Welding
Level 1

Trainee Guide
Fifth Edition

PEARSON

Boston   Columbus   Indianapolis   New York   San Francisco   Amsterdam
Cape Town Dubai London Madrid Milan Munich Paris Montreal Toronto Delhi
Mexico City Sao Paulo Sydney Hong Kong Seoul Singapore Taipei Tokyo

NCCER

President: Don Whyte
Director of Product Development: Daniele Dixon
Welding Project Manager: Jamie Carroll
Senior Manager of Production: Tim Davis
Quality Assurance Coordinator: Debie Hicks

Desktop Publishing Coordinator: James McKay
Permissions Specialists: Adrienne Payne
Production Specialist: Adrienne Payne
Editor: Tanner Yea

Writing and development services provided by Topaz Publications, Liverpool, NY

Lead Writer/Project Manager: Troy Staton
Desktop Publisher: Joanne Hart
Art Director: Alison Richmond

Permissions Editor: Tonia Burke
Writers: Troy Staton, Carol Herbert, Thomas Burke

Pearson Education, Inc.

Director, Global Employability Solutions: Jonell Sanchez
Head of Associations: Andrew Taylor
Editorial Assistant: Kelsey Kissner
Program Manager: Alexandrina B. Wolf
Operations Supervisor: Deidra M. Skahill
Art Director: Diane Ernsberger
Digital Product Strategy Manager: Maria Anaya
Digital Studio Project Managers: Heather Darby,
  Tanika Henderson
Directors of Marketing: David Gesell, Margaret Waples
Field Marketers: Brian Hoehl

Composition: NCCER
Printer/Binder: LSC Communications
Cover Printer: LSC Communications
Text Fonts: Palatino and Univers

Credits and acknowledgments for content borrowed from other sources and reproduced, with permission, in this textbook appear at the end of each module.

5 18

PEARSON

Perfect bound	ISBN10:	0-13-416311-7
	ISBN13:	978-0-13-416311-6
Case bound	ISBN10:	0-13-413110-X
	ISBN13:	978-0-13-413110-8

Preface

Welding is a high-tech industry that can take you to places all over the world. Welds are everywhere — just look around. From ladders to aircraft carriers; from NASCAR to national defense; from the laboratory to sales and repair; the welding trade impacts virtually every industry.

Few career choices offer as many options for employment and opportunities for growth. According to the Department of Labor, "the basic skills of welding are similar across industries, so welders can easily shift from one industry to another, depending on where they are needed most." Welders will continue to be in high demand since there is a continual need to rebuild aging infrastructure such as highways and buildings. In addition, the creation of new power generation facilities will yield new job prospects for trained and skilled welders.

Evolving technology creates more uses for welding in the workplace. For example, new methods are being developed to bond dissimilar materials and nonmetallic materials such as plastics, composites, and new alloys. Also, advances in laser beam and electron beam welding, new fluxes, and other new technologies and techniques all point to an increasing need for educated, skilled welders.

This fifth edition of *Welding Level One* introduces the fundamentals of the welding trade. The four levels of this curriculum present an apprentice approach and will help you be knowledgeable, safe, and effective on the job. The Welding curriculum has been revised by industry subject matter experts from across the nation who have incorporated the latest methods and technology of the trade.

New with *Welding Level One*

This fifth edition of *Welding Level One* boasts a new instructional design that organizes the material into a layout that mirrors the learning objectives. The new format engages trainees and enhances the learning experience by presenting concepts in a clear, concise manner. For example, trade terms are defined at the beginning of each section in which they appear and each section concludes with a brief section review which serves as a study aid. The images and diagrams have been updated to exemplify the most current welding practices with special emphasis on safety.

This edition has been updated to include the latest technology, tools, and techniques of the trade. The suggested teaching hours for *Welding Safety* (29101-15) increased to five hours and *SMAW – Groove Welds with Backing* (29111-15) increased by 20 hours, bringing it to 50 suggested teaching hours. The following modules decreased in suggested teaching hours: *Air-Carbon Arc Cutting and Gouging* (29104-15) was reduced to ten suggested teaching hours; *SMAW – Beads and Fillet Welds* (29109-15) now has 100 suggested teaching hours; and *SMAW – Open-Root Groove Welds – Plate* (29112-15), for-merly called *Open V-Groove*, was reduced by 20 hours, bringing the total suggested teaching hours to 60. Additional changes include a broader emphasis on safety, an easier-to-follow explanation of electrical concepts in *SMAW – Equipment and Setup* (29107-15), and the module formerly called *Shielded Metal Arc Welding –Electrodes* (29108-09) is now *SMAW Electrodes* (29108-15).

This edition also aligns with the new American Welding Society's School Excelling through National Skills Education (SENSE) EG2.0 guidelines for Entry Welder and to the most current AWS standards. This means that, in addition to conforming to NCCER guidelines for credentialing through its Registry, this program can also be used to meet guidelines provided by AWS for Entry Welder training. For more information on the AWS SENSE program, contact AWS at 1-800-443-9353 or visit **www.aws.org**. For information on NCCER's Accreditation and Registry, contact NCCER Customer Service at 1-888-622-3720 or visit **www.nccer.org**.

We invite you to visit the NCCER website at **www.nccer.org** for the latest releases, training information, newsletter, and much more. You can also reference the Pearson product catalog online at **www.nccer.org**. Your feedback is welcome. You may e-mail your comments to **curriculum@nccer.org** or send general comments and inquiries to **info@nccer.org**.

NCCER Standardized Curricula

NCCER is a not-for-profit 501(c)(3) education foundation established in 1995 by the world's largest and most progressive construction companies and national construction associations. It was founded to address the severe workforce shortage facing the industry and to develop a standardized training process and curricula. Today, NCCER is supported by hundreds of leading construction and maintenance companies, manufacturers, and national associations. The NCCER Standardized Curricula was developed by NCCER in partnership with Pearson Education, Inc., the world's largest educational publisher.

Some features of the NCCER Standardized Curricula are as follows:

- An industry-proven record of success
- Curricula developed by the industry for the industry
- National standardization providing portability of learned job skills and educational credits
- Compliance with Office of Apprenticeship requirements for related classroom training (*CFR 29:29*)
- Well-illustrated, up-to-date, and practical information

NCCER also maintains a Registry that provides transcripts, certificates, and wallet cards to individuals who have successfully completed a level of training within a craft in the NCCER Standardized Curricula. *Training programs must be delivered by an NCCER Accredited Training Sponsor in order to receive these credentials.*

Special Features

In an effort to provide a comprehensive user-friendly training resource, we have incorporated many different features for your use. Whether you are a visual or hands-on learner, this book will provide you with the proper tools to get started in the welding industry.

Introduction Page

This page is found at the beginning of each module and lists the Objectives, Performance Tasks, and Trade Terms, for that module. The Objectives list the skills and knowledge you will need in order to complete the module successfully. The Performance Taks give you an opportunity to apply your knowledge to real-world tasks. The list of Trade Terms identifies important terms you will need to know by the end of the module.

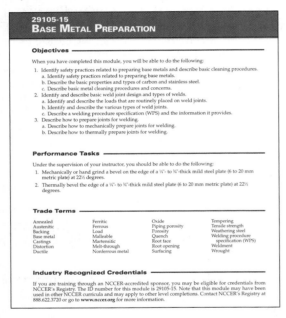

Notes, Cautions, and Warnings

Safety features are set off from the main text in highlighted boxes and organized into three categories based on the potential danger of the issue being addressed. Notes simply provide additional information on the topic area. Cautions alert you of a danger that does not present potential injury but may cause damage to equipment. Warnings stress a potentially dangerous situation that may cause injury to you or a co-worker.

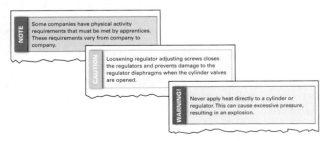

Special Features

Features present technical tips and professional practices from the construction industry. These features often include real-life scenarios similar to those you might encounter on the job site.

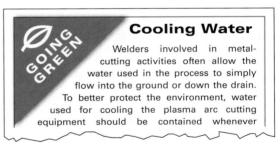

Commonly Available Carbon Electrodes

The carbon electrodes that are usually readily available from general welding suppliers are the copper-coated round and rectangular electrodes used with direct current electrode positive (DCEP) current. Other types and styles, including jointed,

Going Green

Going Green looks at ways to preserve the environment, save energy, and make good choices regarding the health of the planet. Through the introduction of new construction practices and products, you will see how the "greening of the world" has already taken root.

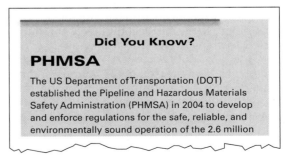

Cooling Water

Welders involved in metal-cutting activities often allow the water used in the process to simply flow into the ground or down the drain. To better protect the environment, water used for cooling the plasma arc cutting equipment should be contained whenever

Did You Know?

The Did You Know? features introduce historical tidbits or modern information about the construction industry. Interesting and sometimes surprising facts about construction are also presented.

Did You Know?

PHMSA

The US Department of Transportation (DOT) established the Pipeline and Hazardous Materials Safety Administration (PHMSA) in 2004 to develop and enforce regulations for the safe, reliable, and environmentally sound operation of the 2.6 million

Case History

Case History features emphasize the importance of safety by citing examples of the costly (and often devastating) consequences of ignoring safety rules and regulations.

Case History
Mexico City Refinery

In Mexico City in 1984, a weld on a liquid propane gas (LPG) tank cracked. The escaping gas cloud drifted toward a nearby residential area until it encountered an ignition source. When the gas ignited, the fireball quickly burned its way back to the leak source. The resulting explosion destroyed the refinery and many residential neighborhoods, leaving 503 people dead and more than 4,000 injured.
The Bottom Line: Welds must be made cor-

Color Illustrations and Photographs

Full-color illustrations and photographs are used throughout each module to provide vivid detail. These figures highlight important concepts from the text and provide clarity for complex instructions. Each figure is denoted in the text in *italic type* for easy reference.

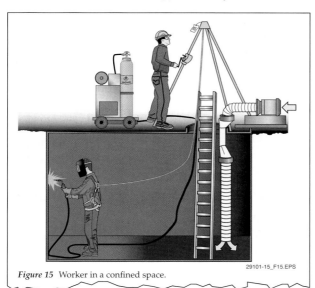

Figure 15 Worker in a confined space.

29101-15_F15.EPS

Step-by-Step Instructions

Step-by-step instructions are used throughout to guide you through technical procedures and tasks from start to finish. These steps show you not only how to perform a task but also how to do it safely and efficiently

Follow these steps to make an overhead fillet weld:

Step 1 Tack two plates together to form a T-joint for the fillet weld coupon.

Step 2 Tack-weld the coupon so it is in the overhead position.

Trade Terms

Each module presents a list of Trade Terms that are discussed within the text, defined in the Glossary at the end of the module, and reinforced with a Trade Terms Quiz. These terms are denoted in the text with **blue bold type** upon their first occurrence. To make searches for key information easier, a comprehensive Glossary of Trade Terms from all modules is found at the back of this book.

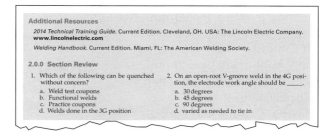

Metals are classified into two basic groups: **ferrous** metals, which are composed mainly of iron, and **nonferrous metals**, which contain very little or no iron. Ferrous metals include all steel, cast iron, **wrought** iron, **malleable** iron, and **ductile**

Section Review

The Section Review features helpful additional resources and review questions related to the objectives in each section of the module.

Additional Resources

2014 Technical Training Guide. Current Edition. Cleveland, OH. USA: The Lincoln Electric Company. www.lincolnelectric.com

Welding Handbook. Current Edition. Miami, FL: The American Welding Society.

2.0.0 Section Review

1. Which of the following can be quenched without concern?
   a. Weld test coupons
   b. Functional welds
   c. Practice coupons
   d. Welds done in the 3G position

2. On an open-root V-groove weld in the 4G position, the electrode work angle should be _____.
   a. 30 degrees
   b. 45 degrees
   c. 90 degrees
   d. varied as needed to tie in

Review Questions

Review Questions are provided to reinforce the knowledge you have gained. This makes them a useful tool for measuring what you have learned.

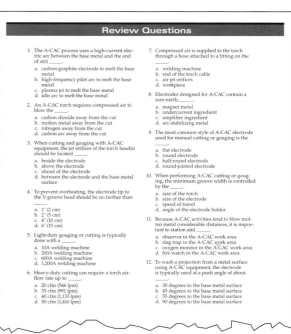

iii

NCCER Standardized Curricula

NCCER's training programs comprise more than 80 construction, maintenance, pipeline, and utility areas and include skills assessments, safety training, and management education.

Boilermaking
Cabinetmaking
Carpentry
Concrete Finishing
Construction Craft Laborer
Construction Technology
Core Curriculum: Introductory
  Craft Skills
Drywall
Electrical
Electronic Systems Technician
Heating, Ventilating, and
  Air Conditioning
Heavy Equipment Operations
Highway/Heavy Construction
Hydroblasting
Industrial Coating and Lining
  Application Specialist
Industrial Maintenance Electrical
  and Instrumentation Technician
Industrial Maintenance Mechanic
Instrumentation
Insulating
Ironworking
Masonry
Millwright
Mobile Crane Operations
Painting
Painting, Industrial
Pipefitting
Pipelayer
Plumbing
Reinforcing Ironwork
Rigging
Scaffolding
Sheet Metal
Signal Person
Site Layout
Sprinkler Fitting
Tower Crane Operator
Welding

Maritime

Maritime Industry Fundamentals
Maritime Pipefitting
Maritime Structural Fitter

Green/Sustainable Construction

Building Auditor
Fundamentals of Weatherization
Introduction to Weatherization
Sustainable Construction
  Supervisor
Weatherization Crew Chief
Weatherization Technician
Your Role in the Green
  Environment

Energy

Alternative Energy
Introduction to the Power Industry
Introduction to Solar Photovoltaics
Introduction to Wind Energy
Power Industry Fundamentals
Power Generation Maintenance
  Electrician
Power Generation I&C
  Maintenance Technician
Power Generation Maintenance
  Mechanic
Power Line Worker
Power Line Worker: Distribution
Power Line Worker: Substation
Power Line Worker: Transmission
Solar Photovoltaic Systems Installer
Wind Turbine Maintenance
  Technician

Pipeline

Control Center Operations, Liquid
Corrosion Control
Electrical and Instrumentation
Field Operations, Liquid
Field Operations, Gas
Maintenance
Mechanical

Safety

Field Safety
Safety Orientation
Safety Technology

Supplemental Titles

Applied Construction Math
Tools for Success

Management

Fundamentals of Crew Leadership
Project Management
Project Supervision

Spanish Titles

Acabado de concreto: nivel uno
Aislamiento: nivel uno
Albañilería: nivel uno
Andamios
Carpintería:
  Formas para carpintería, nivel tres
Currículo básico: habilidades
  introductorias del oficio
Electricidad: nivel uno
Herrería: nivel uno
Herrería de refuerzo: nivel uno
Instalación de rociadores: nivel uno
Instalación de tuberías: nivel uno
Instrumentación: nivel uno, nivel
  dos, nivel tres, nivel cuatro
Orientación de seguridad
Paneles de yeso: nivel uno
Seguridad de campo

Acknowledgments

This curriculum was revised as a result of the farsightedness and leadership of the following sponsors:

Claddagh Enterprises
Central Louisiana Technical College
Exelon Generation
Gulf States, Inc., Michigan Division
KBR Industrial Services
Lee College
Lincoln Electric

Northeast Community College
Northland Pioneer College
Robins & Morton
TIC-The Industrial Company
Toledo Refining Co. LLC
Zachry Industrial, Inc.

This curriculum would not exist were it not for the dedication and unselfish energy of those volunteers who served on the Authoring Team. A sincere thanks is extended to the following:

Tom Ashley
Gerald Bickerstaff
Curtis Casey
Bill Cherry
Rod Hellyer

Frank Johnson
John Knapp
Holley Thomas
Jerry Trainor
Tim Riley

Jason Scales
Dan Sterry
Adam Webb

NCCER Partners

American Fire Sprinkler Association
Associated Builders and Contractors, Inc.
Associated General Contractors of America
Association for Career and Technical Education
Association for Skilled and Technical Sciences
Construction Industry Institute
Construction Users Roundtable
Construction Workforce Development Center
Design Build Institute of America
GSSC – Gulf States Shipbuilders Consortium
ISN
Manufacturing Institute
Mason Contractors Association of America
Merit Contractors Association of Canada
NACE International
National Association of Minority Contractors
National Association of Women in Construction
National Insulation Association
National Technical Honor Society
NAWIC Education Foundation
North American Crane Bureau
North American Technician Excellence
Pearson

Pearson Qualifications International
Prov
SkillsUSA®
Steel Erectors Association of America
U.S. Army Corps of Engineers
University of Florida, M. E. Rinker School of
    Building Construction
Women Construction Owners & Executives,
    USA

Contents

Module Nine

SMAW – Beads and Fillet Welds

Describes the preparation and setup of arc welding equipment and the process of striking an arc. Explains how to detect and correct arc blow. Describes how to make stringer, weave, overlapping beads, and fillet welds. (Module ID 29109-15; 100 hours)

Module Ten

Joint Fit-Up and Alignment

Describes job code specifications. Explains how to use fit-up gauges and measuring devices to check fit-up and alignment and use plate and pipe fit-up and alignment tools to properly prepare joints. Explains how to check for joint misalignment and poor fit. (Module ID 29110-15; 5 hours)

Module Eleven

SMAW – Groove Welds with Backing

Introduces groove welds and explains how to set up welding equipment for making groove welds. Describes how to make groove welds with backing. Provides procedures for making flat, horizontal, vertical, and overhead groove welds. (Module ID 29111-15; 50 hours)

Module Twelve

SMAW – Open-Root Groove Welds – Plate

Introduces various types of groove welds and describes how to prepare for groove welding. Describes the techniques required to produce various open V-groove welds. (Module ID 29112-15; 60 hours)

Glossary

Index

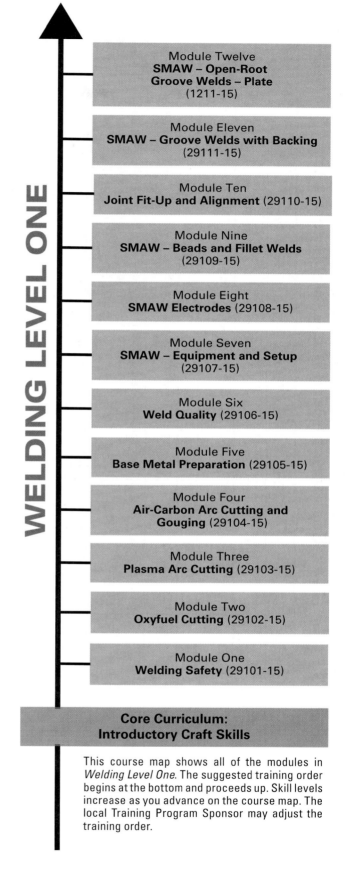

This course map shows all of the modules in *Welding Level One*. The suggested training order begins at the bottom and proceeds up. Skill levels increase as you advance on the course map. The local Training Program Sponsor may adjust the training order.

CORRELATIONS CHART: AWS S.E.N.S.E. EG2.0 GUIDELINES AND NCCER WELDING LEVELS ONE AND TWO

SENSE Guidelines Modules	Passing Score	Visual	Destructive	NCCER / NCCER Standardized Curricula
1 – Occupational Orientation *	No Test	No Test	No Test	Core Curriculum: Module 00107-15, Basic Communication Skills; Module 00108-15, Basic Employability Skills
2 – Safety and Health of Welders *	100%	No Test	No Test	Welding Level 1: Module 29101-15, Welding Safety
3 – Drawing and Welding Symbol Interpretation *	75%	No Test	No Test	Welding Level 2: Module 29201-15, Welding Symbols; Module 29202-15, Reading Welding Detail Drawings
4 – Shielded Metal Arc Welding	75%	Pass /Fail	Pass /Fail	Welding Level 1: Module 29107-15, SMAW – Equipment and Setup; Module 29109-15, SMAW – Beads and Fillet Welds; Module 29111-15, SMAW – Groove Welds with Backing
5 – Gas Metal Arc Welding	75%	Pass /Fail	No Test	Welding Level 2: Module 29205-15, GMAW and FCAW – Equipment and Filler Materials; Module 29209-15, GMAW – Plate
6 – Flux Cored Arc Welding	75%	Pass /Fail	No Test	Welding Level 2: Module 29205-15, GMAW and FCAW – Equipment and Filler Materials; Module 29210-15, FCAW – Plate
7 – Gas Tungsten Arc Welding	75%	Pass /Fail	No Test	Welding Level 2: Module 29207-15, GTAW – Equipment and Filler Materials; Module 29208-15, GTAW – Plate
8 – Thermal Cutting Process * †	75%	No Test	No Test	
Unit 1 Manual Oxyfuel Gas Cutting (OFC) †		Pass /Fail	No Test	Welding Level 1: Module 29102-15, Oxyfuel Cutting
Unit 2 Mechanized Oxyfuel Gas Cutting (OFC) †		Optional	No Test	Welding Level 1: Module 29105-15, Base Metal Preparation
Unit 3 Manual Plasma Arc Cutting—PAC †		Pass /Fail	No Test	Welding Level 1: Module 29103-15, Plasma Arc Cutting
Unit 4 Manual Air Carbon Arc Cutting †		Optional	No Test	Welding Level 1: Module 29104-15, Air-Carbon Arc Cutting and Gouging
9 – Welding Inspection and Testing *	75%	Pass /Fail	No Test	Welding Level 1: Module 29106-15, Weld Quality

* Required module for Level I Entry Welder Completion (plus one welding process module)

† Completion of Units 1 and 3 minimum

29101-15
Welding Safety

OVERVIEW

Qualified welders are in high demand all over the world. Typical welding tasks may include working in heavy construction; joining pipe sections for oil and natural gas pipelines; building ships; and working in a variety of industrial environments such as power plants, refineries, chemical plants, and manufacturing facilities. This module introduces trainees to the all-important topic of safety in the welding trade.

Module One

Trainees with successful module completions may be eligible for credentialing through the NCCER Registry. To learn more, go to **www.nccer.org** or contact us at **1.888.622.3720**. Our website has information on the latest product releases and training, as well as online versions of our *Cornerstone* magazine and Pearson's product catalog.

Your feedback is welcome. You may email your comments to **curriculum@nccer.org**, send general comments and inquiries to **info@nccer.org**, or fill in the User Update form at the back of this module.

This information is general in nature and intended for training purposes only. Actual performance of activities described in this manual requires compliance with all applicable operating, service, maintenance, and safety procedures under the direction of qualified personnel. References in this manual to patented or proprietary devices do not constitute a recommendation of their use.

Objectives

When you have completed this module, you will be able to do the following:

1. Describe basic welding processes, the welding trade, and training/apprenticeship programs.
   a. Describe basic welding processes and the welding trade.
   b. Describe NCCER standardized training and explain apprenticeship programs.
2. Identify and describe personal protective equipment (PPE) related to the welding trade.
   a. Identify and describe body, foot, and hand protective gear.
   b. Identify and describe ear, eye, face, and head protective gear.
3. Identify and describe welding safety practices related to specific hazards or environments.
   a. Describe the importance of welding safety and identify factors related to accidents.
   b. Describe basic welding safety practices related to the general work area.
   c. Describe hot work permits and fire watch requirements.
   d. Describe confined spaces and their related safety practices.
   e. Identify safety practices related to welding equipment.
   f. Identify and describe respiratory hazards, respiratory safety equipment, and ways to ventilate welding work areas.
   g. Explain the purpose of the SDS/MSDS and how it is used.

Performance Tasks

This is a knowledge-based module; there are no Performance Tasks.

Trade Terms

Arc burn
Bonded
Carbon steel
Electric arc
Electrically grounded
Electrode
Flash burn
Flux

Fume plume
Galvanized steel
Immediate danger to life and health (IDLH)
Purge gas
Shielding gas
Stainless steel
Ultraviolet (UV) radiation
Ventricular fibrillation

Industry Recognized Credentials

If you are training through an NCCER-accredited sponsor, you may be eligible for credentials from NCCER's Registry. The ID number for this module is 29101-15. Note that this module may have been used in other NCCER curricula and may apply to other level completions. Contact NCCER's Registry at 888.622.3720 or go to **www.nccer.org** for more information.

Contents

Topics to be presented in this module include:

Figures and Tables

1.0.0 THE WELDING TRADE

Objective

Describe basic welding processes, the welding trade, and training/apprenticeship programs.

a. Describe basic welding processes and the welding trade.
b. Describe NCCER standardized training and explain apprenticeship programs.

Trade Terms

Carbon steel: An alloy of iron combining iron and a small amount of carbon (usually less than 1 percent) to provide hardness.

Electric arc: The flow of an electrical current across an air gap or gaseous space.

Electrode: The point from which a welding arc is produced. In most welding processes, the electrode is a filler metal which is consumed in the process and becomes part of the weld.

Flux: A material used to dissolve or prevent the formation of oxides and other undesirable substances on a weld joint.

Shielding gas: A gas such as argon, helium, or carbon dioxide used to protect the welding electrode wire from contamination in gas metal arc welding (GMAW) and flux core arc welding (FCAW).

Stainless steel: An iron-based alloy usually containing at least 11 percent chromium.

This module is an extension of the Core Curriculum *Basic Safety* module. Its purpose is to identify general safety considerations that apply to various aspects of welding, along with the steps needed to avoid job-related deaths and injuries. Specific equipment and process hazards are covered in more detail in modules that deal with the use of the equipment or process in a particular application.

1.1.0 Welding Work

Welding is the process of joining similar metals using a high-intensity electric arc at temperatures from 6,000°F (3,316°C) to 10,000°F (5,538°C). The arc melts and fuses the base metals. In most cases, a filler metal is melted along with the base metal in order to strengthen the joint. The finished joint is as strong as, or even stronger than, the base metals that have been joined. Welding is commonly done on carbon steel, stainless steel, and aluminum.

Welders are key contributors to a number of construction and industrial crafts, including building and bridge construction; ship and oil rig construction and repair; heavy industry; and pipeline work. Welding is a craft that requires extensive knowledge and skill as well as a commitment to performing high-quality work and, most importantly, a strong commitment to safety.

There is a lot more to welding than simply striking an electric arc on base metals. Welders must know the characteristics of the metals being welded and the special materials and techniques that apply to the different metals. They must also know how to properly handle and store base metals and filler metals; how to prepare metals for welding; and how to interpret welding symbols and specifications.

The most common type of welding is shielded metal arc welding (SMAW). In the SMAW method, a flux-coated, consumable electrode is used to place the weld (*Figure 1*). SMAW is used primarily for welding iron and steel, including stainless steel.

Gas metal arc welding (GMAW), sometimes called MIG, or metal inert gas welding, uses a consumable wire that is fed through the welding gun to make the weld. It also uses a shielding gas. GMAW is used on steel and other metals. It is widely used in industrial applications, especially in the manufacture of automobiles where robotic welding is common. Flux core arc welding (FCAW) is very similar to GMAW. The main differences are that the FCAW filler metal is a tubular flux-cored wire and shielding gas may or may not be used.

Gas tungsten arc welding (GTAW), sometimes called TIG (tungsten inert gas) welding, uses a non-consumable tungsten electrode to produce the weld. GTAW is commonly used in welding stainless steel and light metals such as aluminum and magnesium.

In addition to welding, welders are often required to cut steel using oxyfuel, air carbon-arc, or plasma-arc equipment (*Figure 2*). Oxyfuel cutting is the most common method. In the oxyfuel method, oxygen and a fuel gas are mixed in a cutting torch to produce a high-intensity flame that can cut through thick steel. Acetylene is the most common fuel gas because it produces the most intense heat. Propane, MAPP® gas, and butane are also used for cutting work. Plasma-arc cutting relies on a jet of superheated gas that is passed through an electric arc. It is used to cut steel as well as other metals. Carbon-arc cutting uses

(A) SMAW

(B) GMAW

(C) GTAW

29101-15_F01.EPS

Figure 1 Welding methods.

an electric arc formed between a carbon electrode and the base metal to melt the base metal. Carbon-arc cutting is often used in preparing a base metal for welding.

Welders work on critical structures such as bridges, high-rise buildings, ships, pipelines, and automobiles. As you can imagine, defective welds in any of these applications can lead to a disaster. For that reason, welding work is subjected to intense quality control inspections. The more critical the weld, the more intense the inspection is. Inspectors use a variety of methods to examine welds, including X-ray inspection.

In order to qualify as a welder, a person must undergo extensive training that requires many hours of practice. The training provided by this program will prepare you to weld plate and pipe in a variety of positions and to cut carbon steel plate and pipe using common cutting equipment and techniques.

There are numerous opportunities for welders in heavy construction, manufacturing, ship building, and auto racing. Once you have completed your training and have become a journey-level welder, you can advance in the trade in a variety of ways (see *Figure 3*). The more training and education you complete, the greater the opportunities.

1.2.0 NCCER Standardized Training

NCCER is a not-for-profit education foundation established by the nation's leading construction companies. NCCER was created to provide the industry with standardized construction education materials, the NCCER Standardized Curricula (the welding modules are part of this series), and a system for tracking and recognizing students' training accomplishments—the NCCER Registry.

NCCER also offers accreditation, instructor certification, and craft assessments. NCCER is committed to developing and maintaining a training process that is internationally recognized, standardized, portable, and competency-based.

Working in partnership with industry and academia, NCCER has developed a system for program accreditation that is similar to those found in institutions of higher learning. NCCER's accreditation process ensures that students receive quality training based on uniform standards and criteria. These standards are outlined in NCCER's *Accreditation Guidelines* and must be adhered to by NCCER Accredited Training Sponsors.

(A) OXYFUEL CUTTING

(B) MANUAL PLASMA ARC CUTTING

(C) CARBON ARC GOUGING

29101-15_F02.EPS

Figure 2 Cutting methods.

More than 603 training centers across the United States and eight other countries are proud to be NCCER Accredited Training Sponsors/ Accredited Assessment Centers. Millions of craft professionals and construction managers have received quality construction education through NCCER's network of Accredited Training Sponsors. Every year the number of NCCER Accredited Training Sponsors increases significantly.

A craft instructor is a journey-level craft professional or career and technical educator trained and certified to teach the NCCER Standardized Curricula. This network of certified instructors ensures that NCCER training programs meet the standards of instruction set by the industry. There are more than 5,919 master trainers and 61,698 craft instructors within the NCCER instructor network. More information is available at **www.nccer.org**.

The Welding curriculum developed by NCCER provides participants with portable, industry-recognized credentials. These credentials include transcripts, certificates, and wallet cards that are tracked through NCCER's Registry.

This curriculum provides trainees with industry-driven training and education using a competency-based learning approach. This means that trainees must show the instructor that they possess the knowledge and skills needed to safely perform the hands-on tasks that are covered in each module.

When the instructor is satisfied that a trainee has demonstrated through written exams and hands-on performance testing that he or she has the required knowledge and skills for a given module, that information is sent to NCCER and kept in the Registry. The NCCER Registry can then confirm training and skills for workers as they move from state-to-state, company-to-company, or position-to-position.

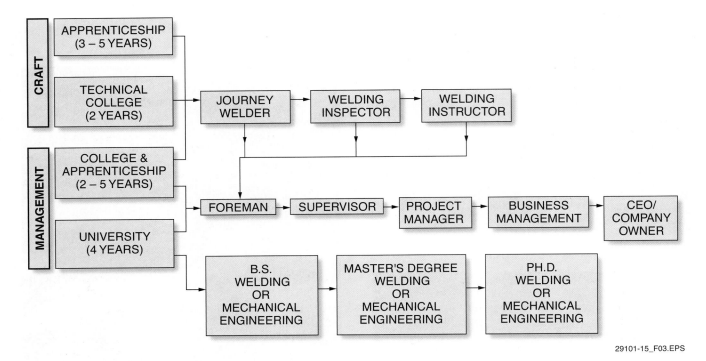

29101-15_F03.EPS

Figure 3 Opportunities in the welding trade.

1.2.1 Apprenticeship Training

Apprentice training goes back thousands of years. Its basic principles have not changed over time. First, it is a means for a person entering the craft to learn from those who have mastered the craft. Second, it focuses on learning by doing; real skills versus theory. Some theory is presented in the classroom. However, it is always presented in a way that helps the apprentice understand the purpose behind the skill that is to be learned.

The US Department of Labor's Office of Apprenticeship sets the minimum standards for training programs across the country. These programs rely on mandatory classroom instruction and on-the-job learning (OJL). They require at least 144 hours of classroom instruction per year and 2,000 hours of OJL per year. In a typical four-year apprenticeship program, trainees spend 576 hours in classroom instruction and 8,000 hours in OJL before receiving certificates issued by registered apprenticeship programs.

NCCER uses the Department of Labor apprenticeship standards as a foundation for comprehensive training that provide trainees with in-depth classroom and OJL experience.

All apprenticeship standards prescribe certain work-related or on-the-job learning. This OJL is broken down into specific tasks in which the apprentice receives hands-on training. In addition, a specified number of hours is required in each task. In a competency-based program, it may be possible to shorten the time required by testing out of specific tasks through a series of written and performance tests.

In a traditional program, the required OJL may be acquired in increments of 2,000 hours per year. Layoffs or illness may affect the duration. The apprentice must log all work time and turn the log in to the apprenticeship committee so that accurate records are maintained.

The classroom instruction and work-related training will not always run concurrently due to such reasons as layoffs, the type of work needed to be done in the field, and other factors. Apprentices with special job experience or coursework may obtain credit toward their classroom requirements. This reduces the total time required in the classroom while maintaining the total 8,000-hour on-the-job training requirement. These special cases depend on the type of program and the regulations and standards under which it operates.

Informal OJL provided by employers is usually less thorough than that provided through a formal apprenticeship program. The degree of training and supervision in this type of program often depends on the size of the employing firm. A small contractor may provide training in only one area, while a large company may be able to provide training in several areas.

For those entering an apprenticeship program, a high school or technical school education is desirable. Courses in shop, mechanical drawing, and

4

NCCER – *Welding Level One*   29101-15

general mathematics are helpful. Manual dexterity, good physical condition, and quick reflexes are important. The ability to solve problems quickly and accurately and to work closely with others is essential. You must also have high respect for safety and its place in the construction environment.

> **NOTE**
>
> Some companies have physical activity requirements that must be met by apprentices. These requirements vary from company to company.

1.2.2 Youth Training and Apprenticeship Programs

Youth apprenticeship programs are available that allow students to begin their apprenticeship or craft training while still in high school. A student entering the program in the 11th grade may complete as much as one year of the training program by high school graduation. In cooperation with local construction industry employers, students may be able to work in the craft and earn money while still in school. Upon graduation, students can enter the industry at a higher level and with more pay than someone just starting in a training program.

Students participating in apprenticeship training may enter the second level or year of the program wherever it is offered. They may also have the option of applying credits at two-year or four-year colleges that offer degree or certificate programs in their selected field of study.

Additional Resources

Arc Welding Safety E205. Latest Edition. Cleveland, OH: Lincoln Electric.

Safety Topics in Welding, Cutting, and Brazing. **www.osha.gov**.

AWS F32M/F32, Ventilation Guide for Weld Fumes. Latest Edition. Miami, FL: American Welding Society.

1.0.0 Section Review

1. Flux core arc welding (FCAW) is similar to _____.
   a. SMAW
   b. GMAW
   c. GTAW
   d. carbon-arc

2. Before graduation, a high school student entering an apprenticeship program in the 11th grade may complete _____.
   a. as much as one year of the program
   b. nearly two years of the program
   c. more than two years of the program
   d. three years of the program

SECTION TWO

2.0.0 PERSONAL PROTECTIVE EQUIPMENT

Objective

Identify and describe personal protective equipment (PPE) related to the welding trade.

a. Identify and describe body, foot, and hand protective gear.
b. Identify and describe ear, eye, face, and head protective gear.

Trade Terms

Arc burn: Burn to the skin produced by brief exposure to intense radiant heat and ultraviolet light.

Flash burn: Burns to the eyes sometimes called welder's flash; caused by exposure to intense radiant heat and ultraviolet light.

Ultraviolet (UV) radiation: Invisible rays capable of causing burns. UV rays from the sun are the cause of sunburn.

29101-15_F04.EPS

Figure 4 Welder with proper protective clothing.

Each person in the shop environment wears general work clothing. Extra protection is needed for each person that is in direct contact with hot materials. Depending on the specific job and conditions, protective equipment can include the following:

- Body protection
- Foot protection
- Hand protection
- Ear protection
- Eye, face, and head protection

2.1.0 Body, Foot, and Hand Protection

Welding produces hazards from molten metal and sparks. For that reason, welders and anyone working in the welding environment must have proper protection.

2.1.1 Body Protection

Basic clothing for welders, as shown in *Figure 4*, should offer protection from flying sparks, heat, and **ultraviolet (UV) radiation** from electric arcs. Shirts should be made of a tight-weave fabric, have long sleeves and pocket flaps, and be worn with the collar buttoned. Pants must not have cuffs and should fit so they hang straight down the leg, touching the shoe tops without creases. Cuffs and creases can catch sparks, which can cause fires. Never wear polyester or other synthetic fibers; sparks will melt these materials, causing serious burns. Wear wool and cotton, which are more resistant to sparks. Cotton clothing, if worn, should be treated with a fire-retardant.

It is important to follow the manufacturer's recommendations for laundering fire-resistant clothing. Fabric softeners may reduce the effectiveness of the treatment, increasing the flammability. Some fire-resistant clothing may be ruined by repeated washings or incorrect washing. When purchasing fire-resistant clothing, look for fire-resistant labeling on the garment itself, rather than on the packaging alone.

Additional protective coverings are required for the following:

- Cutting operations
- Out-of-position or overhead welding/cutting
- Welding operations that produce spattering molten metal and sparks or large amounts of heat

Figure 5 shows a welder wearing protective clothing and equipment. Protective coverings, usually made of leather because of its durability, are often referred to as leathers. However, because leather is heavy and hot to wear, lightweight fire- and heat-resistant clothing is also available. *Figure 6* shows a variety of leather and fire-resistant items that can be worn in combinations to provide various degrees of covering.

Full jackets offer the most protection. A cape is cooler, but offers only shoulder, chest, and arm protection. Leather pants, in combination with a jacket, will protect a welder's lap and legs. If the weather is hot and full leather pants are uncomfortable, leggings, sometimes called chaps, are available. They are strapped to the legs, leaving the back open. However, an apron alone is cooler and will protect a welder's lap and most of the leg area when squatting, sitting, or bending over a table. In some cases, if the welding is done at or below waist level, only full- or half-sleeve arm covers are required in combination with an

29101-15_F05.EPS

Figure 5 Personal protective equipment (PPE).

Terms

Don't confuse the terms *fire-retardant* and *fireproof*. A product impregnated with a fire-retardant delays ignition. A fireproof product is not supposed to burn. Clothing treated with a fire-retardant may lose some of its protective ability if washed, so it may need retreating after washing.

apron. The welding position and type of welding operation typically dictate the appropriate personal protective equipment (PPE), in conjunction with company or site policies.

WARNING!

Sparks of molten metal from welding and metal cutting can reach temperatures greater than 9,900°F (5,500°C) and can travel a significant distance. They can land on the clothing or exposed skin of the welder or others in the immediate area, potentially resulting in burns. Sparks can also land on flammable material a significant distance away, resulting in a fire. For this reason, welders and nearby workers must wear the proper protective clothing. Flammable materials nearby must be covered with non-flammable blankets.

2.1.2 Foot Protection

The Occupational Safety and Health Administration (OSHA) requires that protective footwear be worn when working where falling, rolling, or sharp objects pose a danger of foot injuries and where feet are exposed to electrical hazards. For welders, sparks and droplets of molten metal provide additional hazards. Footwear with leather-reinforced soles or innersoles of flexible metal are recommended. High-top safety shoes or boots should be worn. Ensure that the pant leg covers the tongue and lace area of the footwear. If the tongue and lace area is exposed, wear leather spats (as shown in *Figure 6*) under the pants or leggings and over the front top of the footwear to protect them from sparks or falling molten metal. Spats will prevent sparks from burning through the front of lace-up boots. Protective footwear must comply with *ANSI Z49.1*. Sneakers, tennis shoes, and similar types of footwear must never be worn on the job site.

(A) LEATHER JACKET

(B) LEATHER APRON

(C) FIRE-RESISTANT ARM COVER

(D) LEATHER BOOT OR SHOE PROTECTION (SPATS)

(E) CAPE (FASTENS TO AN APRON)

(F) LEATHER CHAPS

29101-15_F06.EPS

Figure 6 Leather and fire-resistant protective coverings.

2.1.3 Hand Protection

Gloves are the primary type of hand protection. Gloves must be selected on the basis of the hazards involved in doing the work. Gauntlet-type welding gloves must be worn when welding or cutting to protect against UV rays from an electric arc and heat from any thermal welding/cutting process. The most common glove is the standard leather welding glove shown in *Figure 7*. For heavy-duty welding or cutting, special heat-resistant or heat-reflective gloves are also available.

> **WARNING!**
> Don't weld or cut metal with gloves that have been used to handle petroleum products, especially when oxygen is being used. Also consider that the transfer of petroleum residue to the welding rod can affect the weld quality.

2.2.0 Protection Above the Neck

Special attention must be given to eye and ear safety when welding or cutting. Severe burns to the eyes can result from viewing the welding arc without the proper protection. Ears can be damaged by noise and sparks. The face, head, and neck must also be protected from sparks and heat.

2.2.1 Hearing Protection

Welding areas can be very noisy. If overhead work is being performed, hot sparks can cause burns to the ears and ear canals unless a leather hood is used. Such burns are extremely painful, so it is important to wear ear plugs when welding (*Figure 8*). Disposable earplugs are the most common form of hearing protection used in welding. These devices usually have an outer layer of pliable foam and a core layer of acoustical fiber that filters out harmful noise, yet allows you to hear normal conversation. To use disposable earplugs, simply roll each plug into a cone and insert the tapered end into the ear canal while pulling up on the upper portion of your ear. The earplugs will expand, filling the ear canal and creating a proper fit.

> **WARNING!**
> Plain cotton placed in the ear is not acceptable protection. Only approved ear protection is to be used.

2.2.2 Eye, Face, and Head Protection

The heat and light produced by cutting or welding operations can damage the skin and eyes. Injury to the eyes may result in permanent loss of vision. Oxyfuel cutting and welding can cause eye fatigue and mild burns to the skin because of the infrared heat radiated by the process. Welding or cutting operations involving an electric arc of any kind produce UV light. The UV light can cause a severe arc burn to exposed skin and flash burns to unprotected eyes. (The sun also produces UV light, some of which is not filtered out by the Earth's atmosphere and can also dam-

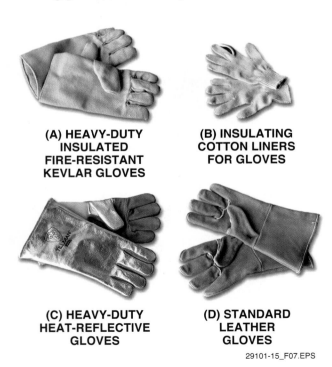

(A) HEAVY-DUTY
INSULATED
FIRE-RESISTANT
KEVLAR GLOVES

(B) INSULATING
COTTON LINERS
FOR GLOVES

(C) HEAVY-DUTY
HEAT-REFLECTIVE
GLOVES

(D) STANDARD
LEATHER
GLOVES

29101-15_F07.EPS

Figure 7 Welding gloves.

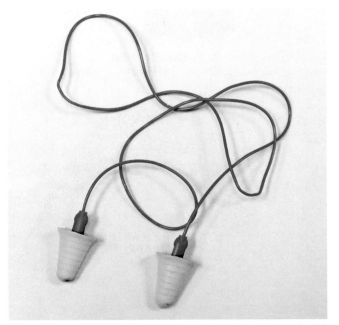

29101-15_F08.EPS

Figure 8 Ear plugs.

age the eyes.) A flash burn can very easily cause permanent damage to the retina of the eye. Welders should never view an electric arc directly or indirectly without wearing a properly tinted lens designed for electric arc use.

> **NOTE**
>
> Goggles are normally used with a leather soft cap to protect the face and head from UV rays.

As specified in *ANSI/AWS Z49.1*, protective spectacles with side shields, arc goggles, or other approved eye protection must be worn by welders in addition to a helmet or face shield with a filtered lens (*Figure 9*). The spectacles or goggles may have either clear or filtered lenses, depending on the amount of exposure to welding or cutting radiation (see *Table 1*). For overhead oxyfuel cutting operations, a leather hood may be used in place of the face shield to provide protection from sparks and molten metal. Goggles can be paired with a leather soft-cap for cutting and burning. The combination face shield with flip-up lenses can be used at some sites with a lens that is dark enough for the application.

For electric arc operations, a welding helmet with a properly tinted lens (shades 9 to 14) must be worn over safety glasses to provide proper protection. Many varieties of helmets are available; typical styles are shown in *Figure 10*. Some of the helmets are available with additional side-view lenses in a lighter tint so that welders can sense, by peripheral vision, any activities occurring beside them. A tinted lens that flips up to reveal a clear safety lens is a good feature. The clear lens enables the welder to continue using the helmet for protection during chipping and grinding.

Most welding and cutting tasks require the use of safety glasses, chemical-resistant goggles, dust goggles, or face shields. Always check your company's policy for the welding or cutting process being used to find out what type of eye protection is needed.

(A) TINTED GOGGLES WITH HEADBAND

(B) CLEAR ELASTIC-STRAP GOGGLES

(C) TINTED ELASTIC-STRAP GOGGLES

(D) TINTED FACE SHIELD (CUTTING)

(E) CLEAR FACE SHIELD

(F) COMBINATION FACE SHIELD WITH FLIP-UP LENSES

29101-15_F09.EPS

Figure 9 Oxyfuel welding/cutting goggles and face shield combinations.

Table 1 Guide for Shade Numbers

From *AWS F2.2 Lens Shade Selector*. Shade numbers are given as a guide only and may be varied to suit individual needs.

Process	Electrode Size in (mm)	Arc Current (Amperes)	Minimum Protective Shade	Suggested* Shade No. (Comfort)
Shielded Metal Arc Welding (SMAW)	Less than 3⁄32 (2.4)	Less than 60	7	–
	3⁄32–5⁄32 (2.4–4.0)	60–160	8	10
	5⁄32–1⁄4 (4.0–6.4)	160–250	10	12
	More than 1⁄4 (6.4)	250–550	11	14
Gas Metal Arc Welding (GMAW) and Flux Cored Arc Welding (FCAW)		Less than 60	7	–
		60–160	10	11
		160–250	10	12
		250–500	10	14
Gas Tungsten Arc Welding (GTAW)		Less than 50	8	10
		50–150	8	12
		150–500	10	14
Air Carbon Arc Cutting (CAC-A)	(Light)	Less than 500	10	12
	(Heavy)	500–1000	11	14
Plasma Arc Welding (PAW)		Less than 20	6	6 to 8
		20–100	8	10
		100–400	10	12
		400–800	11	14
Plasma Arc Cutting (PAC)		Less than 20	4	4
		20–40	5	5
		40–60	6	6
		60–80	8	8
		80–300	8	9
		300–400	9	12
		400–800	10	14
Torch Brazing (TB)		–	–	3 or 4
Torch Soldering (TS)		–	–	2
Carbon Arc Welding (CAW)		–	–	14

Process	Plate Thickness		Suggested* Shade No. (Comfort)
	in	mm	
Oxyfuel Gas Welding (OFW)			
Light	Under 1⁄8	Under 3	4 or 5
Medium	1⁄8 to 1⁄2	3 to 13	5 or 6
Heavy	Over 1⁄2	Over 13	6 or 8
Oxygen Cutting (OC)			
Light	Under 1	Under 25	3 or 4
Medium	1 to 6	25 to 150	4 or 5
Heavy	Over 6	Over 150	5 or 6

*As a rule of thumb, start with a shade that is too dark to see the weld zone. Then go to a lighter shade which gives sufficient view of the weld zone without going below the minimum. In oxyfuel gas welding, cutting, or brazing where the torch and/or the flux produces a high yellow light, it is desirable to use a filter lens that absorbs the yellow or sodium line of the visible light spectrum.

29105-15_T01.EPS

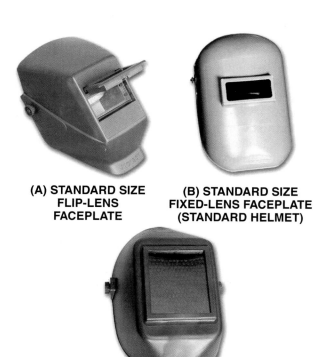

(A) STANDARD SIZE FLIP-LENS FACEPLATE

(B) STANDARD SIZE FIXED-LENS FACEPLATE (STANDARD HELMET)

(C) LARGE-LENS FACEPLATE

29101-15_F10.EPS

Figure 10 Arc-welding helmets.

Lightweight Welding Gloves

Soft, flexible, leather gloves are used for light-duty work and for operations involving GMAW, GTAW, brazing, soldering, and oxyfuel welding where free hand movement is required. These types of welding produce less heat and spatter than others. Lightweight gloves provide more flexibility for the required motions and pulling the trigger of the welding torch.

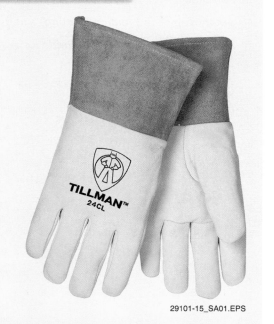

29101-15_SA01.EPS

Auto-Darkening Helmets

Helmets are available with an adjustable auto-darkening lens that automatically changes shades the instant an electric arc starts to occur. Typical shade ranges are 5-14 and 9-13. A sensor just above the lens detects the start of an electric arc and electronically causes the shade of the lens to change in less than $\frac{1}{25,000}$ of a second. The use of this type of lens eliminates the need for the welder to raise the helmet or flip up the faceplate to view the work before striking an arc. The helmet can also be left down for chipping and grinding safety. Auto-darkening helmets cost more than standard helmets, but many welders consider the improvement in quality and productivity to be worth the additional cost. Some require periodic battery replacement and testing to ensure proper operation, but others are solar-powered.

29101-15_SA02.EPS

Additional Resources

ANSI/AWS Z49.1, Safety in Welding, Cutting, and Allied Processes. Miami, FL: American Welding Society.

Arc Welding Safety E205, Latest Edition. Cleveland, OH: Lincoln Electric.

Personal Protective Equipment (PPE) for Welding and Cutting. Miami, FL: American Welding Society.

Safety Topics in Welding, Cutting, and Brazing. **www.osha.gov**.

AWS F32M/F32, Ventilation Guide for Weld Fumes. Latest Edition. Miami, FL: American Welding Society.

2.0.0 Section Review

1. Which of the following fabrics is the best choice for welding clothing?
   a. Rayon
   b. Polyester
   c. Nylon
   d. Cotton

2. The most common form of ear protection is _____ .
   a. disposable ear plugs
   b. ear muffs
   c. wads of cotton
   d. washable ear plugs

3.0.0 WELDING SAFETY PRACTICES

Objective

Identify and describe welding safety practices related to specific hazards or environments.

   a. Describe the importance of welding safety and identify factors related to accidents.
   b. Describe basic welding safety practices related to the general work area.
   c. Describe hot work permits and fire watch requirements.
   d. Describe confined spaces and their related safety practices.
   e. Identify safety practices related to welding equipment.
   f. Identify and describe respiratory hazards, respiratory safety equipment, and ways to ventilate welding work areas.
   g. Explain the purpose of the SDS/MSDS and how it is used.

Trade Terms

Bonded: The permanent joining of metallic parts to form an electrically conductive path that will assure electrical continuity and the capacity to safely conduct any current likely to be imposed on it.

Electrically grounded: Connected to Earth or to some conducting body that serves in place of the earth.

Fume plume: The fumes, gases, and particles from the consumables, base metal, and base metal coating during the welding process.

Galvanized steel: Carbon steel dipped in zinc to inhibit corrosion.

Immediate danger to life and health (IDLH): As defined by OSHA, this term refers to an atmosphere that poses an immediate threat to life, would cause irreversible adverse health effects, or would impair an individual's ability to escape from a dangerous atmosphere.

Purge gas: An inert gas, such as nitrogen, used to drive oxygen away from a weld site.

Ventricular fibrillation: Irregular heart rhythm characterized by a rapid fluttering heartbeat.

There are a number of special hazards associated with welding, including intense heat, flame, and dangers of exposure to a high-intensity electric arc. Metal cutting brings additional hazards to the workplace, including explosive and flammable gases and the intense flame used to cut metal. These conditions create the need for specialized safety practices that must be followed by those performing the work, as well as those in the vicinity of the work.

3.1.0 Welding Safety Hazards

The welding industry is fortunate to have a well-developed standard for safety: *American National Standards Institute (ANSI) Z49.1, Safety in Welding, Cutting, and Allied Processes*. Everyone involved with welding, cutting, and related processes should be familiar with this document. The document discusses personal protective equipment, ventilation, and fire prevention, as well as welding in confined spaces. It contains precautionary information that is important to welders, supervisors, and managers. The document also details specific welding process safety procedures.

There are many causes of accidents. They can usually be divided into two broad categories: personal and physical. Workers must be aware of the factors that can cause accidents so they can understand the consequences of their actions on the job site.

3.1.1 Personal Factors That Cause Accidents

Accidents can often be traced to personal factors such as poor health, lack of experience, and the improper use of alcohol and medications. Something as simple as excessive use of colognes, after-shaves, or perfumes that contain alcohol can represent a burn hazard. A good general rule is: whenever you feel burning or excessive heat, immediately stop work and check it out.

People who are ill or injured may not be able to concentrate on their work. In addition, they may be physically weakened and unable to handle strenuous work. A person doing work that is inherently dangerous is more likely to have an accident when sick or injured. Mental stress can also play a role in accidents; concentration is the issue. An employee who is worried about a serious personal problem is likely to be distracted and unable to focus on work. Workers need to realize that they endanger themselves and those around them when they are not 100 percent effective. As a worker, you need to be able to recognize when others may not be up to par and take appropriate precautions.

Age and inexperience often play a role in accidents. Insurance company studies show that a person who lacks experience is more likely to take risks that cause accidents. Sometimes accidents occur because someone hasn't had enough experience to learn how to avoid them. In fact, insurance company statistics show that more accidents occur in the under-18 age group than in any other. An inexperienced person often is not able to foresee the outcome of an action, or lacks the knowledge to know what works and what doesn't.

People who consume alcohol or use illegal drugs while working risk their lives and the lives of their co-workers. A drug or alcohol hangover is nearly as bad. Legal prescription and over-the-counter drugs can also cause problems. Some cold remedies and cough medicines contain alcohol or other substances that will make people drowsy. Alcohol and drugs affect a person's coordination, alertness, and decision-making ability. Never use these substances while you are working, and don't work with people who use them on the job.

3.1.2 Physical Factors That Cause Accidents

Accidents are often caused by conditions at the job site. For example, foot and vehicle traffic congestion increases at starting and quitting times. As quitting time approaches, workers are more likely to hurry, possibly taking risks they might not otherwise take. People often slow down right after a meal, so workers may be less attentive after lunch than at other times.

People who learned to put their toys away when they were young probably keep a neat, safe work site as adults. The others are likely to have tools, equipment, scrap material, and other items lying around their work site that could trip people. If you keep a messy work site, eventually you or someone else will be injured because of it. Put away your tools, dispose of scrap materials, and secure your equipment and cables when not in use. If you have flammable or hazardous material at your work site, keep it properly contained and covered. When you're done with it, return it to the designated location for disposal or storage.

Damaged or defective welding equipment is dangerous. For example, a frayed or damaged welding cable could give someone a lethal shock or start a fire. Any damaged or defective equipment should be so tagged and either repaired or replaced. All equipment should be periodically inspected for damage. Make sure that periodic maintenance procedures recommended by the manufacturer are performed on schedule.

3.2.0 Work Area Safety

An important factor in area safety is good housekeeping. The work area should be picked up and swept clean. The floors and workbenches should be free of dirt, scrap metal, grease, oil, and anything that is not essential to accomplishing the given task. Collections of steel, welding electrode stubs, wire, hoses, and cables are difficult to work around and easy to trip over. An electrode caddy can be used to hold the electrodes and stubs. Hooks can be made to hold hoses and cables, and scrap steel should be thrown into scrap bins.

The ideal welding shop should have bare concrete floors and electrically grounded and bonded bare metal walls and ceilings to reduce the possibility of fire. Never weld or cut over wood floors, as this increases the possibility of fire.

It is important to keep flammable liquids as well as rags, wood scraps, piles of paper, and other combustibles out of the welding area.

If you must weld in an enclosed building, make every effort to eliminate anything that could trap a spark. Sparks can smolder for hours and then burst into flames. Regardless of where you are welding, be sure to have a fire extinguisher nearby. Also, keep a bucket of water handy to cool off hot metal (other than a test coupon or functional weld) and quickly douse small fires. If a piece of hot metal must be left unattended, use soapstone to write the word HOT on it before leaving. This procedure can also be used to warn people of hot tables, vises, firebricks, and tools.

Perform cutting activities in the workshop in a properly prepared cutting area. Be aware that welding and cutting sparks can ignite any type of flammable vapors. Prevent hot droplets from landing directly on the torch hoses, as they could melt a hole in the hose.

Whenever welding must be done outside a welding booth, use portable screens to protect other personnel from the bright glare of the welding arc (Figure 11). The portable screen also prevents drafts of air from interfering with the stability of the arc. Welding blankets (Figure 12) should be used to cover material and equipment that could be burned or otherwise damaged.

The most common welding accident is burned hands and arms. Keep first-aid equipment nearby to treat burns in the work area. Eye injuries can also occur if you are careless. Post emergency phone numbers in a prominent location.

Figure 11 A typical welding screen.

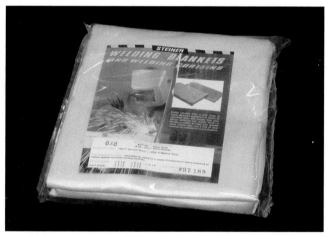

3,000°F (1,620°C) INTERMITTENT, 1,500°F (820°C) CONTINUOUS SILICON DIOXIDE CLOTH

Figure 12 A welding blanket.

The following are some work-area reminders:

- Eliminate tripping hazards by coiling cables and keeping clamps and other tools off the floor.
- Don't get entangled in cables, loose wires, or clothing while you work. You must be able to move freely, especially should your clothing ignite or some other accident occur.
- Clean up oil, grease, or other agents that may ignite and splatter off surfaces while welding.
- Shut off the welding machine and disconnect the power before performing any service or maintenance. A circuit breaker or power disconnect switch should be provided for each machine when hard-wired. Others may be equipped with only a cord and plug.

- Keep the floor free of electrodes and electrode stubs. They could easily cause a slip or fall. All used electrodes should be disposed of in a proper metal container. Throwing electrodes on the floor can result in disciplinary action or even dismissal, due to the safety hazard and poor housekeeping attitude it represents.
- Always remove a SMAW electrode from the electrode holder before laying it down. Leaving a stub in the electrode holder allows it to easily arc to nearby surfaces.
- Work in a dry area, booth, or other shielded area whenever possible.
- Make sure there are no open doors or windows through which sparks may travel to flammable materials.
- Don't coil electrical leads over your body while working on scaffolding.

Conditions on a job site can change quickly. When performing welding or cutting work, it is a good idea to stop occasionally and check your surroundings.

3.3.0 Hot Work Permits and Fire Watches

A hot work permit (*Figure 13*) is an official authorization from the site manager to perform work that may pose a fire hazard. The purpose of the permit is to ensure that a survey process takes place to confirm that the area is safe for cutting or welding, while also establishing accountability for the safety of the job. The permit includes information such as the time, location, and type of work being done. The hot work permit system promotes the development of standard fire safety guidelines. Hot work permits and fire watches are required by OSHA. Permits also help managers keep records of who is working where and at what time. This information is essential in the event of an emergency or at other times when personnel need to be evacuated.

During a fire watch, a person other than the welder or cutting operator must constantly scan the work area for fires. Fire watch personnel must have ready access to fire extinguishers and alarms and know how to use them. When a person is standing fire watch, it should be their only duty.

Cutting operations must never be performed without a fire watch. Whenever oxyfuel cutting equipment is used, there is a great danger of fire. Hot work permits and fire watches are used to minimize this danger. Most sites require the use of hot work permits and fire watches. When they are violated, severe penalties are imposed.

HOT WORK PERMIT

For Cutting, Welding, or Soldering with Portable Gas or ARC Equipment

(References: 1997 Uniform Fire Code Article 49 & National Fire Protection Association Standard NFPA 51B.)

Job Date_____ Start Time_____ Expiration_____ WO #_____

Name of Applicant_____ Company_____ Phone_____

Supervisor_____ Phone_____

Location / Description of work _____

IS FIRE WATCH REQUIRED?

1. _____ (yes or no) Are combustible materials in building construction closer than 35 feet to the point of operation?

2. _____ (yes or no) Are combustibles more than 35 feet away but could be easily ignited by sparks?

3. _____ (yes or no) Are wall or floor openings within 35 foot radius exposing combustible material in adjacent areas, including concealed spaces in floors or walls?

4. _____ (yes or no) Are combustible materials adjacent to the other side of metal partitions, walls, ceilings, or roofs which could be ignited by conduction or radiation?

5. _____ (yes or no) Does the work necessitate disabling a fire detection, suppression, or alarm system component?

YES to any of the above indicates that a qualified fire watch is required.

Fire Watcher Name(s) _____ Phone_____

NOTIFICATIONS

Notify the following groups at least 72 hours prior to work and 30 minutes after work is completed.
Write in names of persons contacted.

Notify in person OR by phone ONLY if question #5 above is answered "yes":

• Facilities Management Fire Alarm Supervisor

Notify by phone or in person: (If by phone, write down name of person and send them a completed copy of this permit.)

• Facilities Management Fire Protection Group
• Environmental Health & Safety Industrial Hygiene Group

SIGNATURES REQUIRED

University Project Manager_____ Date _____ Phone_____

I understand and will abide by the conditions described in this permit. I will implement the necessary precautions which are outlined on both sides of this permit form. Thirty minutes after each hot work session, I will reinspect work areas and adjacent areas to which spark and heat might have spread to verify that they are fire safe, and contact Facilities Management Alarm Technicians to have any disabled fire protection systems reactivated.

_____   _____   Date _____ Phone_____
    Permit Applicant   Company or Department

1/17/15

29101-15_F13.EPS

Figure 13 Hot work permit.

3.4.0 Confined Spaces

A confined space refers to a relatively small or restricted space, such as a storage tank, boiler, or pressure vessel, or small compartments, such as underground utility vaults, small rooms, or the unventilated corners of a room. *Figure 14* shows an example of a confined space entry permit typically found on a job site.

OSHA 29 *CFR* 1910.146 defines a confined space as a space that:

- Is large enough and so configured that an employee can bodily enter and perform assigned work
- Has a limited or restricted means of entry or exit; for example, tanks, vessels, silos, storage bins, hoppers, vaults, and pits
- Is not designed for continuous employee occupancy

OSHA 29 *CFR* 1910.146 further defines a permit-required confined space as a space that:

- Contains or has the potential to contain a hazardous atmosphere

- Contains a material that has the potential for engulfing an entrant
- Has an internal configuration such that an entrant could be trapped or asphyxiated by inwardly converging walls or by a floor that slopes downward and tapers to a smaller cross-section
- Contains any other recognized serious safety or health hazard

For safe working conditions, the oxygen level in a confined space atmosphere must range between 19.5 percent and 23.5 percent by volume, with 21.5 percent being considered the normal level (*Table 2*). Oxygen concentrations below 19.5 percent by volume are considered deficient.

Before operations are started, the wheels of heavy portable equipment must be securely blocked to prevent accidental movement. Where a welder must enter a confined space through a manhole or other opening, all means must be provided for quickly removing the worker in case of emergency (*Figure 15*). When safety belts and lifelines are used for this purpose, they must be attached to the welder's body so that he or she

Oxyfuel Gas Welding Safety Precautions

- Always light the oxyfuel gas torch flame using an approved torch lighter to avoid burning your fingers.
- Never point the torch tip at anyone when lighting it or using it.
- Never point the torch at the cylinders, regulators, hoses, or anything else that may be damaged and cause a fire or explosion.
- Never lay a lighted torch down on the bench or workpiece, and do not hang it up while it is lighted. If the torch is not in the welder's hands, it must be off.
- To prevent possible fires and explosions, check valves and flashback arrestors must be installed in all oxyfuel gas welding and cutting outfits.
- When cutting with oxyfuel gas equipment, clear the area of all combustible materials.
- Skin contact with liquid oxygen can cause frostbite. Be careful when handling liquid oxygen.
- Never use oxygen as a substitute for compressed air.
- All oxygen cylinders with leaky valves or safety fuse plugs and discs should be set aside and marked for the attention of the supplier. Do not tamper with or attempt to repair oxygen cylinder valves. Do not use a hammer or wrench to open the valves.

Master Card No._____

1. Work Description
Area_____   Equipment Location_____
Work to be done:

2. Gas Test		Results	Recheck	Recheck
Required	☐ Instrument Check			
☐ Yes	☐ Oxygen % 20.8 Min.			
☐ No	☐ Combustible % LFL			
		Date/Time/Sig.	Date/Time/Sig.	Date/Time/Sig.

3. Special Instructions: ☐ Check issuer before beginning work   ☐ None

4. Hazardous Materials: ☐ None   What did the line / equipment last contain?

5. Special Protection Required: ☐ None   ☐ Forced Air Ventilation

☐ Avoid Skin Contact   ☐ Gloves _____   ☐ Suit _____

☐ Goggles or Face Shield   ☐ Respirator _____   ☐ Safety Harness

☐ Self-Contained Breathing Equipment   ☐ Hoseline Breathing Equipment

☐ Other, Specify _____   ☐ Standby - Name: _____

6. Fire Protection Required: ☐ None   ☐ Portable Fire Extinguisher   ☐ Fire Hose and Nozzle

☐ Fire Watch   ☐ Other, specify:

7. Condition of Area and Equipment

Required Yes	No		THESE KEY POINTS MUST BE CHECKED
		a.	Lines disconnected & blinded or where disconnecting is not possible, blinds installed? (Includes drains, vents and instrument leads) and appropriate valves locked out?
		b.	Equipment cleaned, washed, purged, ventilated?
		c.	Low voltage or GFCI-protected electrical equipment provided?
		d.	Explosion-proof electrical equipment provided?
		e.	Life lines required to be attached to safety harnesses?

Comments

rev.6/10/15

29101-15_F14A.EPS

Figure 14A Confined space entry permit (sheet 1 of 2).

8. Approval			Permit Authorization Area Supervisor	Permit Acceptance Maint./Contractor Supervisor/Engineer	Date	Time
	Date	Time				
Issued by						
Endorsed by						
Endorsed by						
Endorsed by						

9. Individual Review

I have been instructed on proper Safety Procedures and proper Confined Space Entry Procedures. I have signed in on the appropriate Master Card and have affixed personal locks on energy isolation devices as appropriate.

Signature of all personnel covered by this permit.

Forward to Production Superintendent 7 days after completion of work.

rev.6/10/15

29101-15_F14B.EPS

Figure 14B Confined space entry permit (sheet 2 of 2).

Table 2 Effects of an Increase or Decrease in Oxygen Levels

Oxygen Level	Effects
> 23.5%	Easy ignition of flammable material such as clothes
19.5% – 21.5%*	Normal
17%	Deterioration of night vision, increased breathing volume, accelerated heartbeat
14% – 16%	Very poor muscular coordination, rapid fatigue, intermittent respiration
6% – 10%	Nausea, vomiting, inability to perform, unconsciousness
< 6%	Spasmodic breathing, convulsive movements, and death in minutes

*There is a safety zone of 2%. Above 23.5% is considered enriched.

cannot be jammed in a small exit opening. An attendant with a pre-planned rescue procedure must be stationed outside to observe the welder at all times and must be capable of putting the rescue operations into effect.

When welding or cutting operations are suspended for any substantial period of time, such as during lunch or overnight, all electrodes must be removed from the holders, and the holders must be carefully located so that accidental contact cannot occur. The welding machines must also be disconnected from the power source.

In order to eliminate the possibility of gas escaping through leaks or improperly closed valves when gas welding or cutting, the gas and oxygen supply valves must be closed, the regulators

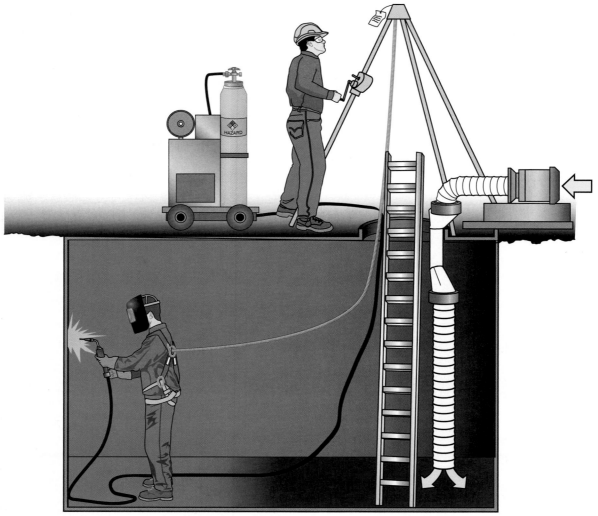

29101-15_F15.EPS

Figure 15 Worker in a confined space.

released, the gas and oxygen lines bled, and the valves on the torch shut off when the equipment will not be used for a substantial period of time. Where practical, the torch and hose must also be removed from the confined space. After welding operations are completed, the welder must mark the hot metal or provide some other means of warning other workers.

> **WARNING!**
>
> Nitrogen and argon are used as backing and purge gases for some welding applications. These gases will displace all the oxygen in the space where they are used.

3.5.0 Welding Equipment Safety

Anyone working with welding equipment must be aware of the equipment-related hazards. Of particular concern are the hazards associated with the voltages and electrical currents produced by welding machines. There are additional hazards related to falling equipment and equipment powered by gasoline engines. *Figure 16* shows some examples of safety notices typically found in operator manuals for welding units.

3.5.1 Electrical Hazards

Electric shock from welding and cutting equipment can result in death, injury, and severe burns. Injury can also result if a fall occurs because of the shock. The amount of current that passes through the human body determines the outcome of an electric shock. The higher the voltage, the greater the chance for a fatal shock. Electric current flows along the path of least resistance to return to its source. If you come in contact with a live conductor, you become a load in the circuit and the current will flow though you. *Figure 17* shows how much resistance the human body presents under various circumstances and how this converts into amps or milliamps when the voltage is 110V. Note

INSTALLATION SAFETY PRECAUTIONS

⚠ WARNING

ELECTRIC SHOCK CAN KILL.
- Do not touch electrically live parts or electrode with skin or wet clothing.
- Insulate yourself from work and ground.
- Always wear dry insulating gloves.

FUMES AND GASES CAN BE DANGEROUS.
- Keep your head out of fumes.
- Use ventilation or exhaust to remove fumes from breathing zone.

WELDING SPARKS CAN CAUSE FIRE OR EXPLOSION.
- Keep flammable material away.
- Do not weld on closed containers.

ARC RAYS CAN BURN EYES AND SKIN
- Wear eye, ear, and body protection.

29101-15_F16.EPS

Figure 16 Safety hazard notices.

that the potential for shock increases dramatically if the skin is damp. A cut will also reduce resistance and increase shock potential. Currents of less than 1 amp can severely injure and even kill a person. Contact with a live electrical circuit can also cause severe burns. Under the right conditions, even a small current can cause serious injury or death. The Construction Employers Association has issued the following guidelines regarding the seriousness of electric shock:

- *Greater than 3 milliamps (mA)* – Indirect accident
- *Greater than 10 mA* – Muscle contraction
- *Greater than 30 mA* – Lung paralysis, usually temporary
- *Greater than 50 mA* – Possible ventricular fibrillation
- *100 mA to 4A* – Certain ventricular fibrillation
- *Greater than 4A* – Heart paralysis, severe burns

Did You Know?

Oxygen-Deficient Air

People don't realize that entering an oxygen-deficient area can result in immediate death. Even a single breath of oxygen-deficient air can prevent muscles from responding and prevent you from escaping. Persons with physical problems such as heart ailments are at the greatest risk. At a refinery in Corpus Christi, Texas, while the welder was at lunch, a helper entered a pipe in which argon had been used as a backing gas. When the welder returned, he found the helper dead.

Nitrogen is lighter than air and, if allowed, will float upward. Argon, on the other hand, is heavier than air and will settle in low-lying areas. When removing nitrogen from a confined space, open the top to allow it to float out. When removing argon from a confined space, provide some way for it to escape out the bottom or push/pull it out using air and mechanical equipment.

DRY SKIN
100,000
TO
600,000 Ω

$$\frac{110 \text{ VOLTS}}{350,000 \ \Omega} = \text{LESS THAN} \atop \text{1 MILLIAMP}$$

WET SKIN
1,000 Ω

$$\frac{110 \text{ VOLTS}}{1,000 \ \Omega} = 110 \text{ MILLIAMPS}$$

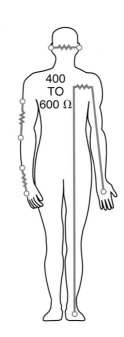

400
TO
600 Ω

EAR TO EAR
$$\frac{110 \text{ VOLTS}}{100 \ \Omega} = 1.1 \text{ AMP}$$

HAND TO FOOT
$$\frac{110 \text{ VOLTS}}{500 \ \Omega} = 220 \text{ MILLIAMPS}$$

29101-15_F17.EPS

Figure 17 Typical body resistances and current.

Each welder and operator of the equipment must be trained and able to recognize the dangers associated with each particular type of equipment to avoid injuries, fatalities, and other electrical accidents.

It is a good idea for all involved to know first-aid procedures, such as cardiopulmonary resuscitation (CPR) for treating electric shock. Treat electrical burns like any other type of burn by applying ice or a cold compress over the burned area. Prevent infection and cover with a clean, dry dressing. Seek medical attention if necessary.

Welding-related electrical hazards come from two sources. The first, which we will call primary hazards, come from the source voltage that powers the welding machine. This voltage is usually 230 or 460 volts. These are potentials that create the possibility of serious, even fatal, shocks. It is important to know that this voltage exists inside the welding machine any time the machine is plugged in. The only way to disable it is to unplug the machine or turn off the power at the disconnect switch. The welding machine must also be properly grounded. This is done by connecting the ground terminal on the machine to a known ground.

> **WARNING!**
>
> Be sure to follow applicable lockout/tagout procedures when disabling the power to the unit. Also, have a qualified electrician install the welding machine so that it will be correctly wired and grounded. If the welding power supply does not work properly, have a qualified technician repair it. Never open the unit yourself.

Secondary hazards exist in the welding process itself. This type of shock can occur if a person touches part of the electrode circuit, such as a bare spot on the electrode cable, and touches the metal work material with another part of the body at the same time. Such a condition can occur if the welder is leaning against the work while welding. Here are some things to remember that will help keep you safe:

- The voltage is highest at the electrode when you are not welding. Due to the high current flow through the leads, the voltage drops somewhat during welding.
- The jaws and screws on the electrode holder, as well as the stick electrode itself, are electrically energized. Make sure they do not touch your skin or clothing, especially if the clothing is damp or wet.

- Never touch an electrode to a grounded conductive surface because these surfaces will become electrically energized. The electrode and work circuit is electrically energized when the welding machine switch is on.
- Never operate arc-welding equipment on a wet or damp floor.
- Wear dry gloves that are in good condition when welding. Keep an extra pair on hand.
- Keep dry insulation between your body, including your arms and legs, and the metal being welded or a ground such as a metal floor or wet earth. Use plywood or a rubber mat between you and the ground or floor.
- Keep the welding cables and electrode holder in good condition.

If you must work in conditions where you might come into contact with the work while welding, use one of the following types of machines:

- Semiautomatic DC constant-voltage welder
- DC manual (stick) welder
- AC welder with reduced voltage control
- Use a ground fault circuit interrupter (GFCI) for power tools that plug in

3.5.2 Gasoline-Powered Units

Gasoline- or diesel-powered welding machines are often used on sites where there is no direct access to 230- or 460-volt electrical outlets. In fact, this type of unit is very common. They range from small, portable units that look like the portable generators you would see at hardware stores, to trailer-mounted units like the one shown in *Figure 18*. In addition to the electrical hazards previously covered, gasoline- and diesel-powered units have their own particular hazards:

Shock Prevention

- Floors must be dry at all times. Use wooden platforms or rubberized carpet/floor coverings or other insulated material.
- Only qualified electricians may work on electric arc welding machine power connections.
- Workbenches must be grounded.
- With the power off, cable connections must be checked for cracked insulators, loose contacts, and other signs of damage.

- Gasoline-powered units produce deadly carbon monoxide just like automobile engines. They must never be operated indoors or in an enclosed space unless the exhaust is vented to the outdoors.
- Gasoline is highly flammable, so it must be kept in approved containers and stored away from hot work.
- When refueling a generator, turn off any welding or cutting flames. Allow the engine to cool to prevent the gasoline from contacting hot engine parts.

3.6.0 Fumes and Gases

The gases, fumes, and dust produced by welding and cutting processes can be hazardous. Adequate ventilation must be provided to prevent workers from breathing these products. There are three methods available to protect personnel against fumes and gases:

- *Natural ventilation* – The movement of air through the workplace caused by natural forces is often enough to remove fumes and gases.
- *Mechanical ventilation* – If natural ventilation is not adequate, portable or fixed fans can be used to provide the necessary ventilation.
- *Source extraction* – This method uses a mechanical device to capture welding fumes at or near the arc.

**4-WHEEL TRAILER WITH
MULTI-PROCESS WELDER**

29101-15_F18.EPS

Figure 18 Gasoline-powered welding machine.

Regardless of the ventilation method used, it is important to avoid breathing in the welding fumes. This means keeping your head up and away from the fume plume, which is the term used to describe the fumes, gases, and particles created by the welding process. The heated fumes and smoke generally rise straight up or move towards the ventilation system opening, where a negative pressure exists. Observe the column of smoke and position yourself to avoid it. A small fan may also be used to divert the smoke, but take care to keep the fan from blowing directly on the work area. The gases from flux and the shielding gases controlled by the welding machine must be present at the welding arc in order to protect the molten metal from the surrounding air. A fan directed at the weld will disturb the process and create flaws.

In addition to the fume plume itself, some consumables and base metals contain toxic materials that require special ventilation. These materials include the following:

- Barium
- Cadmium (found in some steel and fasteners)
- Chromium (found in stainless steel)
- Base metal coatings such as paint
- Cobalt
- Copper
- Manganese
- Nickel
- Silica
- Zinc (found in galvanized metals)

Always check to determine what material you are welding or cutting. Some of the materials listed here produce fumes that are known to cause cancer or can lead to serious respiratory illness. Long-term exposure can even result in death. Some consumables are designated as special ventilation products as a result of the hazards they represent. When welding or cutting these materials, wear approved respiratory equipment and/or ensure that source extraction is used.

Chromium 6, also known as hexavalent chrome, is used in stainless steel. It gives off toxic fumes when heated. Manganese is another hazardous metal used in stainless steel. In addition, it is used in some welding rods, so welders have a greater risk of exposure. Exposure to manganese fumes and dust have been linked to Parkinson's Disease.

Galvanized steel, which is very common, is dipped in zinc to inhibit rust. When flame-cut or welded, it can produce a condition known as zinc chills, which causes flu-like symptoms. For occasional short-term exposure to fumes from zinc- or copper-coated materials, an approved respirator must be used.

In many cases, natural ventilation provides the required level of protection. However, it may be necessary to have a specialist determine if the ventilation is adequate. Some guidelines used in making this determination are as follows:

- The welding area must contain at least 10,000 cubic feet. This represents an area of 22' × 22' × 22' for each welder.
- The ceiling height must be no less than 16 feet.
- Cross-ventilation must not be blocked by partitions, equipment, or other barriers.
- The area must not be designated a confined space.

3.6.1 Source Extraction Equipment

Source extraction, also called local exhaust, can be provided by portable, mobile, or stationary systems such as those shown in *Figure 19*. As a general rule, the farther the extraction hose is from the arc, the greater the volume of air required. There are two classes of source extraction equipment: low vacuum/high volume and high vacuum/low volume.

Low vacuum/high volume equipment can be used in situations where the extraction hose is 6 to 15 inches (≈15 to 38 cm) from the arc (*Figure 20*). It uses extraction hoses, or hose and tubing combinations, with a diameter of 6 to 8 inches (≈15 to 20 cm) to accommodate the higher air volume of 560 to 860 cubic feet per minute (≈16 to 24 cubic meters per minute). The mobile unit shown in *Figure 19* is an example of a low vacuum/high volume system.

High vacuum/low volume systems are designed for close-proximity work and are effective in difficult to reach or confined work spaces. In this case, close proximity means the inlet opening needs to be 2" to 4" (≈5 to 10 cm) from the fume source. There are two types of high vacuum/low volume systems. The fume extraction gun type has a fume extraction nozzle built into the welding gun. Suction nozzles—the second type—are positioned near the weld, usually within 4 inches (10 cm) of the work. Adequate ventilation can be a problem in tight or cramped working quarters. To ensure adequate room ventilation in such cases, local exhaust ventilation should be used to capture fumes. Positioning is very important with these systems, to ensure the immediate area around the weld is not disturbed.

The exhaust hood should be kept close enough to the source of the fumes to ensure their capture.

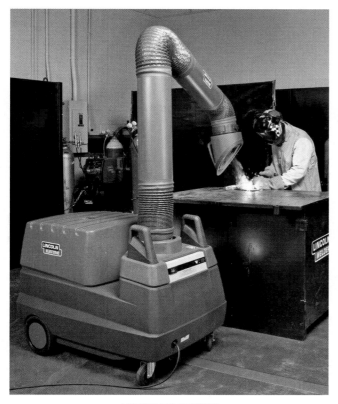

(A) MOBILE UNIT

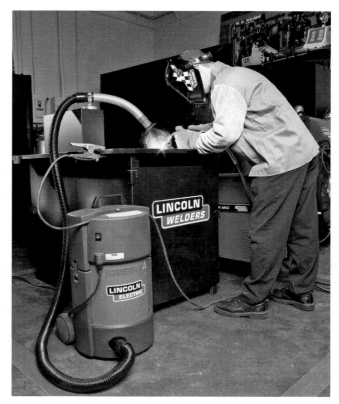

(B) PORTABLE UNIT

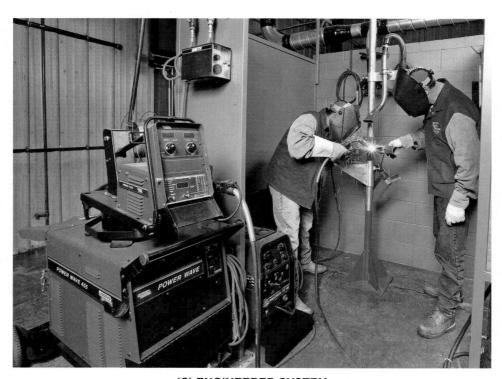

(C) ENGINEERED SYSTEM

29101-15_F19.EPS

Figure 19 Source extraction equipment.

Figure 20 Low vacuum/high volume fume extraction.

Welders should recognize that fumes of any type, regardless of their source, should not be inhaled. The best way to avoid problems is to provide adequate ventilation and keep your face out of the fume plume. Respirators must be used. *Figure 21* shows a welder using both a fume extraction system and a powered air-purifying respirator unit for increased safety.

Figure 21 Personal filtration and an extraction system combined.

3.6.2 Respirators

Special metals require the use of respirators to protect welders from harmful fumes. Respirator selection is based on the type and concentration of the contaminants. Respirators are grouped into three main types based on how they work to protect the wearer from contaminants. The types are the following:

- Air-purifying respirators
- Supplied-air respirators (SARs)
- Self-contained breathing apparatus (SCBA)

If respirators are used, your employer must offer worker training on respirator fitting and use. Medical screenings should also be provided.

Air-Purifying Respirators – Air-purifying respirators provide the lowest level of protection. They are made for use only in atmospheres that have enough oxygen to sustain life (at least 19.5 percent). Air-purifying respirators use special filters and cartridges to remove specific gases, vapors, and particles from the air. The respirator cartridges contain charcoal, which absorbs certain toxic vapors and gases. When the wearer detects any taste or smell, the charcoal's absorption capacity has been reached and the cartridge can no longer remove the contaminant. The respirator filters remove particles such as dust, mists, and metal fumes by trapping them in the filter material. Filters should be changed when it becomes difficult to breathe. Depending on the contaminants, cartridges can be used alone or in combination with a filter/pre-filter and filter cover. Air-purifying respirators should be used for protection only against the types of contaminants listed on the filters and cartridges and on the National Institute for Occupational Safety and Health (NIOSH) approval label affixed to each respirator carton and replacement filter/cartridge carton. Respirator manufacturers typically classify air-purifying respirators into four groups:

- No-maintenance
- Low-maintenance
- Reusable
- Powered air-purifying respirators (PAPRs)

No-maintenance and low-maintenance respirators are typically used for residential or light commercial work that does not call for constant and heavy respirator use. No-maintenance respirators are typically half-mask respirators with permanently attached cartridges or filters. The entire respirator is discarded when the cartridges or filters are spent. Low-maintenance respirators generally are also half-mask respirators that use replaceable cartridges and filters. However, they are not designed for constant use.

Reusable respirators (*Figure 22*) are made in half-mask and full facepiece styles. These respirators require the replacement of cartridges, filters, and respirator parts. Their use also requires a complete respirator maintenance program.

Powered air-purifying respirators (PAPRs) are made in half-mask, full facepiece, and hood styles. They use battery-operated blowers to pull outside air through the cartridges and filters attached to the respirator. The blower motors can be either mask- or belt-mounted. Depending on the cartridges used, they can filter particulates, dusts, fumes, and mists, along with certain gases and vapors. PAPRs like the one shown in *Figure 23* have a belt-mounted, powered air-purifier unit connected to the mask by a breathing tube. The hood style of PAPR can be seen in *Figure 21*. Many models also have an audible and visual alarm that is activated when airflow falls below the required minimum level. This feature gives an immediate indication of a loaded filter or low battery-charge condition. Units with the blower mounted in the mask do not use the breathing

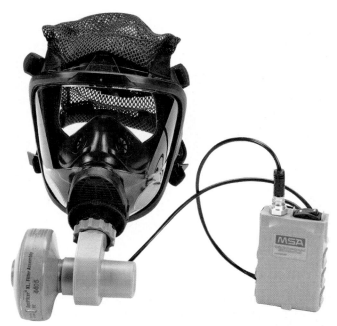

29101-15_F23.EPS

Figure 23 Full facepiece powered air-purifying respirator (PAPR).

tube and belt-mounted package. Since they are more compact, they typically have less filtration capacity.

Supplied-Air Respirators – Supplied-air respirators (*Figure 24*) provide a supply of breathable air for extended periods of time via a high-pressure hose that is connected to an external source, such as a compressor, compressed-air cylinder, or pump. They provide a higher level of protection in atmospheres where air-purifying respirators are

(A) FULL FACEPIECE MASK

(B) HALF MASK

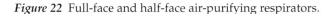

29101-15_F22.EPS

Figure 22 Full-face and half-face air-purifying respirators.

Figure 24 Supplied-air respirator.

not capable of capturing the hazardous fumes. Supplied-air respirators are typically used in toxic atmospheres. Some can be used in atmospheres that are classified as **immediately dangerous to life and health (IDLH)** as long as they are equipped with an air cylinder for emergency escape. An atmosphere is considered IDLH if it poses an immediate hazard to life or produces immediate, irreversible, adverse effects on health. There are two types of supplied-air respirators: continuous-flow and pressure-demand. The unit shown in *Figure 24* is a continuous-flow model.

The continuous-flow supplied-air respirator provides air to the user in a constant stream. One or two hoses may be used to deliver the air from the air source to the facepiece. Unless the compressor or pump is especially designed to filter the air or a portable air-filtering system is used, the unit must be located where there is breathable air (Grade D or better as described in *Compressed Gas Association [CGA] Commodity Specification G-7.1*). Continuous-flow respirators are made with tight-fitting half-masks or full facepieces. They are also made with hoods. The flow of air to the user may be adjusted either at the air source (fixed flow) or on the unit's regulator (adjustable

flow). Pressure-demand supplied-air respirators are similar to the continuous-flow type except that they supply air to the user's facepiece via a pressure-demand valve as the user inhales and fresh air is required. They typically have a two-position exhalation valve that allows the worker to switch between pressure-demand and negative-pressure modes to permit entry into, movement within, and exit from a work area.

Self-Contained Breathing Apparatus (SCBA) – SCBAs (*Figure 25*) can be used in oxygen-deficient atmospheres (below 19.5 percent oxygen); in poorly ventilated or confined spaces; and in IDLH atmospheres. These respirators provide a supply of air for 30 to 60 minutes from a compressed-air cylinder worn on the user's back.

An emergency escape breathing apparatus (EEBA) is a smaller version of an SCBA cylinder. EEBA units are used for escape from hazardous environments and generally provide only a five- to ten-minute supply of air. This type is usually kept in strategic locations where they can be used in the event of an emergency. In facilities that require that they be available, all workers must become familiar with their exact location.

3.6.3 Respiratory Program

In the United States, OSHA regulations, along with your company's procedures, must be followed when selecting the proper type of respirator for a particular job (*Figure 26*). Respirator selection is based on the contaminant present and its concentration level. This determination should be made by a trained safety specialist. The respirator must be properly fitted and used in accordance with the manufacturer's instructions. It must be worn during all times of exposure. Regardless of

Figure 25 Self-contained breathing apparatus (SCBA).

BEFORE USING A RESPIRATOR YOU MUST DETERMINE THE FOLLOWING:

1. THE TYPE OF CONTAMINANT(S) FOR WHICH THE RESPIRATOR IS BEING SELECTED
2. THE CONCENTRATION LEVEL OF THAT CONTAMINANT
3. WHETHER THE RESPIRATOR CAN BE PROPERLY FITTED ON THE WEARER'S FACE

YOU MUST READ AND UNDERSTAND ALL RESPIRATOR INSTRUCTIONS, WARNINGS, AND USE LIMITATIONS CONTAINED ON EACH PACKAGE BEFORE USE.

29101-15_F26.EPS

Figure 26 Use the right respirator for the job.

the kind of respirator needed, OSHA regulations require employers to have a respirator protection program consisting of the following:

- Standard operating procedures for selection and use
- Employee training
- Regular cleaning and disinfecting
- Sanitary storage
- Regular inspection
- Annual fit testing
- Pulmonary function testing

As an employee, you are responsible for wearing respiratory protection when needed. Both vapors and fumes can be eliminated in certain concentrations by the use of air-purifying devices, as long as oxygen levels are acceptable. Examples of fumes are smoke billowing from a fire, or the fumes generated when welding. Always check the cartridge on your respirator to make sure it is the correct type to use for the air conditions and contaminants found on the job site.

When selecting a respirator to wear while working with specific materials, first identify the hazardous ingredients contained in the material and their exposure levels, then choose the proper respirator to protect yourself at these levels. Always read the product's safety data sheet (SDS) or material safety data sheet (MSDS). It identifies the hazardous ingredients and should

list the type of respirator and cartridge recommended for use with the product.

Limitations that apply to all half-mask (air-purifying) respirators are as follows:

- These respirators do not completely eliminate exposure to contaminants, but they will reduce the level of exposure to below-hazardous levels.
- These respirators do not supply oxygen and must not be used in areas where the oxygen level is below 19.5 percent.
- These respirators must not be used in areas where chemicals provide weak warning signs of their presence, such as a lack of taste or a unique odor.

If your breathing becomes difficult; if you become dizzy or nauseated; if you smell or taste the chemical; or if you have other noticeable effects, leave the area immediately, return to a fresh air area, and seek assistance.

All respirators are useless unless properly fit-tested to each individual. To obtain the best protection from your respirator, you must perform positive and negative fit checks each time you wear it. These fit checks must be done until you have obtained a good face seal.

To perform the positive fit check, do the following:

Step 1 Adjust the facepiece for the best fit, then adjust the head and neck straps to ensure good fit and comfort.

> **WARNING!**
>
> Do not overtighten the head and neck straps. Tighten them only enough to stop leakage. Overtightening can cause facepiece distortion and dangerous leaks.

Step 2 Block the exhalation valve with your hand or other material.

Step 3 Breathe out into the mask to increase the pressure inside.

Step 4 Check for air leakage around the edges of the facepiece.

Step 5 If the facepiece puffs out slightly for a few seconds, a good face seal has been obtained.

To perform a negative fit check, do the following:

Step 1 Block the inhalation valve with your hand or other material.

Step 2 Attempt to inhale.

Step 3 Check for air leakage around the edges of the facepiece.

Step 4 If the facepiece caves in slightly for a few seconds, a good face seal has been obtained.

3.7.0 Safety Data Sheets

Safety data sheets (SDSs) and material safety data sheets (MSDSs) are fact sheets prepared by the chemical manufacturer or importer. All manufacturers of potentially hazardous materials must provide detailed information regarding possible hazards resulting from the use of their product. An SDS/MSDS must be made available to anyone using the product or anyone working in the area where the product is in use. Each product used on a construction site must have an SDS/MSDS readily available for examination by workers. *Figure 27* shows a sample MSDS storage box where the information is available on site. An SDS/MSDS describes the substance, its hazards, safe handling, first aid, and emergency spill procedures.

The SDS replaces the MSDS that has been used for a number of years. The primary difference in the two is the arrangement of the entries; both provide the same basic information and have the same function. The Hazard Communication (HAZCOM) standard requires a new, internationally standardized format for the SDS. The sections of the SDS form include the following:

- *Section 1, Product identification, manufacturer information, recommended uses and restrictions*
- *Section 2, Hazard identification*
- *Section 3, Composition/information on ingredients*
- *Section 4, First aid measures*
- *Section 5, Firefighting information*
- *Section 6, Accidental release measures*
- *Section 7, Handling and storage*
- *Section 8, Exposure controls/personal protection*
- *Section 9, Physical and chemical properties*
- *Section 10, Stability and reactivity*
- *Section 11, Toxicological properties*
- *Section 12, Ecological properties*
- *Section 13, Disposal considerations*
- *Section 14, Transport information*
- *Section 15, Regulatory information*
- *Section 16, Other information*

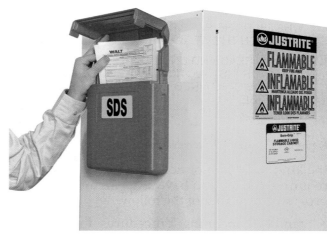

29101-15_F27.EPS

Figure 27 SDS/MSDS storage box.

Respirator Inspection, Care, and Maintenance

A respirator must be clean and in good condition, and all of its parts must be in place for it to give you proper protection. Respirators must be cleaned every day. Failure to do so will limit their effectiveness and offer little or no protection. For example, suppose you wore the respirator yesterday and did not clean it. The bacteria from breathing into the respirator, plus the airborne contaminants that managed to enter the facepiece, will have made the inside of your respirator unsanitary. Continued use may cause you more harm than good. Remember, only a clean and complete respirator will provide you with the necessary protection. Follow these guidelines:

- Inspect the condition of your respirator before and after each use.
- Do not wear a respirator if the facepiece is distorted or if it is worn and cracked. You will not be able to get a proper face seal.
- Do not wear a respirator if any part of it is missing. Replace worn straps or missing parts before using.
- Do not expose respirators to excessive heat or cold, chemicals, or sunlight.
- Clean and wash your respirator after every time you use it. Remove the cartridge and filter, hand wash the respirator using mild soap and a soft brush, and let it air dry overnight.
- Sanitize your respirator each week. Remove the cartridge and filter, then soak the respirator in a sanitizing solution for at least two minutes. Thoroughly rinse with warm water and let it air dry overnight.
- Store the clean and sanitized respirator in its resealable plastic bag. Do not store the respirator face down. This will cause distortion of the facepiece.

The SDS format is gradually replacing the MSDS and is required for all new products. During the transition, you may find that the data sheets for many existing products are still in the MSDS format.

Figure 28 shows a sample SDS for solvent cement used with PVC pipe products. The most important things to look for on an SDS are the specific hazards, personal protection, handling procedures, and first aid information. Most SDSs have a 24-hour emergency-response number.

Look for the information you would need to use the cement described on the SDS form in *Figure 28*. First locate the hazards. Section 2 shows that the adhesive is flammable and an eye and skin irritant that can cause respiratory irritation, dizziness, and drowsiness.

Next, find out how to minimize these hazards. Section 7 gives general handling and storage information. It indicates that ventilation is needed to reduce hazardous vapors. This can be as simple as a fan in an open window. If ventilation is not enough, respiratory protection is needed. Section 8 tells you how to protect your eyes and skin.

Section 4 lists the first-aid measures for eye contact, skin contact, or inhalation. Section 5 explains fire hazards and firefighting measures. You now have the information you need in case of an emergency.

WARNING!

The SDSs must be kept in the work area and be readily accessible to all workers. Hazardous chemical producers are required to provide an SDS/MSDS in the official language(s) of each country or market where it is being produced and sold. Employers should ensure that workers are capable of reading and understanding the contents of an SDS/MSDS, and take steps to provide the information in a language they can understand. The company's safety officer or competent person should always review the SDS/MSDS before the hazardous material is used.

All SDSs/MSDSs must be kept on site. Ask your supervisor to tell you where they are located and to point out the sections that relate to your job. The health and safety of you and your co-workers depends on it. Companies often post these sheets on a bulletin board or put them in a convenient place near the work area.

At this writing, the MSDS is still in use for many products. The layout of the MSDS may vary somewhat between products and/or manufacturers. This is one of the reasons for the change to the SDS format. *Figure 29* shows a portion of an MSDS for Excalibur 7018-1 welding electrodes. As a welder, it is important to be familiar with the safety precautions related to the consumable products in use. Remember that some consumables contain potentially hazardous substances.

GHS SAFETY DATA SHEET

WELD-ON® 705™ Low VOC Cements for PVC Plastic Pipe

Date Revised: **DEC 2011**
Supersedes: **FEB 2010**

SECTION 1 - PRODUCT AND COMPANY IDENTIFICATION

PRODUCT NAME: WELD-ON® 705™ Low VOC Cements for PVC Plastic Pipe
PRODUCT USE: Low VOC Solvent Cement for PVC Plastic Pipe
SUPPLIER:

MANUFACTURER: IPS Corporation
17109 South Main Street, Carson, CA 90248-3127
P.O. Box 379, Gardena, CA 90247-0379
Tel. 1-310-898-3300

EMERGENCY: Transportation: CHEMTEL Tel. 800.255-3924, 813-248-0585 (International) **Medical:** Tel. 800.451.8346, 760.602.8703 3E Company (International)

SECTION 2 - HAZARDS IDENTIFICATION

GHS CLASSIFICATION:

Health		Environmental		Physical	
Acute Toxicity:	Category 4	Acute Toxicity:	None Known	Flammable Liquid	Category 2
Skin Irritation:	Category 3	Chronic Toxicity:	None Known		
Skin Sensitization:	NO				
Eye:	Category 2B				

GHS LABEL: OR

Signal Word: Danger

WHMIS CLASSIFICATION: CLASS B, DIVISION 2

Hazard Statements	Precautionary Statements
H225: Highly flammable liquid and vapor	P210: Keep away from heat/sparks/open flames/hot surfaces – No smoking
H319: Causes serious eye irritation	P261: Avoid breathing dust/fume/gas/mist/vapors/spray
H332: Harmful if inhaled	P280: Wear protective gloves/protective clothing/eye protection/face protection
H335: May cause respiratory irritation	P304+P340: IF INHALED: Remove victim to fresh air and keep at rest in a position comfortable for breathing
H336: May cause drowsiness or dizziness	P403+P233: Store in a well ventilated place. Keep container tightly closed
EUH019: May form explosive peroxides	P501: Dispose of contents/container in accordance with local regulation

SECTION 3 - COMPOSITION/INFORMATION ON INGREDIENTS

	CAS#	EINECS #	REACH Pre-registration Number	CONCENTRATION % by Weight
Tetrahydrofuran (THF)	109-99-9	203-726-8	05-2116297729-22-0000	25 - 50
Methyl Ethyl Ketone (MEK)	78-93-3	201-159-0	05-2116297728-24-0000	5 - 36
Cyclohexanone	108-94-1	203-631-1	05-2116297718-25-0000	15 - 30

All of the constituents of this adhesive product are listed on the TSCA inventory of chemical substances maintained by the US EPA, or are exempt from that listing.
* Indicates this chemical is subject to the reporting requirements of Section 313 of the Emergency Planning and Community Right-to-Know Act of 1986 (40CFR372).
\# indicates that this chemical is found on Proposition 65's List of chemicals known to the State of California to cause cancer or reproductive toxicity.

SECTION 4 - FIRST AID MEASURES

Contact with eyes:	Flush eyes immediately with plenty of water for 15 minutes and seek medical advice immediately.
Skin contact:	Remove contaminated clothing and shoes. Wash skin thoroughly with soap and water. If irritation develops, seek medical advice.
Inhalation:	Remove to fresh air. If breathing is stopped, give artificial respiration. If breathing is difficult, give oxygen. Seek medical advice.
Ingestion:	Rinse mouth with water. Give 1 or 2 glasses of water or milk to dilute. Do not induce vomiting. Seek medical advice immediately.

SECTION 5 - FIREFIGHTING MEASURES

			HMIS	NFPA	
Suitable Extinguishing Media:	Dry chemical powder, carbon dioxide gas, foam, Halon, water fog.				0-Minimal
Unsuitable Extinguishing Media:	Water spray or stream.	Health	2	2	1-Slight
Exposure Hazards:	Inhalation and dermal contact	Flammability	3	3	2-Moderate
Combustion Products:	Oxides of carbon, hydrogen chloride and smoke	Reactivity	0	0	3-Serious
		PPE	B		4-Severe
Protection for Firefighters:	Self-contained breathing apparatus or full-face positive pressure airline masks.				

SECTION 6 - ACCIDENTAL RELEASE MEASURES

Personal precautions:	Keep away from heat, sparks and open flame.
	Provide sufficient ventilation, use explosion-proof exhaust ventilation equipment or wear suitable respiratory protective equipment. Prevent contact with skin or eyes (see section 8).
Environmental Precautions:	Prevent product or liquids contaminated with product from entering sewers, drains, soil or open water course.
Methods for Cleaning up:	Clean up with sand or other inert absorbent material. Transfer to a closable steel vessel.
Materials not to be used for clean up:	Aluminum or plastic containers

SECTION 7 - HANDLING AND STORAGE

Handling: Avoid breathing of vapor, avoid contact with eyes, skin and clothing.
Keep away from ignition sources, use only electrically grounded handling equipment and ensure adequate ventilation/fume exhaust hoods.
Do not eat, drink or smoke while handling.

Storage: Store in ventilated room or shade below 44°C (110°F) and away from direct sunlight.
Keep away from ignition sources and incompatible materials: caustics, ammonia, inorganic acids, chlorinated compounds, strong oxidizers and isocyanates.
Follow all precautionary information on container label, product bulletins and solvent cementing literature.

SECTION 8 - PRECAUTIONS TO CONTROL EXPOSURE / PERSONAL PROTECTION

EXPOSURE LIMITS:

Component	ACGIH TLV	ACGIH STEL	OSHA PEL	OSHA STEL:
Tetrahydrofuran (THF)	50 ppm	100 ppm	200 ppm	
Methyl Ethyl Ketone (MEK)	200 ppm	300 ppm	200 ppm	
Cyclohexanone	20 ppm	50 ppm	50 ppm	

Engineering Controls: Use local exhaust as needed.
Monitoring: Maintain breathing zone airborne concentrations below exposure limits.
Personal Protective Equipment (PPE):
Eye Protection: Avoid contact with eyes, wear splash-proof chemical goggles, face shield, safety glasses (spectacles) with brow guards and side shields, etc. as may be appropriate for the exposure.
Skin Protection: Prevent contact with the skin as much as possible. Butyl rubber gloves should be used for frequent immersion.
Use of solvent-resistant gloves or solvent-resistant barrier cream should provide adequate protection when normal adhesive application practices and procedures are used for making structural bonds.
Respiratory Protection: Prevent inhalation of the solvents. Use in a well-ventilated room. Open doors and/or windows to ensure airflow and air changes. Use local exhaust ventilation to remove airborne contaminants from employee breathing zone and to keep contaminants below levels listed above.
With normal use, the Exposure Limit Value will not usually be reached. When limits approached, use respiratory protection equipment.

29101-15_F28A.EPS

Figure 28A Solvent Cement SDS (sheet 1 of 2).

SECTION 9 - PHYSICAL AND CHEMICAL PROPERTIES

Appearance:	Clear or gray, medium syrupy liquid		
Odor:	Ketone	**Odor Threshold:**	0.88 ppm (Cyclohexanone)
pH:	Not Applicable		
Melting/Freezing Point:	-108.5°C (-163.3°F) Based on first melting component: THF	**Boiling Range:**	66°C (151°F) to 156°C (313°F)
Boiling Point:	66°C (151°F) Based on first boiling component: THF	**Evaporation Rate:**	> 1.0 (BUAC = 1)
Flash Point:	-20°C (-4°F) TCC based on THF	**Flammability:**	Category 2
Specific Gravity:	0.9611 @23°C (73°F)	**Flammability Limits:**	**LEL:** 1.1% based on Cyclohexanone
Solubility:	Solvent portion soluble in water. Resin portion separates out.		**UEL:** 11.8% based on THF
Partition Coefficient n-octanol/water:	Not Available	**Vapor Pressure:**	129 mm Hg @ 20°C (68°F)based on THF
Auto-ignition Temperature:	321°C (610°F) based on THF	**Vapor Density:**	>2 (Air = 1)
Decomposition Temperature:	Not Applicable	**Other Data: Viscosity:**	Medium bodied
VOC Content:	When applied as directed, per SCAQMD Rule 1168, Test Method 316A,VOC content is: $\leq$ 510 g/l.		

SECTION 10 - STABILITY AND REACTIVITY

Stability:	Stable
Hazardous decomposition products:	None in normal use. When forced to burn, this product gives off oxides of carbon, hydrogen chloride and smoke.
Conditions to avoid:	Keep away from heat, sparks, open flame and other ignition sources.
Incompatible Materials:	Oxidizers, strong acids and bases, amines, ammonia

SECTION 11 - TOXICOLOGICAL INFORMATION

Likely Routes of Exposure: Inhalation, Eye and Skin Contact

Acute symptoms and effects:

Inhalation:	Severe overexposure may result in nausea, dizziness, headache. Can cause drowsiness, irritation of eyes and nasal passages.
Eye Contact:	Vapors slightly uncomfortable. Overexposure may result in severe eye injury with corneal or conjunctival inflammation on contact with the liquid.
Skin Contact:	Liquid contact may remove natural skin oils resulting in skin irritation. Dermatitis may occur with prolonged contact.
Ingestion:	May cause nausea, vomiting, diarrhea and mental sluggishness.
Chronic (long-term) effects:	None known to humans

Toxicity:

	LD_{50}	LC_{50}
Tetrahydrofuran (THF)	Oral: 2842 mg/kg (rat)	Inhalation 3 hrs. 21,000 mg/m³ (rat)
Methyl Ethyl Ketone (MEK)	Oral: 2737 mg/kg (rat), Dermal: 6480 mg/kg (rabbit)	Inhalation 8 hrs. 23,500 mg/m³ (rat)
Cyclohexanone	Oral: 1535 mg/kg (rat), Dermal: 948 mg/kg (rabbit)	Inhalation 4 hrs. 8,000 PPM (rat)

Reproductive Effects	**Teratogenicity**	**Mutagenicity**	**Embryotoxicity**	**Sensitization to Product**	**Synergistic Products**
Not Established	Not Established	Not Established	Not Established	Not Established	Not Established

SECTION 12 - ECOLOGICAL INFORMATION

Ecotoxicity:	None Known
Mobility:	In normal use, emission of volatile organic compounds (VOC's) to the air takes place, typically at a rate of $\leq$ 510 g/l.
Degradability:	Biodegradable
Bioaccumulation:	Minimal to none.

SECTION 13 - WASTE DISPOSAL CONSIDERATIONS

Follow local and national regulations. Consult disposal expert.

SECTION 14 - TRANSPORT INFORMATION

Proper Shipping Name:	Adhesives
Hazard Class:	3
Secondary Risk:	None
Identification Number:	UN 1133
Packing Group:	PG II
Label Required:	Class 3 Flammable Liquid
Marine Pollutant:	NO

EXCEPTION for Ground Shipping
DOT Limited Quantity: Up to 5L per inner packaging, 30 kg gross weight per package.
Consumer Commodity: Depending on packaging, these quantities may qualify under DOT as "ORM-D" .

TDG INFORMATION	
TDG CLASS:	FLAMMABLE LIQUID 3
SHIPPING NAME:	ADHESIVES
UN NUMBER/PACKING GROUP:	UN 1133, PG II

SECTION 15 - REGULATORY INFORMATION

Precautionary Label Information: Highly Flammable, Irritant

Symbols: F, Xi

Ingredient Listings: USA TSCA, Europe EINECS, Canada DSL, Australia AICS, Korea ECL/TCCL, Japan MITI (ENCS)

Risk Phrases:
R11: Highly flammable.
R20: Harmful by inhalation.
R36/37: Irritating to eyes and respiratory system.
R66: Repeated exposure may cause skin dryness or cracking
R67: Vapors may cause drowsiness and dizziness

Safety Phrases:
S9: Keep container in a well-ventilated place.
S16: Keep away from sources of ignition - No smoking.
S25: Avoid contact with eyes.
S26: In case of contact with eyes, rinse immediately with plenty of water and seek medical advice.
S33: Take precautionary measures against static discharges.
S46: If swallowed, seek medical advise immediately and show this container or label.

SECTION 16 - OTHER INFORMATION

Specification Information:

Department issuing data sheet:	IPS, Safety Health & Environmental Affairs
E-mail address:	<EHSinfo@ipscorp.com>
Training necessary:	Yes, training in practices and procedures contained in product literature.
Reissue date / reason for reissue:	12/14/2011 / Updated GHS Standard Format
Intended Use of Product:	Solvent Cement for PVC Plastic Pipe

All ingredients are compliant with the requirements of the European Directive on RoHS (Restriction of Hazardous Substances).

This product is intended for use by skilled individuals at their own risk. The information contained herein is based on data considered accurate based on current state of knowledge and experience. However, no warranty is expressed or implied regarding the accuracy of this data or the results to be obtained from the use thereof.

29101-15_F28B.EPS

Figure 28B Solvent Cement SDS (sheet 2 of 2).

Date:	1/21/2013	MSDS No.:	US-M292
Trade Name:	Excalibur 7018-1 MR		
Sizes:	All		
Supersedes:	8/29/2011		

MATERIAL SAFETY DATA SHEET
For Welding Consumables and Related Products
Conforms to Hazard Communication Standard 29CFR 1910.1200 Rev. October 1988

SECTION I - IDENTIFICATION

Manufacturer/ Supplier:	The Lincoln Electric Company 22801 St. Clair Avenue Cleveland, OH 44117-1199 (216) 481-8100	Product Type:	Covered Electrode
		Classification:	AWS E7018-1H4R

SECTION II - HAZARDOUS MATERIAL (1)

IMPORTANT!
This section covers the materials from which this product is manufactured. The fumes and gases produced during welding with the normal use of this product are covered by Section V; see it for industrial hygiene information.
CAS Number shown is representative for the ingredients listed. All ingredients listed may not be present in all sizes.
(1) The term "hazardous" in "Hazardous Materials" should be interpreted as a term required and defined in the Hazards Communication Standard and does not necessarily imply the existence of any hazard. All materials are listed on the TSCA inventory.

Ingredients:	CAS No.	Wt.%	TLV mg/m^3	PEL mg/m^3
Iron	7439-89-6	15	10*	15*
Limestone and/or calcium carbonate	1317-65-3	10	10*	15
Titanium dioxides	13463-67-7	< 5	10	15
Fluorides (as F)	7789-75-5	< 5	2.5	2.5
Silicates and other binders	1344-09-8	< 5	10*	15*
Manganese and/or manganese alloys and compounds (as Mn)*****	7439-96-5	< 5	0.2	5 (c)
Mineral silicates	1332-58-7	< 5	5**	5**
Silicon and/or silicon alloys and compounds (as Si)	7440-21-3	1	10*	15*
Zirconium alloys and compounds (as Zr)	12004-83-0	1	5	5
Cellulose and other carbohydrates	65996-61-4	0.5	10*	15*
Quartz	14808-60-7	< 0.5	#0.025**	#0.1**
Molybdenum alloys (as Mo)	7439-98-7	< 0.5	10	10
Lithium compounds (as Li)	554-13-2	< 0.5	10*	15*
Carbon steel core wire	7439-89-6	55	10*	15*

Supplemental Information:

(*) Not listed. The OSHA PEL for nuisance particles is 15 milligrams per cubic meter. The ACGIH guideline for total particulate is 10 milligrams per cubic meter. PEL value for iron oxide is 10 milligrams per cubic meter. TLV value for iron oxides is 5 milligrams per cubic meter.

(**) As respirable dust.

(*****) Subject to the reporting requirements of Sections 311, 312, and 313 of the Emergency Planning and Community Right-to-Know Act of 1986 and of 40CFR 370 and 372.

(c) Value is for manganese fume. Present PEL is 5 milligrams per cubic meter (ceiling value). Values proposed by OSHA in 1989 were 1.0 milligrams per cubic meter TWA and 3.0 milligrams per cubic meter STEL (Short Term Exposure Limit).

(#) Crystalline silica (quartz) is on the IARC (International Agency for Research on Cancer) and NTP (National Toxicology Program) lists as posing a carcinogenic risk to humans.

SECTION III - HAZARD DATA

Non Flammable; Welding arc and sparks can ignite combustibles and flammable products. See Z49.1 referenced in Section VI.
Product is inert, no special handling or spill procedures required. Not regulated by DOT.

Rev 1/13

(CONTINUED ON SIDE TWO)

29101-15_F29A.EPS

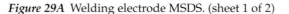

Figure 29A Welding electrode MSDS. (sheet 1 of 2)

| Product: | Excalibur 7018-1 MR |
| Date: | 1/21/2013 |

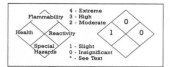

	4 - Extreme
Flammability	3 - High
	2 - Moderate
Health	Reactivity
Special	1 - Slight
Hazards	0 - Insignificant
	* - See Text

SECTION IV - HEALTH HAZARD DATA

Threshold Limit Value: The ACGIH recommended general limit for Welding Fume NOS - (Not Otherwise Specified) is 5 mg/m³.
ACGIH-1999 preface states that the TLV-TWA should be used as guides in the control of health hazards and should not be used as fine lines between safe and dangerous concentrations. See Section V for specific fume constituents which may modify this TLV. Threshold Limit Values are figures published by the American Conference of Government Industrial Hygienists. Units are milligrams per cubic meter of air.

Effects of Overexposure: Electric arc welding may create one or more of the following health hazards:
Fumes and Gases can be dangerous to your health. Common entry is by inhalation. Other possible routes are skin contact and ingestion.

Short-term (acute) overexposure to welding fumes may result in discomfort such as metal fume fever, dizziness, nausea, or dryness or irritation of nose, throat, or eyes. May aggravate pre-existing respiratory problems (e.g. asthma, emphysema).

Long-term (chronic) overexposure to welding fumes can lead to siderosis (iron deposits in lung) and may affect pulmonary function. Manganese overexposure can affect the central nervous system, resulting in impaired speech and movement. Bronchitis and some lung fibrosis have been reported. Repeated exposure to fluorides may cause excessive calcification of the bone and calcification of ligaments of the ribs, pelvis and spinal column. May cause skin rash. Titanium dioxide is listed by the IARC (International Agency for Research on Cancer) as a Group 2B carcinogen (possibly carcinogenic to humans based on animal studies). Respiratory exposure to the crystalline silica present in this welding electrode is not anticipated during normal use. Respiratory overexposure to airborne crystalline silica is known to cause silicosis, a form of disabling pulmonary fibrosis which can be progressive and may lead to death. Crystalline silica is on the IARC (International Agency for Research on Cancer) and NTP (National Toxicology Program) lists as posing a cancer risk to humans. WARNING: This product, when used for welding or cutting, produces fumes or gases which contain chemicals known to the State of California to cause birth defects and, in some cases, cancer. (California Health & Safety Code Section 25249.5 et seq.)

Arc Rays can injure eyes and burn skin. *Skin cancer has been reported.*
Electric Shock can kill. If welding must be performed in damp locations or with wet clothing, on metal structures or when in cramped positions such as sitting, kneeling or lying, or if there is a high risk of unavoidable or accidental contact with workpiece, use the following equipment: Semiautomatic DC Welder, DC Manual (Stick) Welder, or AC Welder with Reduced Voltage Control.
Emergency and First Aid Procedures: Call for medical aid. Employ first aid techniques recommended by the American Red Cross.
IF BREATHING IS DIFFICULT give oxygen. IF NOT BREATHING employ CPR (Cardiopulmonary Resuscitation) techniques.
IN CASE OF ELECTRICAL SHOCK, turn off power and follow recommended treatment. In all cases call a physician.

SECTION V - REACTIVITY DATA

Hazardous Decomposition Products: Welding fumes and gases cannot be classified simply. The composition and quantity of both are dependent upon the metal being welded, the process, procedure and electrodes used.

Other conditions which also influence the composition and quantity of the fumes and gases to which workers may be exposed include: coatings on the metal being welded (such as paint, plating, or galvanizing), the number of welders and the volume of the worker area, the quality and amount of ventilation, the position of the welder's head with respect to the fume plume, as well as the presence of contaminants in the atmosphere (such as chlorinated hydrocarbon vapors from cleaning and degreasing activities.)

When the electrode is consumed, the fume and gas decomposition products generated are different in percent and form from the ingredients listed in Section II. Decomposition products of normal operation include those originating from the volatilization, reaction, or oxidation of the materials shown in Section II, plus those from the base metal and coating, etc., as noted above.

Reasonably expected fume constituents of this product would include: Primarily iron oxide and fluorides; secondarily complex oxides of manganese, potassium, silicon, sodium, and titanium.

Maximum fume exposure guideline for this product (based on manganese content) is 4.0 milligrams per cubic meter.

Gaseous reaction products may include carbon monoxide and carbon dioxide. Ozone and nitrogen oxides may be formed by the radiation from the arc.

Determine the composition and quantity of fumes and gases to which workers are exposed by taking an air sample from inside the welder's helmet if worn or in the worker's breathing zone. Improve ventilation if exposures are not below limits. See ANSI/AWS F1.1, F1.2, F1.3 and F1.5, available from the American Welding Society, 550 N.W. LeJeune Road, Miami, FL 33126.

SECTION VI AND VII
CONTROL MEASURES AND PRECAUTIONS FOR SAFE HANDLING AND USE

Read and understand the manufacturer's instruction and the precautionary label on the product. Request Lincoln Safety Publication E205. See American National Standard Z49.1, "Safety In Welding, Cutting and Allied Processes" published by the American Welding Society, 550 N.W. LeJeune Road, Miami, FL, 33126 (both available for free download at http://www.lincolnelectric.com/community/safety/) and OSHA Publication 2206 (29CFR1910), U.S. Government Printing Office, Superintendent of Documents, P.O. Box 371954, Pittsburgh, PA 15250-7954 for more details on many of the following:
Ventilation: Use enough ventilation, local exhaust at the arc, or both to keep the fumes and gases from the worker's breathing zone and the general area. Train the welder to keep his head out of the fumes. *Keep exposure as low as possible.*
Respiratory Protection: Use respirable fume respirator or air supplied respirator when welding in confined space or general work area when local exhaust or ventilation does not keep exposure below TLV.
Eye Protection: Wear helmet or use face shield with filter lens shade number 12 or darker. Shield others by providing screens and flash goggles.

Protective Clothing: Wear hand, head, and body protection which help to prevent injury from radiation, sparks and electrical shock. See Z49.1.
At a minimum this includes welder's gloves and a protective face shield, and may include arm protectors, aprons, hats, shoulder protection, as well as dark substantial clothing. Train the welder not to permit electrically live parts or electrodes to contact skin . . . or clothing or gloves if they are wet. Insulate from work and ground.
Disposal Information: Discard any product, residue, disposable container, or liner as ordinary waste in an environmentally acceptable manner according to Federal, State and Local Regulations unless otherwise noted. No applicable ecological information available.

29101-15_F29B.EPS

Figure 29B Welding electrode MSDS. (sheet 2 of 2)

Additional Resources

ANSI/AWS Z49.1, Safety in Welding, Cutting, and Allied Processes. Miami, FL: American Welding Society.

Arc Welding Safety E205. Latest Edition. Cleveland, OH: Lincoln Electric.

Personal Protective Equipment (PPE) for Welding and Cutting. Miami, FL. American Welding Society.

Safety Topics in Welding, Cutting, and Brazing. **www.osha.gov**.

AWS F32M/F32, Ventilation Guide for Weld Fumes. Latest Edition. Miami, FL: American Welding Society.

3.0.0 Section Review

1. The AWS standard for welding and cutting safety is _____ .
   a. Z49.1
   b. CFR 1929
   c. CFR 1910
   d. Z45.5

2. The primary purpose of a welding screen is to _____ .
   a. keep non-welders from watching welders at work
   b. protect others from the glare of the welding arc
   c. prevent insects from affecting the welding arc
   d. keep nearby material from catching fire

3. When a worker is assigned the responsibility of providing a fire watch, the worker _____ .
   a. cannot do another task
   b. can retrieve additional consumables for the welder from the shop
   c. can also weld or cut to remain productive
   d. may conduct weld testing

4. Argon is used as a _____ .
   a. breathing gas
   b. cutting fuel gas
   c. solvent
   d. purge gas

5. Which of the following is a *correct* statement about electrical hazards associated with welding?
   a. The jaws of the electrode holder do not carry electrical current.
   b. GFCIs should not be used when welding.
   c. Electrical shocks can cause severe burns.
   d. Welding machines typically operate on 120-volt house current.

6. Which of these metals is *least* likely to produce toxic fumes when heated?
   a. Manganese
   b. Hexavalent chrome
   c. Carbon steel
   d. Galvanized steel

7. In which section of an SDS are hazards identified?
   a. Section 1
   b. Section 2
   c. Section 5
   d. Section 12

SUMMARY

Welders join metal using a high-intensity electric arc to melt and fuse the base metals. Welders may work on critical structures such as pipelines, high-rise buildings, ships, and bridges, so the trade requires a high level of skill and concentration, which can only be achieved through extensive training and practice. Welders also are called upon to cut and gouge metal using equipment that produces a high-intensity flame. The most common cutting method is oxyfuel cutting, in which oxygen is mixed with a fuel gas such as acetylene to produce the cutting flame.

Safety is everyone's responsibility. Proper clothing, footwear, and eye protection are required for safe welding and cutting. Workers who fail to comply with safety rules are subject to dismissal. All welding shops must have established plans for dealing with accidents. Take the time to learn the proper procedures for accident response and reporting before you need to respond to an emergency.

Accidents are very harmful for both employees and employers, and they are often caused by poor behavior and unsafe conditions. Most accidents can be prevented. By knowing and avoiding the behaviors that cause accidents and by keeping working conditions safe, it may be possible to avoid injuries and reduce hazards.

Trying to bluff your way through a job you do not understand is asking for trouble. Even if you think you know the correct procedures, a review may bring out an important part of the job that you may have forgotten. Do not be afraid to ask questions. The responses you receive may help new or less experienced co-workers get answers to questions they may be embarassed to ask.

Materials, fumes, welding radiation exposure, and storing and handling cylinders present particular hazards for the welder. Developing an attitude of safety is an excellent way for every worker to avoid or reduce all of these hazards. Practicing good safety attitudes means that you do the following:

- Report all unsafe conditions and behaviors immediately.
- Keep work areas clean and orderly at all times.
- Immediately report all accidents and injuries, no matter how minor.
- Be certain you completely understand the instructions given before starting work.
- Know how and where medical help may be obtained.
- Wear the required protective devices when working in a hazardous operation area.
- Do not use alcohol or drugs. If you are ill and must take prescribed medication, notify your supervisor immediately.

1. The most common type of welding is ____ .
   a. GTAW
   b. GMAW
   c. FCAW
   d. SMAW

2. A competency-based curriculum is one in which ____ .
   a. the focus is on classroom training
   b. students must demonstrate knowledge and skills
   c. instructors must have four-year degrees
   d. all the training is done on the job

3. Youth apprenticeship programs are available that allow students to begin their apprenticeship or craft training ____ .
   a. in middle school
   b. in high school
   c. instead of their traditional subjects
   d. instead of English classes

4. Body protection while welding is best provided by wearing ____ .
   a. pants with cuffs
   b. polyester clothing
   c. cotton clothing
   d. loose-fitting shirts

5. Which of the following is a *correct* statement?
   a. It is not necessary to wear other eye protection if you are wearing a welding helmet.
   b. The same shade of lens can be used for almost all welding and cutting tasks.
   c. The recommended shade for electric arc operations is 9–14.
   d. Safety glasses with side shields should not be used because they block peripheral vision.

6. Accidents are more likely for employees in the age group ____ .
   a. 18 and under
   b. 19 to 25
   c. 26 to 45
   d. 60 and over

7. A good work area safety practice is ____ .
   a. welding near an open window
   b. welding over wooden floors
   c. using cardboard to control welding/grinding sparks
   d. writing HOT on hot metal before leaving it unattended

8. A hot work permit ____ .
   a. authorizes the performance of work potentially posing a fire hazard
   b. promotes development of standard fire safety guidelines
   c. records unsafe conditions at a job site
   d. helps the manager keep records of hazardous spaces

9. Cutting operations should never be performed without which of these items in the area?
   a. A bucket of sand
   b. A bucket of water
   c. A fire watch
   d. A fire hose

10. A confined space is one that ____ .
    a. has a flammable atmosphere
    b. has unrestricted means of entry or exit
    c. is designed for continuous employee occupancy
    d. is large enough that a worker can enter and work

11. An atmosphere is considered oxygen deficient when the oxygen level is below ____ .
    a. 19.5 percent
    b. 20 percent
    c. 21.5 percent
    d. 23.5 percent

12. The potential for electric shock ____ .
    a. decreases when the skin is damp
    b. remains the same when the skin is damp
    c. increases when the skin is damp
    d. decreases when the skin is cut

13. A potentially toxic coating that is used in galvanized metal is _____ .
    a. zinc
    b. chromium
    c. barium
    d. manganese

14. To make sure a respirator provides proper protection, what must be done each time it is worn?
    a. Only a positive fit check must be performed.
    b. Positive and negative fit checks must be performed.
    c. Only a negative fit check must be performed.
    d. The head and neck straps must be tightened as much as possible

15. An SDS is a form used to _____ .
    a. file a worker's compensation claim with an insurance company
    b. list the contents, hazards, and precautions that pertain to a chemical or material
    c. record unsafe conditions that exist at a job site
    d. recommend changes to an employer safety program

Trade Terms Quiz

Fill in the blank with the correct term that you learned from your study of this module.

1. An electric circuit that is connected to the earth is said to be _____ .

2. Helium or argon is used as a(n) _____ in some welding and cutting operations.

3. The cloud of gases and other materials that rises during the welding process is called the _____ .

4. An irregular heart rhythm characterized by a fluttering heartbeat is called _____.

5. A condition that poses an immediate or delayed threat to life or health is referred to as _____ .

6. When metal parts are joined together to form an electrically conductive path, they are said to be _____ .

7. When skin is exposed to intense ultraviolet light, it can cause a(n) _____ .

8. When carbon is added to steel to increase hardness, the result is _____ .

9. Steel dipped in zinc is referred to as _____ .

10. If your eyes are exposed to intense ultraviolet light, the injury is referred to as a(n) _____ .

11. When nitrogen is used to drive oxygen away from a weld site, it is called a(n) _____ .

12. Chromium, manganese, and other materials are added to steel to produce _____ .

13. The flow of an electrical current across an air gap is called a(n) _____ .

14. The point from which a welding arc is produced is called a(n) _____ .

15. A(n) _____ is a material used to dissolve or prevent the formation of oxides and other undesirable substances on a weld joint.

16. The invisible rays produced by arc welding that can cause burns to the eye are _____ .

Trade Terms

Arc burn 7
Bonded 6
Carbon steel 8
Electric arc 13
Electrically grounded 1

Electrode 14
Flash burn 10
Flux 15
Fume plume 3
Galvanized steel 9

Immediate danger to life and health (IDLH) 5
Purge gas 11
Shielding gas 2
Stainless steel 12

Ultraviolet (UV) radiation 16
Ventricular fibrillation 4

Bill D. Cherry

Zachry Industrial, Inc.
Manager — Weld Testing

How did you choose a career in the construction industry?
I liked looking at a finished product that resulted from my efforts and knowing that it made an impact on people's lives. Most, if not all, of the structures I welded are still in use today. This gives me a strong sense of accomplishment.

Who inspired you to enter the industry?
I was inspired by the father of a school friend who owned a private construction company.

How important are education and training in construction?
They help me maintain my current position and also to teach the trades I have experience with.

What kinds of work have you done in your career?
After graduating from high school, I started out in the road-boring business. This is where I learned how to weld. My next job was at a coal-fired power plant in San Antonio, Texas where I was exposed to many different welding processes and procedures. After that, I was given the opportunity to learn welding inspection and training. I was able to achieve Certified Welding Instructor (CWI), Certified Welding Educator (CWE), and NCCER Master Welding Instructor certifications. I have also trained students to become welders since that time.

Tell us about your present job.
I primarily test welders for our construction and maintenance jobsites. I also have been training welders for 25 years.

What factors have contributed most to your success?
I attribute my success to one person, who gave me the opportunity to advance my career. Without that one individual, I might still be in an unfulfilling career. One person can make a big difference in your life. I started as a welding helper at Zachary Industrial. I eventually became a certified welder, and then a welding inspector, and now manager of weld testing.

What types of training have you been through?
The majority of my welding skills were self-taught. I did receive formal training to become a Certified Welding Instructor (CWI) and have been through the NCCER Instructor Certification Training Program (ICTP). Of course, on-the-job training also played a major role.

Would you suggest construction as a career to others? Why?
Absolutely. Obviously it is not for everyone, but it can be a very rewarding and lucrative career.

What advice would you give to those new to the field?
My advice to those new to the field is never stop learning and always strive to better yourself. Never just sit back and be content with what you are currently doing. The opportunities are always there. You have to actively pursue them.

Tell us an interesting career-related fact or accomplishment.
I hold the only known certification from NCCER as a Master Welding Instructor. Also, I have done many different types of construction. I have administered more than 10,000 weld tests to date. Another thing that means a lot to me is having former students stop by my shop just to say hello and let me know how well they are doing in the welding field.

How do you define craftsmanship?
I define it as the ability to do quality work the first time around.

Trade Terms Introduced in This Module

Arc burn: Burn to the skin produced by brief exposure to intense radiant heat and ultraviolet light.

Bonded: The permanent joining of metallic parts to form an electrically conductive path that will assure electrical continuity and the capacity to safely conduct any current likely to be imposed on it.

Carbon steel: An alloy of iron combining iron and usually less than 1 percent carbon to provide hardness.

Electric arc: The flow of an electrical current across an air gap or gaseous space.

Electrically grounded: Connected to Earth or to some conducting body that serves in place of the earth.

Electrode: The point from which a welding arc is produced.

Flash burn: Burns to the eyes sometimes called welder's flash; caused by exposure to intense radiant heat and ultraviolet light.

Flux: A material used to dissolve or prevent the formation of oxides and other undesirable substances on a weld joint.

Fume plume: The fumes, gases, and particles from the consumables, base metal, and base metal coating during the welding process.

Galvanized steel: Carbon steel dipped in zinc to inhibit corrosion.

Immediate danger to life and health (IDLH): As defined by OSHA, this term refers to an atmosphere that poses an immediate threat to life, would cause irreversible adverse health effects, or would impair an individual's ability to escape from a dangerous atmosphere.

Purge gas: An inert gas, such as nitrogen, used to drive oxygen away from a weld site.

Shielding gas: A gas such as argon, helium, or carbon dioxide used to protect the welding electrode wire from contamination in GMAW and FCAW welding.

Stainless steel: An iron-based alloy usually containing at least 11 percent chromium.

Ultraviolet (UV) radiation: Invisible rays capable of causing burns. UV rays from the sun are the causes of sunburn.

Ventricular fibrillation: Irregular heart rhythm characterized by a rapid fluttering heartbeat.

Additional Resources

This module presents thorough resources for task training. The following resource material is suggested for further study.

ANSI/AWS Z49.1, Safety in Welding, Cutting, and Allied Processes. Miami, FL: American Welding Society.
Arc Welding Safety E205. Latest Edition. Cleveland, OH: Lincoln Electric.
Personal Protective Equipment (PPE) for Welding and Cutting. Miami, FL: American Welding Society.
Safety Topics in Welding, Cutting, and Brazing. **www.osha.gov**.
AWS F32M/F32, Ventilation Guide for Weld Fumes. Latest Edition. Miami, FL: American Welding Society.

Figure Credits

Justin Poland, Robins and Morton, Module Opener, Figure 1A

The Lincoln Electric Company, Cleveland, OH, USA, Figures 1B, 1C, 2A, 2B, 4, 5, 6A–6C, 18–21, 29, SA02

Zachry Industrial, Inc., Figure 2C

Topaz Publications, Inc., Figures 6D–6F, 7, 8, 9A–9E, 10, 12

John Tillman Company, SA01

AWS F2.2:2001, Lens Shade Selector, reproduced with permission from the American Welding Society (AWS), Miami, FL, USA, Table 1

Sellstrom Manfacturing, Figures 9F, 11

North Safety Products USA, Figure 22A

MSA The Safety Company, Figures 22B, 23–25

Photo courtesy of Justrite Mfg. Co. LLC, Figure 27

Weld-On Adhesives, Inc., a division of IPS Corporation, Figure 28

Section Review Answer Key

Answer	Section Reference	Objective
Section One		
1. b	1.1.0	1a
2. a	1.2.2	1b
Section Two		
1. d	2.1.1	2a
2. a	2.2.1	2b
Section Three		
1. a	3.1.0	3a
2. b	3.2.0	3b
3. a	3.3.0	3c
4. d	3.4.0	3d
5. c	3.5.1	3e
6. c	3.6.0	3f
7. b	3.7.0	3g

NCCER CURRICULA — USER UPDATE

NCCER makes every effort to keep its textbooks up-to-date and free of technical errors. We appreciate your help in this process. If you find an error, a typographical mistake, or an inaccuracy in NCCER's curricula, please fill out this form (or a photocopy), or complete the online form at **www.nccer.org/olf**. Be sure to include the exact module ID number, page number, a detailed description, and your recommended correction. Your input will be brought to the attention of the Authoring Team. Thank you for your assistance.

Instructors – If you have an idea for improving this textbook, or have found that additional materials were necessary to teach this module effectively, please let us know so that we may present your suggestions to the Authoring Team.

NCCER Product Development and Revision

13614 Progress Blvd., Alachua, FL 32615

Email: curriculum@nccer.org
Online: www.nccer.org/olf

❏ Trainee Guide   ❏ Lesson Plans   ❏ Exam   ❏ PowerPoints   Other _____

Craft / Level: _____   Copyright Date: _____

Module ID Number / Title: _____

Section Number(s): _____

Description: _____

_____

_____

_____

_____

Recommended Correction: _____

_____

_____

_____

_____

Your Name: _____

Address: _____

_____

Email: _____   Phone: _____

29102-15
Oxyfuel Cutting

OVERVIEW

Oxyfuel cutting is a method for cutting metal that uses an intense flame produced by burning a mixture of a fuel gas and pure oxygen. It is a versatile metal cutting method that has many uses on job sites. Because of the flammable gases and open flame involved, there is a danger of fire and explosion when oxyfuel equipment is used. However, these risks can be minimized when the operator is well-trained and knowledgeable about the function and operation of each part of an oxyfuel cutting outfit.

Module Two

Trainees with successful module completions may be eligible for credentialing through the NCCER Registry. To learn more, go to **www.nccer.org** or contact us at **1.888.622.3720**. Our website has information on the latest product releases and training, as well as online versions of our *Cornerstone* magazine and Pearson's product catalog.

Your feedback is welcome. You may email your comments to **curriculum@nccer.org**, send general comments and inquiries to **info@nccer.org**, or fill in the User Update form at the back of this module.

This information is general in nature and intended for training purposes only. Actual performance of activities described in this manual requires compliance with all applicable operating, service, maintenance, and safety procedures under the direction of qualified personnel. References in this manual to patented or proprietary devices do not constitute a recommendation of their use.

Objectives

When you have completed this module, you will be able to do the following:

1. Describe oxyfuel cutting and identify related safe work practices.
   a. Describe basic oxyfuel cutting.
   b. Identify safe work practices related to oxyfuel cutting.
2. Identify and describe oxyfuel cutting equipment and consumables.
   a. Identify and describe various gases and cylinders used for oxyfuel cutting.
   b. Identify and describe hoses and various types of regulators.
   c. Identify and describe cutting torches and tips.
   d. Identify and describe other miscellaneous oxyfuel cutting accessories.
   e. Identify and describe specialized cutting equipment.
3. Explain how to set up, light, and shut down oxyfuel equipment.
   a. Explain how to properly prepare a torch set for operation.
   b. Explain how to leak test oxyfuel equipment.
   c. Explain how to light the torch and adjust for the proper flame.
   d. Explain how to properly shut down oxyfuel cutting equipment.
4. Explain how to perform various oxyfuel cutting procedures.
   a. Identify the appearance of both good and inferior cuts and their causes.
   b. Explain how to cut both thick and thin steel.
   c. Explain how to bevel, wash, and gouge.
   d. Explain how to make straight and bevel cuts with portable oxyfuel cutting machines.

Performance Tasks

Under the supervision of your instructor, you should be able to do the following:

1. Set up oxyfuel cutting equipment.
2. Light and adjust an oxyfuel torch.
3. Shut down oxyfuel cutting equipment.
4. Disassemble oxyfuel cutting equipment.
5. Change empty gas cylinders.
6. Cut shapes from various thicknesses of steel, emphasizing:
   - Straight line cutting
   - Square shape cutting
   - Piercing
   - Beveling
   - Cutting slots
7. Perform washing.
8. Perform gouging.
9. Use a track burner to cut straight lines and bevels.

Trade Terms

Backfire
Carburizing flame
Drag lines
Dross
Ferrous metals
Flashback
Gouging

Kerf
Neutral flame
Oxidizing flame
Pierce
Soapstone
Washing

Industry Recognized Credentials

If you are training through an NCCER-accredited sponsor, you may be eligible for credentials from NCCER's Registry. The ID number for this module is 29102-15. Note that this module may have been used in other NCCER curricula and may apply to other level completions. Contact NCCER's Registry at 888.622.3720 or go to **www.nccer.org** for more information.

Contents ———————

Topics to be presented in this module include:

Contents (continued)

Figures and Tables

Figures and Tables (continued)

SECTION ONE

1.0.0 OXYFUEL CUTTING BASICS

Objective

Describe oxyfuel cutting and identify related safe work practices.

a. Describe basic oxyfuel cutting.
b. Identify safe work practices related to oxyfuel cutting.

Trade Terms

Dross: The material (oxidized and molten metal) that is expelled from the kerf when cutting using a thermal process. It is sometimes called slag.

Ferrous metals: Metals containing iron.

Soapstone: Soft, white stone used to mark metal.

29102-15_F01.EPS

Figure 1 Oxyfuel cutting.

In order to perform oxyfuel cutting in a safe and effective manner, it is critical to understand the basic principles of oxyfuel cutting and be thoroughly familiar with safe work practices associated with the process.

1.1.0 The Oxyfuel Cutting Process

Oxyfuel cutting, also called flame cutting or burning, is a process that uses the flame and oxygen from a cutting torch to cut **ferrous metals**. The flame is produced by burning a fuel gas mixed with pure oxygen. The flame heats the metal to be cut to the kindling temperature (a cherry-red color); then a stream of high-pressure pure oxygen is directed from the torch at the metal's surface. This causes the metal to instantaneously oxidize or burn away. The cutting process results in oxides that mix with molten iron and produce **dross**, which is blown from the cut by the jet of cutting oxygen. This oxidation process, which takes place during the cutting operation, is similar to an accelerated rusting process. *Figure 1* shows an operator performing oxyfuel cutting.

The oxyfuel cutting process is usually performed only on ferrous metals such as straight carbon steels, which oxidize rapidly. This process can be used to quickly cut, trim, and shape ferrous metals, including the hardest steel.

Oxyfuel cutting can be used for certain metal alloys, such as stainless steel; however, the process requires higher preheat temperatures (white

heat) and about 20 percent more oxygen for cutting. In addition, sacrificial steel plate or rod may have to be placed on top of the cut to help maintain the burning process. Other methods, such as carbon arc cutting, powder cutting, inert gas cutting, and plasma arc cutting, are much more practical for cutting steel alloys and nonferrous metals. Some of these methods are covered in other modules.

1.2.0 Oxyfuel Safety Summary

Cutting activities present unique hazards depending upon the material being cut and the fuel used to power the equipment. The proper safety equipment and precautions must be used when working with oxyfuel equipment because of the potential danger from the high-pressure flammable gases and high temperatures used. The following is a summary of safety procedures and practices that must be observed while cutting or welding. Keep in mind that this is just a summary. Complete safety coverage is provided in the *Welding Safety* module. Trainees who have not completed that module should do so before continuing.

1.2.1 Protective Clothing and Equipment

Oxyfuel cutting produces intense light and heat. It can also produce flying sparks and toxic fumes. To avoid injury, operators must wear appropriate personal protective equipment (PPE) when performing oxyfuel cutting operations. The following is a list of safety practices related to protective clothing and equipment that operators should follow:

- Always use safety glasses with a full face shield or a helmet (*Figure 2*). The glasses, face shield, or helmet lens must have the proper light-reducing tint for the type of cutting to be performed.
- Wear proper protective leather and/or flame-retardant clothing along with welding gloves that protect from flying sparks and molten metal as well as heat.
- Wear high-top safety shoes or boots. Make sure that the tongue and lace area of the footwear will be covered by a pant leg. If the tongue and lace area is exposed or the footwear must be protected from burn marks, wear leather spats under the pants or chaps and over the top of the footwear.
- Wear a 100-percent cotton cap with no mesh material included in its construction. The bill of the cap points to the rear. If a hard hat is required for the environment, use one that allows the attachment of rear deflector material and a face shield. A hard hat with a rear deflector is generally preferred when working overhead, and may be required by some employers and job sites.

WARNING!

Do not wear a cap with a button in the middle. The conductive metal button beneath the fabric represents a safety hazard.

- Wear a face shield over safety glasses for cutting. Either the face shield or the lenses of the safety glasses must be an approved shade for the application. A welding hood equipped with a properly tinted lens is also acceptable. A shade 3 to 6 filter is recommended, depending on the thickness of the metal being cut, as required by *ANSI Z49.1*.
- Wear earplugs to protect ear canals from sparks. Wear hearing protection to protect against the consistent sound of the torch.

WARNING!

Ear protection is essential to protect ears from the noise of the torch. Other personal protective equipment (PPE) must be worn to protect the operator from hot metal and slag.

EAR PLUGS

SAFETY GLASSES

SHADED FACE SHIELD 3-5 SHADE

GAUNTLET-TYPE WELDING GLOVES

COTTON OR WOOL OUTER GARMENTS

PANT LEG EXTENDS ALL THE WAY TO THE INSTEP OF THE BOOT (NO CUFF)

LEATHER BOOT OR SHOE PROTECTION (SPATS)

29102-15_F02.EPS

Figure 2 Personal protective equipment (PPE) for oxyfuel cutting.

NCCER – *Welding Level One* 29102-15

- Cutting operations involving materials or coatings containing cadmium, mercury, lead, zinc, chromium, and beryllium result in toxic fumes. For long-term cutting of such materials, always wear an approved full face, supplied-air respirator (SAR) that uses breathing air supplied externally of the work area. For occasional, very short-term exposure to zinc or copper fumes, a high-efficiency particulate arresting (HEPA)-rated or metal-fume filter may be used on a standard respirator.

1.2.2 Fire/Explosion Prevention

Most welding environment fires occur during oxyfuel gas welding or cutting. To minimize fire and explosion hazards, all cutting should be done in designated areas of the shop if possible. These areas should be made safe for cutting operations with concrete floors, arc filter screens, protective drapes, and fire extinguishers. A welding blanket (*Figure 3*) can be used to protect items in the area that would otherwise be damaged. No combustibles should be stored nearby. The work area should be kept neat and clean, and any metal scrap or dross must be cold before disposal.

Operators should be well-trained in the function and operation of each part of an oxyfuel gas welding or cutting station. In addition, it is often required that at least one fire watch be posted with an extinguisher to watch for possible fires.

The following list contains other steps that operators should follow to help prevent fires and explosions.

- Never carry matches or gas-filled lighters. Sparks can cause the matches to ignite or the lighter to explode, resulting in serious injury.
- Always comply with any site requirement for a hot-work permit and/or a fire watch.
- Never use oxygen to blow off clothing. The oxygen can remain trapped in the fabric for a time. If a spark hits the clothing during this time, the clothing can burn rapidly and violently out of control.
- Never release a large amount of oxygen or use oxygen in place of compressed air. Its presence around flammable materials or sparks can cause rapid and uncontrolled combustion. Keep pure oxygen away from oil, grease, and other petroleum products.
- Make sure that any flammable material in the work area is moved or shielded by a fire-resistant covering.

3,000°F (1,649°C) INTERMITTENT, 1,500°F (816°C) CONTINUOUS SILICON DIOXIDE CLOTH

29102-15_F03.EPS

Figure 3 Welding blanket.

- Approved fire extinguishers must be available before attempting any heating, welding, or cutting operations. Make sure the extinguisher is charged, the inspection tag is valid, and any individual that may be required to operate it knows how to do so.
- Never release a large amount of fuel gas, especially acetylene. Propylene and propane tend to concentrate in and along low areas and can ignite at a considerable distance from the release point. Acetylene is lighter than air but is even more dangerous. When mixed with air or oxygen, it will explode at much lower concentrations than any other fuel gas.
- To prevent fires, maintain a neat and clean work area, and make sure that any metal scrap or slag is cold before disposal.

Before cutting containers such as tanks or barrels, check to see if they have contained any explosive, hazardous, or flammable materials, including petroleum products, citrus products, or chemicals that decompose into toxic fumes when heated. Proper procedures for cutting or welding hazardous containers are described in the *American Welding Society (AWS) F4.1, Safe Practices for the Preparation of Containers and Piping for Welding and Cutting,* and *ANSI Z49.1*. As a standard practice, always clean and then fill any tanks or barrels with water, or purge them with a flow of inert gas such as nitrogen to displace any oxygen.

Containers must be cleaned by steam cleaning, flushing with water, or washing with detergent until all traces of the material have been removed.

After cleaning the container, fill it with water (*Figure 4*) or a purge gas, such as carbon dioxide, argon, or nitrogen to displace the explosive fumes. Air, which contains oxygen, is displaced from inside the container by the water or inert gas. Without oxygen, combustion cannot take place.

A water-filled vessel is the best alternative. When using water, position the container to minimize the air space. When using an inert gas, provide a vent hole so the inert gas can push the air and other vapors out to the atmosphere. Keep in mind that these precautions do not guarantee

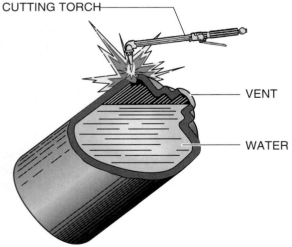

CUTTING TORCH

VENT

WATER

NOTE: ANSI Z49.1 AND AWS STANDARDS SHOULD BE FOLLOWED.

29102-15_F04.EPS

Figure 4 Eliminating/minimizing oxygen in a container.

the absence of flammable materials inside. For that reason, these types of activities should not be done without adequate supervision and the use of proper testing methods.

1.2.3 Work Area Ventilation

Cutting operations should always be performed in a well-ventilated area. This greatly reduces the risks associated with toxic fumes. This rule is especially true for confined spaces. Confined-space procedures must be followed before cutting begins. Proper ventilation must be provided before any cutting procedures take place, but never use oxygen for ventilation purposes.

Work area safety rules apply to all workers in the area. In a typical work area, there might be numerous workers performing various tasks. Always remain aware of other personnel in the area and take the necessary precautions to ensure their safety as well as your own.

1.2.4 Cylinder Handling and Storage

Operators must be aware of the hazards involved in the use of fuel gas, oxygen, and shielding gas cylinders and know how to store these cylinders are stored safely. One basic rule is that only compressed gas cylinders containing the correct gas for the process should be used. Regulators must be correct for the gas and pressure and must function properly. All hoses, fittings, and other parts must be suitable and maintained in good condition.

Any cylinder that leaks, bad valves, or damaged threads must be identified and reported to the supplier. Use a piece of **soapstone** or a tag to write the problem on the cylinder. If closing the cylinder valve cannot stop the leak, move the cylinder outdoors to a safe location, away from any source of ignition, and notify the supplier. Post a warning sign and then slowly release the pressure.

In its gaseous form, acetylene is extremely unstable and explodes easily. For this reason it must remain at pressures below 15 pounds per square inch (psi) or 103 kilopascals (kPa). If an acetylene cylinder is tipped over, stand it upright and wait at least an hour before using it. If the cylinder has spent a significant period of time laying on its side,

Oxygen Consumption

Two-thirds of the oxygen in a neutral flame comes from the air you breathe inside the work space. This could be an issue in a heavily occupied, tight work space.

it is best to allow at least two hours to ensure all of the liquid has drained down to the base of the cylinder. Acetylene cylinders contain liquid acetone. If the liquid is withdrawn from a cylinder, it will foul the safety check valves and regulators and decrease the stability of the acetylene stored in the cylinder. For this reason, acetylene must never be withdrawn at a per-hour rate that exceeds one-seventh of the volume of the cylinder(s) in use. Acetylene cylinders in use should be opened no more than one and one-half turns and, preferably, no more than three-fourths of a turn.

Other precautions associated with gas cylinders include the following:

- Keep cylinders in the upright position and securely chained to an undercarriage or fixed support so that they cannot be knocked over accidentally. Even though they are more stable, cylinders attached to a gas distribution manifold should be chained or otherwise confined, as should cylinders stored in a special room used only for cylinder storage.
- Keep cylinders that contain combustible gases in one area of the building for safety. Cylinders must be at a safe distance from arc welding or cutting operations and any other source of heat, sparks, or flame.
- Cylinder storage areas must be located away from halls, stairwells, and exits so that in case of an emergency they will not block an escape route. Storage areas should also be located away from heat, radiators, furnaces, and welding sparks. The location of storage areas should be where unauthorized people cannot tamper with the cylinders. A warning sign that reads "Danger—No Smoking, Matches, or Open Lights", or similar wording, should be posted in the storage area.
- Oxygen and fuel gas cylinders or other flammable materials must be stored separately. The storage areas must be separated by 20 feet (6.1 m) or by a wall 5 feet (1.5 m) high with at least a 30-minute burn rating. The purpose of the distance or wall is to keep the heat of a small fire from causing the oxygen cylinder safety valve to release. If the safety valve releases the oxygen, a small fire could quickly become an inferno.
- Inert gas cylinders may be stored separately or with either fuel cylinders or oxygen cylinders. Empty cylinders must be stored separately from full cylinders, although they may be stored in the same room or area. All cylinders must be stored vertically and have the protective caps screwed on firmly.
- Never allow a welding electrode, electrode holder, or any other electrically energized parts to come in contact with the cylinder.
- When opening a cylinder valve, operators should stand with the valve stem between themselves and the regulator.
- If a cylinder is frozen to the ground, use warm water (not boiling) to loosen it.

Cylinders equipped with a valve protection cap must have the cap in place unless the cylinder is in use. The protection cap prevents the valve from being broken off if the cylinder is knocked over. If the valve of a full high-pressure cylinder (such as argon or oxygen) is broken off, the cylinder can take flight like a missile if it has not been secured properly. Never lift a cylinder by the safety cap or valve. The valve can easily break off or be damaged. When moving cylinders, the valve protection cap must be replaced, especially if the cylinders are mounted on a truck or trailer. Cylinders must never be dropped or handled roughly.

WARNING!

Using a wrench inserted through the cap can open the valve. If the cap is stuck, use a strap wrench to remove, or call the gas supplier.

Additional Resources

ANSI Z49.1, Safety in Welding, Cutting, and Allied Processes. Miami, FL: American Welding Society.

Uniweld Products, Inc. Numerous videos are available at **www.uniweld.com/en/uniweld-videos**. Last accessed: November 30, 2014.

The Harris Products Group, a division of Lincoln Electric. Numerous videos are available at **www.harrisproductsgroup.com/en/Expert-Advice/videos.aspx**. Last accessed: November 30, 2014.

1.0.0 Section Review

1. The oxyfuel cutting process creates oxides that mix with molten iron and produce dross, which is blown from the cut by a jet of _____ .
   a. pure nitrogen
   b. pure oxygen
   c. carbon dioxide
   d. cooling water

2. Carbon dioxide, argon, and nitrogen are all gases that are suitable for _____ .
   a. mixing with a fuel gas to enable oxyfuel cutting
   b. oxidizing dross to create reusable slag
   c. filling and pressurizing confined spaces
   d. purging explosive fumes from containers

SECTION TWO

2.0.0 OXYFUEL CUTTING CONSUMABLES AND EQUIPMENT

Objective

Identify and describe oxyfuel cutting equipment and consumables.

  a. Identify and describe various gases and cylinders used for oxyfuel cutting.
  b. Identify and describe hoses and various types of regulators.
  c. Identify and describe cutting torches and tips.
  d. Identify and describe other miscellaneous oxyfuel cutting accessories.
  e. Identify and describe specialized cutting equipment.

Trade Terms

Backfire: A loud snap or pop as a torch flame is extinguished.

Flashback: The flame burning back into the tip, torch, hose, or regulator, causing a high-pitched whistling or hissing sound.

Gouging: The process of cutting a groove into a surface.

Kerf: The gap produced by a cutting process.

Washing: A term used to describe the process of cutting out bolts, rivets, previously welded pieces, or other projections from the metal surface.

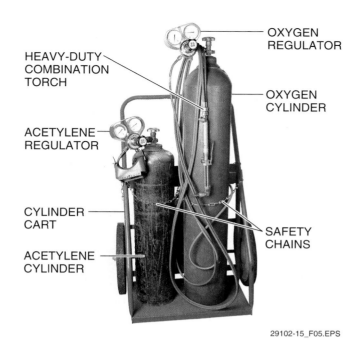

29102-15_F05.EPS

Figure 5 Typical oxyfuel welding/cutting outfit.

2.1.1 Oxygen

Oxygen (O_2) is a colorless, odorless, tasteless gas that supports combustion. It is not considered a fuel gas, but is necessary for combustion. Combined with burning material, pure oxygen causes a fire to flare and burn out of control. When mixed with fuel gases, oxygen produces the high-temperature flame required to flame-cut metals.

Oxygen is stored at more than 2,000 psi (13,790 kPa) in hollow steel cylinders. The cylinders come in a variety of sizes based on different international standards. Specific details about the cylinders are typically regulated by the government agency for the country in which the cylinders are transported, such as the US Department of Transportation (DOT), Transport Canada (TC), and the Department for Transport (DfT) in Europe. *Figure 6* shows high-pressure oxygen cylinder markings and capacities in cubic feet based on US DOT specifications. The smallest standard cylinder holds about 85 cubic feet (2.4 cubic meters) of oxygen, and the largest ultra-high-pressure cylinder holds about 485 cubic feet (13.7 cubic meters). The most common oxygen cylinder size used for cutting operations is the 227-cubic foot (6.4-cubic meter) cylinder. It is more than 4' (1.2 m) tall and 9" (10.2 cm) in diameter. Regardless of their size and capacity, oxygen cylinders must be tested every 10 years in the United States. Other locations may have a different testing standard.

The equipment used to perform oxyfuel cutting includes oxygen and fuel gas cylinders, oxygen and fuel gas regulators, hoses, and a cutting torch. A typical movable oxyfuel (oxy-acetylene) cutting outfit is shown in *Figure 5*.

2.1.0 Cutting Gases

Many oxyfuel cutting outfits use oxygen and acetylene. However, other fuel gases including natural gas and liquefied gases are also used with oxygen for cutting. This section examines some of the more common cutting gases.

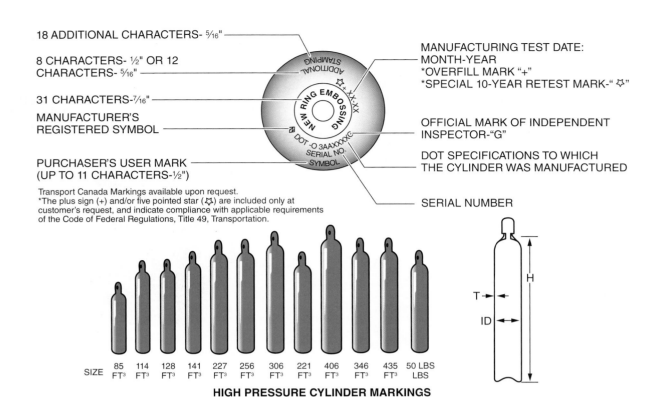

18 ADDITIONAL CHARACTERS- 5/16"

8 CHARACTERS- ½" OR 12 CHARACTERS- 5/16"

31 CHARACTERS-7/16"

MANUFACTURER'S REGISTERED SYMBOL

PURCHASER'S USER MARK (UP TO 11 CHARACTERS-½")

Transport Canada Markings available upon request.
*The plus sign (+) and/or five pointed star (☆) are included only at customer's request, and indicate compliance with applicable requirements of the Code of Federal Regulations, Title 49, Transportation.

MANUFACTURING TEST DATE: MONTH-YEAR
*OVERFILL MARK "+"
*SPECIAL 10-YEAR RETEST MARK-" ☆ "

OFFICIAL MARK OF INDEPENDENT INSPECTOR-"G"

DOT SPECIFICATIONS TO WHICH THE CYLINDER WAS MANUFACTURED

SERIAL NUMBER

| SIZE | 85 FT³ | 114 FT³ | 128 FT³ | 141 FT³ | 227 FT³ | 256 FT³ | 306 FT³ | 221 FT³ | 406 FT³ | 346 FT³ | 435 FT³ | 50 LBS LBS |

HIGH PRESSURE CYLINDER MARKINGS

DOT SPECIFICATIONS	O₂ CAPACITY (FT³)		WATER CAPACITY (IN³)		NOMINAL DIMENSIONS (IN)			NOMINAL WEIGHT (LB)	PRESSURE (PSI)	
	AT RATED SERVICE PRESSURE	AT 10% OVERCHARGE	MINIMUM	MAXIMUM	AVG. INSIDE DIAMETER "ID"	HEIGHT "H"	MINIMUM WALL "T"		SERVICE	TEST

STANDARD HIGH PRESSURE CYLINDERS[1]

3AA2015	85	93	960	1040	6.625	32.50	0.144	48	2015	3360
3AA2015	114	125	1320	1355	6.625	43.00	0.144	61	2015	3360
3AA2265	128	140	1320	1355	6.625	43.00	0.162	62	2265	3775
3AA2015	141	155	1630	1690	7.000	46.00	0.150	70	2015	3360
3AA2015	227	250	2640	2710	8.625	51.00	0.184	116	2015	3360
3AA2265	256	281	2640	2710	8.625	51.00	0.208	117	2265	3775
3AA2400	306	336	2995	3060	8.813	55.00	0.226	140	2400	4000
3AA2400	405	444	3960	4040	10.060	56.00	0.258	181	2400	4000

ULTRALIGHT® HIGH PRESSURE CYLINDERS[1]

E-9370-3280	365	NA	2640	2710	8.625	51.00	0.211	122	3280	4920
E-9370-3330	442	NA	3181	3220	8.813	57.50	0.219	147	3330	4995

ULTRA HIGH PRESSURE CYLINDERS[2]

3AA3600	347[3]	374	2640	2690	8.500	51.00	0.336	170	3600	6000
3AA6000	434[3]	458	2285	2360	8.147	51.00	0.568	267	6000	10000
E-10869-4500	435[3]	NA	2750	2890	8.813	51.00	0.260	148	4500	6750
E-10869-4500	485[3]	NA	3058	3210	8.813	56.00	0.260	158	4500	6750

1. Regulators normally permit filling these cylinders with 10% overcharge, provided certain other requirements are met.
2. Under no circumstances are these cylinders to be filled to a pressure exceeding the marked service pressure at 70°F.
3. Nitrogen capacity at 70°F.

All cylinders normally furnished with ¾" NGT internal threads, unless otherwise specified.
Nominal weights include neck ring but exclude valve and cap, add 2 lbs. (.91 kg) for cap and 1½ lb. (.8 kg) for valve.
Cap adds approximately 5 in. (127 mm) to height.
Cylinder capacities are approximately 5 in. (127 mm) to height.
Cylinder capacities are approximately at 70°F. (21°C).

29102-15_F06.EPS

Figure 6 High-pressure oxygen cylinder markings and sizes.

Oxygen cylinders have bronze cylinder valves on top (*Figure 7*). The cylinder valve controls the flow of oxygen out of the cylinder. A safety plug on the side of the cylinder valve allows oxygen in the cylinder to escape if the pressure in the cylinder rises too high. Although the escaping oxygen presents a hazard, the risk of explosion represents an even more significant hazard. Oxygen cylinders are usually equipped with Compressed Gas Association (CGA) Valve Type 540 valves for service up to 3,000 pounds per square inch gauge (psig), or 20,684 kPa. Some cylinders are equipped with CGA Valve Type 577 valves for up to 4,000 psig (27,579 kPa) oxygen service or CGA Valve Type 701 valves for up to 5,500 psig (34,473 kPa) service. Each CGA valve type is for a specific type of gas and pressure rating. Use care when handling oxygen cylinders because oxygen is stored at such high pressures. When it is not in use, always cover the cylinder valve with the protective steel safety cap and tighten it securely (*Figure 8*).

> **WARNING!**
> Do not remove the protective cap unless the cylinder is secured. If the cylinder falls over and the valve assembly breaks off, the cylinder will be propelled like a rocket, causing severe injury or death to anyone in its path.

2.1.2 Acetylene

Acetylene gas (C_2H_2), a compound of carbon and hydrogen, is lighter than air. It is formed by dissolving calcium carbide in water. It has a strong, distinctive, garlic-like odor, which is added to the gas intentionally so that it can be detected. In its

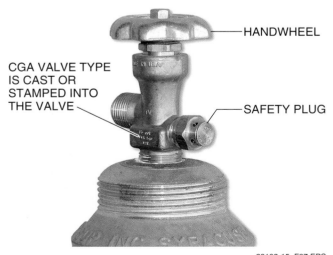

29102-15_F07.EPS

HANDWHEEL

CGA VALVE TYPE IS CAST OR STAMPED INTO THE VALVE

SAFETY PLUG

Figure 7 Oxygen cylinder valve.

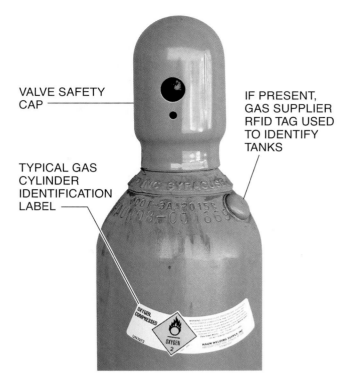

VALVE SAFETY CAP

IF PRESENT, GAS SUPPLIER RFID TAG USED TO IDENTIFY TANKS

TYPICAL GAS CYLINDER IDENTIFICATION LABEL

29102-15_F08.EPS

Figure 8 Oxygen cylinder with standard safety cap.

gaseous form, acetylene is extremely unstable and explodes easily. Because of this instability, it cannot be compressed at pressures of more than 15 psi (103 kPa) when in its gaseous form. At higher pressures, acetylene gas breaks down chemically, producing heat and pressure that could result in a violent explosion. When combined with oxygen, acetylene creates a flame that burns hotter than 5,500°F (3,037°C), one of the hottest gas flames. Acetylene can be used for flame cutting, welding, heating, flame hardening, and stress relieving.

Because of the explosive nature of acetylene gas, it cannot be stored above 15 psi (103 kPa) in a hollow cylinder. To solve this problem, acetylene cylinders are specially constructed to store acetylene. The acetylene cylinder is filled with a porous material that creates a solid cylinder, instead of a hollow cylinder as used for all other common gases. The porous material is soaked with liquid acetone, which absorbs the acetylene gas, stabilizing it and allowing for storage at pressures above 15 psi (103 kPa).

Because of the liquid acetone inside the cylinder, acetylene cylinders must always be used in an upright position. If the cylinder is tipped over, stand the cylinder upright and wait at least one hour before using it. If liquid acetone is withdrawn from a cylinder, it will foul the safety check valves and regulators. It will also cause extremely unstable torch operation. Always take care to control the acetylene gas pressure leaving a cylinder at pressures less than 15 psig (103.4 kPa) and at hourly rates that do not exceed one-seventh of the cylinder capacity. This can easily happen if a torch with a large nozzle is connected to a cylinder that is too small for the task. High rates of discharge may cause liquid acetone to be caught up in the gas stream and be carried out with the gas.

Acetylene cylinders have safety fuse plugs in the top and bottom of the cylinder (*Figure 9*) that melt at 212°F (100°C). A fuse, or fusible, plug is a type of pressure relief device. It is not a valve however; once it is activated, it cannot be reclosed and must be replaced. Fuse plugs, also known as rupture disks, are often used in place of relief valves for low-pressure applications like this one. In the event of a fire, the fuse plugs will release the acetylene gas, preventing the cylinder from exploding.

As with oxygen cylinders, acetylene cylinders are available in a variety of sizes and are typically regulated by government agencies. *Figure 10* shows high-pressure acetylene cylinder markings and capacities in cubic feet based on US DOT specifications. The smallest standard cylinder holds about 10 cubic feet (0.28 cubic meters) of gas. The largest standard cylinder holds about 420 cubic feet (11.9 cubic meters) of gas. A cylinder that holds about 850 cubic feet (24.1 cubic meters) is also available. Like oxygen cylinders, acetylene cylinders used in the United States must be tested every 10 years.

Acetylene cylinders are usually equipped with a CGA 510 brass cylinder valve. The valve controls the flow of acetylene from the cylinder into a regulator. Some acetylene cylinders are equipped with an alternate CGA 300 valve. Some obsolete valves still in use require a special long-handled wrench with a square socket end to operate the valve.

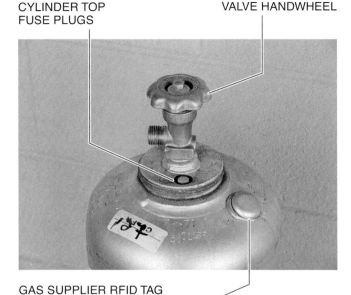

CYLINDER TOP FUSE PLUGS

VALVE HANDWHEEL

GAS SUPPLIER RFID TAG FOR CYLINDER IDENTIFICATION

CYLINDER BOTTOM FUSE PLUGS

29102-15_F09.EPS

Figure 9 Standard acetylene cylinder valve and fuse plugs.

Lifetime Cylinder Management

Gas cylinders may be fitted with radio-frequency identification (RFID) tags so that they can be readily identified and tracked. The RFID tag is electronically scanned and a coded number is matched against the records for identification and tracking. This aids in quickly determining the identity of the purchaser or user of the cylinder, where it has been, and for determining the testing and maintenance records of the cylinder. The RFID tag may appear to be a button like the one shown here. However, it can also be concealed in the walls of the cylinder neck and covered for protection, or have a tight-fitting collar that snaps around the cylinder valve neck.

29102-15_SA01.EPS

Alternate High-Pressure Cylinder Valve Cap

High-pressure cylinders can be equipped with a clamshell cap that can be closed to protect the cylinder valve with or without a regulator installed on the valve. This enables safe movement of the cylinder after the cylinder valve is closed. This type of cap is usually secured to the cylinder body cap threads when it is installed so that it cannot be removed. When the clamshell is closed, it can also be padlocked to prevent unauthorized operation of the cylinder valve.

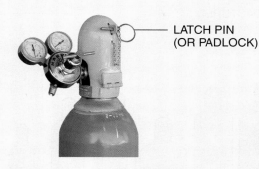

LATCH PIN
(OR PADLOCK)

CLAMSHELL OPEN
TO ALLOW CYLINDER
VALVE OPERATION

CLAMSHELL CLOSED FOR
MOVEMENT OR PADLOCKED
TO PREVENT OPERATION OF
CYLINDER VALVE

CLAMSHELL CLOSED
FOR TRANSPORT

29102-15_SA02.EPS

Alternate Acetylene Cylinder Safety Cap

Acetylene cylinders can be equipped with a ring guard cap that protects the cylinder valve with or without a regulator installed on the valve. This enables safe movement of the cylinder after the cylinder valve is closed. This type of cap is usually secured to the cylinder body cap threads when it is installed so that it cannot be removed. Some other types of cylinders, such as propane cylinders, may also be fitted with this type of guard.

29102-15_SA03.EPS

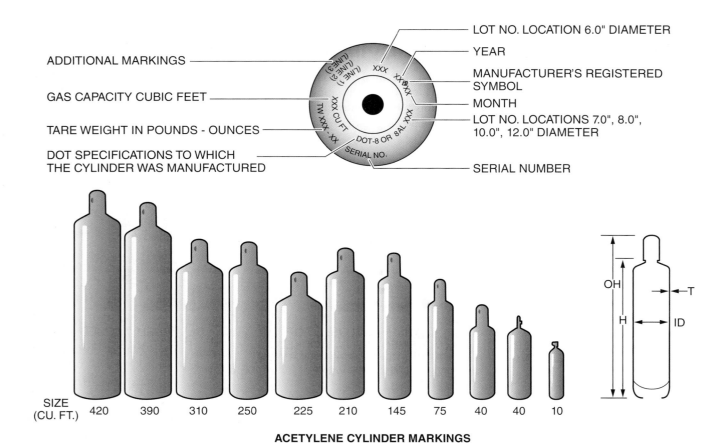

ADDITIONAL MARKINGS

GAS CAPACITY CUBIC FEET

TARE WEIGHT IN POUNDS - OUNCES

DOT SPECIFICATIONS TO WHICH THE CYLINDER WAS MANUFACTURED

LOT NO. LOCATION 6.0" DIAMETER

YEAR

MANUFACTURER'S REGISTERED SYMBOL

MONTH

LOT NO. LOCATIONS 7.0", 8.0", 10.0", 12.0" DIAMETER

SERIAL NUMBER

SIZE (CU. FT.)   420   390   310   250   225   210   145   75   40   40   10

ACETYLENE CYLINDER MARKINGS

DOT SPECIFICATIONS	CAPACITY			NOMINAL DIMENSIONS (IN)				ACETONE (LB - OZ)	APPROXIATE TARE WEIGHT WITH VALVE WITHOUT CAP (LB)
	ACETYLENE	MIN. WATER							
	(FT³)	(IN³)	(LB.)	AVG. INSIDE DIAMETER "ID"	HEIGHT W/OUT VALVE OR CAP "H"	HEIGHT W/VALVE AND CAP "OH"	MINIMUM WALL "T"		
8 AL[1]	10	125	4.5	3.83	13.1375	14.75	0.0650	1-6	8
8[1]	40	466	16.8	6.00	19.8000	23.31	0.0870	5-7	25
8[2]	40	466	16.8	6.00	19.8000	28.30	0.0870	5-7	28
8[3]	75	855	30.8	7.00	25.5000	31.25	0.0890	9-8	45
8	100	1055	38.0	7.00	30.7500	36.50	0.0890	12-2	55
8	145	1527	55.0	8.00	34.2500	40.00	0.1020	18-10	76
8	210	2194	79.0	10.00	32.2500	38.00	0.0940	25-13	105
8AL	225	2630	94.7	12.00	27.5000	32.75	0.1280	29-6	110
8	250	2606	93.8	10.00	38.0000	43.75	0.0940	30-12	115
8AL	310	3240	116.7	12.00	32.7500	38.50	0.1120	39-5	140
8AL	390	4151	150.0	12.00	41.0000	46.75	0.1120	49-14	170
8AL	420	4375	157.5	12.00	43.2500	49.00	0.1120	51-14	187
8	60	666	24.0	7.00	25.79 OH		0.0890	7-11	40
8	130	1480	53.3	8.00	36.00 OH		0.1020	17-2	75
8AL	390	4215	151.8	12.00	46.00 OH		0.1120	49-14	180

1. Tapped for 3/8" valve but are not equipped with valve protection caps.
2. Includes valve protection cap.
3. Can be tared to hold 60 ft³ (1.7 m³) of acetylene gas.
   Standard tapping (except cylinders tapped for 3/8") 3/4"-14 NGT.

Weight includes saturation gas, filler, paint, solvent, valve, fuse plugs.
Does not include cap of 2 lb. (.91 kg.)
Cylinder capacities are based upon commercially pure acetylene gas at 250 psi (17.5 kg/cm²), and 70°F (15°C).

29102-15_F10.EPS

Figure 10 Acetylene cylinder markings and sizes.

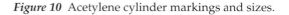

The smallest standard acetylene cylinder, which holds 10 cubic feet (0.28 cubic meters), is equipped with a CGA 200 small series valve, and 40 cubic foot (1.13 cubic meter) cylinders use a CGA 520 small series valve. As with oxygen cylinders, place a protective valve cap on the acetylene cylinders during transport (*Figure 11*).

> **WARNING!**
>
> Do not remove the protective cap unless the cylinder is secured. If the cylinder falls over and the nozzle breaks off, the cylinder will release highly explosive gas.

2.1.3 Liquefied Fuel Gases

Many fuel gases other than acetylene are used for cutting. They include natural gas and liquefied fuel gases such as propylene and propane. Their flames are not as hot as acetylene, but they have higher British thermal unit (Btu) ratings and are cheaper and safer to use. Job site policies typically determine which fuel gas to use. *Table 1* compares the flame temperatures of oxygen mixed with various fuel gases.

Propylene mixtures are hydrocarbon-based gases that are stable and shock-resistant, making them relatively safe to use. They are purchased under trade names such as High Purity Gas (HPG™), Apachi™, and Prestolene™. These gases and others have distinctive odors to make leak detection easier. They burn at temperatures around 5,193°F (2,867°C), hotter than natural gas and propane. Propylene gases are used for flame cutting, scarfing, heating, stress relieving, brazing, and soldering.

Propane is also known as liquefied petroleum (LP) gas. It is stable and shock-resistant, and it has a distinctive odor for easy leak detection. It burns at 4,580°F (2,526°C), which is the lowest

Table 1 Flame Temperatures of Oxygen With Various Fuel Gases

Type of Gas	Flame Temperature
Acetylene	More than 5,500°F (3,038°C)
Propylene	5,130°F (2,832°C)
Natural gas	4,600°F (2,538°C)
Propane	4,580°F (2,527°C)

temperature of any common fuel gas. Propane has a slight tendency toward backfire and flashback and is used quite extensively for cutting procedures.

Natural gas is delivered by pipeline rather than by cylinders. Manifolds must be available on site for the connection of regulators and hoses. It burns at about 4,600°F (2,537°C). Natural gas is relatively stable and shock-resistant and has a slight tendency toward backfire and flashback. Because of its recognizable odor, leaks are easily detectable. Natural gas is used primarily for cutting on job sites with permanent cutting stations.

Liquefied fuel gases are shipped in hollow steel cylinders (*Figure 12*). When empty, they are much lighter than acetylene cylinders.

The hollow steel cylinders for liquefied fuel gases come in various sizes. They can hold from 30 to 225 pounds (13.6 to 102.1 kilograms) of fuel gas. As the cylinder valve is opened, the

29102-15_F11.EPS

Figure 11 Acetylene cylinder with standard valve safety cap.

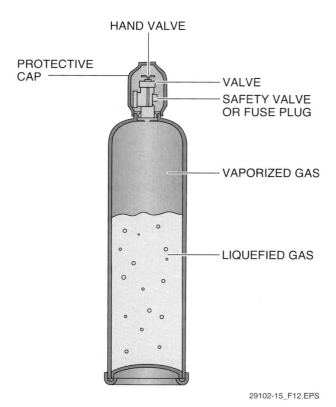

29102-15_F12.EPS

Figure 12 Liquefied fuel gas cylinder.

vaporized gas is withdrawn from the cylinder. The remaining liquefied gas absorbs heat and releases additional vaporized gas. The pressure of the vaporized gas varies with the outside temperature. The colder the outside temperature, the lower the vaporized gas pressure will be. If high volumes of gas are removed from a liquefied fuel gas cylinder, the pressure drops, and the temperature of the cylinder will also drop. A ring of frost can form around the base of the cylinder due to the cooling effect as the liquid vaporizes and absorbs heat. If high withdrawal rates continue, the regulator may also start to ice up. If high withdrawal rates are required, special regulators with electric heaters should be used.

> **WARNING!**
>
> Never apply heat directly to a cylinder or regulator. This can cause excessive pressure, resulting in an explosion.

The pressure inside a liquefied fuel gas cylinder is not an indicator of how full or empty the cylinder is. The weight of a cylinder determines how much liquefied gas is left. Liquefied fuel gas cylinders are equipped with CGA 510, 350, or 695 valves, depending on the fuel and storage pressures.

> **WARNING!**
>
> Do not remove the protective cap on liquefied fuel gas cylinders unless the cylinder is secured. If the cylinder falls over and the nozzle breaks off, the cylinder will release highly explosive gas. Cylinders containing a liquid such as propane must be kept in an upright position. If the valve is broken or is opened with the cylinder horizontal, the fuel can emerge as a liquid that will shoot a long distance before it vaporizes. If it is ignited, it produces an extremely dangerous, uncontrolled flame.

2.2.0 Regulators and Hoses

Regulators (*Figure 13*) are attached to the oxygen and fuel gas cylinder valves. They reduce the high cylinder pressures to the required lower working pressures and maintain a steady flow of gas from the cylinder.

A regulator's pressure-adjusting screw controls the gas pressure. Turned clockwise, it increases the pressure of gas. Turned counterclockwise, it reduces the pressure of gas. When turned counterclockwise until loose (released), it stops the

Handling and Storing Liquefied Gas Cylinders

Liquefied fuel gas cylinders have a safety valve built into the valve at the top of the cylinder. The safety valve releases gas if the pressure begins to rise. Use care when handling fuel gas cylinders because the gas in cylinders is stored at significant pressures. Cylinders should never be dropped or hit with heavy objects, and they should always be stored in an upright position. When not in use, the cylinder valve must always be covered with the protective steel cap.

flow of gas. When the adjusting screw feels very loose, it is an indication that the end of the screw is no longer in contact with the regulating spring. However, the regulator adjusting screw should never be considered a shut-off valve. When shut-off is desired, use the cylinder valve. The regulator can remain at its set position, allowing any gas remaining between the cylinder valve and the regulator to escape.

Most regulators contain two gauges. The high-pressure or cylinder-pressure gauge indicates the actual cylinder pressure (upstream); the low-pressure gauge indicates the pressure of the gas leaving the regulator (downstream).

Oxygen regulators differ from fuel gas regulators. Oxygen regulators may have green markings and always have right-hand threads on all connections. The oxygen regulator's high-pressure gauge generally reads up to 3,000 psi (20,684 kPa) and includes a second scale that shows the amount of oxygen in the cylinder in terms of cubic feet. The low-pressure or working-pressure gauge may read 100 psi (689 kPa) or higher.

Fuel gas regulators may have red markings and usually have left-hand threads on all the connections. As a reminder that the regulator has left-hand threads, a V-notch may be cut into the corners of the fitting nut. These notches are visible on the fitting nut of the fuel gas regulator shown in *Figure 13*. The fuel gas regulator's high-pressure gauge usually reads up to 400 psi (2,758 kPa). The low-pressure or working-pressure gauge may read up to 40 psi (276 kPa). Acetylene gauges, however, are always red-lined at 15 psi (103 kPa) as a reminder that acetylene pressure should not be increased beyond that point.

Single-stage and two-stage regulators will be discussed in the following sections.

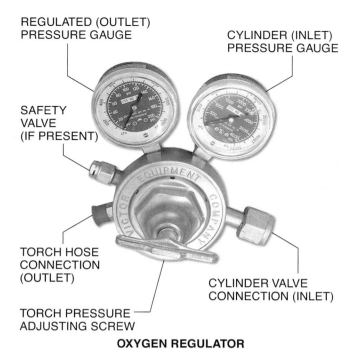

REGULATED (OUTLET) PRESSURE GAUGE

CYLINDER (INLET) PRESSURE GAUGE

SAFETY VALVE (IF PRESENT)

TORCH HOSE CONNECTION (OUTLET)

CYLINDER VALVE CONNECTION (INLET)

TORCH PRESSURE ADJUSTING SCREW

OXYGEN REGULATOR

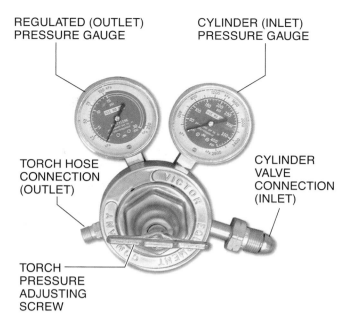

REGULATED (OUTLET) PRESSURE GAUGE

CYLINDER (INLET) PRESSURE GAUGE

TORCH HOSE CONNECTION (OUTLET)

CYLINDER VALVE CONNECTION (INLET)

TORCH PRESSURE ADJUSTING SCREW

FUEL GAS REGULATOR

29102-15_F13.EPS

Figure 13 Oxygen and acetylene regulators.

WARNING!

To prevent injury and damage to regulators, always follow these guidelines:

- Never subject regulators to jarring or shaking, as this can damage the equipment beyond repair.
- Always check that the adjusting screw is fully released before the cylinder valve is turned on and when the welding has been completed.
- Always open cylinder valves slowly and stand with the valve stem between you and the regulator.
- Never use oil to lubricate a regulator. This can result in an explosion when the regulator is in use.
- Never use fuel gas regulators on oxygen cylinders, or oxygen regulators on fuel gas cylinders.
- Never work with a defective regulator. If it is not working properly, shut off the gas supply and have the regulator repaired by someone who is qualified to work on it.
- Never use large wrenches, pipe wrenches, pliers, or slip-joint pliers to install or remove regulators.

2.2.1 Single-Stage Regulators

Single-stage, spring-compensated regulators reduce pressure in one step. As gas is drawn from the cylinder, the internal pressure of the cylinder decreases. A single-stage, spring-compensated regulator is unable to automatically adjust for this decrease in internal cylinder pressure. Therefore, it becomes necessary to adjust the spring pressure periodically to modify the output gas pressure as the gas in the cylinder is consumed. These regulators are the most commonly used because of their low cost and high flow rates.

2.2.2 Two-Stage Regulators

The two-stage, pressure-compensated regulator reduces pressure in two steps. It first reduces the input pressure from the cylinder to a predetermined intermediate pressure. This intermediate pressure is then adjusted by the pressure-adjusting screw. With this type of regulator, the delivery pressure to the torch remains constant, and no readjustment is necessary as the gas in the cylinder is consumed. Standard two-stage regulators (*Figure 14*) are more expensive than single-stage regulators and have lower flow rates. There are also heavy-duty types with higher flow rates that are usually preferred for thick material and/or continuous-duty cutting operations.

2.2.3 Check Valves and Flashback Arrestors

Check valves and flashback arrestors (*Figure 15*) are safety devices for regulators, hoses, and torches. Check valves allow gas to flow in one direction only. Flashback arrestors stop fire from being able to travel backwards through the hose.

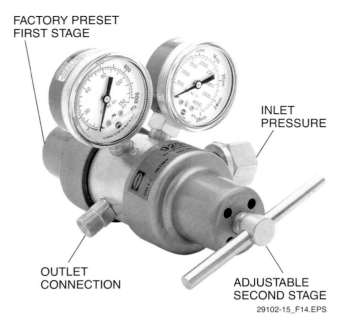

FACTORY PRESET
FIRST STAGE

INLET
PRESSURE

OUTLET
CONNECTION

ADJUSTABLE
SECOND STAGE

29102-15_F14.EPS

Figure 14 Two-stage regulator.

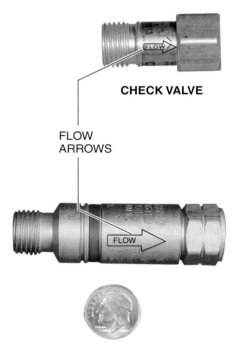

CHECK VALVE

FLOW
ARROWS

FLOW

**FLASHBACK ARRESTOR WITH
INTERNAL CHECK VALVE**

29102-15_F15.EPS

Figure 15 Add-on check valve and flashback arrestor.

Check valves consist of a ball and spring that open inside a cylinder. The valve allows gas to move in one direction but closes if the gas attempts to flow in the opposite direction. When a torch is first pressurized or when it is being shut off, back-pressure check valves prevent the entry and mixing of acetylene with oxygen in the oxygen hose or the entry and mixing of oxygen with acetylene in the acetylene hose.

Flashback arrestors prevent flashbacks from reaching the hoses and/or regulator. They have a flame-retarding filter that will allow heat, but not flames, to pass through. Most flashback arrestors also contain a check valve.

Add-on check valves and flashback arrestors are designed to be attached to the torch handle connections and to the regulator outlets. At a minimum, flashback arrestors with check valves should be attached to the torch handle connections. Both devices have arrows on them to indicate flow direction. When installing add-on check valves and flashback arrestors, be sure the arrow matches the desired gas flow direction.

The fittings for oxyfuel equipment are brass or bronze, and certain components are often fitted with soft, flexible, O-ring seals. The seal surfaces of the fittings or O-rings can be easily damaged by over-tightening with standard wrenches. For that reason, only a torch wrench (sometimes called a gang wrench) should be used to install regulators, hose connections, check valves, flashback arrestors, torches, and torch tips. Longer wrenches that provide more leverage should be avoided.

The universal torch wrench shown in *Figure 16* is equipped with various size wrench cutouts for use with a variety of equipment and standard CGA components. The length of a torch wrench is limited to reduce the chance of damage to fittings because of excessive torque. In some cases,

Serious Cutting

The cutting power of oxyfuel equipment can be very surprising. This 6" (15.2 cm) carbon steel block is no match for a worker with the right torch.

29102-15_SA04.EPS

Figure 16 Universal torch wrench.

manufacturers specify only hand-tightening for certain fitting connections of a torch set (tips or cutting/welding attachments, for example). In any event, follow the manufacturer's specific instructions when connecting the components of a torch set.

2.2.4 Gas Distribution Manifolds

In some applications, cutting gases are used on a large scale. Instead of using a cylinder for each operator, a large bank of cylinders connected to a common manifold is used. *Figure 17* is a representation of such an arrangement. High-pressure hoses called pigtails are used to connect the gas supply tanks to the manifold. Regulators are provided at both the source and the workstation hookups.

Figure 18 shows a manifold setup that might be used in a pipe fabrication shop. Operators would tie-in with their hoses at the drops on the manifolds. Each of these manifolds has four drops. Each drop would be provided with a pressure gauge like the one shown. One function of this gauge is to provide an indication if there is a leak

Figure 18 Operator hookups for cutting gases.

in a hose or in the torch. Hose length must be considered when using a manifold distribution system because the pressure drop increases with

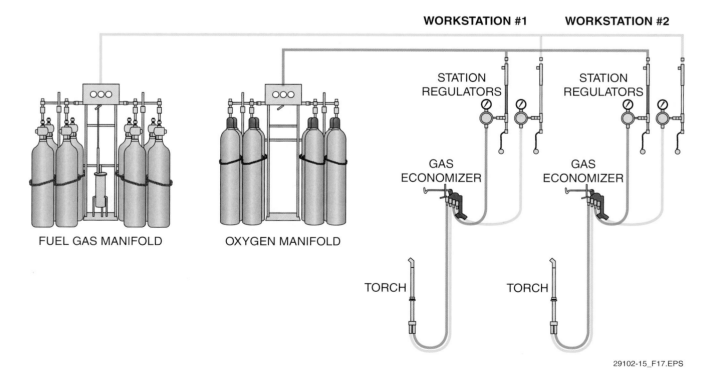

Figure 17 Gas distribution to stations through a manifold.

the length of the hoses. Hoses with a larger inside diameter are needed for long runs.

OSHA provides the following safety precautions specifically for manifold systems:

- Fuel gas and oxygen manifolds must bear the name of the substance they contain.
- Fuel gas and oxygen manifolds must not be placed in confined spaces; they must be placed in safe, well ventilated, and accessible locations.
- Hose connections must be designed so that they cannot be interchanged between fuel gas and oxygen manifolds and supply header connections. Adapters may not be used to interchange hoses.
- Hose connections must be kept free of grease and oil.
- Manifold and header hose connections must be capped when not in use.
- Nothing may be placed on a manifold that will damage the manifold or interfere with the quick closing of the valves.

2.2.5 Hoses

Hoses transport gases from the regulators to the torch. Oxygen hoses are usually green with right-hand threaded connections. Hoses for fuel gas are usually red and have left-hand threaded connections. The fuel gas connection fittings are grooved as a reminder that they have left-hand threads.

Proper care and maintenance of the hose is important for maintaining a safe, efficient work area. Remember the following guidelines for hoses:

- Protect the hose from molten dross or sparks, which will burn the exterior. Although some hoses are flame retardant, they will burn.
- Do not place the hoses under the metal being cut. If the hot metal falls on the hose, the hose will be damaged. Keep hoses as far away from the cutting activity as possible.
- Frequently inspect and replace hoses that show signs of cuts, burns, worn areas, cracks, or damaged fittings. The hoses are tough and durable, but not indestructible.
- Never use pipe-fitting compounds or lubricants around hose connections. These compounds often contain oil or grease, which ignite and burn or explode in the presence of oxygen.

Propane and propylene require hoses designed for the mixture of hydrocarbons present in these fuels. Ensure that any hoses used are appropriate for the fuel gas. Hose and fuel gas providers can help provide the correct hoses.

2.3.0 Torches and Tips

Cutting torches mix oxygen and fuel gas for the torch flame and control the stream of oxygen necessary for the cutting jet. Depending on the job site, either a one-piece or a combination cutting torch may be used.

2.3.1 One-Piece Hand Cutting Torch

The one-piece hand cutting torch, sometimes called a demolition torch or a straight torch, contains the fuel gas and oxygen valves that allow the gases to enter the chambers and then flow

Fuel and Oxygen Cylinder Separation for Fixed Installations

For fixed installations involving one or more cylinders coupled to a manifold, fuel and oxygen cylinders must be separated by at least 20' (6.10 m) or be divided by a wall 5' (1.52 m) or higher with a 30-minute burn rating, per American National Standards Institute Z49.1. This also applies to cylinders in storage. Special wheeled cradles designed to distribute gas from multiple cylinders are available.

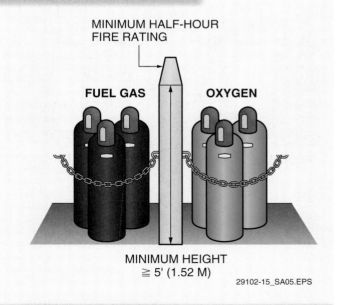

MINIMUM HALF-HOUR FIRE RATING

FUEL GAS   OXYGEN

MINIMUM HEIGHT ≧ 5' (1.52 M)

29102-15_SA05.EPS

into the part of the torch where they are mixed. The main body of the torch is called the handle. The torch valves control the fuel gas and oxygen flow needed for preheating the metal to be cut. The cutting oxygen lever, which is spring-loaded, controls the jet of cutting oxygen. Hose connections are located at the end of the torch body behind the valves.

Figure 19 shows a three-tube, one-piece hand cutting torch in which the preheat fuel and oxygen are mixed in the tip. These torches are designed for heavy-duty cutting and little else. They have long supply tubes from the torch handle to the torch head to reduce radiated heat to the operator's hands. Cutting torches are generally available with sufficient capacity to cut steel up to 12" (≈30 cm) thick. Larger-capacity torches, with the ability to cut steel up to 36" (≈90 cm) thick, can also be obtained. Torches with this kind of capacity require a significant oxygen and fuel gas supply to perform.

Two different types of oxyfuel cutting torches are in general use. The positive-pressure torch (*Figure 19*) is designed for use with fuel supplied through a regulator from pressurized fuel storage cylinders. The injector torch is designed to use a vacuum created by the oxygen flow to draw the necessary amount of fuel from a very-low-pressure fuel source, such as a natural gas line or acetylene generator. The injector torch, when used, is most often found in continuous-duty, high-volume manufacturing applications. Both types may employ one of two different fuel-mixing methods:

- Torch-handle, or supply-tube, mixing
- Torch-head or tip mixing

The two methods can normally be distinguished by the number of supply tubes from the torch handle to the torch head. Torches that use three tubes from the handle to the head mix the preheat fuel and oxygen at the torch head or tip. This method tends to help eliminate any flashback damage to the torch head supply tubes and torch handle. One tube carries fuel gas to the head. The other two tubes carry oxygen; one carries the oxygen for the preheat flame while the other carries the oxygen for cutting.

The cutting torch with two tubes usually mixes the preheat fuel and oxygen in a mixing chamber in the torch body or in one of the supply tubes (*Figure 20*). Injector torches usually have the injector located in one of the supply tubes, and the mixing occurs in the tube between the injector and the torch head. Some older torches that have only two visible tubes are actually three-tube torches that mix the preheat fuel and oxygen in the torch head or tip. This is accomplished by using a separate preheat fuel tube inside a larger preheat oxygen tube.

2.3.2 Combination Torch

The combination torch consists of a cutting torch attachment that fits onto a welding torch handle. These torches are normally used in light-duty or medium-duty applications. Fuel gas and oxygen valves are on the torch handle. The cutting attachment has a cutting oxygen lever and another oxygen valve to control the preheat flame. When the cutting attachment is screwed onto the torch handle, the torch handle oxygen valve is opened all the way, and the preheat oxygen is controlled by an oxygen valve on the cutting attachment. When the cutting attachment is removed, brazing and heating tips can be screwed onto the torch handle. *Figure 21* shows a two-tube combination torch in which preheat mixing is accomplished in a supply tube. These torches are usually positive-pressure torches with mixing occurring in the attachment body, supply tube, head, or tip. These torches may be equipped with built-in flashback arrestors and check valves.

2.3.3 Cutting Torch Tips

Cutting torch tips, or nozzles, fit into the cutting torch and are either screwed in or secured with a tip nut. There are one- and two-piece cutting tips (*Figure 22*).

One-piece cutting tips are made from a solid piece of copper. Two-piece cutting tips have a separate external sleeve and internal section.

Torch manufacturers supply literature explaining the appropriate torch tips and gas pressures for various applications. *Table 2* shows a sample cutting tip chart that lists recommended tip sizes and gas pressures for use with acetylene fuel gas and a specific manufacturer's torch and tips.

> **CAUTION**
>
> Do not use the cutting tip chart from one manufacturer for the cutting tips of another manufacturer. The gas flow rate of the tips may be different, resulting in excessive flow rates. Different gas pressures may also be required. The cutting torch tip to be used depends on the base metal thickness and fuel gas being used. Special-purpose tips are also available for use in such operations as gouging and grooving.

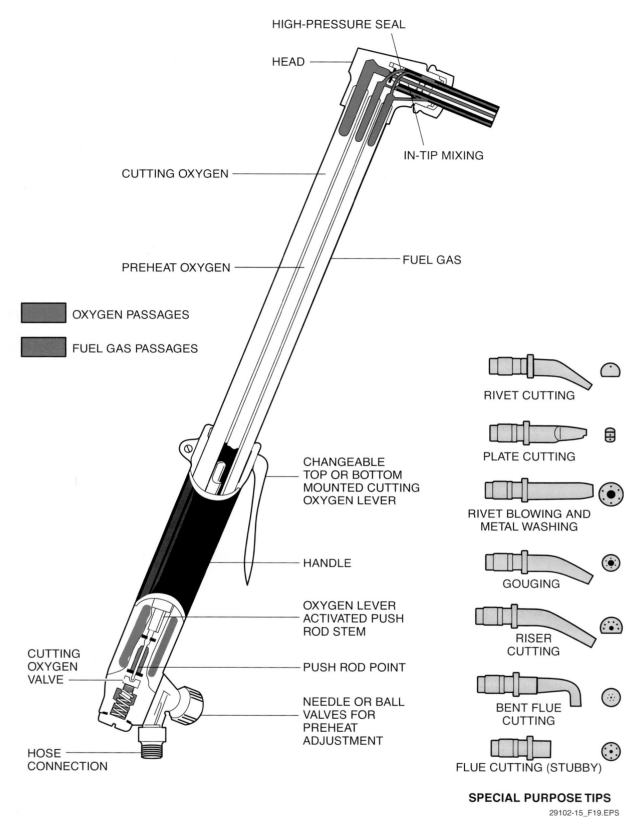

HIGH-PRESSURE SEAL

HEAD

CUTTING OXYGEN

IN-TIP MIXING

PREHEAT OXYGEN

FUEL GAS

OXYGEN PASSAGES

FUEL GAS PASSAGES

CHANGEABLE TOP OR BOTTOM MOUNTED CUTTING OXYGEN LEVER

HANDLE

OXYGEN LEVER ACTIVATED PUSH ROD STEM

CUTTING OXYGEN VALVE

PUSH ROD POINT

NEEDLE OR BALL VALVES FOR PREHEAT ADJUSTMENT

HOSE CONNECTION

RIVET CUTTING

PLATE CUTTING

RIVET BLOWING AND METAL WASHING

GOUGING

RISER CUTTING

BENT FLUE CUTTING

FLUE CUTTING (STUBBY)

SPECIAL PURPOSE TIPS

29102-15_F19.EPS

Figure 19 Heavy-duty three-tube one-piece positive-pressure hand cutting torch.

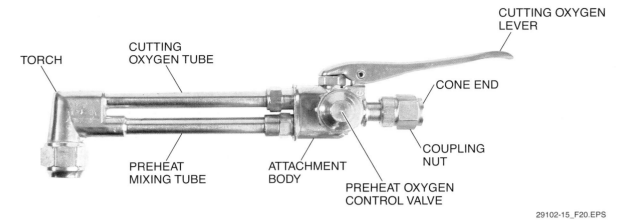

Figure 20 Cutting torch attachment.

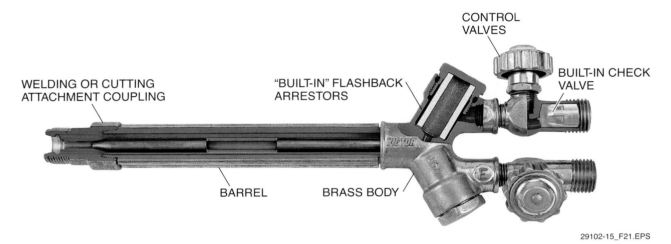

Figure 21 Combination cutting torch

The cutting torch tip to be used depends on the base metal thickness and fuel gas being used. Special-purpose tips are also available for such operations as gouging and grooving.

One-piece torch tips are generally used with acetylene cutting because of the high temperatures involved. They can have four, six, or eight preheat holes in addition to the single cutting hole. *Figure 23* shows the arrangement of typical acetylene torch cutting tips.

Tips used with liquefied fuel gases must have at least six preheat holes (*Figure 24*). Because fuel gases burn at lower temperatures than acetylene, more holes are necessary for preheating. Tips used with liquefied fuel gases can be one- or two-piece cutting tips. *Figure 25* shows a typical two-piece cutting tip used with liquefied fuel gases.

Special-purpose tips are available for special cutting jobs such as cutting sheet metal, rivets, risers, and flues, as well as washing and gouging. *Figure 26* shows special-purpose torch cutting tips, which are described as follows.

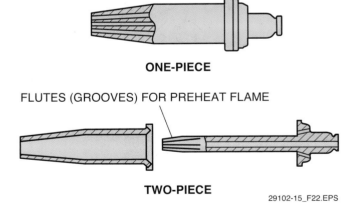

Figure 22 One- and two-piece cutting tips.

- The sheet metal cutting tip has only one preheat hole. This minimizes the heat and prevents distortion in the sheet metal. These tips are normally used with a motorized carriage, but can also be used for hand cutting.
- Rivet cutting tips are used to cut off rivet heads, bolt heads, and nuts.

Table 2 Sample Acetylene Cutting Tip Chart

| Cutting Tip Series 1-101, 3-101, and 5-101 | | | | | | | | | | | | | |
| Metal Thickness | | Tip Size | Cutting Oxygen Pressure* | | Preheat Oxygen* | | Acetylene Pressure* | | Speed | | Kerf Width | |
(in)	(mm)		(psig)	(kPa)	(psig)	(kPa)	(psig)	(kPa)	(in/ min)	(cm/ min)	(in)	(mm)
⅛	3.18	000	20-25	138-172	3-5	21-34	3-5	21-34	20-30	51-76	0.04	01.02
¼	6.35	00	20-25	138-172	3-5	21-34	3-5	21-34	20-28	51-71	0.05	01.27
⅜	9.52	0	25-30	172-207	3-5	21-34	3-5	21-34	18-26	46-66	0.06	01.52
½	12.70	0	30-35	207-241	3-6	21-41	3-5	21-34	16-22	41-56	0.06	01.52
¾	19.05	1	30-35	207-241	4-7	28-48	3-5	21-34	15-20	38-51	0.07	1.78
1	25.40	2	35-40	241-276	4-8	28-55	3-6	21-41	13-18	33-46	0.09	02.29
2	50.80	3	40-45	276-310	5-10	34-69	4-8	28-55	10-12	25-30	0.11	02.79
3	76.20	4	40-50	276-345	5-10	34-69	5-11	34-76	8-10	20-25	0.12	03.05
4	101.60	5	45-55	310-379	6-12	41-83	6-13	41-90	6-9	15-23	0.15	03.81
6	152.40	6	45-55	310-379	6-15	41-103	8-14	55-97	4-7	10-18	0.15	03.81
10	254.00	7	45-55	310-379	6-20	41-138	10-15	69-103	3-5	8-13	0.34	08.64
12	304.80	8	45-55	310-379	7-25	48-172	10-15	69-103	3-4	8-10	0.41	10.41

*The lower side of the pressure listings is for hand cutting and the higher side is for machine cutting.

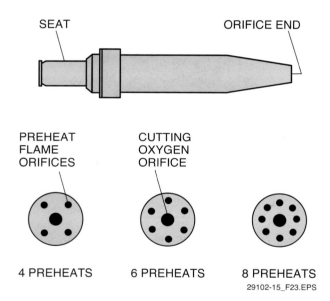

Figure 23 Orifice-end views of one-piece acetylene torch cutting tips.

- Riser cutting tips are similar to rivet cutting tips and can also be used to cut off rivet heads, bolt heads, and nuts. They have extra preheat holes to cut risers, flanges, or angle legs faster. They can be used for any operation that requires a cut close to and parallel to another surface, such as in removing a metal backing.
- Rivet blowing and metal washing tips are heavy-duty tips designed to withstand high heat. They are used for coarse cutting and for removing such items as clips, angles, and brackets.

- Gouging tips are used to groove metal in preparation for welding.
- Flue cutting tips are designed to cut flues inside boilers. They also can be used for any cutting operation in tight quarters where it is difficult to get a conventional tip into position.

Nearly all manufacturers use different tip-to-torch mounting designs, sealing surfaces, and diameters. In addition, tip sizes and flow rates are usually not the same between manufacturers even though the model number designations may be the same. This makes it impossible to safely interchange cutting tips between torches from different manufacturers. Even though some tips from different manufacturers may appear to be the same, do not interchange them. The sealing surfaces are very precise, and serious leaks may occur that

Acetylene Flow Rates

Manufacturers provide listings of the maximum fuel flow rate for each acetylene tip size in addition to recommended acetylene pressures. When selecting a tip, make sure that its maximum flow rate (in cubic feet or cubic meters per hour) does not exceed one-seventh of the total fuel capacity for the acetylene cylinder in use. Multiple cylinders must be manifolded together if the flow rate exceeds the cylinder(s) in use in order to prevent withdrawal of acetone along with acetylene.

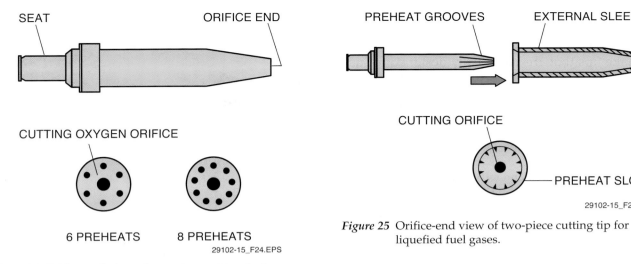

Figure 24 Orifice-end view of one-piece cutting tip for liquefied fuel gases.

Figure 25 Orifice-end view of two-piece cutting tip for liquefied fuel gases.

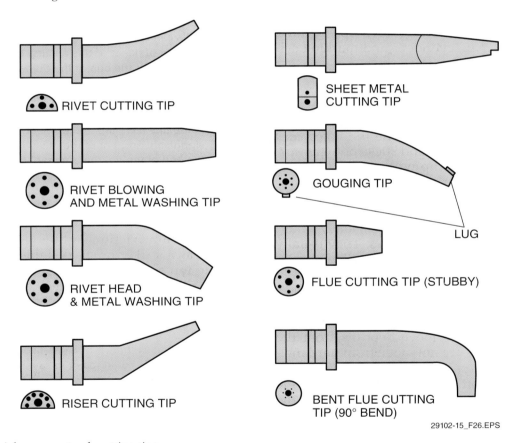

Figure 26 Special-purpose torch cutting tips.

could result in a dangerous fire or flashback. Each torch tip must be properly matched with the handle to which it is attached. Some manufacturers do make tips that are specifically designed for use with other manufacturer's torch handle. If these are used, ensure that the tip is listed as a precise match for the handle model in use.

> **CAUTION**
>
> Do not mix torch tips and handles from different manufacturers, unless they are specifically identified as being compatible in the manufacturer's documentation or catalog.

2.4.0 Other Accessories

In addition to the components that make up the actual oxyfuel cutting equipment, there are numerous accessories that are used along with the equipment. This section examines some accessories that are used for cleaning torch tips, lighting torches, transporting cylinders, and marking cylinders and metal workpieces.

2.4.1 Tip Cleaners and Tip Drills

With use, cutting tips become dirty. Carbon and other impurities build up inside the holes, and molten metal often sprays and sticks onto the surface of the tip. A dirty tip will result in a poor-quality cut with an uneven kerf and excessive dross buildup. To ensure good cuts with straight kerfs and minimal dross buildup, clean cutting tips with tip cleaners or tip drills (*Figure 27*).

Tip cleaners are tiny round files. They usually come in a set with files to match the diameters of the various tip holes. In addition, each set usually includes a file that can be used to lightly recondition the face of the cutting tip. Tip cleaners are inserted into the tip hole and moved back and forth a few times to remove deposits from the hole. The small files are not made from hard metals. Torch tips are typically made of brass, which is also relatively soft. Using an aggressive tip file made from hard metals could easily damage the precise opening.

Tip drills are used for major cleaning and for holes that are plugged. Tip drills are tiny drill bits that are sized to match the diameters of tip holes. The drill fits into a drill handle for use. The handle is held, and the drill bit is turned carefully inside the hole to remove debris. They are more brittle than tip cleaners, making them more difficult to use. If a torch tip is properly cared for, tip drills are rarely needed.

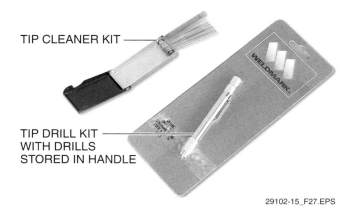

Figure 27 Tip cleaner and drill kits.

29102-15_F27.EPS

> **CAUTION**
>
> Tip cleaners and tip drills are brittle. Care must be taken to prevent these devices from breaking off inside a hole. Broken tip cleaners are difficult to remove. Improper use of tip cleaners or tip drills can enlarge the tip, causing improper burning of gases. If this occurs, the tip must be discarded. If the end of the tip has been partially melted or deeply gouged, do not attempt to cut it off or file it flat. The tip should be discarded and replaced with a new tip. This is because some tips have tapered preheat holes, and if a significant amount of metal is removed from the end of the tip, the preheat holes will become too large.

2.4.2 Friction Lighters

Always use a friction lighter (*Figure 28*), also known as a striker or spark-lighter, to ignite the cutting torch. The friction lighter works by rubbing a piece of flint on a steel surface to create sparks.

> **WARNING!**
>
> Do not use a match or a gas-filled lighter to light a torch. This could result in severe burns and/or could cause the lighter to explode.

2.4.3 Cylinder Cart

The cylinder cart, or bottle cart, is a modified hand truck that has been equipped with chains or straps to hold cylinders firmly in place. Bottle carts help ensure the safe transportation of gas cylinders. *Figure 29* shows two cylinder carts used for oxyfuel cylinders. Some carts are equipped with tool trays or boxes as well as rod holders.

TRIGGER OPERATED STRIKER

COMMON CUP-TYPE STRIKER

29102-15_F28.EPS

Figure 28 Friction lighters.

2.4.4 Soapstone Markers

Because of the heat involved in cutting operations, along with the tinted lenses that are required, ordinary pen or pencil marking for cutting lines or welding locations is not effective. The oldest and most common material used for marking is soapstone in the form of sticks or cylinders (*Figure 30*). Soapstone is soft and feels greasy and slippery. It is actually steatite, a dense, impure form of talc that is heat resistant. It also shows up well through a tinted lens under the illumination of an electric arc or gas cutting flame. Some welders prefer to use silver-graphite pencils (*Figure 30*) for marking dark materials and red-graphite pencils for aluminum or other bright metals. Graphite is also highly heat resistant. A few manufacturers also market heat-resistant paint/dye markers for cutting and welding.

2.5.0 Specialized Cutting Equipment

In addition to the common hand cutting torches, other types of equipment are used in oxyfuel cutting applications. This equipment includes mechanical guides used with a hand cutting torch, various types of motorized cutting machines, and oxygen lances. All of the motorized units use special straight body machine cutting or welding torches with a gear rack attached to the torch body to set the tip distance from the work.

2.5.1 Mechanical Guides

On long, circular, or irregular cuts, it is very difficult to control and maintain an even kerf with a hand cutting torch. Mechanical guides can help maintain an accurate and smooth kerf along the cutting line. For straight line or curved cuts, use a one- or two-wheeled accessory that clamps on the torch tip in a fixed position. The wheeled accessory maintains the proper tip distance while the tip is guided by hand along the cutting line.

The torch tip fits through and is secured to a rotating mount between two small metal wheels. The wheel heights are adjustable so that the tip distance from the work can be set. The radius of

Cup-Type Striker

When using a cup-type striker to ignite a welding torch, hold the cup of the striker slightly below and to the side of the tip, parallel with the fuel gas stream from the tip. This prevents the ignited gas from deflecting back from the cup and reduces the amount of carbon soot in the cup. Note that the flint in a striker can be replaced.

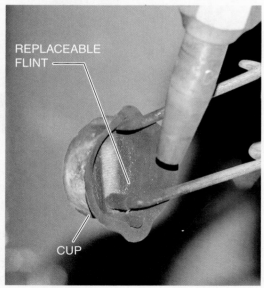

REPLACEABLE FLINT

CUP

29102-15_SA06.EPS

(A)

(B)

29102-15_F29.EPS

Figure 29 Cylinder carts.

SOAPSTONE STICK
AND HOLDER

SOAPSTONE CYLINDER
AND HOLDER

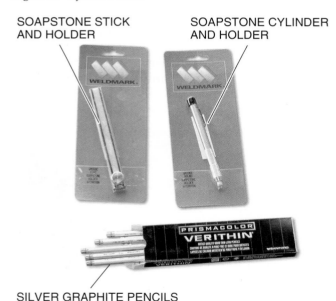

SILVER GRAPHITE PENCILS

29102-15_F30.EPS

Figure 30 Soapstone and graphite markers.

the circle is set by moving the pivot point on a radius bar. After a starting hole is cut (if needed), the torch tip is placed through and secured to the circle cutter rotating mount. Then the pivot point is placed in a drilled hole or a magnetic holder at the center of the circle. When the cut is restarted, the torch can be moved in a circle around the cut, guided by the circle cutter. The magnetic guide is used for straight line cuts. The magnets hold it securely in place during cutting operations.

When large work with an irregular pattern must be cut, a template is often used. The torch is drawn around the edges of the template to trace the pattern as the cut is made. If multiple copies must be cut, a metal pattern held in place and designed to allow space for tip distance from the pattern is usually used. For a one- or two-time

Sharpening Soapstone Sticks

The most effective way to sharpen a soapstone stick marker is to shave it on one side with a file. By leaving one side flat, accurate lines can be drawn very close to a straightedge or a pattern.

29102-15_SA07.EPS

copy, a heavily weighted Masonite or aluminum template that is spaced off the workpiece could be carefully used and discarded.

2.5.2 Motor-Driven Equipment

A variety of fixed and portable motorized cutting equipment is available for straight and curved cutting/welding. The computer-controlled plate cutting machine (*Figure 31*) is a fixed-location machine used in industrial manufacturing applications. The computer-controlled machine can be programmed to **pierce** and then cut any pattern from flat metal stock. There is also an optical pattern-tracing machine that follows lines on a drawing using a light beam and an optical detector. They both can be rigged to cut multiple items using multiple torches operated in parallel. Both units have a motor-driven gantry that travels the length of a table and a transverse motor-driven torch head that moves back and forth across the table. Both units are also equipped to use oxyfuel or plasma cutting torches.

Other types of pattern-tracing machines use metal templates that are clamped in the machine. A follower wheel traces the pattern from the template. The pattern size can be increased or decreased by electrical or mechanical linkage to a moveable arm holding one or more cutting torches that cut the pattern from flat metal stock.

Portable track cutting machines, or track burners, can be used in the field for straight or curved cutting and beveling. *Figures 32* and *33* show units driven by a variable-speed motor. The unit shown in *Figure 32* is available with track extensions for any length of straight cutting or beveling, along with a circle cutting attachment. Some models use a single, centered track.

Machine oxyfuel gas cutters or track burners are basic guidance systems driven by a variable speed electric motor to enable the operator to cut or bevel straight lines at any desired speed. The device (*Figure 33*) is usually mounted on a track or used with a circle-cutting attachment to enable the operator to cut various circle diameters up to 96" (≈244 cm). It consists of a heavy-duty tractor unit fitted with an adjustable torch mount and gas hose attachments. It is also equipped with an On/Off switch, a Low/High speed range switch, a Forward/Reverse directional switch, and a speed-adjusting dial calibrated in inches (or metric equivalent) per minute.

The device shown in *Figure 33* offers the following operational features:

- Makes straight-line cuts of any length
- Makes circle cuts up to 96" (≈244 cm) in diameter
- Makes bevel or chamfer cuts
- Has an infinitely variable cutting speed from 1" to 110" (2.5 cm to 279 cm) per minute
- Has dual speed and directional controls to enable operation of the machine from either end

29102-15_F31.EPS

Figure 31 Computer-controlled plate cutting machine.

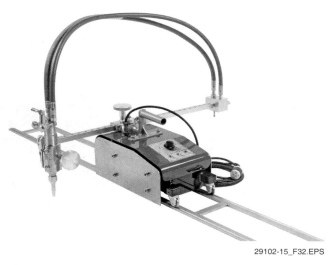

29102-15_F32.EPS

Figure 32 Track burner with oxyfuel machine torch.

Figure 34 shows the location of the following controls:

- *Power on/power off* – Turn the machine on and off by toggling the On/Off toggle switch.
- *Speed range control* – Set the machine's speed range by toggling the switch up or down. The use of two speed ranges allows for more precise speed control.
- *Directional control* – Set the machine's direction.
- *Speed control* – Turn the large knob to adjust the cutting speed based on the percentage of the selected speed range.

A portable, motor-driven band track or hand-cranked ring gear cutter/beveler can be set up in the field for cutting and beveling pipe with oxyfuel or plasma machine torches (*Figures 35*

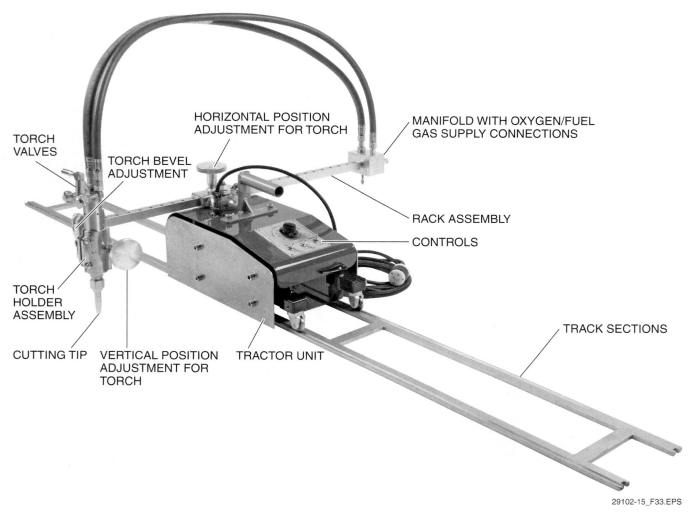

TORCH
VALVES

TORCH BEVEL
ADJUSTMENT

HORIZONTAL POSITION
ADJUSTMENT FOR TORCH

MANIFOLD WITH OXYGEN/FUEL
GAS SUPPLY CONNECTIONS

RACK ASSEMBLY

CONTROLS

TORCH
HOLDER
ASSEMBLY

CUTTING TIP

VERTICAL POSITION
ADJUSTMENT FOR
TORCH

TRACTOR UNIT

TRACK SECTIONS

29102-15_F33.EPS

Figure 33 Track burner features.

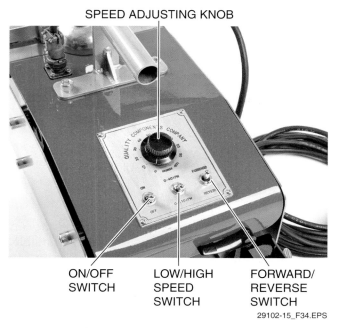

SPEED ADJUSTING KNOB

ON/OFF
SWITCH

LOW/HIGH
SPEED
SWITCH

FORWARD/
REVERSE
SWITCH

29102-15_F34.EPS

Figure 34 Track burner controls.

and *36*). The stainless steel band track cutter uses a chain and motor sprocket drive to rotate the machine cutting torch around the pipe a full 360 degrees. The all-aluminum ring gear type of cutter/beveler is positioned on the pipe, and then the saddle is clamped in place. In operation, the ring gear and the cutting torch rotate at different rates around the saddle for a full 360-degree cut.

2.5.3 *Exothermic Oxygen Lances*

Exothermic (combustible) oxygen lances are a special oxyfuel cutting tool usually used in heavy industrial applications and demolition work. The lance is a steel pipe that contains magnesium- and aluminum-cored powder or rods (fuel). In opera-

29102-15_F35.EPS

Figure 35 Band-track pipe cutter/beveler.

tion, the lance is clamped into a holder (*Figure 37*) that seals the lance to a fitting that supplies oxygen to it through a hose at pressures of 75 to 80 psi (517 to 552 kPa). With the oxygen turned on, the end of the lance is ignited with an acetylene torch or flare. As long as the oxygen is applied, the lance will burn and consume itself. The oxygen-fed flame of the burning magnesium, aluminum, and steel pipe creates temperatures approaching 10,000°F (5,538°C). At this temperature, the lance will rapidly cut or pierce any material, including steel, metal alloys, and cast iron, even under water. The lances for the holder shown in *Figure 37* are 10' (≈3 m) long and range in size from ⅜" (9.5 mm) to 1" (25.4 mm) in diameter. The larger sizes can be coupled to obtain a longer lance.

Shop-Made Straight-Line Cutting Guide

A simple solution for straight-line cutting is to clamp a piece of angle iron to the work and use a band clamp around the cutting torch tip to maintain the cutting tip distance from the work. When the cut is started, the band clamp rests on the top of the vertical leg of the angle iron, and the torch is drawn along the length of the angle iron at the correct cutting speed.

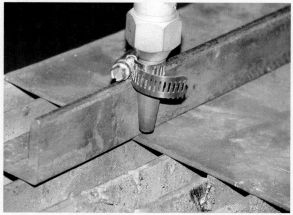

29102-15_SA08.EPS

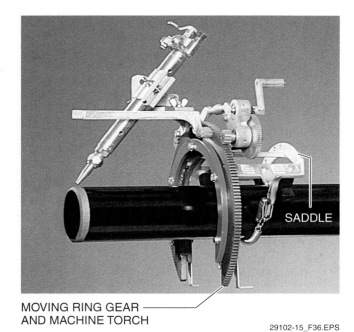

MOVING RING GEAR
AND MACHINE TORCH

SADDLE

29102-15_F36.EPS

Figure 36 Ring gear pipe cutter/beveler.

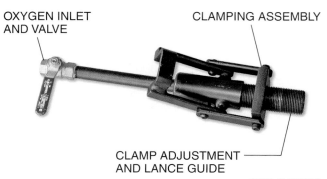

OXYGEN INLET
AND VALVE

CLAMPING ASSEMBLY

CLAMP ADJUSTMENT
AND LANCE GUIDE

29102-15_F37.EPS

Figure 37 Oxygen lance holder.

A small pistol-grip heat-shielded unit that can be used with an electric welder is also available. This small unit uses lances from ¼" (6.35 mm) to ⅜" (9.52 mm) in diameter that are 22" to 36" (55.9 to 91.4 cm) long and that cut very rapidly at a maximum burning time of 60 to 70 seconds. The small unit is primarily used to burn out large frozen pins and frozen headless bolts or rivets. Like a large lance, it can be used to cut any material, including concrete-lined pipe. Both units are relatively inexpensive and can be set up in the field with only an oxygen cylinder, hose, and ignition device.

2.0.0 Section Review

1. A lighter-than-air compound of carbon and hydrogen that is formed by dissolving calcium carbide in water is _____ .

   a. argon
   b. acetylene
   c. propane
   d. propylene

2. Safety devices used on regulators, hoses, and torches to prevent reverse gas flow and protect against fires and explosions are _____ .

   a. check valves and flashback arrestors
   b. orifice plates and torch tips
   c. HEPA filters and flashback valves
   d. distribution manifolds and friction strikers

3. What basic type of oxyfuel cutting torch is designed for use with fuel that is supplied through a regulator from a pressurized fuel storage cylinder?

   a. Motor-controlled torch
   b. Vacuum-injector torch
   c. Neutral-flame torch
   d. Positive-pressure torch

4. The most common material used for marking cutting lines on metal is a form of heat-resistant talc called _____ .

   a. soapstone
   b. Masonite
   c. graphite
   d. flint

5. A type of oxyfuel cutting tool that uses a steel pipe containing magnesium- and aluminum-cored powder or rods for heavy industrial cutting applications and demolition work is a(n) _____ .

   a. circle cutting accessory
   b. plasma machine torch
   c. exothermic oxygen lance
   d. ring gear beveler

3.0.0 OXYFUEL EQUIPMENT SETUP AND SHUTDOWN

Objective

Explain how to set up, light, and shut down oxyfuel equipment.

  a. Explain how to properly prepare a torch set for operation.
  b. Explain how to leak test oxyfuel equipment.
  c. Explain how to light the torch and adjust for the proper flame.
  d. Explain how to properly shut down oxyfuel cutting equipment.

Performance Tasks

1. Set up oxyfuel equipment.
2. Light and adjust an oxyfuel torch.
3. Shut down oxyfuel cutting equipment.
4. Disassemble oxyfuel cutting equipment.
5. Change empty gas cylinders.

Trade Terms

Carburizing flame: A flame burning with an excess amount of fuel; also called a reducing flame.

Neutral flame: A flame burning with correct proportions of fuel gas and oxygen.

Oxidizing flame: A flame burning with an excess amount of oxygen.

Operators should be trained and tested in the correct methods of safely preparing, starting, testing for leaks, and shutting down an oxyfuel cutting station. This part of the module examines procedures for setting up oxyfuel equipment, leak testing the equipment, lighting and adjusting the torch, and shutting down the equipment after use.

3.1.0 Setting Up Oxyfuel Equipment

When setting up oxyfuel equipment, follow procedures to ensure that the equipment operates properly and safely. The following sections explain the procedures for setting up oxyfuel equipment.

3.1.1 Transporting and Securing Cylinders

Cylinders should be transported to the workstation in an upright position on an appropriate hand truck or bottle cart. Once the cylinders are at the workstation, they must be secured with chain to a fixed support so that they cannot be knocked over accidentally. Leaving the cylinders in a proper cylinder cart is common; removal of the cylinders from the cart is not required. Then the protective cap from each cylinder can be removed and the outlet nozzles inspected to make sure that the seat and threads are not damaged. Place the protective caps where they will not be lost and where they will be readily available for reinstallation.

> **WARNING!**
> Always handle cylinders with care. They are under high pressure and should never be dropped, knocked over, rolled, or exposed to heat in excess of 140°F (60°C). When moving cylinders, always be certain that the valve caps are in place. Use a cylinder cage to lift cylinders. Never use a sling or electromagnet for cylinder lifting.

3.1.2 Cracking Cylinder Valves

To crack open a cylinder valve, start by ensuring that the cylinder is fully secured. Then crack open the cylinder valve momentarily to remove any dirt from the valve opening (*Figure 38*).

> **WARNING!**
> Operators should always stand with the valve stem between themselves and the regulator when opening valves to avoid injury from dirt that may be lodged in the valve. If a cloth is used during the cleaning process, it must not have any oil or grease on it. Oil or grease mixed with compressed oxygen can cause an explosion.

Hoisting Cylinders

Never attempt to lift a cylinder using the holes in a safety cap. Always use a lifting cage. Make sure that the cylinder is secured in the cage. Cages designed for storing and lifting cylinders are available in various sizes. The model shown here includes a partition to separate oxygen cylinders from fuel gas cylinders.

29102-15_SA09.EPS

> **CAUTION**
>
> Use care not to over-tighten connections. The brass connections used will strip if over-tightened. Repair or replace equipment that does not seal properly. If the torch valves leak, try lightly tightening the packing nut at the valves. If a cutting attachment or torch tip leaks, disassemble and check the sealing surfaces and/or O-rings for damage. If the equipment or torch tip is new and the sealing surfaces do not appear damaged, over-tighten the connection slightly to seat the sealing surfaces; then loosen and retighten normally. If the leaks persist or the equipment is visibly damaged, replace the torch, cutting attachment (if used), or cutting tip as necessary.

OUTLET FACING AWAY —

29102-15_F38.EPS

Figure 38 Cracking a cylinder valve.

3.1.3 Attaching Regulators

To attach the regulators, first check that the regulator is closed (adjustment screw is backed out and loose/turns with no resistance).

Check the regulator fittings to ensure that they are free of oil and grease (*Figure 39*).

Connect and tighten the oxygen regulator to the oxygen cylinder using a torch wrench (*Figure 40*).

> **WARNING!**
>
> Do not work with a regulator that shows signs of damage, such as cracked gauges, bent thumbscrews, or worn threads. Set it aside for repairs. Operators should never attempt to repair regulators. When tightening the connections, always use a torch wrench.

CHECK THAT FITTINGS ARE CLEAN

29102-15_F39.EPS

Figure 39 Checking connection fittings.

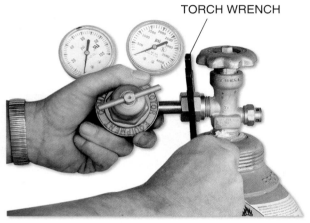

TORCH WRENCH

29102-15_F40.EPS

Figure 40 Tightening regulator connection.

Connect and tighten the fuel gas regulator to the fuel gas cylinder. Remember that most fuel gas fittings have left-hand threads. Next, clean the outlet connection of the regulator. Crack the cylinder valve slightly and turn the pressure adjustment screw clockwise until you feel some resistance and the regulator begins to allow gas to pass through and expel any debris. Shut the cylinder valve and close the regulator (*Figure 41*).

3.1.4 *Installing Flashback Arrestors or Check Valves*

The installation of flashback arrestors or check valves is important and easy to accomplish if they are not already installed. Attach a flashback arrestor and/or check valve to the hose connection on the oxygen regulator or torch handle (*Figure 42*) and tighten with a torch wrench. Then attach and tighten a flashback arrestor and/or check valve to the fuel gas regulator or torch handle.

OPEN REGULATOR TO CLEAN OUTLET

29102-15_F41.EPS

Figure 41 Cleaning the regulator.

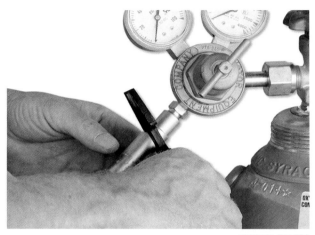

29102-15_F42.EPS

Figure 42 Attaching a flashback arrestor.

Once again, remember that fuel gas fittings have left-hand threads.

> **WARNING!**
>
> At least one flashback arrestor must be used with each hose. The absence of a flashback arrestor could result in flashback. Flashback arrestors can be attached either to the regulator, the torch, or both; however, flashback arrestors installed at the torch handle are preferred if only one is being used.

3.1.5 *Connecting Hoses to Regulators*

New hoses contain talc and possibly loose bits of rubber. These materials must be blown out of the hoses before the torch is connected. If they are not blown out, they will clog the tiny torch needle valves or tip openings.

To connect the hoses to the regulators, first inspect both the oxygen and fuel gas hoses for any damage, burns, cuts, or fraying. Replace any damaged hoses. Then connect the oxygen hose to the oxygen regulator flashback arrestor or check valve (*Figure 43*), and connect the fuel gas hose to the fuel gas regulator flashback arrestor and/or check valve. Complete the installation by opening the cylinder valves and regulators and purging the hoses until they are clear.

3.1.6 Attaching Hoses to the Torch

To attach the hoses to the torch, first attach flashback arrestors to the oxygen and fuel gas hose connections on the torch body—unless the torch has built-in flashback arrestors and check valves. Attach and tighten the oxygen hose to the oxygen fitting on the flashback arrestor or torch (*Figure 44*). Then attach and tighten the hose to the fuel gas fitting on the flashback arrestor or torch.

3.1.7 Connecting Cutting Attachments (Combination Torch Only)

If cutting attachments are being connected to a combination torch, be sure to check the torch manufacturer's instructions for the correct installation method. Then connect the attachment and tighten by hand as required.

3.1.8 Installing Cutting Tips

Before installing a cutting tip in a cutting torch, first identify the thickness of the material to be cut. Then identify the proper size cutting tip from the manufacturer's recommended tip size chart for the fuel being used.

> **WARNING!**
>
> If acetylene fuel is being used, make sure that the maximum fuel flow rate per hour of the tip does not exceed one-seventh of the fuel cylinder capacity. If a purplish flame is observed when the torch is operating, the fuel rate is too high and acetone is being withdrawn from the acetylene cylinder along with the acetylene gas.

Once the cutting tip has been selected, inspect the cutting tip sealing surfaces and orifices for damage or plugged holes. If the sealing surfaces are damaged, discard the tip. If the orifices are plugged, clean them with a tip cleaner or drill.

Check the torch manufacturer's instructions for the correct method of installing cutting tips. Then install the cutting tip and secure it with a torch wrench or by hand as required (*Figure 45*).

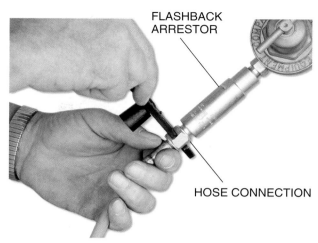

FLASHBACK ARRESTOR

HOSE CONNECTION

29102-15_F43.EPS

Figure 43 Connecting hose to regulator flashback arrestor.

NOTE THAT THIS TORCH HAS BUILT-IN FLASHBACK ARRESTORS AND CHECK VALVES

HOSE CONNECTION

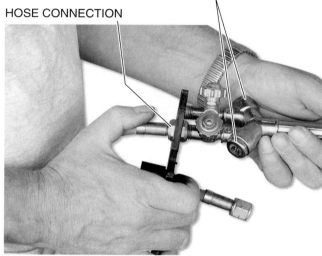

29102-15_F44.EPS

Figure 44 Connecting hoses to torch body.

3.1.9 Closing Torch Valves and Loosening Regulator Adjusting Screws

Closing the torch valves and loosening the regulator adjusting screws (*Figure 46*) are done before opening either cylinder valve. First, check the fuel and oxygen valves on the torch to be sure they are closed. Then check both the oxygen and fuel gas regulator adjusting screws to be sure they are loose (backed out).

> **CAUTION**
>
> Loosening regulator adjusting screws closes the regulators and prevents damage to the regulator diaphragms when the cylinder valves are opened.

Figure 45 Installing a cutting tip.

REGULATOR ADJUSTING SCREWS

TORCH VALVES

29102-15_F46.EPS

Figure 46 Torch valves and regulator adjusting screws.

3.1.10 Opening Cylinder Valves

To open cylinder valves (*Figure 47*), stand on the opposite side of the cylinder from the regulator and crack open the oxygen cylinder valve until the pressure on the regulator gauge rises and stops. The pressure in the cylinder and the pressure at the inlet of the regulator are now equal. Now open the oxygen cylinder valve all the way.

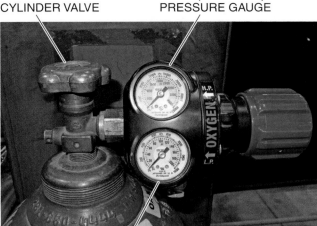

CYLINDER VALVE

INLET (CYLINDER) PRESSURE GAUGE

OUTLET (REGULATED) PRESSURE GAUGE

29102-15_F47.EPS

Figure 47 Cylinder valve and gauges.

Oxygen cylinder valves must be opened all the way until the valve seats at the top. Seating the valve at the fully open position prevents high-pressure leaks at the valve stem.

Once the oxygen cylinder valve is open, slowly open the fuel gas cylinder valve until the cylinder pressure gauge indicates the cylinder pressure, but no more than one and a half turns. This allows it to be quickly closed in case of a fire. This is especially important with acetylene.

3.1.11 Purging the Torch and Setting the Working Pressures

After the oxygen and fuel gas tank valves have been opened, the torch valves are opened to purge the torch and set the working pressures on the regulators.

Fully open the oxygen valve on the torch. Then depress and hold the cutting oxygen lever. Turn the oxygen regulator adjusting screw clockwise until the working pressure gauge shows the correct working pressure with the gas flowing. Allow the gas to flow for five to ten seconds to purge the torch and hoses of air or fuel gas. Then release the cutting lever and close the oxygen valve.

Note that, on single-stage regulators, the working pressure shown on the gauge will likely rise when the cutting lever is released. You will see the pressure gauge fall somewhat each time the lever is opened; this is normal operation. Set the pressure on the oxygen regulator while the gas is flowing to ensure it will be as desired when working.

Open the fuel valve on the torch about one eighth of a turn. Turn the fuel regulator adjusting screw clockwise until the working pressure gauge shows the correct fuel gas working pressure with the gas flowing. Allow the gas to flow for five to ten seconds to purge the hoses and torch of air. Then close the torch fuel valve. If acetylene is used, check that the acetylene static pressure does not rise above 15 psig (103 kPa). If it does, immediately open the torch fuel valve and reduce the regulator output pressure as needed. Because of its instability, acetylene cannot be used at pressures of more than 15 psig (103kPa) when in gaseous form. At higher pressures, acetylene gas breaks down chemically, producing heat and pressure that could result in a violent explosion inside the hose.

> **WARNING!**
>
> The working pressure gauge readings on single-stage regulators will rise after the torch valves are turned off. This is normal. However, if acetylene is being used as the fuel gas, make sure that the static pressure does not rise above 15 psig (103 kPa). Make sure that equipment is purged and leak tested in a well-ventilated area to avoid creating an explosive concentration of gases.

3.2.0 Testing for Leaks

Equipment must be tested for leaks immediately after it is set up and periodically thereafter. The torch should be checked for leaks before each use. Leaks could cause a fire or explosion. To test for leaks, apply a commercially prepared leak-testing formula (*Figure 48*) or a solution of detergent and water to each potential leak point. If bubbles form, a leak is present. Be aware though, that solutions made from common household soaps are not usually as effective as commercially prepared leak detection solutions.

> **WARNING!**
>
> If a detergent is used for leak testing, make sure the detergent contains no oil. In the presence of oxygen, oil can cause fires or explosions.

There are numerous leak points to test, including the following:

- Oxygen cylinder valve
- Fuel gas cylinder valve
- Oxygen regulator and regulator inlet and outlet connections
- Fuel gas regulator and regulator inlet and outlet connections

Portable Oxyacetylene Equipment

Most oxyfuel equipment used on large job sites and in shops is very heavy and is usually transported using a special hand truck. This type of equipment is typically used when extensive cutting must be accomplished. However, for small tasks in unusual locations, such as a commercial building rooftop, portable equipment that can be hand carried by one person is often used.

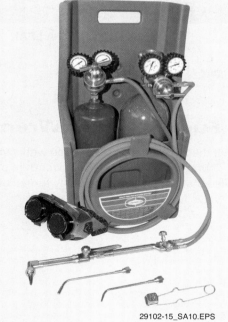

29102-15_SA10.EPS

APPLYING THE SOLUTION

REACTION TO A LEAK

29102-15_F48.EPS

Figure 48 Leak detection fluid use.

- Hose connections at the regulators, check valves/flashback arrestors, and torch
- Torch valves and cutting oxygen lever valve
- Cutting attachment connection (if used)
- Cutting tip

If there is a leak at the fuel gas cylinder valve stem, attempt to stop it by tightening the packing gland at the base of the stem. If this does not stop the leak, mark and remove the cylinder from service and notify the supplier. For other leaks, tighten the connections slightly with a wrench. If this does not stop the leak, turn off the gas pressure, open all connections, and inspect the fitting for damage.

> **WARNING!**
> Do not use Teflon® tape or pipe dope on these fittings since they do not provide a good seal. Make sure that equipment is purged and leak tested in a well-ventilated area to avoid creating an explosive concentration of gases.

3.2.1 *Initial and Periodic Leak Testing*

Initial and periodic leak testing is performed during initial equipment setup and periodically thereafter. First, set the equipment to the correct working pressures with the torch valves turned off. Then, using a leak-test solution, check for leaks at the cylinder valves, regulator relief ports, and regulator gauge connections (*Figure 49*). Also, check for leaks at hose connections, regulator connections, and check valve/flame arrestor connections up to the torch.

Fuel Cylinder Wrench

If the fuel cylinder is equipped with a valve requiring a T-wrench, always leave the wrench in place on the valve so that the fuel can be quickly turned off. This type of valve is obsolete but still in use.

29102-15_SA11.EPS

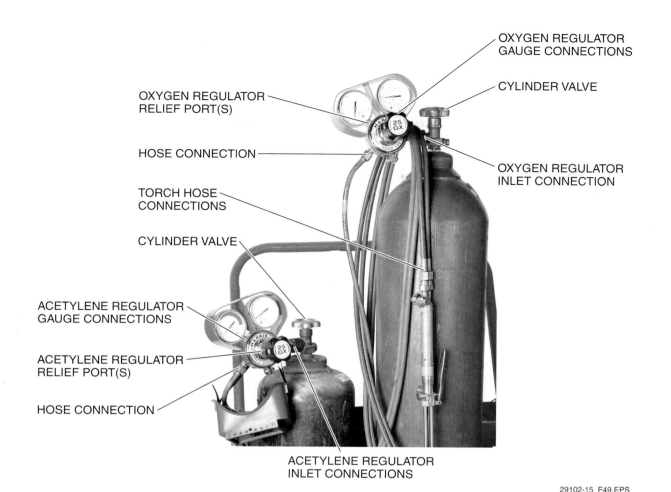

OXYGEN REGULATOR GAUGE CONNECTIONS

CYLINDER VALVE

OXYGEN REGULATOR RELIEF PORT(S)

HOSE CONNECTION

OXYGEN REGULATOR INLET CONNECTION

TORCH HOSE CONNECTIONS

CYLINDER VALVE

ACETYLENE REGULATOR GAUGE CONNECTIONS

ACETYLENE REGULATOR RELIEF PORT(S)

HOSE CONNECTION

ACETYLENE REGULATOR INLET CONNECTIONS

29102-15_F49.EPS

Figure 49 Typical initial and periodic leak-test points.

3.2.2 *Leak-Down Testing of Regulators, Hoses, and Torch*

Before the torch is ignited for use, the regulators, hoses, and torch should be quickly tested for leaks. First, set the equipment to the correct working pressures with the torch valves turned off. Then loosen both regulator adjusting screws. Check the working pressure gauges after a minute or two to see if the pressure drops. If the pressure drops, check the hose connection and regulators for leaks; otherwise, proceed with the test.

Place a thumb or finger over the cutting tip orifices and press tightly to block them (*Figure 50*). Turn on the torch oxygen valve and then depress and hold the cutting oxygen lever down. After the gauge pressure drops slightly, observe the oxygen working pressure gauge for a minute to see if the pressure continues to drop. If the pressure keeps dropping, perform the leak test described in the following section to determine the source of the leak. If the pressure does not change, close the torch oxygen valve and release the pressure at the cutting tip.

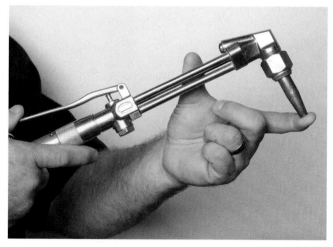

29102-15_F50.EPS

Figure 50 Blocking cutting tip for a leak test.

Next, block the tip again and turn on the torch fuel valve. After the gauge pressure drops slightly, carefully observe the fuel working pressure gauge for a minute. If the pressure continues to drop, perform the leak test described in the following section to determine the source of the leak. If the pressure does not change, close the torch fuel valve and release the pressure at the cutting tip.

If no leaks are apparent during the leak-down test, set the equipment to the correct working pressures.

3.2.3 Full Leak Testing of a Torch

Performing a full leak test involves testing for and isolating torch leaks in several places, as shown in *Figure 51*.

Start by setting the equipment to the correct working pressures with the torch valves turned off. Then, place a thumb or finger over the cutting tip orifices and press to block the orifices.

Turn on the torch oxygen valve and then depress the cutting oxygen lever. With the cutting tip blocked, check for leaks using a leak-test solution at the torch oxygen valve, cutting oxygen lever valve, cutting attachment connection to the handle (if used), preheat oxygen valve (if present), and cutting tip seal at the torch head. If no leaks are found, release the cutting oxygen lever, close the torch oxygen valve, and release the pressure at the cutting tip.

Next, with the cutting tip again blocked, open the torch fuel valve. Using a leak-test solution, check for leaks at the torch fuel valve, cutting attachment connection to the handle (if used), and cutting tip seal at the torch head. If no leaks are detected, close the torch fuel valve and release the pressure at the cutting tip.

3.3.0 Controlling the Oxyfuel Torch Flame

To be able to safely use a cutting torch, the operator must understand the flame and be able to adjust it and react to unsatisfactory conditions. The following sections will explain the oxyfuel flame and how to control it safely.

3.3.1 Oxyfuel Flames

There are three types of oxyfuel flames: neutral flame, carburizing flame, and oxidizing flame.

- *Neutral flame* – A neutral flame burns proper proportions of oxygen and fuel gas. The inner cones will be light blue in color, surrounded by a darker blue outer flame envelope that results when the oxygen in the air combines with the super-heated gases from the inner cone. A neutral flame is used for all but special cutting applications.
- *Carburizing flame* – A carburizing flame has a white feather created by excess fuel. The length of the feather depends on the amount of excess fuel present in the flame. The outer flame envelope is longer than that of the neutral flame, and it is much brighter in color. The excess fuel in the carburizing flame (especially acetylene) produces large amounts of carbon. The carbon will combine with red-hot or molten metal, making the metal hard and brittle. The carburizing flame is cooler than a neutral flame and is never used for cutting. It is used for some special heating applications.

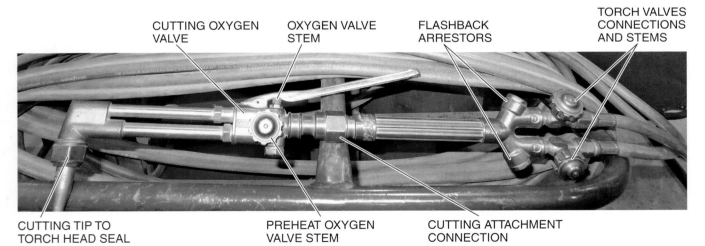

CUTTING OXYGEN VALVE   OXYGEN VALVE STEM   FLASHBACK ARRESTORS   TORCH VALVES CONNECTIONS AND STEMS

CUTTING TIP TO TORCH HEAD SEAL   PREHEAT OXYGEN VALVE STEM   CUTTING ATTACHMENT CONNECTION

29102-15_F51.EPS

Figure 51 Torch leak-test points.

- *Oxidizing flame* – An oxidizing flame has an excess of oxygen. The inner cones are shorter, much bluer in color, and more pointed than a neutral flame. The outer flame envelope is very short and often fans out at the ends. An oxidizing flame is the hottest flame. A slightly oxidizing flame is recommended with some special fuel gases, but in most cases it is not used. The excess oxygen in the flame can combine with many metals, forming a hard, brittle, low-strength oxide. However, the preheat flames of a properly adjusted cutting torch will be slightly oxidizing when the cutting oxygen is shut off.

Figure 52 shows the various flames that occur at a cutting tip for both acetylene and LP gas.

Acetylene Burning in Atmosphere
Open fuel gas valve until smoke clears from flame.

Carburizing Flame
(Excess acetylene with oxygen)
Preheat flames require more oxygen.

Neutral Flame
(Acetylene with oxygen) Temperature 5589°F (3087°C).
Proper preheat adjustment when cutting.

Neutral Flame with Cutting Jet Open
Cutting jet must be straight and clean.
If it flares, the pressure is too high for the tip size.

Oxidizing Flame
(Acetylene with excess oxygen) Not recommended for average cutting. However, if the preheat flame is adjusted for neutral with the cutting oxygen on, then this flame is normal after the cutting oxygen is off.

OXYACETYLENE FLAME

LP Gas Burning in Atmosphere
Open fuel gas valve until flame begins to leave tip end.

Reducing Flame
(Excess LP-gas with oxygen) Not hot enough for cutting.

Neutral Flame
(LP-gas with oxygen) For preheating prior to cutting.

Oxidizing Flame with Cutting Jet Open
Cutting jet stream must be straight and clean.

Oxidizing Flame without Cutting Jet Open
(LP-gas with excess oxygen) The highest temperature flame for fast starts and high cutting speeds.

OXYPROPANE FLAME

29102-15_F52.EPS

Figure 52 Acetylene and LP (propane) gas flames.

3.3.2 Backfires and Flashbacks

When the torch flame goes out with a loud pop or snap, a backfire has occurred. Backfires are usually caused when the tip or nozzle touches the work surface or when a bit of hot dross briefly interrupts the flame. When a backfire occurs, relight the torch immediately. Sometimes the torch even relights itself. If a backfire recurs without the tip making contact with the base metal, shut off the torch and find the cause. Possible causes are the following:

- Improper operating pressures
- A loose torch tip
- Dirt in the torch tip seat or a bad seat

When the flame goes out and burns back inside the torch with a hissing or whistling sound, a flashback is occurring. Immediately shut off the oxygen valve on the torch; the flame is burning inside the torch. If the flame is not extinguished quickly, the end of the torch will melt off. The flashback will stop as soon as the oxygen valve is closed. Therefore, quick action is crucial. Flashbacks can cause fires and explosions within the cutting rig and, therefore, are very dangerous. Flashbacks can be caused by the following:

- Equipment failure
- Overheated torch tip
- Dross or spatter hitting and sticking to the torch tip
- Oversized tip (tip is too large for the gas flow rate being used)

After a flashback has occurred, wait until the torch has cooled. Then, blow oxygen (not fuel gas) through the torch for several seconds to remove soot that may have built up in the torch during the flashback before relighting it. If the torch makes a hissing or whistling sound after it is reignited or if the flame does not appear to be normal, shut off the torch immediately and have the torch serviced by a qualified technician.

3.3.3 Igniting the Torch and Adjusting the Flame

After the cutting equipment has been properly set up and purged, the torch can be ignited and the flame adjusted for cutting. The procedure for igniting the torch starts with choosing the appropriate cutting torch tip according to the base metal thickness being cut and the fuel gas being used. Always inspect the cutting tip sealing surfaces and orifices prior to installation. Attach the tip to the cutting torch by placing it on the end of the torch and tightening the nut. Some manufac-turers recommend tightening the nut with a torch wrench, while others recommend tightening the nut by hand. Check the manufacturer's documentation for the equipment in use to ensure the tip is being installed correctly.

> **NOTE**
>
> Refer to the manufacturer's charts. Depending on the tip selected, the oxygen and fuel gas pressure may have to be adjusted.

Prior to opening the oxygen and fuel gas valves, be sure to put on the proper PPE. Also ensure that you are not depressing the oxygen cutting lever. If present, close the preheat oxygen valve and open the torch oxygen valve fully. Open the fuel gas valve on the torch handle about one-quarter turn. Then, holding the friction lighter near the side and to the front of the torch tip, ignite the torch.

> **WARNING!**
>
> Hold the friction lighter near the side of the tip, rather than directly in front of it, to prevent the ignited gas from being deflected backwards. Always use a friction lighter. Never use matches or cigarette lighters to light the torch because this could result in severe burns and/or could cause the lighter to explode. Always point the torch away from yourself, other people, equipment, and flammable material.

Once the torch is lit, adjust the torch fuel gas flame by adjusting the flow of fuel gas with the fuel gas valve. Increase the flow of fuel gas until the flame stops smoking or pulls slightly away from the tip. Decrease the flow until the flame returns to the tip. Open the preheat oxygen valve (if present) or the oxygen torch valve very slowly and adjust the torch flame to a neutral flame. Then press the cutting oxygen lever all the way down and observe the flame. It should have a long, thin, high-pressure oxygen cutting jet up to 8" (≈20 cm) long, extending from the cutting oxygen hole in the center of the tip. If it does not, do the following:

- Check that the working pressures are set as recommended on the manufacturer's chart.
- Clean the cutting tip. If this does not clear up the problem, change the cutting tip.

With the cutting oxygen on, observe the preheat flame. If it has changed slightly to a carburizing flame, increase the preheat oxygen until the flame is neutral. After this adjustment, the preheat flame will change slightly to an acceptable oxidizing flame when the cutting oxygen is shut off.

3.3.4 Shutting Off the Torch

Shutting off the torch itself is done by releasing the cutting oxygen lever and then closing the torch or preheat oxygen valves. After that, quickly close the torch fuel gas valve to extinguish the flame.

> **WARNING!**
>
> Always turn off the oxygen flow first to prevent a possible flashback into the torch.

3.4.0 Shutting Down Oxyfuel Cutting Equipment

When a cutting job is completed and the oxyfuel equipment is no longer needed, it must be shut down. *Figure 53* identifies the order in which various pieces of the oxyfuel cutting equipment are shut down.

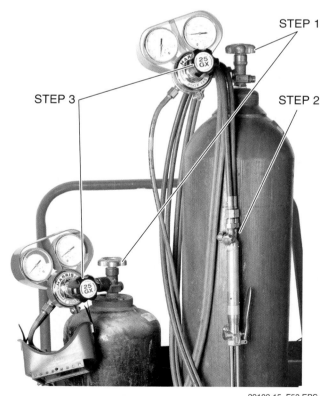

29102-15_F53.EPS

Figure 53 Shutting down oxyfuel cutting equipment.

Step 1 To begin the shutdown, first close the fuel gas and oxygen cylinder valves. Leave the regulators at their present setting.

Step 2 Open the fuel gas valve on the torch to allow all remaining gas to escape, and then close it. Next, open the oxygen valve on the torch to allow all remaining gas to escape, and then close it. These actions relieve the gas pressure in the hose and regulators, all the way back to the cylinder valve. Do not proceed to the next step until all pressure is released and all regulator gauges — both inlet and outlet — read zero.

Step 3 Turn the fuel gas and oxygen regulator adjusting screws counterclockwise to back them out, until they are loose.

Step 4 Coil and secure the hose and torch to prevent damage.

3.4.1 Disassembling Oxyfuel Equipment

In some situations, it may be necessary to disassemble the oxyfuel equipment after it has been used. Before any disassembly takes place, make sure that the equipment has been properly shut down. This includes checking that the cylinder valves are closed and all pressure gauges read zero.

Remove both hoses from the torch assembly and then detach the hoses from the regulators. Remove both regulators from the cylinders and reinstall the protective caps on the cylinders. The cylinders should now be returned to their proper storage place.

> **WARNING!**
>
> Always transport and store gas cylinders in the upright position. Be sure they are properly secured (chained) and capped. Regardless of whether the cylinders are empty or full, never store fuel gas cylinders and oxygen cylinders together without providing the required separation distance or fire-rated partition.

Obtaining Maximum Fuel Flow

Increasing the fuel flow until the flame pulls away from the tip and then decreasing the flow until the flame returns to the tip sets the maximum fuel flow for the tip size in use.

3.4.2 Changing Cylinders

Empty is a relative term when discussing gas cylinders. These cylinders should never be completely emptied because reverse flow could occur. Oxygen tanks, for example, should never get below the required working pressure, or about 25 psi (172 kPa). Once the cylinder pressure drops near the working pressure value, the torch will stop performing properly and the cylinder will have to be replaced. As a result, some residual pressure is typically present. Follow these procedures to change a cylinder:

> **WARNING!**
>
> When moving cylinders, always be certain that they are in the upright position and the valve caps are secured in place. Never use a sling or electromagnet to lift cylinders. To lift cylinders, use a cylinder cage.

Step 1 Begin by making sure that the equipment has been properly shut down. This includes checking that the cylinder valves are closed and all pressure gauges read zero.

Step 2 Remove the regulator from the empty cylinder and replace the protective cap on the cylinder.

Step 3 Mark MT (empty) and the date (or the accepted site notation for indicating an empty cylinder) near the top of the cylinder using soapstone (*Figure 54*).

29102-15_F54.EPS

Figure 54 Typical empty cylinder marking.

Step 4 Transport the empty cylinder from the workstation to the storage area. Place the empty cylinder in the empty cylinder section of the storage area for the type of gas it contained.

Marking and Tagging Cylinders

Do not use permanent markers on cylinders; use soapstone or another temporary marker. If a cylinder is defective, place a warning tag on it.

Additional Resources

ANSI Z49.1, Safety in Welding, Cutting, and Allied Processes. Miami, FL: American Welding Society.

Uniweld Products, Inc. Numerous videos are available at **www.uniweld.com/en/uniweld-videos**. Last accessed: November 30, 2014.

The Harris Products Group, a division of Lincoln Electric. Numerous videos are available at **www.harrisproductsgroup.com/en/Expert-Advice/videos.aspx**. Last accessed: November 30, 2014.

3.0.0 Section Review

1. Gas cylinders can be lifted to height _____ .
   a. with a strong electromagnet
   b. using a cylinder cage
   c. with a sling routed through the openings in the cap
   d. with a cable routed through the openings in the cap

2. Immediately after setup and periodically thereafter, oxyfuel equipment must be tested for _____ .
   a. purging
   b. carburizing
   c. flashbacks
   d. leaks

3. An oxidizing torch flame is one that has a(n) _____ .
   a. excess of fuel gas
   b. proper oxygen/fuel gas mix
   c. excess of oxygen
   d. abnormally low temperature

4. When oxyfuel cutting equipment is being shut down, how should the fuel gas and oxygen regulator adjusting screws be positioned after bleeding the remaining gas pressure in the hoses?
   a. Both screws tight
   b. Fuel gas screw tight; oxygen screw loose
   c. Fuel gas screw loose; oxygen screw tight
   d. Both screws loose

4.0.0 PERFORMING CUTTING PROCEDURES

Objective

Explain how to perform various oxyfuel cutting procedures.

   a. Identify the appearance of both good and inferior cuts and their causes.
   b. Explain how to cut both thick and thin steel.
   c. Explain how to bevel, wash, and gouge.
   d. Explain how to make straight and bevel cuts with portable oxyfuel cutting machines.

Performance Tasks

   6. Cut shapes from various thicknesses of steel, emphasizing:
      • Straight line cutting
      • Square shape cutting
      • Piercing
      • Beveling
      • Cutting slots
   7. Perform washing.
   8. Perform gouging.
   9. Use a track burner to cut straight lines and bevels.

Trade Terms

Drag lines: The lines on the edge of the material that result from the travel of the cutting oxygen stream into, through, and out of the metal.

The following sections explain how to recognize good and bad cuts, how to prepare for cutting operations, and how to perform straight-line cutting, piercing, bevel cutting, washing, and gouging.

4.1.0 Preparing for Oxyfuel Cutting with a Hand Cutting Torch

Before metal can be cut, the equipment must be set up and the metal prepared. One important step is to properly lay out the cut by marking it with soapstone or punch marks. The few minutes this takes will result in a quality job, reflecting craftsmanship and pride. The following procedures describe how to prepare to make a cut.

Prepare the metal to be cut by cleaning any rust, scale, or other foreign matter from the surface. If possible, position the work so that it can be cut comfortably. Mark the lines to be cut with soapstone or a scriber. Then select the correct cutting torch tip according to the thickness of the metal to be cut, the type of cut to be made, the amount of preheat needed, and the type of fuel gas to be used. Ignite the torch and use the procedures outlined in the following sections for performing specific types of cutting operations.

4.1.1 Inspecting the Cut

Before attempting to make a cut, operators must be able to recognize good and bad cuts and know what causes bad cuts. This is explained in the following list and illustrated in *Figure 55*:

• A good cut features a square top edge that is sharp and straight, not ragged. The bottom edge can have some dross adhering to it but not an excessive amount. What dross there is should be easily removable with a chipping hammer. The drag lines should be near vertical and not very pronounced.
• When preheat is insufficient, bad gouging results at the bottom of the cut because of slow travel speed.
• Too much preheat will result in the top surface melting over the cut, an irregular cut edge, and an excessive amount of dross.
• When the cutting oxygen pressure is too low, the top edge will melt over because of the resulting slow cutting speed.
• Using cutting oxygen pressure that is too high will cause the operator to lose control of the cut, resulting in an uneven kerf.
• A travel speed that is too slow results in bad gouging at the bottom of the cut and irregular drag lines.
• When the travel speed is too fast, there will be gouging at the bottom of the cut, a pronounced break in the drag line, and an irregular kerf.
• A torch that is held or moved unsteadily across the metal being cut can result in a wavy and irregular kerf.
• When a cut is lost and then not restarted carefully, bad gouges will result at the point where the cut is restarted.

A square kerf face with minimal notching not exceeding $\frac{1}{16}$" (1.6 mm) deep is expected and, in fact, required in the Performance Accreditation Tasks for this module.

DIRECTION OF TRAVEL

GOOD CUT

PREHEAT
INSUFFICIENT

TOO MUCH
PREHEAT

CUTTING PRESSURE
TOO LOW

OXYGEN PRESSURE
TOO HIGH AND
UNDERSIZE TIP

TRAVEL SPEED
TOO SLOW

TRAVEL SPEED
TOO FAST

TORCH HELD OR
MOVED UNSTEADILY

CUT NOT RESTARTED
CAREFULLY, CAUSING
GOUGES AT RESTARTING
POINTS (CIRCLED)

© American Welding Society (AWS) *Welding Handbook*

29102-15_F55.EPS

Figure 55 Examples of good and bad cuts.

The tasks in the sections that follow are designed to develop skills with a cutting torch. Each task should be practiced until there is thorough familiarity with the procedure. After each task is completed, it should be taken to the instructor for evaluation. Do not proceed to the next task until the instructor says to continue.

4.2.0 Cutting Steel

The effectiveness of cutting steel with an oxyfuel cutting outfit depends on factors such as the thickness of the steel, the cutting tip that is being used, and the skill of the operator.

4.2.1 *Cutting Thin Steel*

Thin steel is considered material ³⁄₁₆" (≈5 mm) thick or less. A major concern when cutting thin steel is distortion caused by the heat of the torch and the

cutting process. To minimize distortion, move as quickly as possible without losing the cut.

To begin the process for cutting thin steel, first prepare the metal surface. Then light the torch and hold it so that the tip is pointing in the direction the torch is traveling at a 15- to 20-degree angle. Make sure that a preheat orifice and the cutting orifice are centered on the line of travel next to the metal (*Figure 56*).

> **CAUTION**
>
> Holding the tip upright (perpendicular to the metal) when cutting thin steel will overheat the metal, causing distortion. Maintain the 15 to 20-degree push angle as shown.

Preheat the metal to a dull red. Use care not to overheat thin steel because this will cause distortion. The edge of the tip can be lightly rested on

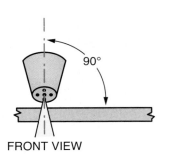

FRONT VIEW

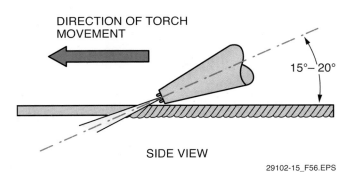

DIRECTION OF TORCH MOVEMENT

15°–20°

SIDE VIEW

29102-15_F56.EPS

Figure 56 Cutting thin steel.

the surface of the metal being cut and then slid along the surface when making the cut in thin metal. Press the cutting oxygen lever to start the cut, and then move quickly along the line. To minimize distortion, move as quickly as possible without losing the cut.

4.2.2 Cutting Thick Steel

Most oxyfuel cutting is done on steel that is more than ³⁄₁₆" (≈5 mm) thick. Whenever heat is applied to metal, distortion is a problem, but as the steel gets thicker, it becomes less of a problem.

To cut thick steel with a cutting torch, start by preparing the metal surface. Then light the torch and adjust the torch flame. Follow the number sequence shown in *Figure 57* to perform the cut.

The torch can be moved from either right to left or left to right. Choose the direction that allows the best visibility of the cut. When cutting begins, the tips of the preheat flame should be held ¹⁄₁₆" to ⅛" (1.6 to 3.2 mm) above the workpiece. For steel up to ⅜" (10 mm) thick, the first and third procedures can usually be omitted.

4.2.3 Piercing a Plate

Before holes or slots can be cut in a plate, the plate must be pierced. Piercing puts a small hole through the metal where the cut can be started. Because more preheat is necessary on the surface

of a plate than at the edge, choose the next-larger cutting tip than is recommended for the thickness to be pierced. When piercing steel that is more than 3" (≈8 cm) thick, it may help to first preheat the bottom side of the plate directly under the spot to be pierced. The following steps describe how to pierce a plate for cutting. *Figure 58* provides a visual reference.

Step 1 Start by preparing the metal surface and the torch for cutting.

Step 2 Ignite the torch and adjust the flame.

Step 3 Hold the torch tip ¼" to ⁵⁄₁₆" (6.4 mm to 7.9 mm) above the spot to be pierced until the surface is a bright cherry red.

Step 4 Raise the tip about ½" (12.7 mm) above the metal surface and tilt the torch slightly so that molten metal does not blow directly back into the tip as the oxygen lever is depressed. Depress the oxygen lever.

Step 5 Maintain the tipped position until a hole burns through the plate. Then rotate the torch back to the vertical position (perpendicular to the plate).

Step 6 Lower the torch back to the initial distance from the plate and continue to cut outward from the original hole to the line to be cut. Then follow the line.

4.3.0 Beveling, Washing, and Gouging

While oxyfuel cutting equipment is commonly associated with cutting through metal plate, it is also well suited for cutting angles in the edge of steel plate, removing bolts and rivets, and cutting groves in metal surfaces.

4.3.1 Cutting Bevels

Bevel cutting is often performed to prepare the edge of steel plate for welding. The procedure for bevel cutting is illustrated in *Figure 59*.

Step 1 Prepare the metal surface and the torch.

Step 2 Ignite the torch and adjust the flame.

Step 3 Hold the torch so that the tip faces the metal at the desired bevel angle. Using a piece of angle iron as a cutting guide as shown in *Figure 59* will result in a 45-degree bevel angle. Angle iron can be used as a guide for any angle, as long as the operator consciously maintains the torch at the proper bevel angle.

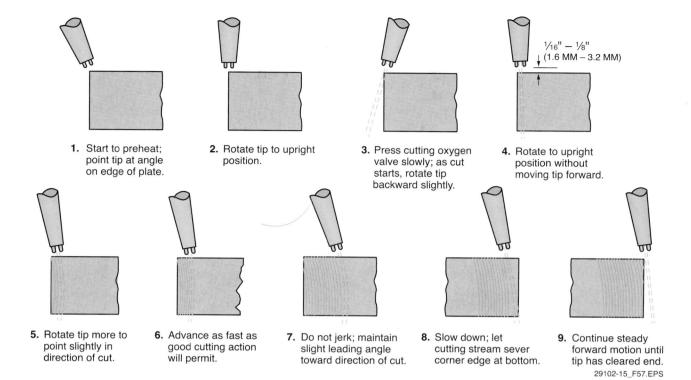

1. Start to preheat; point tip at angle on edge of plate.

2. Rotate tip to upright position.

3. Press cutting oxygen valve slowly; as cut starts, rotate tip backward slightly.

4. Rotate to upright position without moving tip forward.

5. Rotate tip more to point slightly in direction of cut.

6. Advance as fast as good cutting action will permit.

7. Do not jerk; maintain slight leading angle toward direction of cut.

8. Slow down; let cutting stream sever corner edge at bottom.

9. Continue steady forward motion until tip has cleared end.

29102-15_F57.EPS

Figure 57 Flame cutting with a hand torch.

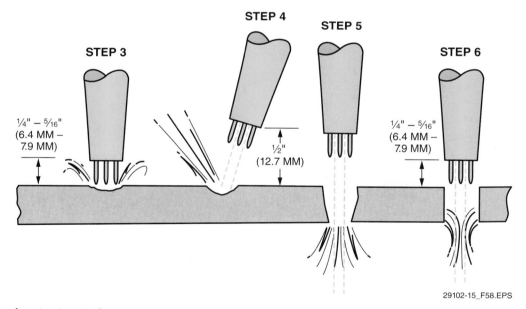

29102-15_F58.EPS

Figure 58 Steps for piercing steel.

Step 4 Preheat the edge to a bright cherry red.

Step 5 Press the cutting oxygen lever to start the cut.

Step 6 As cutting begins, move the torch tip at a steady rate along the line to be cut. Pay particular attention to the torch angle to ensure it creates a uniform bevel along the entire length of the cut.

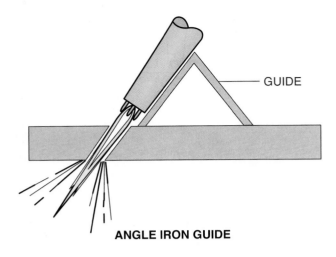

GUIDE

ANGLE IRON GUIDE

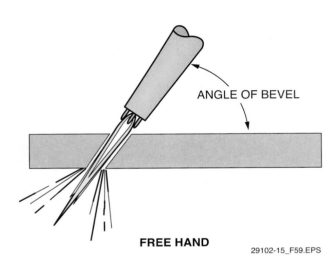

ANGLE OF BEVEL

FREE HAND

29102-15_F59.EPS

Figure 59 Cutting a bevel.

4.3.2 Washing

Washing is a term used to describe the process of cutting out bolts, rivets, previously welded pieces, or other projections from the surface. Washing operations use a special tip with a large cutting hole that produces a low-velocity stream of oxygen. The low-velocity oxygen stream helps prevent cutting into the surrounding base metal. *Figure 60* is a simplified illustration of a washing procedure.

Step 1 Prepare the metal surface and torch.

Step 2 Ignite the torch and adjust the flame.

Step 3 Preheat the metal to be cut until it is a bright cherry red.

Step 4 Rotate the cutting torch tip to roughly a 55-degree angle to the metal surface.

Step 5 At the top of the material, press the cutting oxygen lever to begin cutting the material to be removed. Continue moving back and forth across the material while

WASHING TIP

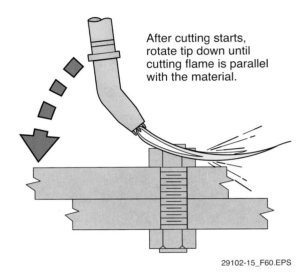

After cutting starts, rotate tip down until cutting flame is parallel with the material.

29102-15_F60.EPS

Figure 60 Washing.

rotating the tip to a position parallel with the material. Move the tip back and forth and down to the workpiece surface. Take care not to cut into it.

> **CAUTION**
>
> As the surrounding metal heats up, there is a greater danger of cutting into it. Try to complete the washing operation as quickly as possible. If the surrounding metal gets too hot, stop and let it cool down.

4.3.3 Gouging

Gouging (*Figure 61*) is the process of cutting a groove into a surface. Gouging operations use a special curved tip that produces a low-velocity stream of oxygen that curves up, allowing the operator to control the depth and width of the groove. It is an effective means to gouge out cracks or weld defects for welding. Gouging tips can also be used to remove steel backing from welds or to wash off bolt or rivet heads. However, gouging tips are not as effective as washing tips for removing the shank of a bolt or rivet.

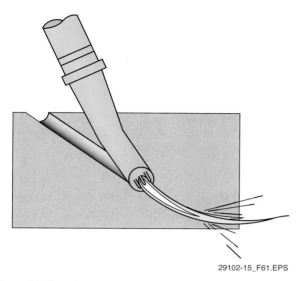

Figure 61 Gouging.

The travel speed and torch angle are very important when gouging. If the travel speed or torch angle is incorrect, the gouge will be irregular and there will be a buildup of dross inside the gouge. Practice until the gouge is clean and even, with a consistent depth.

Step 1 Prepare the metal surface and torch.

Step 2 Ignite the torch and adjust the flame.

Step 3 Holding the torch so that the preheat holes are pointed directly at the metal, preheat the surface until it becomes a bright cherry red.

Step 4 When the steel has been heated to a bright cherry red, slowly roll the torch away from the metal so that the holes are at an angle that will cut the gouge to the correct depth. While rolling the torch away, depress the cutting oxygen lever gradually.

Step 5 Continue to move the cutting torch along the line of the gouge while rocking it back and forth to create a gouge of the required depth and width.

4.4.0 Operating Oxyfuel Track Burners

Oxyfuel track burners, such as the one shown in *Figure 62*, provide a convenient way for operators to make straight cuts, curved cuts, and beveled cuts in the field. They can also enhance precision and uniformity in cuts. When a number of cuts are needed, track burners significantly increase productivity. Most models are portable, allowing them to be set up on the job site.

4.4.1 Torch Adjustment

The rack assembly on the track burner permits the torch holder assembly to move toward or away from the tractor unit. The torch holder allows vertical positioning of the torch. The torch bevel adjustment allows torch positioning at any commonly required angle. After adjusting the torch to the desired position, tighten all clamping screws to prevent the torch from making any unexpected movements.

4.4.2 Straight-Line Cutting

The following provides some basic information regarding the set-up of track burners. Although most track burners have a great deal in common, be sure to follow the manufacturer's operating procedures for the system in use.

To perform straight-line cutting with an oxyfuel track burner, first place the machine track on the workpiece and line it up before placing the machine on the track. Be sure the track is long enough for the cut to be made. If not, install additional track. Connect track sections carefully and ensure it is properly supported. When properly connected, the machine should travel smoothly from one track section to the next. If the cut is long, the track may have to be clamped at both ends beyond the cut to keep the track from moving during the cutting process.

> **WARNING!**
>
> Many cutting machines are not designed to detect the end of their track or workpiece. Take care that an unattended machine does not fall from an elevated workpiece while in operation.

Once the track has been positioned, place the machine on the track. Be sure that the supply gas hoses and the power lines are long enough and free to move with the machine so that it can complete the cut properly. Move the machine to the approximate point where the cut will start. Then set the Low/High speed switch to the desired cutting speed or speed range. Set the On/Off switch to the Off position. Next, plug the power cord into an appropriate power supply outlet. Ensure that all clamping screws are properly tightened. Now ignite and properly adjust the torch, and preheat the start of the cut. Set the Forward/Reverse switch to the desired direction of travel. Simultaneously turn on the cutting oxygen and rotate the cutting speed control knob to the desired rate of travel. When the cut is completed, stop the machine and shut off the torch.

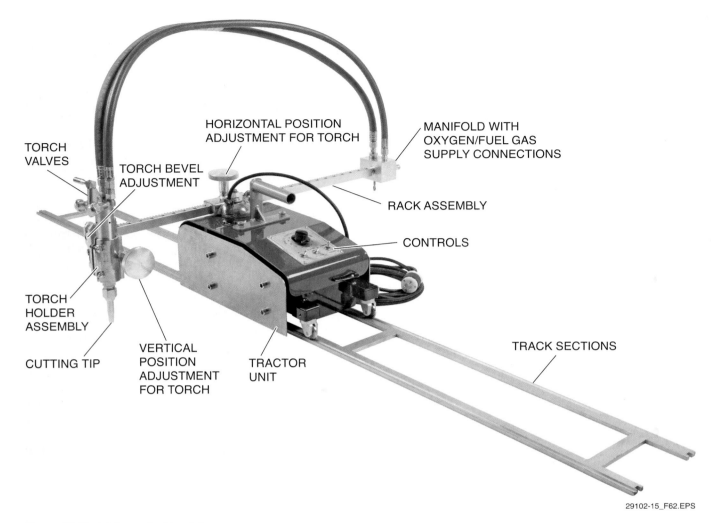

TORCH VALVES

TORCH BEVEL ADJUSTMENT

HORIZONTAL POSITION ADJUSTMENT FOR TORCH

MANIFOLD WITH OXYGEN/FUEL GAS SUPPLY CONNECTIONS

RACK ASSEMBLY

CONTROLS

TORCH HOLDER ASSEMBLY

CUTTING TIP

VERTICAL POSITION ADJUSTMENT FOR TORCH

TRACTOR UNIT

TRACK SECTIONS

29102-15_F62.EPS

Figure 62 Portable oxyfuel track burner.

4.4.3 Bevel Cutting

Bevel cutting can also be performed with a portable oxyfuel track burner. As before, first place the machine track on the workpiece and line it up before placing the machine on the track. The track must be long enough for the cut to be made. If it is not, install additional track. Connect track sections carefully. Extend the track on both sides of the cut and support the track. When properly connected, the machine should travel smoothly from one track section to the next. If the cut is long, the track may have to be clamped at both ends beyond the cut to keep the track from moving during the cut.

With the track properly positioned, place the machine on the track. Make sure the supply gas hoses and the power lines are long enough and free to move with the machine so that it can complete the cut properly. Loosen the bevel adjusting knob, set the torch angle to the desired bevel angle, and then tighten the bevel adjusting knob.

Move the machine to the approximate point where the cut will start. Then set the Low/High speed switch to the desired cutting speed. Set the On/Off switch to the Off position. Next, plug the power cord into a 115 alternating current (AC), 60 Hertz (Hz) power outlet. Ensure that all clamping screws are properly tightened. Now ignite and properly adjust the torch, and preheat the start of the cut. Set the Forward/Reverse switch to the desired direction of travel. Simultaneously turn on the cutting oxygen and rotate the cutting speed control knob to the desired rate. When the cut is completed, stop the machine and shut off the torch.

Additional Resources

ANSI Z49.1, Safety in Welding, Cutting, and Allied Processes. Miami, FL: American Welding Society.

Uniweld Products, Inc. Numerous videos are available at **www.uniweld.com/en/uniweld-videos**. Last accessed: November 30, 2014.

The Harris Products Group, a division of Lincoln Electric. Numerous videos are available at **www.harrisproductsgroup.com/en/Expert-Advice/videos.aspx**. Last accessed: November 30, 2014.

4.0.0 Section Review

1. An oxyfuel cut that has gouging at the bottom, a pronounced break in the drag line, and an irregular kerf is most likely to occur when the ____ .

   a. travel speed is too fast
   b. preheat is insufficient
   c. cutting oxygen pressure is too low
   d. cutting oxygen pressure is too high

2. The process of burning a small hole through the metal where an oxyfuel cut can be started is called ____ .

   a. gouging
   b. beveling
   c. piercing
   d. washing

3. A washing operation uses a special torch tip with a large cutting hole that produces a ____ .

   a. high-velocity stream of oxygen
   b. grooved cut in the base metal surface
   c. cleansing layer of dross in the cut
   d. low-velocity stream of oxygen

4. On a portable oxyfuel track burner, the cutting torch can be angled in a plane that is perpendicular to the track by using the torch ____ .

   a. clamping screws
   b. bevel adjustment
   c. kerf positioner
   d. holder assembly

SUMMARY

Oxyfuel cutting has many uses on many different job sites. It can be used to cut metal plate and shapes to size, prepare joints for welding, clean metals or welds, and disassemble structures. Oxyfuel cutting equipment can range in size from small, portable sets to large, automated, fixed-position machines. In all cases, high pressures and flammable gases are involved. For that reason, there is always the danger of fire and explosion when using oxyfuel equipment. These risks can be minimized when the operator is well trained and knowledgeable. By understanding the safety precautions and equipment fundamentals presented in this module, operators will be better prepared to use oxyfuel cutting equipment and cutting techniques in their workplaces.

1. Oxyfuel cutting is best suited for use on ____.

   a. steel alloys
   b. ferrous metals
   c. nonferrous metals
   d. stainless steel

2. The stream of high-pressure cutting oxygen that is directed from an oxyfuel torch causes the metal to instantaneously ____ .

   a. solidify
   b. de-kerf
   c. magnitize
   d. oxidize

3. The recommended range of tinting for either the face shield or the safety glass lenses used during an oxyfuel cutting operation is ____ .

   a. 1 to 2
   b. 3 to 6
   c. 7 to 8
   d. 9 to 10

4. Most welding environment fires occur during ____ .

   a. oxyfuel gas equipment transporting
   b. carbon arc welding activities
   c. oxyfuel gas welding or cutting
   d. acetylene tank installation

5. When preparing a tank or vessel that might have contained flammable materials to be cut with an oxyfuel cutting torch, the best approach is to ____ .

   a. fill it with water
   b. purge it with oxygen
   c. fill it with air
   d. purge it with acetylene

6. When pure oxygen is combined with a fuel gas, it produces a ____ .

   a. high-pressure jet of cutting oxygen
   b. non-explosive, non-flammable vapor
   c. colorless, odorless, and tasteless gas
   d. high-temperature flame for cutting

7. The most common size of oxygen cylinder used in oxyfuel cutting applications is ____ .

   a. 85 cubic feet (2.4 cubic meters)
   b. 227 cubic feet (6.4 cubic meters)
   c. 350 cubic feet (9.9 cubic meters)
   d. 485 cubic feet (13.7 cubic meters)

8. If an acetylene cylinder is found lying on its side, what should be done before it is used?

   a. Contact the supplier for instructions.
   b. Release a small amount of the gas to atmosphere.
   c. Stand it upright and wait at least one hour.
   d. Add a liquid stabilizer to the gas.

9. The maximum hourly rate at which acetylene gas can be withdrawn from a cylinder is ____ .

   a. one half of the cylinder's capacity
   b. one third of the cylinder's capacity
   c. one fifth of the cylinder's capacity
   d. one seventh of the cylinder's capacity

10. The fuel gas that burns with the lowest flame temperature is ____ .

    a. propane
    b. propylene
    c. acetylene
    d. butane

11. Most gas pressure regulators contain two gauges—one that indicates the cylinder pressure and one that indicates the ____ .

    a. ideal cylinder pressure for the conditions
    b. pressure of the gas at the regulator outlet
    c. pressure in the accompanying cylinder
    d. maximum pressure for the cylinder

12. Oxyfuel cutting equipment that has left-hand threads and a V-notch in the nut are most likely to be ____ .

    a. oxygen fittings
    b. purging gas fittings
    c. aftermarket fittings
    d. fuel gas fittings

13. The type of cutting torch that uses a vacuum created by oxygen flow to draw in fuel from a very-low-pressure fuel source is a(n) _____ .
    a. siphon torch
    b. inert torch
    c. injector torch
    d. suspension torch

14. A friction lighter produces sparks when its steel surface is rubbed with a piece of _____ .
    a. flint
    b. soapstone
    c. graphite
    d. steatite

15. A computer-controlled plate cutting machine and an optical pattern-tracing machine are examples of _____ .
    a. portable oxyfuel cutters
    b. exothermic machines
    c. manual guide cutters
    d. fixed-location machines

16. Cracking the cylinder valves during the setup of oxyfuel equipment is a way of _____ .
    a. leak testing the valve regulators
    b. equalizing the cylinder pressures
    c. removing dirt from the valves
    d. venting residual gas from the cylinders

17. Before opening cylinder valves, verify that the adjusting screws on the oxygen and fuel gas regulators have been _____ .
    a. tightened
    b. closed
    c. loosened
    d. purged

18. Oxyfuel cutting equipment is typically leak tested using a(n) _____ .
    a. supervisor's sense of smell
    b. solution that produces bubbles
    c. lit match or a candle
    d. ultrasonic detector

19. An oxyfuel cutting flame that has an excess of fuel is called a(n) _____ .
    a. hot lean flame
    b. neutral flame
    c. oxidizing flame
    d. carburizing flame

20. When disassembling oxyfuel equipment, verify that all pressure gauges are _____ .
    a. reading zero
    b. open and showing atmospheric pressure
    c. chained securely to each other
    d. marked MT for storage

21. Gas cylinders should never be completely emptied because of the risk of _____ .
    a. reverse flow
    b. valve cracking
    c. disproportional mixing
    d. cylinder implosion

22. When a cut has been made with oxyfuel cutting equipment, the drag lines of the cut should be close to _____ .
    a. thirty degrees with minimal notching
    b. forty-five degrees with a wavy kerf
    c. horizontal and notched at the kerf
    d. vertical and not very pronounced

23. A major concern when cutting thin steel with an oxyfuel cutting torch is _____ .
    a. flashback
    b. sparking
    c. distortion
    d. backfire

24. An oxyfuel cutting process that is often used for cutting off bolts, rivets, and other projections is called _____ .
    a. beveling
    b. gouging
    c. piercing
    d. washing

25. What part of an oxyfuel track burner allows the vertical positioning of the torch to be controlled?
    a. The clamping screws
    b. The torch holder
    c. The kerf regulator
    d. The bevel adjustment

Trade Terms Quiz

Fill in the blank with the correct term that you learned from your study of this module.

1. The gap produced by a cutting process is called a(n) _____ .

2. A loud snap or pop that can be heard as a torch flame is extinguished is called a(n) _____ .

3. The soft, white material that is commonly used to mark metal is _____ .

4. Metals that contain iron are called _____ .

5. Creating a groove in the surface of a workpiece is a process called _____ .

6. A flame that is burning with too much fuel is called a(n) _____ .

7. A flame that is burning with too much oxygen is called a(n) _____ .

8. When correct portions of fuel gas and oxygen are fed to a flame, the flame is said to be a(n) _____ .

9. When a thermal process is used for cutting, the material expelled from the kerf is called _____ .

10. Cutting off projections such as bolts, rivets, and previous welded pieces is a process referred to as _____ .

11. The name used to describe what occurs when an oxyfuel cutting torch penetrates a metal plate is _____ .

12. When a flame burns back into the tip of a torch and causes a high-pitched whistling sound, the condition is called _____ .

13. The lines on the edge of a cut that result from the cutting oxygen streaming into, through, and out of the metal are called _____ .

Trade Terms

Backfire 2
Carburizing flame 6
Drag lines 13
Dross 9
Ferrous metals 4

Flashback 12
Gouging 5
Kerf 1
Neutral flame 8
Oxidizing flame 7

Pierce 11
Soapstone 3
Washing 10

Appendix

PERFORMANCE ACCREDITATION TASKS

The American Welding Society (AWS) School Excelling through National Skills Standards Education (SENSE) program is a comprehensive set of minimum Standards and Guidelines for Welding Education programs. The following performance accreditation is aligned with and designed around the SENSE program.

The Performance Accreditation Tasks (PATs) correspond to and support the learning objectives in *AWS EG2.0, Guide for the Training and Qualification of Welding Personnel: Entry-Level Welder*.

Note that in order to satisfy all learning objectives in *AWS EG2.0*, the instructor must also use the PATs contained in the second level of the NCCER Welding curriculum.

PATs 1 and 2 correspond to *AWS EG2.0, Module 8 – Thermal Cutting Processes, Unit 1 – Manual OFC Principles*, Key Indicators 5, 6, and 7.

PAT 3 corresponds to *AWS EG2.0, Module 8 – Thermal Cutting Processes, Unit 1 – Manual OFC Principles*, Key Indicators 3 and 4.

PATs provide specific acceptable criteria for performance and help to ensure a true competency-based welding program for students.

The following tasks are designed to test your competency with an oxyfuel cutting torch. Do not perform these cutting tasks until directed to do so by your instructor.

SETTING UP, IGNITING, ADJUSTING, AND SHUTTING DOWN OXYFUEL EQUIPMENT

Using oxyfuel equipment that has been completely disassembled, demonstrate how to:

- Set up oxyfuel equipment
- Ignite and adjust the flame
  - Carburizing
  - Neutral
  - Oxidizing
- Shut off the torch
- Shut down the oxyfuel equipment

Criteria for Acceptance:

- Set up the oxyfuel equipment in the correct sequence   _____
- Demonstrate that there are no leaks   _____
- Properly adjust all three flames   _____
- Shut off the torch in the correct sequence   _____
- Shut down the oxyfuel equipment   _____

CUTTING A SHAPE

Using a carbon steel plate, lay out and cut the shape and holes shown in the figure. If available, use a machine track cutter to straight cut the longer dimension.

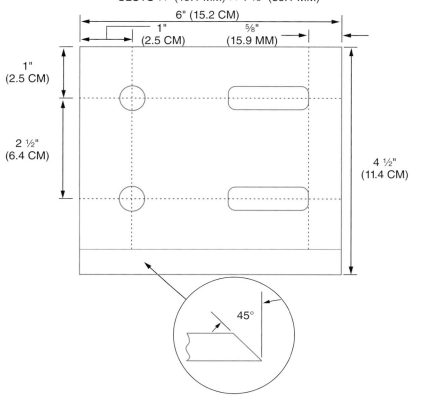

NOTE: MATERIAL – CARBON STEEL ¼" (>6 MM) THICK OR GREATER
HOLES ¾" (19.1 MM) DIAMETER
SLOTS ¾" (19.1 MM) × 1 ½" (38.1 MM)

29102-15_A01.EPS

Criteria for Acceptance:

- Perform this task in the flat position (1G)   _____
- Outside dimensions ±⅛" (3.2 mm)   _____
- Inside dimensions (holes and slots) ±⅛" (3.2 mm)   _____
- Square ±5 degrees   _____
- Minimal amount of dross sticking to plate which can be easily removed   _____
- Square kerf face with minimal notching not exceeding ¹⁄₁₆" (1.6 mm) deep   _____

CUTTING A SHAPE

Using a carbon steel plate, lay out and cut the shape and holes shown in the figure. If available, use a machine track cutter to bevel and straight cut the longer dimension.

NOTE: MATERIAL – CARBON STEEL ¼" (>6 MM) THICK OR GREATER
HOLES ¾" (19.1 MM) DIAMETER
SLOTS ¾" (19.1 MM) × 1 ½" (38.1 MM)

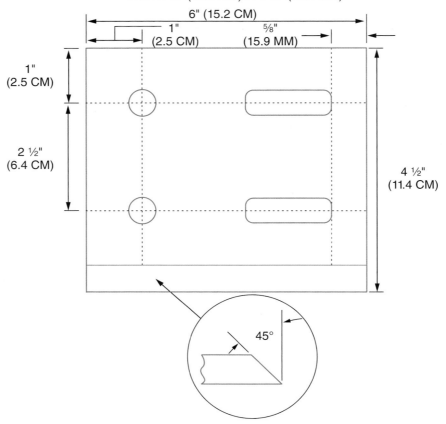

29102-15_A02.EPS

Criteria for Acceptance:

- Perform this task in the horizontal position (2G)   _____
- Outside dimensions ±⅛" (3.2 mm)   _____
- Inside dimensions (holes and slots) ±⅛" (3.2 mm)   _____
- Square ±5 degrees   _____
- Bevel ±2 degrees   _____
- Minimal amount of dross sticking to plate which can be easily removed   _____
- Square kerf face with minimal notching not exceeding 1⁄16" (1.6 mm) deep   _____

Trade Terms Introduced in This Module

Backfire: A loud snap or pop as a torch flame is extinguished.

Carburizing flame: A flame burning with an excess amount of fuel; also called a reducing flame.

Drag lines: The lines on the edge of the material that result from the travel of the cutting oxygen stream into, through, and out of the metal.

Dross: The material (oxidized and molten metal) that is expelled from the kerf when cutting using a thermal process. It is sometimes called slag.

Ferrous metals: Metals containing iron.

Flashback: The flame burning back into the tip, torch, hose, or regulator, causing a high-pitched whistling or hissing sound.

Gouging: The process of cutting a groove into a surface.

Kerf: The gap produced by a cutting process.

Neutral flame: A flame burning with correct proportions of fuel gas and oxygen.

Oxidizing flame: A flame burning with an excess amount of oxygen.

Pierce: To penetrate through metal plate with an oxyfuel cutting torch.

Soapstone: Soft, white stone used to mark metal.

Washing: A term used to describe the process of cutting out bolts, rivets, previously welded pieces, or other projections from the metal surface.

Additional Resources

This module presents thorough resources for task training. The following resource material is suggested for further study.

ANSI Z49.1, Safety in Welding, Cutting, and Allied Processes. Miami, FL: American Welding Society.

Plasma Cutters Handbook: Choosing Plasma Cutters, Shop Safety, Basic Operation, Cutting Procedures, ANSI Z49.1, Safety in Welding, Cutting, and Allied Processes. Miami, FL: American Welding Society.

The Harris Products Group, a division of Lincoln Electric. Numerous videos are available at **www.harrisproductsgroup.com/en/Expert-Advice/videos.aspx**. Last accessed: November 30, 2014.

Uniweld Products, Inc. Numerous videos are available at **www.uniweld.com/en/uniweld-videos**. Last accessed: November 30, 2014.

Figure Credits

The Lincoln Electric Company, Cleveland, OH, USA, Module Opener, Figures 1, 2, 5, 14, 32–34, 48–50, 53, 60 (photo), 62, SA04

Topaz Publications, Inc., Figures 3, 7–9, 11, 13, 15, 16, 18, 27, 28, 29A, 30, 37–45, SA02, SA03, SA06–SA08, SA11

Courtesy of Uniweld Products, Figure 20, SA10

Victor Technologies, Figure 21

Vestil Manufacturing, Figure 29B

Koike Aronson, Inc. – Worldwide manufacturer of cutting, welding and positioning equipment, Figure 31

Courtesy of H & M Pipe Beveling Machine Company, Inc., Figures 35, 36

Zachry Industrial, Inc., Figures 46, 47, 51

Courtesy of Smith Equipment, Figure 52

© American Welding Society (AWS) *Welding Handbook* 1991, Welding Processes Volume No. 2, Edition No. 8, Miami: American Welding Society, Figure 55

Xerafy, SA01

Courtesy of Saf-T-Cart, SA09

Section Review Answer Key

Answer	Section Reference	Objective
Section One		
1. b	1.1.0	1a
2. d	1.2.2	1b
Section Two		
1. b	2.1.2	2a
2. a	2.2.3	2b
3. d	2.3.1	2c
4. a	2.4.4	2d
5. c	2.5.3	2e
Section Three		
1. b	3.1.1	3a
2. d	3.2.0	3b
3. c	3.3.1	3c
4. d	3.4.0	3d
Section Four		
1. a	4.1.1	4a
2. c	4.2.3	4b
3. d	4.3.2	4c
4. b	4.4.1	4d

NCCER CURRICULA — USER UPDATE

NCCER makes every effort to keep its textbooks up-to-date and free of technical errors. We appreciate your help in this process. If you find an error, a typographical mistake, or an inaccuracy in NCCER's curricula, please fill out this form (or a photocopy), or complete the online form at **www.nccer.org/olf**. Be sure to include the exact module ID number, page number, a detailed description, and your recommended correction. Your input will be brought to the attention of the Authoring Team. Thank you for your assistance.

Instructors – If you have an idea for improving this textbook, or have found that additional materials were necessary to teach this module effectively, please let us know so that we may present your suggestions to the Authoring Team.

NCCER Product Development and Revision
13614 Progress Blvd., Alachua, FL 32615

Email: curriculum@nccer.org
Online: www.nccer.org/olf

❏ Trainee Guide   ❏ Lesson Plans   ❏ Exam   ❏ PowerPoints   Other _____

Craft / Level: _____   Copyright Date: _____

Module ID Number / Title: _____

Section Number(s): _____

Description: _____

_____

_____

_____

Recommended Correction: _____

_____

_____

_____

Your Name: _____

Address: _____

_____

Email: _____   Phone: _____

29103-15
Plasma Arc Cutting

OVERVIEW

Welders must be familiar with the task of plasma arc cutting. This module presents information related to the plasma arc cutting process, including safety procedures, setup, gas types, flow rates, and equipment.

Module Three

Trainees with successful module completions may be eligible for credentialing through the NCCER Registry. To learn more, go to **www.nccer.org** or contact us at **1.888.622.3720**. Our website has information on the latest product releases and training, as well as online versions of our *Cornerstone* magazine and Pearson's product catalog.

Your feedback is welcome. You may email your comments to **curriculum@nccer.org**, send general comments and inquiries to **info@nccer.org**, or fill in the User Update form at the back of this module.

This information is general in nature and intended for training purposes only. Actual performance of activities described in this manual requires compliance with all applicable operating, service, maintenance, and safety procedures under the direction of qualified personnel. References in this manual to patented or proprietary devices do not constitute a recommendation of their use.

PLASMA ARC CUTTING

Objectives

When you have completed this module, you will be able to do the following:

1. Explain plasma arc cutting processes and identify related safety precautions.
   a. Describe the plasma arc cutting processes.
   b. Identify safety practices related to plasma arc cutting.
2. Identify and describe plasma arc cutting equipment.
   a. Identify and describe plasma arc power units.
   b. Identify and describe plasma arc torches and accessories.
   c. Identify and describe plasma arc cutting gases and gas control devices.
3. Describe how to set up, safely operate, and care for plasma arc cutting equipment.
   a. Describe how to set up plasma arc cutting equipment and the adjacent work area.
   b. Describe how to safely operate plasma arc cutting equipment.
   c. Describe how to care for plasma arc cutting equipment.

Performance Tasks

Under the supervision of your instructor, you should be able to do the following:

1. Set up plasma arc cutting equipment.
2. Set the amperage and gas pressures or flow rates for the type and thickness of metal to be cut using plasma arc equipment.
3. Square-cut metal using plasma arc equipment.
4. Bevel-cut metal using plasma arc equipment.
5. Pierce and cut slots in metal using plasma arc equipment.
6. Dismantle and store the equipment.

Trade Terms

Amperage
Dross
Duty cycle
Phase

Plasma
Potential
Solenoid valve

Industry Recognized Credentials

If you are training through an NCCER-accredited sponsor, you may be eligible for credentials from NCCER's Registry. The ID number for this module is 29103-15. Note that this module may have been used in other NCCER curricula and may apply to other level completions. Contact NCCER's Registry at 888.622.3720 or go to **www.nccer.org** for more information.

Contents

Topics to be presented in this module include:

Figures and Tables

1.0.0 PLASMA ARC CUTTING PROCESSES

Objective

Explain plasma arc cutting processes and identify related safety precautions.

a. Describe plasma arc cutting processes.
b. Identify safety practices related to plasma arc cutting.

Trade Terms

Dross: A waste byproduct of molten metal.

Plasma: A fourth state of matter (not solid, liquid, or gas) created by heating a gas to such a high temperature that it boils electrons off the gas molecules (ionization).

(A) MANUAL PLASMA ARC CUTTING

(B) MECHANIZED PLASMA ARC CUTTING

29103-15_F01.EPS

Figure 1 Plasma arc cutting (PAC).

Plasma arc cutting (PAC) uses a jet of plasma to pierce, cut, and gouge metal. The plasma is created by superheating gas in an electric arc. The temperature of the plasma jet is hot enough to melt almost any metal. The process can be set up for manual or mechanized operation (*Figure 1*) and, in most cases, is faster and more efficient than any other cutting method. PAC offers the following advantages:

- Cuts both ferrous and nonferrous metals
- Produces minimal dross
- Creates a minimal heat-affected zone
- Causes little or no distortion
- Possesses high cutting speeds
- Produces a very narrow kerf

The plasma arc cutting setup (*Figure 2*) requires an electrical power supply and a cutting gas, which can be compressed air. Depending on the equipment, a separate shielding or cooling gas may also be required. PAC torches convert the cutting gas into plasma by passing the gas through an electric arc inside the torch. The torch has a relatively small opening (orifice), which constricts the arc and high-pressure gas flow. This results in the plasma stream exiting the torch as a supersonic jet hotter than any flame. Depending on the current flow, the plasma cutting jet may reach temperatures of 30,000°F (16,650°C) or higher. When the jet contacts metal, it transfers its tremendous heat, causing the metal to melt instantly. The high-velocity jet blasts the molten metal away, forming a hole, groove, or gouge. Be advised that this high-velocity jet blast can carry molten material 40 feet (12 m) away from the workpiece.

1.1.0 Plasma Arc Cutting Process

There are two types of plasma arc cutting processes: transferred arc and non-transferred arc.

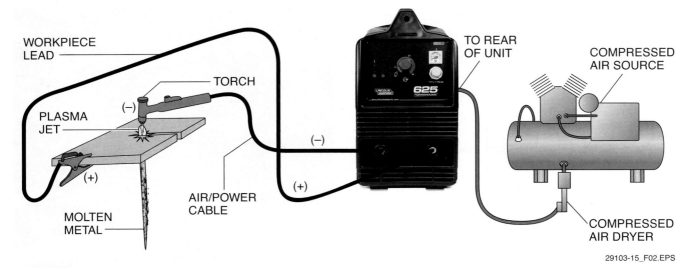

Figure 2 Plasma arc cutting setup diagram.

The transferred arc process is the most common and is used to cut materials that are electrically conductive, such as metals. The non-transferred arc process is less common and is used to cut materials that are not electrically conductive, such as ceramics and concrete.

1.1.1 Transferred Arc Process

In the transferred-arc process, the workpiece is part of the electrical circuit. The initial arc is established between the electrode and the workpiece via an arc transfer process. A number of different arc-initiation methods are used for the transfer process. These include contact starting, discharge-starting to the workpiece, and discharge-starting using a pilot arc. Of these methods, discharge-starting using a pilot arc is the most common. In this method, a low-voltage, low-current, high-frequency (HF) arc is established between the electrode and the torch nozzle inside the torch. This small arc between the torch electrode and torch nozzle is called a pilot arc. The pilot arc is used to ionize the air jet, resulting in plasma generation. When the torch is near the workpiece, a higher-voltage, current-cutting arc is transferred and established between the torch electrode and the work via the plasma. The higher-voltage, current-cutting arc maintains the plasma jet and the pilot arc is then shut off. For the transferred-arc process to work, one lead of the power supply (the positive lead) must be connected to the metal to be cut. *Figure 3* shows a schematic of the transferred-arc process.

1.1.2 Non-Transferred Arc Process

In the non-transferred arc process, the entire arc is established within the torch, between the torch electrode and the torch nozzle. The material being cut is not electrically connected to the arc circuit, since it is typically non-conductive material. The non-transferred arc process is not normally found in facilities that cut only metal. *Figure 4* shows a schematic of the non-transferred arc process.

1.2.0 Safety Practices

The following is a summary of safety procedures and practices that must be observed while cutting. Keep in mind that this is a summary only. Complete safety coverage is provided in the *Welding Safety* module. If you have not completed that module, do so before continuing. Above all, be sure to wear appropriate protective clothing and equipment when welding or cutting.

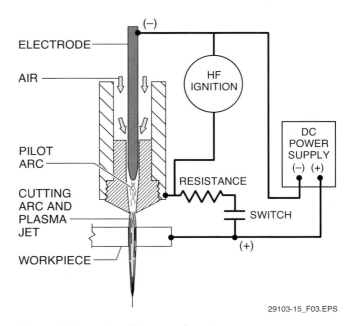

Figure 3 Schematic of the transferred-arc process.

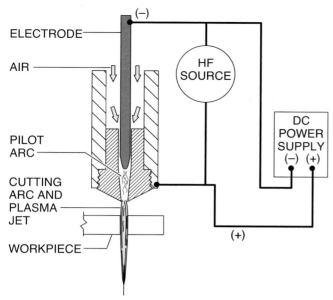

Figure 4 Schematic of the non-transferred arc process.

A plasma arc cutting process that uses air or nitrogen and is set up correctly is generally safer than oxyfuel processes. There is no chance of flashbacks and other dangers caused by flammable gases such as those used in oxyfuel processes. However, the voltages used, the extremely high temperatures of the cutting jet, and the intense ultraviolet radiation of the arc are hazardous. In addition to the safety practices provided in the following section, the following practices should also be observed:

- Make sure that all sources of energy to the PAC equipment are turned off before disassembling a plasma torch to replace any consumable components. Open-circuit DC voltages as high as 400 volts may be present between the torch components and the workpiece when they are connected.
- The extremely high temperatures of the plasma cutting jet can quickly and easily cut through gloves and flesh, resulting in severe injuries.

WARNING!

Working with energized PAC equipment is extremely hazardous. Always disconnect the power supply and all other sources of energy before servicing or otherwise working with plasma cutting equipment. Never place your hands under the workpiece near the cutting jet. Even with protective clothing, you can be severely injured.

1.2.1 *Protective Clothing and Equipment*

Welding activities can cause injuries unless you wear all of the protective clothing and equipment that is designed specifically for the welding industry. The following information includes important safety guidelines about protective clothing and equipment:

- Always use safety glasses with a full face shield or a helmet. The glasses, face shield, or helmet lens must have the proper light-reducing tint for the type of welding or cutting being performed. Use corrective lenses if needed; some companies will not allow contact lenses to be worn during cutting or welding activities. Never view an electric arc directly or indirectly without using a properly tinted lens.
- Wear proper protective leather and/or flame-retardant clothing along with welding gloves that will protect you from flying sparks and molten metal.
- Wear high-top safety shoes or boots. Make sure that the tongue and lace area of the footwear will be covered by a pant leg or other protective material. Boots with no laces at all are generally preferred for both welding and cutting activities. If the tongue and lace area is exposed, or if the footwear must be protected from burn marks, wear leather spats under your pants or chaps and over the top front area of the footwear.
- Wear a 100-percent cotton cap with no mesh material included in its construction. The bill of the cap points to the rear. If a hard hat is required for the environment, use one that allows the attachment of rear deflector material and a face shield. A hard hat with a rear deflector is generally preferred when working overhead, and may be required by some employers and job sites.

WARNING!

Do not wear a cap with a button in the middle. The conductive metal button beneath the fabric represents a safety hazard.

- Wear a face shield over snug-fitting cutting goggles or safety glasses for gas welding or cutting. Either the face shield or the lenses of the welding goggles must be an approved shade for the application. A welding hood equipped with a properly tinted lens is also acceptable. See *Table 1* for a list of lens shades based on arc current.

Table 1 Plasma Arc Cutting – Guide to Lens Shade Numbers

From *AWS F2.2 Lens Shade Selector*. Shade numbers are given as a guide only and may be varied to suit individual needs.

Process	Arc Current (Amperes)	Minimum Protective Shade	Suggested* Shade No. (Comfort)
Plasma Arc Cutting (PAC)	Less than 20	4	4
	20–40	5	5
	40–60	6	6
	60–80	8	8
	80–300	8	9
	300–400	9	12
	400–800	10	14

*As a rule of thumb, start with a shade that is too dark to see the weld zone. Then go to a lighter shade which gives sufficient view of the weld zone without going below the minimum. In oxyfuel gas welding, cutting, or brazing where the torch and/or the flux produces a high yellow light, it is desirable to use a filter lens that absorbs the yellow or sodium line of the visible light spectrum.

29103-15_T01.EPS

- Wear earplugs to protect ear canals from sparks. Wear hearing protection to protect against the consistent sound of the torch.

> **WARNING!**
> Ear protection is essential to protect ears from the noise of the torch. Other personal protective equipment (PPE) must be worn to protect the operator from hot metal and slag.

1.2.2 Fire/Explosion Prevention

Welding activities usually involve the use of fire or extreme heat to melt metal. Whenever fire is used, it must be controlled and contained. Welding or cutting activities are often performed on vessels that may once have contained flammable or explosive materials. Residues from those materials can catch fire or explode when a welder begins work on such a vessel. The following are fire and explosion prevention guidelines associated with welding:

- Never carry matches or gas-filled lighters in your pockets. Sparks can cause the matches to ignite or the lighter to explode, causing serious injury.
- Always comply with all site and/or employer requirements for a hot-work permit and a fire watch.
- Never use oxygen to blow dirt or dust off clothing. The oxygen can remain trapped in the fabric for a time. If a spark hits the oxygen in the fabric, the clothing can burn rapidly and violently.
- Make sure that any flammable material in the work area is moved or shielded by a fire-resistant covering.

- Approved fire extinguishers must be available before attempting any heating, welding, or cutting operations. Make sure the extinguisher is charged, the inspection tag is valid, and any individual that may be required to operate it knows how to do so.
- Never release a large amount of oxygen or use oxygen in place of compressed air. The presence of oxygen around flammable materials or sparks can cause rapid and uncontrolled combustion. Keep oxygen away from oil, grease, and other petroleum products.
- Never release a large amount of fuel gas, especially acetylene. Methane and propane are heavier than air and tend to migrate to and concentrate in low areas. As a result, they can ignite at a considerable distance from the release point. Acetylene is lighter than air but is even more dangerous than methane; when mixed with air or oxygen, it will explode at much lower concentrations than any other common fuel gas.
- To prevent fires, maintain a neat and clean work area, and make sure that any metal scrap or slag is cold before disposal.

Before cutting or welding containers such as tanks or barrels, find out if they contained any explosive, hazardous, or flammable materials, including petroleum products, citrus products, or chemicals that decompose into toxic fumes when heated. Proper procedures for cutting or welding hazardous containers are described in the *American Welding Society (AWS) F4.1, Safe Practices for the Preparation of Containers and Piping for Welding and Cutting,* and *ANSI Z49.1.* As a standard practice, always clean and then fill any tanks or barrels with water, or purge them with a flow of inert gas such as nitrogen to displace any oxygen.

NCCER – *Welding Level One* 29103-15

Containers must be cleaned by steam cleaning, flushing with water, or washing with detergent until all traces of the material have been removed.

After cleaning the container, fill it with water or a purging gas, such as carbon dioxide, argon, or nitrogen to displace the explosive fumes. Air, which contains oxygen, is displaced from inside the container by the water or inert gas. Without oxygen, combustion cannot take place.

A water-filled vessel is the best alternative. When using water, position the container to minimize the air space. When using an inert gas, provide a vent hole so the inert gas can push the air and other vapors out to the atmosphere. Keep in mind, though, that even these precautions do not guarantee the absence of flammable materials inside. For that reason, these types of activities should not be done without proper supervision and the use of proper testing methods.

1.2.3 Work Area Ventilation

Vapors and fumes tend to rise in the air from their sources. Welders often have to work above the welding area where the fumes are being created. Welding fumes can cause personal injuries. Good work area ventilation helps to remove the vapors and to protect the welder. The following is a list of work area ventilation guidelines to consider before and during welding activities:

- Make sure confined-space procedures are followed before conducting any welding or cutting in the confined space.
- Always perform cutting or welding operations in a well-ventilated area. Cutting or welding operations involving materials such as coatings that contain cadmium, lead, zinc, or chromium will result in toxic fumes. For cutting or welding of such materials, always wear an approved respirator with an appropriate filter as directed by your employer.
- Make sure confined spaces are ventilated properly for cutting or welding purposes.
- Never use pure oxygen from cylinders for the purpose of ventilation or breathing air.

Career Opportunities

The growth of the automobile racing industry has created career opportunities for workers who are skilled at welding and cutting metal. In addition to fabrication of race cars at the shop, there is a demand for at-the-track repair work. Lincoln Electric offers a Motorsports Welding School to address the special needs of the racing community.

29103-15_SA01.EPS

Additional Resources

AWS F3.2M/F3.2, Ventilation Guide for Weld Fume. Latest Edition. Miami, FL: American Welding Society.

Plasma Cutters Handbook: Choosing Plasma Cutters, Shop Safety, Basic Operation, Cutting Procedures, Advanced Cutting Tips, CNC Plasma Cutters, Troubleshooting, and Sample Projects. Eddie Paul. New York, NY: Penguin Group.

1.0.0 Section Review

1. The plasma arc cutting process type most commonly used is the ____ .
   a. electrical process
   b. non-transferred arc process
   c. oxyfuel pilot process
   d. transferred arc process

2. When operating plasma arc cutting equipment, it is acceptable to wear a hat made of ____ .
   a. polyester
   b. nylon
   c. 100-percent cotton
   d. wool

2.0.0 PLASMA ARC CUTTING EQUIPMENT

Objective

Identify and describe plasma arc cutting equipment.

   a. Identify and describe plasma arc power units.
   b. Identify and describe plasma arc torches and accessories.
   c. Identify and describe plasma arc cutting gases and gas control devices.

Trade Terms

Amperage: The unit of measure used for the intensity of an electric current; often abbreviated to *amp* or *amps*.

Duty cycle: The percentage of time a plasma arc cutting machine can cut without overheating within a ten-minute period.

Phase: In a three-phase power supply system, the sine waves of the voltage on each of the three separate conductors is displaced from each of the others by 120 degrees, although all conductors are carrying alternating current at the same frequency and are synchronized. Each of the three sine waves is referred to as a phase of the power source.

Potential: The relative electrical voltage difference between two points of reference; the electromotive force or difference in potential between two points, expressed in volts.

Solenoid valve:: A valve used to control the flow of gases or liquids that is opened or closed by the action of an energized electromagnet.

There are many different manufacturers of PAC equipment. The following sections provide general information common to most equipment types. For information on the operation of specific equipment at your site, always refer to the manufacturer's documentation.

Basic plasma arc cutting equipment includes the following items:

- A power unit, where the process is controlled
- Plasma arc cutting torch with torch cable
- Workpiece lead assembly
- Plasma, shielding and cooling gases, and gas control components

2.1.0 PAC Power Units

Plasma-arc cutting is performed using direct-current, electrode-negative (DCEN) power. This means that the electrode of the torch has a negative potential compared to the workpiece lead. The power units are designed specifically for plasma arc cutting, and they are available in many sizes and configurations. The smallest and simplest PAC units are designed for light-duty manual cutting of sheet metal and light gauge plate. They typically plug into a single-phase 115VAC or 230VAC outlet. Typical maximum cutting amperages for these light-duty units range from 14 to 30 amperes. Their duty cycle is usually in the 35 percent range. The simplest units use filtered, compressed air for plasma and cooling gas. They have a high-frequency generator for easy arc initiation. Some of these small air units contain their own air compressor, to supply air for both cooling and plasma generation.

Commercial-grade portable units are available that can operate on various single-phase or three-phase power supplies. These are rated for cutting material up to 1¼" (3.2 cm) thick at duty cycles of 50 to 60 percent. At maximum current and operating at a reduced duty cycles, they are capable of severing material up to 1½" (3.8 cm) thick. These units normally require an external air or gas supply for the torch. The size of the unit needed is determined by the type and thickness of the metal to be cut.

Console controls are relatively simple. They usually include most of the following controls and indicators:

- Power On/Off switch
- Power On and/or unit-ready indicator
- Air or gas On/Off switch or setting (for monitoring and adjusting torch air or gas flow, or for purging lines)
- Output current control
- Power unit trouble indicator(s)

Figure 5 shows a typical portable commercial-grade plasma arc cutting unit.

The Hazards of PAC

Ultraviolet radiation, hot particle matter, and noise are all hazards associated with PAC, but they are manageable with the proper equipment. Devices such as water tables and water mufflers use water to control such hazards in mechanized plasma cutting applications.

RESET SWITCH, PURGE SET, ON/OFF SWITCH

GAS/AIR PRESSURE GAUGE

GAS/AIR CONNECTION (IN REAR)

AIR PRESSURE CONTROL

CURRENT OUTPUT CONTROL

CARRYING STRAP

WORKPIECE CABLE

TORCH CABLE

29103-15_F05.EPS

Figure 5 Commercial-grade plasma arc cutting unit.

Larger industrial units are used for heavy-duty manual cutting and general-duty mechanized cutting. These units usually operate on a 230/460VAC three-phase power supply, producing DC output currents up to 750 amperes. Their duty cycle is usually 100 percent, meaning that they are capable of constant operation. They can cut stainless steel and aluminum up to 2½" (6.4 cm) thick and carbon steel up to 3" (7.6 cm) thick. Some units are water-cooled, often containing their own closed-loop torch cooling system.

The larger PAC units are often dual-flow, mechanized units that operate with several types of gases. One gas is used for cutting, and the other is used as a shield around the cutting area, to improve the quality of the cut. Some units may use only one of the gases for the pilot arc.

Although gouging is not the primary use for plasma arc torches, some commercial and industrial units may be used for gouging when fitted with a gouging torch, or a torch with gouging components. Plasma arc gouging is sometimes referred to using the nonstandard acronym PAG.

Controls on the larger units typically include a power On/Off switch, output current control and ammeter, several gas flow controls and gauges, gas selector controls, local/remote switch, and open-circuit voltmeter. A leather cable cover accessory is often used to protect the torch cable from damage from the resulting hot or molten metal. The covers are typically designed with snaps or Velcro® to secure it around the torch cable.

2.1.1 Workpiece Clamp

The workpiece clamp (*Figure 6*) provides the connection between the end of the workpiece lead and the workpiece. The area on the workpiece where the clamp is being attached must be cleaned well enough for good metal-to-metal contact. Workpiece clamps are mechanically connected to the welding lead, and they come in a variety of shapes and sizes. Workpiece clamps are rated by the current they can carry without overheating. If a workpiece clamp needs to be replaced, be sure to select one that is rated at or above the output capacity of the PAC power unit.

When PAC will be used to cut equipment containing electrical or electronic components, batteries, bearings, or seals, the workpiece clamp must be properly located to prevent damage to associated components. If the cutting current passes through any type of bearing, seal, valve, contacting surface, or electrical or electronic component, it could cause severe damage to the item from arcing and overheating.

Cutting Ratings

Manufacturers may rate the cutting capability of their equipment in different ways. Some may rate it at the listed duty cycle, and others may rate it at the maximum cutting (severing) capability at a much reduced duty cycle. When selecting a power supply, make sure that you understand the manufacturer's rating system.

Control Interlocks

Interlocks may be incorporated into the controls of PAC systems. An interlock common in PAC systems monitors the air supply pressure. If the plasma torch is operated without an adequate supply of air, internal arcing may damage the torch. For this reason, a gas pressure safety switch is included in the circuit. Adequate gas pressure must be present before the torch will operate. The safety switch also shuts down the torch if there is a gas supply failure during cutting. Another interlock is an over-temperature switch that shuts down the unit if the duty cycle is exceeded and the unit overheats. Some units are also equipped with a retaining cup switch that shuts down the unit if the retaining cup holding the tip, nozzle, and swirl ring in the torch is loose.

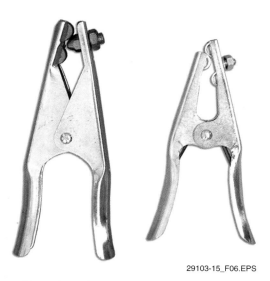

29103-15_F06.EPS

Figure 6 Typical workpiece clamps.

Carefully evaluate the cutting task and position the workpiece clamp so that the cutting current will not pass through any sensitive components or across gaps. If in doubt, ask your supervisor for assistance before proceeding.

> **CAUTION**
>
> To prevent damage from arcing across gapped or contacting surfaces, make sure the path from the workpiece clamp to the torch arc does not pass through these surfaces. Cutting current can severely damage bearings, seals, valves, or contacting surfaces. Position the workpiece clamp to prevent the cutting current from passing through them.
>
> Disconnect the grounded battery lead on any mobile equipment that is being repaired using the plasma arc cutting process. This prevents the cutting current from causing battery explosion or battery damage. Never operate PAC equipment near a battery.

Do not cut near batteries. A spark could cause a battery to explode, showering the area with battery acid. Always remove batteries from the work area.

Also, workpiece leads must never be connected to pipes carrying flammable, volatile, or corrosive materials. The cutting current could cause overheating or sparks, resulting in an explosion or fire.

> **CAUTION**
>
> The cutting current can destroy electronic or electrical equipment. Have an electrician check the equipment and, if necessary, isolate the system before cutting.

2.2.0 PAC Torches

The most common plasma arc cutting torch uses a pilot arc and the transferred arc process. A high-frequency pilot arc is first started between the electrode and the nozzle. A small amount of gas is injected into the arc chamber where it is heated to a plasma and escapes through the nozzle as a fine jet. However, the torch current circuit is designed so that the full voltage, or potential, and current are not available in the torch between the internal electrode and nozzle. The base metal to be cut is connected to the power unit by the workpiece clamp to make a complete circuit. This gives the workpiece a higher potential difference than the nozzle. When the torch is moved near the workpiece, the higher-current arc extends (transfers) to the workpiece because of its higher potential. The control unit automatically increases the amperage and gas flow(s) to produce the longer cutting arc, and disables the pilot arc. *Figure 7* shows the result of the plasma transfer process that takes place between the torch and the workpiece.

> **GOING GREEN**
>
> ## Cooling Water
>
> Welders involved in metal-cutting activities often allow the water used in the process to simply flow into the ground or down the drain. To better protect the environment, water used for cooling the plasma arc cutting equipment should be contained whenever possible and sent to a waste treatment facility.

Figure 7 A transferred plasma cutting arc.

Plasma arc torches may be handheld for manual cutting or machine mounted for mechanized cutting. The heaviest duty torches are usually mechanized and water-cooled. Handheld torches are usually light-duty to heavy-duty, with heavy-duty models typically being water-cooled. Some torches are equipped with separate passages and orifices for cooling and shielding gases. Torches have a button or lever on the torch handle to start the pilot arc. A foot pedal can also be used to start the arc. When a foot pedal is used, the switch on the control panel is set for remote operation.

The typical PAC torch usually contains a replaceable retaining cup, a copper electrode with a hafnium or thoriated tungsten insert, a gas swirl ring, and a nozzle. Nozzles have various-sized holes for cutting metals of different thicknesses. The torch also contains coolant passages for gas or water cooling. *Figure 8* shows a typical handheld PAC torch and its consumable components.

Mechanized PAC torches are designed to be operated on automated carriers. They can usually be used in the same carriers designed for oxyfuel cutting heads. However, any carrier used must be capable of moving at the much greater speeds required for effective plasma arc cutting.

Water-injection PAC torches inject pressurized water to create a whirlpool around the jet, below the point where the arc narrows in the torch's arc cavity. The water swirls around the plasma jet at high velocity and compresses the jet even more. The narrower jet cuts with a more uniform melt rate and leaves the kerfs of stainless steel and titanium free of dross. This type of torch is normally mechanized.

Several styles of shrouds, heat shields, nozzles, and electrodes are usually available for a given model of PAC torch. These components vary in style and size to adapt to different metal types and thicknesses. Nozzles and other components for gouging are offered to fit some of the heavier duty torches. Electrode types and styles vary with torch manufacturers.

The most common parts that are changed in a torch are the electrode and the nozzle. The electrode is changed when it becomes contaminated. The nozzle is changed when it becomes worn and no longer true to its original shape or size, or when metal of a different type or thickness needs to be cut. When performing low-amperage cutting, it is possible to drag the torch on the workpiece using a drag shield or standoff post guide without causing damage or sacrificing the cut. Cutting at current levels above 60 amps or for an extended time usually requires a standoff guide, which attaches to the cup and holds the tip at the required distance from the workpiece. *Figure 9* shows two common styles. The standoff distance, or the distance between the torch tip and the workpiece, should be kept between ¹⁄₁₆" and ¼" (1.6 to 6.4 mm) depending on the thickness of the workpiece.

2.3.0 PAC Gases and Controls

All PAC units require one or more types of gases. The specific type of gases required and the controls necessary to set and adjust them vary with the type, size, and manufacturer of the PAC equipment.

Auto-Voltage Adjustments

Some newer PAC units sense the input voltage and frequency when they are connected to a power source. The units continuously and automatically reconfigure themselves to operate properly with the connected power. This eliminates improper initial power connections and provides continuous adjustment to changing input voltage so that the PAC arc current remains stable during cutting operations. This feature also increases the flexibility of the unit to operate in various locations.

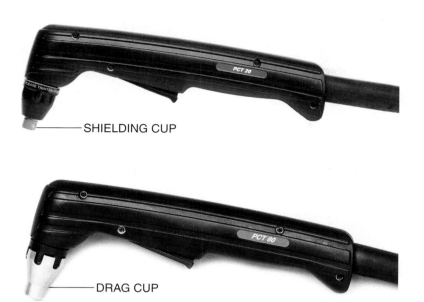

SHIELDING CUP

DRAG CUP

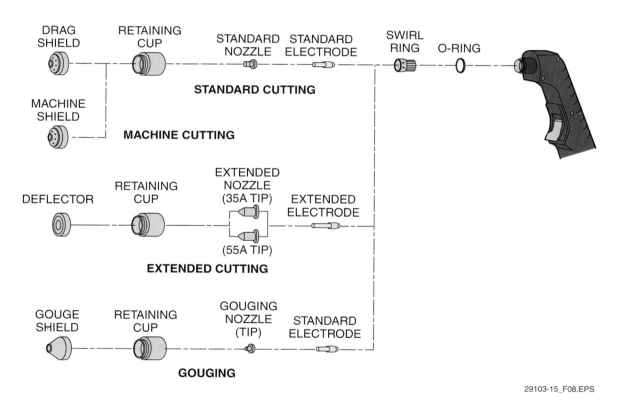

Figure 8 Handheld plasma torches.

29103-15_F08.EPS

(A) STANDOFF ROLLER GUIDE

(B) STANDOFF POST GUIDE

29103-15_F09.EPS

Figure 9 Standoff guides.

2.3.1 PAC Gases

Several different gases and gas mixtures are used with plasma arc cutting units. Gases used include air, nitrogen, oxygen, argon and argon mixtures, hydrogen, and carbon dioxide. The simplest units use clean compressed air to cut carbon steel, stainless steel, and aluminum up to about ³⁄₁₆" (4.8 mm) thick. Most heavy-duty air plasma units are rated to cut carbon steel and stainless steel up to about 1¼" (32 mm) thick and possibly sever material up to 1¾" (44.5 mm) thick. Dual-flow PAC units can operate with two different gases. One gas may be used for the pilot arc and for cutting, and another for shielding or cooling. *Table 2* provides general recommendations for plasma and shield gases.

Specific gas requirements for a particular PAC unit or system are specified in the manufacturer's operating instructions manual and/or the information tags on the unit. Refer to these specifications, and use only the gases specified for the type of equipment being used.

2.3.2 PAC Gas Controls

Plasma arc cutting gases are used to generate plasma, cool the torch, and shield the cut against corrosion. Most gas sources must have their pres-

Replacing Consumables

The torch nozzle and electrode are considered consumables, which are components requiring frequent replacement due to deterioration, and both should be replaced at the same time. It is a good idea to always have a replacement torch nozzle and electrode handy. This will reduce downtime when cutting with the plasma unit.

Contact-Start or Discharge-Start Torches

High-frequency pilot arc starting of a PAC torch can interfere with nearby electronic circuits, including computers and microprocessor controllers. To eliminate the use of a high-frequency start, some newer PAC torches use an internal contact or capacitor-discharge pilot arc starting system. One such contact-start system has a spring-loaded electrode that is in contact with the nozzle when the torch is off. When a trigger is pressed, the PAC unit senses the shorted electrode via high-current flow and opens the gas solenoid. Gas pressure then forces the electrode away from the nozzle. This action draws an arc that establishes the plasma flow. The arc instantly transfers to the workpiece because of the higher voltage difference between the electrode and the workpiece.

Table 2 Recommended Plasma/Shield Gas Combinations

Material	Air/Air	O_2/Air	N_2/CO_2	N_2/Air	H35/N_2
Carbon Steel	Most Economical Good Cut Quality Good Speed Good Gouging Good Weldability	Best Cut Quality Maximum Cut Speed Best Weldability	Some Dross Long Electrode Life	Not Recommended	Best Gouging Long Electrode Life Some Dross
Stainless Steel	Most Economical Good Speed Some Dross	Not Recommended	Good Cut Quality Good Gouging Minimal Dross Long Electrode Life	Long Electrode Life Lowest Shield Gas Cost	Best Cut Quality Best Gouging Minimal Dross Long Electrode Life Cuts Thicker Material
Aluminum	Most Economical Good Speed Some Dross	Not Recommended	Good Cut Quality Good Gouging Minimal Dross Long Electrode Life	Not Recommended	Best Cut Quality Best Gouging Minimal Dross Long Electrode Life Cuts Thicker Material

O_2 = Oxygen
N_2 = Nitrogen
CO_2 = Carbon Dioxide
H35 = Mixture of 35% Hydrogen and 65% Oxygen

29103-15_T02.EPS

sures reduced before they can be used. The type of regulator used to control the gas pressure depends on the type of PAC equipment being used and the base metal being cut.

When a compressed-air supply is being used, a heavy-duty air filter and dryer is required to eliminate oil and moisture from the pressurized air. The high-pressure shop air must be reduced by an external regulator, or internally by the PAC, to the required working pressure. Also, depending on the equipment being used, the gas must be provided at a given flow rate for the task at hand, generally measured in cubic feet of gas per hour (cfh). Always check to be sure that the air supply will provide the flow rate required for the PAC equipment being used. If the airflow rate is too low, it will cause poor cuts and overheating of the torch. Refer to the manufacturer's specifications for the required flow rate for the equipment being used. To conserve consumables, make sure to have air dryers in the compressed air system as well. Moisture in the compressed air can damage the equipment.

Gases other than a local compressed air source are normally supplied in pressurized cylinders at 1,500 to 2,000 psi (103 to 138 bars). The high cylinder pressures are reduced by pressure regulators (*Figure 10*) to a lower working pressure. A gas hose connected to the outlet of the regulator delivers the gas to the PAC unit. Most heavy-duty systems require two high-pressure cylinders and two pressure regulators. One cylinder contains the cutting gas and the other a shielding or cooling gas.

Fumes and Gases

GOING GREEN

The various gases used in PAC activities, as well as the fumes generated from these activities, are often released into the atmosphere without much thought about their effect on the environment. To help protect the environment, efforts are needed to better contain the release of PAC gases and fumes. Whenever possible, PAC work should be performed using some type of capture and recovery system so that the fumes and gases are not released into the atmosphere.

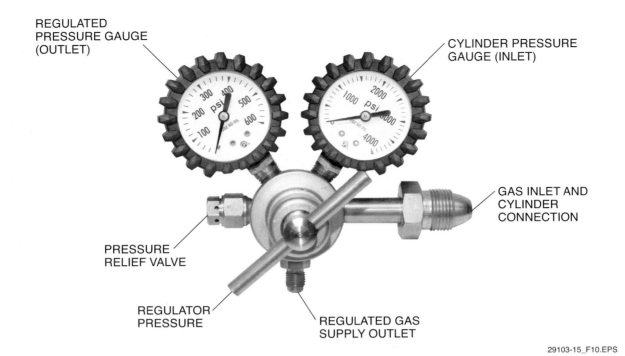

REGULATED
PRESSURE GAUGE
(OUTLET)

CYLINDER PRESSURE
GAUGE (INLET)

GAS INLET AND
CYLINDER
CONNECTION

PRESSURE
RELIEF VALVE

REGULATOR
PRESSURE

REGULATED GAS
SUPPLY OUTLET

29103-15_F10.EPS

Figure 10 Gas cylinder pressure regulator.

The PAC process was originally developed to cut nonferrous metals using inert gases. Advances in technology led to using air and/or oxygen as the plasma gas so steel could be cut using PAC.

Some systems are able to manage two gas sources to enable the user to manage different applications. *Figure 11* shows a unit commonly known as a dual-flow PAC with two internal gas pressure regulators.

Gas flow to the torch is usually controlled by a **solenoid valve**. Solenoid valves are usually mounted inside the unit and are controlled by a pushbutton on the torch. There is one solenoid valve for each gas source.

Many smaller PAC units come equipped with a built-in air compressor. These light-duty machines are portable and run on 110V power, which makes them ideal for use at remote job sites. *Figure 12* shows a light-duty unit.

29103-15_F11.EPS

Figure 11 Industrial dual-flow PAC unit.

Plasma Torch Cutting Guides

Many manufacturers furnish cutting guide accessories that can be used on several models of their torches. A circle-cutting guide kit can also be used as a freehand standoff roller guide.

29103-15_SA02.EPS

29103-15_F12.EPS

Figure 12 Light-duty PAC unit in use.

Additional Resources

Plasma Cutters Handbook: Choosing Plasma Cutters, Shop Safety, Basic Operation, Cutting Procedures, Advanced Cutting Tips, CNC Plasma Cutters, Troubleshooting, and Sample Projects. Eddie Paul. New York, NY: Penguin Group.

2.0.0 Section Review

1. The duty-cycle range for a light-duty unit is typically _____ .

    a. 35 percent
    b. 50 percent
    c. 75 percent
    d. 90 percent

2. The typical plasma arc cutting torch usually contains a replaceable retaining cup, comprised of a copper electrode with a(n) _____ .

    a. magnesium tungsten insert
    b. aluminum or thoriated tungsten insert
    c. steel tungsten alloy insert
    d. hafnium or thoriated tungsten insert

3. Some types of PAC torches inject water to create a whirlpool of water around the jet to _____ .

    a. cut with a more uniform melt rate
    b. ensure the cut will have a wider kerf
    c. allow the torch to be used on a carrier
    d. minimize the amount of dross from the cut

Section Three

3.0.0 Plasma Arc Cutting Equipment Operation and Care

Objective

Describe how to set up, safely operate, and care for plasma arc cutting equipment.

a. Describe how to set up plasma arc cutting equipment and the adjacent work area.
b. Describe how to safely operate plasma arc cutting equipment.
c. Describe how to care for plasma arc cutting equipment.

Performance Tasks

1. Set up plasma arc cutting equipment.
2. Set the amperage and gas pressures or flow rates for the type and thickness of metal to be cut.
3. Square-cut metal using plasma arc cutting equipment.
4. Bevel-cut metal using plasma arc equipment.
5. Pierce and cut slots in metal using plasma arc cutting equipment.
6. Dismantle and store the equipment.

The following section explains how to correctly set up PAC equipment and use it for cutting. Various types of cutting techniques are also presented.

3.1.0 Setting Up PAC Equipment

PAC equipment is relatively easy to set up because it is supplied as a complete system. However, the following are some things to consider before cutting is attempted:

- The unit must have the rated power and duty cycle to cut the intended material and thickness.
- The required power characteristics of the PAC unit, including the phase, voltage, and amperage must be provided.
- The appropriate nozzle must be installed in the torch for the task at hand.

- The required gas or gases must be available and at the required pressures and flow rates.
- Spare consumable components must be on hand.

3.1.1 Preparing the Work Area for PAC

The PAC control unit and workpiece must be located close enough to each other for the torch cable to comfortably reach the workpiece and still provide maneuverability. If possible, the workpiece should be located at a comfortable height and position for the torch operator. The cutting area should be well ventilated and cleared of all combustible material.

Plasma arc cutting or gouging can spray molten metal for considerable distances, possibly as far as 40 ft (12 m). It is important that everything combustible is removed from the range of the sprayed metal. If necessary, flame-resistant shields or curtains should be erected to protect workers and equipment from ultraviolet arc rays and metal splatter. Some employers and/or job sites may require a hot-work permit to be completed and a fire watch to be posted.

> **WARNING!**
>
> PAC produces fumes and gases that can be harmful to your health. The composition and rate of generation of fumes and gases depend on factors such as arc current, cutting speed, material being cut, and gases used. Various paints and coatings can also create or contribute to the production of harmful fumes. The fume and gas byproducts usually consist of ozone, oxides of the metal being cut, and oxides of nitrogen. These fumes must be removed from the work area through a capture and exhaust or recovery system. Check all applicable local codes; some codes may require that exhausted fumes be filtered before being vented to the atmosphere.

Eye and ear protection are essential when performing PAC as a safeguard against noise and to prevent metal spray from entering the operator's eyes and ears. If the site ventilation is not adequate to keep the smoke and fumes away from the operator, a respirator may also be required.

> **WARNING!**
>
> Ear protection is essential to protect hearing. Other PPE must be worn to protect the operator from hot metal and slag.

3.1.2 Setting the Correct Cutting Amperage

The cutting amperage depends on the type of equipment being used, the type and thickness of the material being cut, and the type of gas being used (*Figure 13*). Light-gauge sheet steel may require as little as 7 amps of current, while 2" (5.1 cm) aluminum plate may require 250 amps. Always refer to the manufacturer's recommendations to identify the correct amperage for the specific equipment and job. Using the incorrect amperage will result in a poor-quality cut. If the amperage is too high, severe damage to the torch may also occur.

3.1.3 Setting Gas Parameters and Installing Gas Cylinders

Several different gases may be used on the same system for cutting the same or different types and thicknesses of metals. In many cases, the same gas or mixture will not be used for both thick and thin metals of the same type. Refer to the manufacturer's instructions or the unit data plate for specific gas recommendations.

Some gases used in cutting can be dangerous if accidentally released. Welders routinely have to remove empty cylinders and replace them with filled cylinders, increasing the potential for a release.

When connecting a gas cylinder to the PAC unit, follow these steps:

Step 1 After identifying the correct gas type, place it adjacent to the unit and secure it in an upright position.

Step 2 Remove the protective valve cap.

WARNING!

If oxygen is being used, be sure any cloth used on or near the fittings or regulators is free of oil or grease. Oil or grease mixed with oxygen can explode, causing personal injury or death.

Step 3 While standing to one side of the cylinder valve outlet, and with safety glasses in place, quickly crack open the valve a small amount to blow out any dirt or debris. Reclose immediately to prevent unnecessary gas loss.

WARNING!

To avoid personal injury, always stand with the cylinder between you and the attached regulator when opening valves.

Step 4 Connect the regulator to the cylinder valve and tighten with the proper wrench. Do not over-tighten. Connect the hose from the PAC unit to the regulator outlet.

Step 5 Open the cylinder valve very slowly at first, then open fully. Observe the regulator inlet gauge to determine the pressure remaining in the cylinder.

Step 6 Slowly adjust the regulator to the desired output pressure. The pressure usually has to fall into a range to satisfy the requirements of the PAC unit. The PAC unit regulates the final pressure to the torch.

CAUTION

Do not use pliers or pipe wrenches to tighten the regulator's connections. Use an adjustable wrench or correctly sized open-end wrench. Also, be careful not to over-tighten the connections because they are made of soft bronze and can be damaged easily.

3.2.0 Operating PAC Equipment

PAC equipment is relatively simple to operate. The cutting unit's operating instructions provide information for properly setting gas pressure and flow rates, output current, torch standoff distance, and torch travel speed (for mechanized cutting) for various metal types and thicknesses.

(A)

(B)

29103-15_F13.EPS

Figure 13 Gas and amperage settings vary according to material type and thickness.

To operate the PAC equipment, follow these general steps. Note that this is a general reference only. Refer to the manufacturer's documentation for specific instructions on the PAC equipment being used.

Step 1 Make sure that the capacity of the PAC power unit matches or exceeds the work to be done. Also ensure that the power source for the unit is sufficient, based on the unit data plate.

Step 2 Identify the location of the primary disconnect for the electrical source to be used in case of emergency.

Step 3 Check to make sure that the required gas(es) is connected. If a pressurized cylinder is used, ensure that it has adequate pressure for the work to be done. If compressed air will be used, ensure that it is also connected and at the appropriate pressure.

> **WARNING!**
> Some stainless steel contains chromium 6, also known as hexavalent chrome. This metal will give off toxic fumes when heated. Always use proper respiratory protection when cutting this type of stainless steel.

Step 4 Plug the unit in (if required), energize the PAC unit power source, and turn the unit on.

Step 5 In accordance with the PAC manufacturer's instructions, adjust the gas regulator pressure and/or flow rate to the values recommended for the type and thickness of metal being cut.

Step 6 Set the output current to the value recommended in the manufacturer's instructions for the type and thickness of metal being cut.

Step 7 Attach the work lead clamp to the workpiece to be cut.

> **CAUTION**
> If cutting on machinery or other assembled equipment, be sure to position the work lead clamp so that the cutting current will not pass through seals, bearings, or other contacting surfaces that could be damaged from heat or arcing. Also isolate any electrical components from the PAC system/workpiece circuit.

Step 9 Prepare the work area as previously described.

> **WARNING!**
> Be advised that various gases are extremely flammable, and PAC processes operate at extreme temperatures. The appropriate fire suppression equipment must be stationed in close proximity for emergency use.

Step 10 Put on the appropriate PPE.

Step 11 Hold the torch at the recommended stand-off distance above the point where the cut is to begin, lower your hood, and press the torch arc-start button or foot pedal. As soon as cutting begins, move the torch at a 0-degree travel angle with the recommended speed along the cutting line.

3.2.1 Square-Cutting Metal

Square edges can be cut with PAC by holding the torch at 90 degrees (also referred to as a 0-degree work angle) to the metal surface while it is advanced smoothly, with no side-to-side movement. A straightedge or metal angle can be used to guide and steady the torch. The straightedge can be aligned and quickly clamped down at each end. When cutting thicker metals, the cut does taper slightly as it narrows with depth. This can

Safety Caps

Safety caps should be kept with cylinders at all times. Immediately after removing a cylinder's regulator assembly, reinstall the safety cap over the cylinder valve. This protects the cylinder valve if the cylinder should fall over. If the cylinder valve is suddenly sheared off, the cylinder may propel itself like a rocket. This could result in serious injury or death.

29103-15_SA03.EPS

result in cuts where one or both faces are not quite square with the plate face. The angle of the face is referred to as the cut angle. To compensate for the cut angle on thick materials, the torch should be slightly angled, leaning toward the scrap portion of the workpiece, to produce a square edge on the desired piece. All the taper will then be occur on the scrap side (*Figure 14[A]*).

3.2.2 Bevel-Cutting Metal

Bevel cuts are made with the same technique used for square cuts, except that the torch is held at the required bevel angle. A length of angle iron, laid on its open face and shimmed as necessary, can be used to support the torch's side at the required angle. *Figure 14(B)* shows how to use a metal angle to hold a bevel angle.

3.2.3 Piercing and Slot-Cutting in Metal

PAC can be used to pierce and cut slots in metal in any position. Full, fire-resistant body protection, including ear protection, is recommended when piercing or cutting metal in situations where out-of-position orientations could cause the metal splash to fall on the torch operator.

To pierce very thin metal, hold the torch directly over the point to be pierced at a 0-degree

Gas Pressure and Flow Rates

PAC unit gas pressures and/or flow rates vary with equipment design, gas type, and cut depth (metal thickness). Typical gas flow rates vary from 15 cfh for nitrogen when cutting 0.1" (2.5 mm) aluminum, carbon steel, or stainless steel, up to 62 cfh of argon and 31 cfh of nitrogen when cutting 3" (7.6 cm) carbon steel, or 2½" (6.4 cm) aluminum or stainless steel. Refer to the manufacturer's instructions for the correct gas pressure and flow rates for each situation.

work angle, and press the arc button. As soon as the jet passes completely through the workpiece, move the torch head in a smooth circle (or other pattern) to produce the desired diameter or shape for the hole. When piercing thicker materials, rotate the torch to a 10-degree work angle. This will help to prevent metal from splashing directly up into the torch. When the jet passes through the workpiece, rotate the torch back to a 0-degree work angle and complete the cut.

When cutting slots or patterns, a template can be used to precisely guide the torch tip.

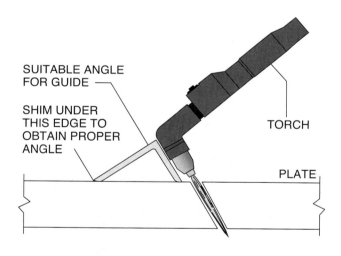

(A) COMPENSATING FOR TAPER
IN THICKER METAL CUTS

(B) USING AN ANGLE IRON TO SET
AND HOLD A BEVEL ANGLE

29103-15_F14.EPS

Figure 14 Compensating for taper and holding bevel angle in thicker metal cuts.

3.2.4 Proper Equipment Storage and Housekeeping

Proper equipment storage and housekeeping are essential for work efficiency and safety. When finished using the PAC equipment, follow these steps to store the equipment and maintain good housekeeping:

Step 1 Turn off the power source.

Step 2 Turn off all gas sources, bleed pressure slowly and carefully, and disconnect the lines from the PAC unit. When using cylinders, be sure to close the bottle valve, remove the regulator assembly, and install the protective cap before moving the cylinder.

> **WARNING!**
> If cylinders are to be removed, be sure to remove the pressure regulators and replace the cylinder caps before releasing and moving the cylinders.

Step 3 Unplug and coil the power unit cord and all gas lines.

Step 4 Coil the torch cable and the workpiece lead.

Step 5 Return the PAC equipment to its proper storage location.

Step 6 Clean off the work area and sweep up any slag and debris in the area.

Figure 15 shows a PAC unit that is properly stored.

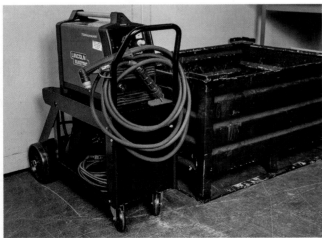

29103-15_F15.EPS

Figure 15 Properly stored PAC unit.

Starting the Torch on a Thick Workpiece

Similar to the oxyfuel cutting of a thick workpiece, it is easier to begin a manual cut at the edge by rotating the PAC torch slightly so that the plasma cutting jet begins cutting at the top edge only. The torch is then rotated vertically so that the cutting jet slices into the remaining thickness before the torch is moved along the cutting line.

3.3.0 Care of Plasma Arc Cutting Equipment

As with all equipment, regular maintenance should be performed according to the manufacturer's recommendations. The cutting equipment should not be altered or modified in any way, except for those changes identified in the manufacturer's literature. The plasma machine requires relatively little maintenance, but its consumables must be inspected regularly and replaced as needed. Minor external repairs usually involve these basic parts only.

3.3.1 Understanding Basic Operating Concerns

If compressed air is used with a PAC unit, shop air may be used as long as it is dry and free of oil. When properly set, the air pressure regulator on the power unit will maintain the correct pressure for operation. The air supply must be dry; only compressed air that has passed through a reliable air dryer should be used. The compressed air supply is typically filtered at the power unit as well.

Swirl Ring Torches

Most swirl rings on PAC torches produce a clockwise tangential plasma swirl as viewed from the torch to the work surface. If the torch is at a 0-degree work angle, the clockwise swirl causes the right side of the kerf to be essentially square and the left side to be angled. For a counterclockwise swirl, the opposite is true.

Your ability to change and replace parts will depend on your knowledge of the procedures contained in the manufacturer's literature for the equipment in question.

Perform the following procedures, and always abide by the manufacturer's guidelines for removal and replacement of worn or damaged parts:

- Check the shield for any signs of wear.
- The shield should be clean and clear of metal debris.
- Unscrew the shield and check the inside for wear. If the holes are blocked, attempt to clean them using a torch tip cleaner.
- If the shield cannot be cleaned, replace it.
- Inspect the retaining cup for damage and replace as necessary.
- Examine the nozzle for wear or damage. If the hole in the nozzle is worn or oval-shaped, replace the nozzle.

- Remove the electrode and check the tip for pitting. If pits exist and are more than $\frac{1}{16}$" (1.6 mm) deep, replace the electrode.
- Inspect the swirl ring, which should be clean, and the holes on the sides free of clogs. If the swirl ring is damaged, replace it.
- Check the O-ring. Apply a small amount of lubricant (provided in a repair kit) to the O-ring before reinstalling it. The O-ring should make a firm seal when all parts are assembled and tight.

3.3.2 Moisture

A significant concern when using PAC equipment is that moisture may enter the unit through the air supply. This can cause the torch to sputter or hiss. If your PAC equipment experiences moisture problems, first drain the air filter/moisture separator bowl. The filter element may also need to be replaced. Refer to the manufacturer's literature for cleaning and replacement of the filter element.

3.3.3 Cooling Filter

Many power units contain an air filter housing to ensure the air supply is clean. Refer to the manufacturer's literature and clean the filter element periodically as required. An air filter that is neglected may cause the PAC unit to overheat and shut down unexpectedly.

Additional Resources

Plasma Cutters Handbook: Choosing Plasma Cutters, Shop Safety, Basic Operation, Cutting Procedures, Advanced Cutting Tips, CNC Plasma Cutters, Troubleshooting, and Sample Projects. Eddie Paul. New York, NY: Penguin Group.

3.0.0 Section Review

1. To remove, attach, or tighten the regulator's connection to a gas cylinder, you would use _____ .

   a. no tools; this connection is to be hand tightened only
   b. a correctly sized wrench with an added extension for leverage to apply maximum force
   c. a properly sized pipe wrench adjusted to fit the regulator fitting
   d. an adjustable wrench or correctly sized open-end wrench

2. To pierce very thin metal with a plasma torch, you should _____ .

   a. hold the torch directly over the point to be pierced at a 0-degree work angle
   b. start from the bottom, piercing the metal from below
   c. hold the torch directly over the point to be pierced at 10-degree work angle
   d. hold the torch directly over the point to be pierced at a 45-degree work angle

3. If the PAC equipment exhibits a moisture problem, you should first _____ .

   a. attach a vacuum pump to the air filter/moisture separator bowl assembly
   b. replace the filter element and air filter/moisture separator bowl assembly
   c. drain the air filter/moisture separator bowl
   d. stop using air and operate the system using argon for at least 12 hours of operation

SUMMARY

Plasma arc cutting is useful for cutting and piercing many types of metals. Unlike oxyfuel cutting, it can be used to cut aluminum, magnesium, copper, nickel, and stainless steel. PAC is usually much faster than oxyfuel cutting, and it produces little or no distortion or alteration zone changes. The cuts are fast and clean, with little or no dross, and there is no chance of carbon inclusions, as can happen with carbon arc cutting. However, because the torch jet produces metal spray, good site fire-prevention practices and the consistent use of appropriate PPE are essential.

With an understanding of metals and the knowledge to correctly set up a PAC unit for the task at hand, workers can quickly become competent in the use of PAC equipment to produce excellent cuts.

1. Plasma arc cutting uses a jet of plasma that can reach temperatures as high as _____ .
   a. 10,000°F (5,537°C)
   b. 20,000°F (11,093°C)
   c. 25,000°F (13,871°C)
   d. 30,000°F (16,650°C)

2. What are the two primary types of plasma arc cutting processes?
   a. Transferred arc and non-transferred arc
   b. Transferred arc and phased arc
   c. Phased arc and non-transferred arc
   d. Oxyfuel pilot arc and non-transferred arc

3. In the transferred arc process, the pilot arc is used to _____ .
   a. cut the metal
   b. ionize the air jet
   c. establish contact with the metal
   d. keep the torch heated

4. In a non-transferred arc process, the material being cut is _____ .
   a. adjacent to the arc circuit
   b. dependent on the arc circuit
   c. electrically connected to the arc circuit
   d. not electrically connected to the arc circuit

5. Which of the following statements regarding PAC safety is true?
   a. If a cap or hat is worn, it should be constructed of 100-percent polyester or nylon.
   b. The electric arc can be safely observed without tinted lenses, but it is not recommended for an extended period.
   c. Boots with no laces at all are generally preferred for both welding and cutting activities.
   d. Flammable materials can remain near the cutting activity, due to the cool nature of the plasma arc.

6. Light-duty PAC equipment typically requires a _____ .
   a. 115VAC, single-phase power source
   b. 220VAC, three-phase power source
   c. 277VAC, single-phase power source
   d. 440VAC, three-phase power source

7. Some heavy-duty PAC systems _____ .
   a. require no torch cooling
   b. are water-cooled
   c. use very little current
   d. weigh less than 5 pounds (2.3 kg)

8. The workpiece clamp is rated by the _____ .
   a. voltage that it can carry without overheating
   b. capacitance that it can carry without overheating
   c. amperage that it can carry without overheating
   d. gas volume that it uses at maximum load

9. Severe damage from arcing and overheating can be caused by the PAC current passing through a _____ .
   a. workpiece
   b. cable
   c. cutting table
   d. bearing

10. Mechanized PAC torches are designed to be _____ .
    a. handheld
    b. mounted in a vise
    c. used at a slow speed
    d. mounted on automated carriers

11. Gases other than compressed air are normally supplied in pressurized cylinders at _____ .
    a. 200 psi (14 bars) to 400 psi (28 bars)
    b. 800 psi (55 bars) to 1,000 psi (69 bars)
    c. 1,000 psi (69 bars) to 1,500 psi (103 bars)
    d. 1,500 psi (103 bars) to 2,000 psi (138 bars)

12. If site ventilation is not adequate to keep fumes away from the operator, a(n) _____ .
    a. fan must be used
    b. respirator may be required
    c. fire watch must be used
    d. airline must be used

13. Light-gauge sheet steel may be cut using a cutting amperage as low as _____ .
    a. 7 amps
    b. 9 amps
    c. 15 amps
    d. 20 amps

14. Square edges can be cut in thin materials by holding the torch at _____ .
    a. 30 degrees to the metal surface
    b. 45 degrees to the metal surface
    c. 75 degrees to the metal surface
    d. 90 degrees to the metal surface

15. When inspecting an electrode before use, it should be replaced if pits exist and are deeper than _____ .
    a. ⅟₃₂" (0.8 mm)
    b. ⅟₁₆" (1.6 mm)
    c. ⅛" (3.2 mm)
    d. ³⁄₁₆" (4.8 mm)

Trade Terms Quiz

Fill in the blank with the correct term that you learned from your study of this module.

1. A fourth state of matter created by heating a gas to such a high temperature that it boils electrons off the gas molecules is called _____ .

2. The relative electrical voltage difference between two points of reference is called _____ .

3. A(n) _____ can control the flow of gases and liquids, and is operated by an electromagnet.

4. The percentage of time a PAC unit can cut without overheating within a 10-minute period is called its _____ .

5. Each of the three sine waves of a power supply that uses three conductors, all carrying alternating current, is referred to as a(n) _____ .

6. The unit of measure for electrical current is _____ .

7. A waste byproduct of molten metal resulting from plasma cutting processes is known as _____ .

Trade Terms

Amperage 6
Dross 7
Duty cycle 4
Phase 5

Plasma 1
Potential 2
Solenoid valve 3

Appendix

PERFORMANCE ACCREDITATION TASKS

The American Welding Society (AWS) School Excelling through National Skills Standards Education (SENSE) program is a comprehensive set of minimum Standards and Guidelines for Welding Education programs. The following performance accreditation tasks are aligned with and designed around the SENSE program.

The Performance Accreditation Tasks (PATs) correspond to and support the learning objectives in *AWS EG2.0, Guide for the Training and Qualification of Welding Personnel: Entry-Level Welder*.

Note that in order to satisfy all learning objectives in *AWS EG2.0*, the instructor must also use the PATs contained in the second level of the NCCER Welding curriculum.

PATs 1 and 2 correspond to *AWS EG2.0, Module 8 – Thermal Cutting Processes, Unit 3 – Manual Plasma Arc Cutting (PAC)*, Key Indicators 3, 4, and 5.

PATs provide specific acceptable criteria for performance and help to ensure a true competency-based welding program for students.

The following tasks are designed to develop your competency with a plasma arc cutting torch. Practice each task until you are thoroughly familiar with the procedure.

As you complete each task, take it to your instructor for evaluation. Do not proceed to the next task until instructed to do so by your instructor.

PLASMA ARC CUTTING

Using electrically conductive material, lay out and cut the shape shown in the figure in the horizontal position. When it is finished, cut your initials into the shape.

NOTE: MATERIAL – CARBON STEEL

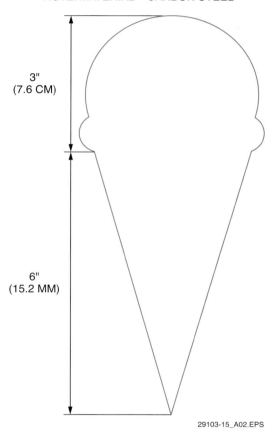

3"
(7.6 CM)

6"
(15.2 MM)

29103-15_A02.EPS

Criteria for Acceptance:

- Dimensions ±¹⁄₁₆"   _____
- Minimal amount of dross sticking to plate which can be easily removed   _____

PLASMA ARC CUTTING

Using electrically conductive material, lay out and cut the shape shown in the figure in the flat position.

NOTE: MATERIAL – CARBON STEEL
  HOLES ¾" (19.1 MM) DIAMETER
  SLOTS ¾" (19.1 MM) × 1 ½" (38.1 MM)

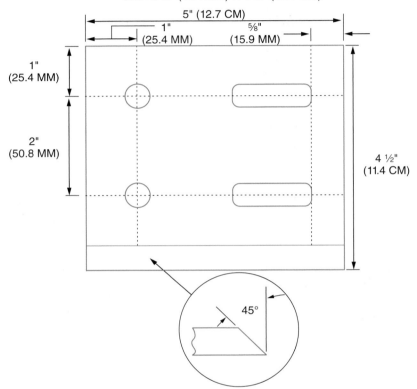

29103-15_A01.EPS

Criteria for Acceptance:

- Outside dimensions ±¹⁄₁₆" _____
- Inside (holes and slots) dimensions ±¹⁄₁₆" _____
- Square ±2° _____
- Bevel ±2° _____
- Minimal amount of dross sticking to plate which can be easily removed _____
- Square kerf face with minimal notching not exceeding ±¹⁄₁₆" deep _____

Cornerstone of Craftsmanship

Curtis Casey
Northland Pioneer College
Emeritus Welding Faculty

Curtis received his welding training in the US Navy, where he was originally slated to be trained as a submarine power plant operator. However, those jobs were filled when he joined, so he chose advanced training as a welder. Curtis welded on SSBN class submarines for six years, working on high-tensile hulls, pressure pipe, and nuclear components. After the Navy, a new opportunity opened up at Bechtel Power Corporation, where he became a Certified Welding Inspector. For the next 15 years, Curtis traveled around the United States helping to build nuclear power plants. He also found the time to earn a Welding Technology degree on the G.I. Bill and to take on supervisory responsibilities. Not long after joining the faculty at Northland Pioneer College, Curtis was appointed chairperson of the Welding Department, which now serves hundreds of students each year.

How did you choose a career in the construction industry?
I took the opportunity to train while in the US Navy's nuclear welding program after graduating from high school. There I got a chance to weld on submarines and other naval vessels.

Who inspired you to enter the industry?
My stepfather inspired me to enlist in the US Navy where I received advanced training in welding. He was a career sailor who always had great stories about seeing the world.

What types of training have you been through?
In the Navy, I received training in the metals trades to be a hull maintenance technician. I finished welding school, learning to weld on both common and exotic materials. After the military I went to college while I worked as a welding inspector in nuclear power plants and got a chance to teach weld inspection as an adjunct professor while attending college. From then on, I was hooked on training and found the opportunity to teach welding full time at Northland Pioneer College. Currently I am a welding consultant wearing many hats within the industry from inspecting to auditing to teaching and consulting.

How important are education and training in construction?
Education is an ongoing process; life-long learning is another way to say it. Education and training are the tools we fit ourselves with to be able to create and mold our careers. Without them our direction in life could easily be aimless and unproductive.

How important are NCCER credentials to your career?
NCCER credentialing gives me an edge over many of my colleagues and peers because of the clout the organization has within the industry. Many companies now are requiring these credentials just to get your foot in their door.

How has training/construction impacted your life?
Simply put, it has given me great sense of satisfaction knowing that I can design and build on ideas that come to my mind. I can take a project from inception to creation because of my training.

What kinds of work have you done in your career?
I started my adult working life as a welder, then worked as a welding inspector, a field welding engineer, a welding business owner, a welding instructor, welding department chair, welding curriculum developer, and welding consultant. I think welding sums it up pretty well. I still maintain my AWS Certified Welding Instructor (CWI) and Certified Welding Educator (CWE) certifications, and I'm actively involved with my students in SkillsUSA, a national student leadership organization.

Tell us about your present job.
I'm currently consulting in the welding arena while occasionally inspecting and auditing other NCCER training and assessment centers.

What do you enjoy most about your job?

My passion has been teaching welding at Northland Pioneer College for the past 15 years, and now in semi-retirement. I love the versatility in what I do. One day I could be doing R & D at a fabrication shop, and the next I could be inspecting welds or auditing programs. It is refreshing to see the newest technology being put to use at certain companies; in this job I see it firsthand.

What factors have contributed most to your success?

Work ethics! I learned at an early age to be responsible for my own successes and failures. I pitched papers on a paper route when I was 11 years old. That taught me to be dependable and gave me exposure to the business world and having to deal with customers, good and bad.

Would you suggest construction as a career to others? Why?

If they were interested in seeing an idea come to fruition from design to completion and if they were willing to get their hands dirty, then I would recommend construction. One can receive great satisfaction and get a sense of worth when they take part in the creation of something that people can actually use in their lives.

What advice would you give to those new to the welding field?

Try it out through a job shadowing program to see if you like it. Then if you dream about it that night, you probably have a passion for it. Then simply follow your passion! Get the right training, surround yourself with like-minded people, keep a good attitude, network a lot, and go to work!

Tell us an interesting career-related fact or accomplishment.

I have had the chance to teach close to 1,000 people the craft that has given me my career, and have seen many former students return to share their success in the industry. That's an emotional paycheck worth the effort.

How do you define craftsmanship?

Developing a good balance between technical and artistic skills.

Trade Terms Introduced in This Module

Amperage: The unit of measure used for the intensity of an electric current; often abbreviated to *amp* or *amps*.

Dross: A waste byproduct of molten metal.

Duty cycle:: The percentage of time a plasma arc cutting machine can cut without overheating within a ten-minute period.

Phase: In a three-phase power supply system, the sine waves of the voltage on each of the three separate conductors is displaced from each of the others by 120 degrees, although all conductors are carrying alternating current at the same frequency and are synchronized. Each of the three sine waves is referred to as a phase of the power source.

Plasma: A fourth state of matter (not solid, liquid, or gas) created by heating a gas to such a high temperature that it boils electrons off the gas molecules (ionization).

Potential: The relative electrical voltage difference between two points of reference; the electromotive force or difference in potential between two points, expressed in volts.

Solenoid valve: A valve used to control the flow of gases or liquids that is opened or closed by the action of an energized electromagnet.

Additional Resources

This module presents thorough resources for task training. The following resource material is suggested for further study.

Plasma Cutters Handbook: Choosing Plasma Cutters, Shop Safety, Basic Operation, Cutting Procedures, Advanced Cutting Tips, CNC Plasma Cutters, Troubleshooting, and Sample Projects. Eddie Paul. New York, NY: Penguin Group.

AWS F3.2M/F3.2, Ventilation Guide for Weld Fume. Latest Edition. Miami, FL: American Welding Society.

Figure Credits

The Lincoln Electric Company, Cleveland, OH, USA, Module Opener, Figures 1A, 2 (photo), 5, 8 (photos), 11, 12, 15, SA01

Courtesy of BUG-O Systems Inc., Figure 1B

AWS F2.2:2001, Lens Shade Selector, reproduced with permission from the American Welding Society (AWS), Miami, FL, USA, Table 1

Topaz Publications, Inc., Figures 6, SA03

ESAB Welding and Cutting Products, Figures 7, 13A

Courtesy of Miller Electric Mfg. Co., Figure 9A

Victor Technologies, Figure 9B, 13B

Courtesy of Uniweld Products, Figure 10

The Eastwood Company, SA02

Section Review Answer Key

Answer	Section Reference	Objective
Section One		
1. d	1.1.0	1a
2. c	1.2.1	1b
Section Two		
1. a	2.1.0	2a
2. d	2.2.0	2b
3. a	2.2.0	2c
Section Three		
1. d	3.1.3	3a
2. a	3.2.3	3b
3. c	3.3.2	3c

NCCER CURRICULA — USER UPDATE

NCCER makes every effort to keep its textbooks up-to-date and free of technical errors. We appreciate your help in this process. If you find an error, a typographical mistake, or an inaccuracy in NCCER's curricula, please fill out this form (or a photocopy), or complete the online form at **www.nccer.org/olf**. Be sure to include the exact module ID number, page number, a detailed description, and your recommended correction. Your input will be brought to the attention of the Authoring Team. Thank you for your assistance.

Instructors – If you have an idea for improving this textbook, or have found that additional materials were necessary to teach this module effectively, please let us know so that we may present your suggestions to the Authoring Team.

NCCER Product Development and Revision

13614 Progress Blvd., Alachua, FL 32615

Email: curriculum@nccer.org
Online: www.nccer.org/olf

❏ Trainee Guide   ❏ Lesson Plans   ❏ Exam   ❏ PowerPoints   Other _____

Craft / Level: _____   Copyright Date: _____

Module ID Number / Title: _____

Section Number(s): _____

Description: _____

_____

_____

_____

Recommended Correction: _____

_____

_____

_____

Your Name: _____

Address: _____

_____

Email: _____   Phone: _____

29104-15

Air-Carbon Arc Cutting and Gouging

OVERVIEW

This module presents information related to air-carbon arc cutting (A-CAC) and gouging. This method is a useful process in which metal materials to be cut and gouged are melted by the heat of a carbon arc. In combination with a jet of compressed air, the molten material is pushed away from the cutting and gouging action.

Module Four

Trainees with successful module completions may be eligible for credentialing through the NCCER Registry. To learn more, go to **www.nccer.org** or contact us at **1.888.622.3720**. Our website has information on the latest product releases and training, as well as online versions of our *Cornerstone* magazine and Pearson's product catalog.

Your feedback is welcome. You may email your comments to **curriculum@nccer.org**, send general comments and inquiries to **info@nccer.org**, or fill in the User Update form at the back of this module.

This information is general in nature and intended for training purposes only. Actual performance of activities described in this manual requires compliance with all applicable operating, service, maintenance, and safety procedures under the direction of qualified personnel. References in this manual to patented or proprietary devices do not constitute a recommendation of their use.

Objectives

When you have completed this module, you will be able to do the following:

1. Define air-carbon arc cutting and identify the related equipment and consumables.
   a. Define air-carbon arc cutting.
   b. Identify and describe air-carbon arc cutting equipment.
   c. Identify and describe various types of electrodes.
   d. Identify safety practices related to air-carbon arc cutting.
2. Describe how to set up, safely operate, and care for air-carbon arc cutting equipment.
   a. Describe how to prepare the equipment and work area for air-carbon arc cutting.
   b. Describe how to wash and gouge metals.
   c. Describe how to care for air-carbon arc cutting equipment.

Performance Tasks

Under the supervision of your instructor, you should be able to do the following:

1. Select and install air-carbon arc cutting electrodes.
2. Prepare the work area and air-carbon arc cutting equipment for safe operation.
3. Use air-carbon arc cutting equipment for washing.
4. Use air-carbon arc cutting equipment for gouging.
5. Perform storage and housekeeping activities for air-carbon arc cutting equipment.

Trade Terms

Carbon-graphite electrode
Concentric cable system

Oxidize
Polarity

Industry Recognized Credentials

If you are training through an NCCER-accredited sponsor, you may be eligible for credentials from NCCER's Registry. The ID number for this module is 29104-15. Note that this module may have been used in other NCCER curricula and may apply to other level completions. Contact NCCER's Registry at 888.622.3720 or go to **www.nccer.org** for more information.

Contents

Topics to be presented in this module include:

Figures and Tables ─────────────

SECTION ONE

1.0.0 AIR-CARBON ARC CUTTING

Objectives

Define air-carbon arc cutting and identify the related equipment and consumables.

   a. Define air-carbon arc cutting.
   b. Identify and describe air-carbon arc cutting equipment.
   c. Identify and describe various types of electrodes.
   d. Identify safety practices related to air-carbon arc cutting.

Performance Task

   1. Select and install air-carbon arc cutting electrodes.

Trade Terms

Carbon-graphite electrode: An electrode composed of a mixture of soft amorphous carbon and hard graphite carbon that may be coated with copper.

Oxidize: To combine with oxygen, such as in burning (rapid oxidation) or rusting (slow oxidation).

A ir-carbon arc cutting (A-CAC) is an arc-based cutting and gouging process, unlike oxyfuel processes. The base metal is melted with an electric arc formed between a carbon electrode and the base metal. A jet of compressed air passing through the electrode holder and its head blows the molten metal away. The process can be used in all positions on any metal that will conduct electricity, such as carbon steel, stainless steel, cast iron, and copper, as well as their alloys. It is primarily used to gouge weld grooves, to back-gouge welds, to gouge out defective welds, and to prepare cracks for welding. Carbon electrode arcs produce cleaner cuts than metal electrode arc cutting methods because no foreign metals are introduced at the arc. In addition, A-CAC is often used in washing and beveling operations. *Figure 1(A)* shows a groove being cut by the A-CAC process. *Figure 1(B)* shows the washing method using the A-CAC process.

This module will explain A-CAC equipment, its operation, and its use for cutting and gouging metals.

1.1.0 A-CAC Process and Equipment

Common applications of air-carbon arc cutting include the following:

- Cutting or gouging out defective metal
- Removing defective welds
- Creating a groove for a weld

The A-CAC process uses a high-current electric arc between the base metal and the end of a carbon-graphite electrode to melt the base metal. The currents used with this process are generally much higher than those required for the shielded metal arc welding (SMAW) process.

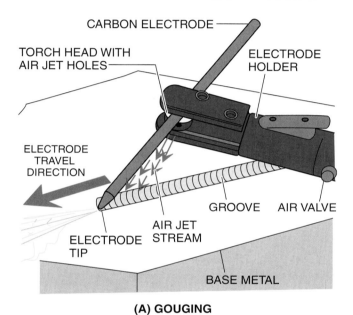

(A) GOUGING

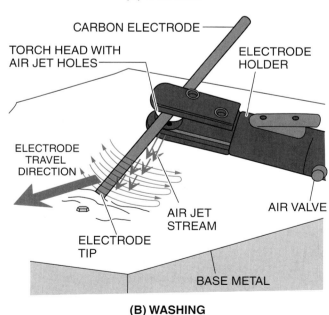

(B) WASHING

29104-15_F01.EPS

Figure 1 Gouging and washing with A-CAC process.

Air-jet orifices are located in the special torch heads. When gouging or beveling, the orifices direct jets of compressed air under the A-CAC electrode, between the electrode and the base metal, to blow the molten metal pool away from the cut. When the electrode is moved along the surface of the base metal, a groove or bevel can be made.

A groove's depth and width are controlled by the following:

- Electrode size
- Arc amperage
- Electrode angle to the base metal
- Electrode advancement rate
- Electrode movement
- Regulated air pressure

1.2.0 A-CAC Equipment

A-CAC equipment includes an A-CAC torch and its attachments, a welding power source, and a supply of compressed air.

1.2.1 A-CAC Torch and Attachments

Typical A-CAC equipment consists of a carbon-graphite electrode gripped in a specialized electrode holder. The electrode holder is commonly referred to as a torch, torch handle, or torch holder. The torch handle is equipped with V-groove heads containing air jet holes, and a combination power cable and compressed air hose is attached to the handle. *Figure 2* shows a typical A-CAC torch. *Figure 3* is a simplified diagram of an A-CAC setup.

The electrode is gripped by the V-groove head. The electrode tip should be within 6" (15 cm) of the V-groove head to prevent overheating. The contacts are mounted in the lower and upper jaws of the electrode holder. The jaws are spring-loaded to clamp onto the electrode. Depending on the torch design, the contact area of one head may be flat and the other V-grooved, or both may be V-grooved. If both are V-grooved, one or both heads will contain air-jet orifices.

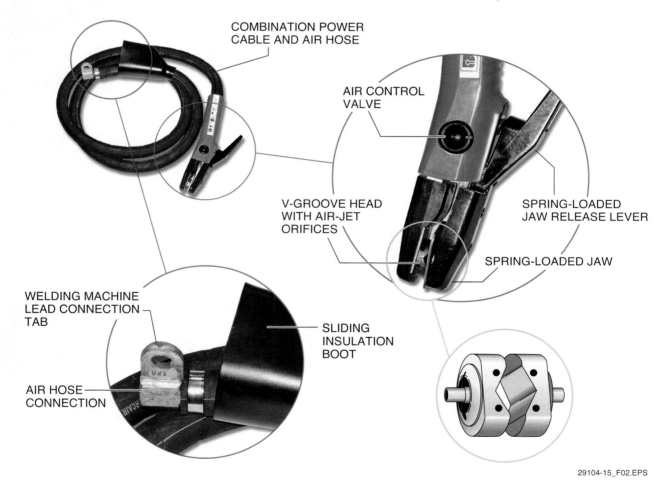

COMBINATION POWER CABLE AND AIR HOSE

AIR CONTROL VALVE

V-GROOVE HEAD WITH AIR-JET ORIFICES

SPRING-LOADED JAW RELEASE LEVER

SPRING-LOADED JAW

WELDING MACHINE LEAD CONNECTION TAB

AIR HOSE CONNECTION

SLIDING INSULATION BOOT

29104-15_F02.EPS

Figure 2 Typical A-CAC torch.

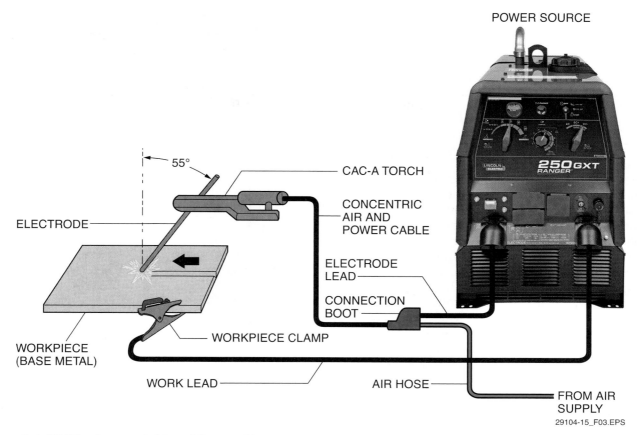

POWER SOURCE

CAC-A TORCH

CONCENTRIC AIR AND POWER CABLE

ELECTRODE LEAD

CONNECTION BOOT

55°

ELECTRODE

WORKPIECE CLAMP

WORKPIECE (BASE METAL)

WORK LEAD

AIR HOSE

FROM AIR SUPPLY

29104-15_F03.EPS

Figure 3 A-CAC torch connected to welding machine.

The air-jet orifices direct jets of air parallel to the electrode to blow the molten metal away. If required, heads may be replaced or head positions exchanged to alter orifice positions for different operations. When cutting and gouging, the jets must be located under the electrode (between the electrode and the base metal surface). For washing, the jet orifices should be located on one or both sides of the electrode. This helps prevent cutting into the base metal surface. *Figure 4* shows different head types and arrangements for a heavy-duty torch. These light- and heavy-duty heads are available in different sizes to accommodate ⅛" (3.2 mm) to 1" (2.5 cm) electrodes.

A-CAC Cutting

Since A-CAC cutting is not dependent on oxidation, it can cut metals that normally cannot be cut with oxyacetylene cutting methods.

At least one V-groove is needed to position the electrode in the center of the head. The V-grooved head provides about twice as much electrical contact surface as the flat head. A flat head is usually fixed, but a V-grooved head is free to rotate in its mount inside the torch handle. This allows the electrode and grooved head to be swiveled within the torch handle for the best or most comfortable work angle. Pressing the lever on the top of the torch handle raises the upper jaw and head, releasing the electrode for repositioning or replacement.

Air passages within the torch handle supply the head orifices with compressed air. A push-push air control valve extends through the torch handle. Pushing it from one side of the torch handle opens the valve and air flows to the head orifices; pushing it back from the other side closes the valve. Air-carbon arc cutting air pressure should be regulated to the lowest air pressure that cleanly blows away the molten metal, usually a range of 40 to 100 pounds (2.8 to 6.9 bars) per square inch gauge (psig).

The connections of the A-CAC torch include both an electrical cable and a compressed air hose. For low-current applications, the torch cable consists of a cable and a hose within a common cover. For high-current applications, the torch cable is a concentric conductor that has an air conduit in the center of a hollow electric cable to provide cooling for the electric cable. Concentric cables may include a swivel on the torch end to give the operator more freedom to maneuver the torch. Some heavy-duty and continuous-duty torches also have additional hoses or lines to carry cooling water to and from the torch, especially with currents exceeding 300A. *Figure 5* shows sections of the two cable types.

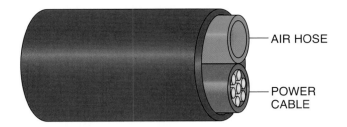

LOW-CURRENT APPLICATIONS

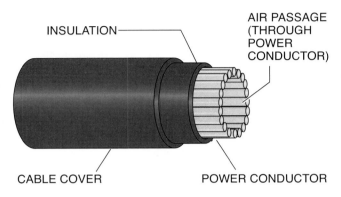

HIGH-CURRENT APPLICATIONS

29104-15_F05.EPS

Figure 5 Torch cable configurations.

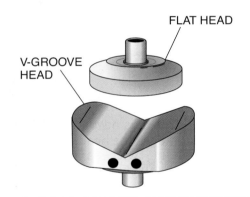

CUTTING, GOUGING, DEFECT REMOVAL

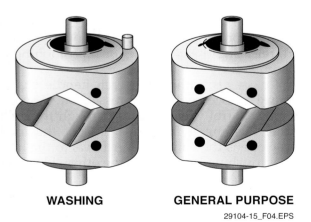

WASHING   **GENERAL PURPOSE**

29104-15_F04.EPS

Figure 4 Head types and arrangements.

1.2.2 A-CAC Cutting Power Supply Options

Either AC or DC constant current welding machines may be used to supply power to A-CAC torches, but DC welding machines are usually preferred. The type of welding current used (AC or DC) depends upon the base metal to be cut or gouged and the electrode type used.

Foundry Torch

The heads shown in *Figure 4* are designed for interchangeable use in a heavy-duty foundry torch, such as the one shown here. This allows the torch to be used for defect removal or pad washing, or as a general fine removal or piercing tool.

29104-15_SA01.EPS

Light-duty gouging and cutting can be done with a 200 amp machine. Medium-duty gouging and cutting will require a 600 amp machine. Heavy-duty cutting may require 1,200- to 2,000-amp equipment.

> **NOTE**
> Avoid using constant voltage (CV) welding machines with A-CAC processes because of their voltage and duty cycle limitations.

Table 1 lists common base metals along with the recommended welding current and polarity to be used with each metal.

1.2.3 A-CAC Torch Air Supply

An A-CAC torch requires regulated compressed air to blow the molten metal from the cut. Compressed air is supplied to the torch through a hose attached to a fitting on the end of the torch cable. While air pressure and volume settings can be over a wide range, they must be high enough that all molten metal will be blown from the cut and dross will not be left in the groove. Light-duty and medium-duty cutting and gouging typically require approximately 8 cubic feet per minute (cfm) [227 liters per minute (lpm)] to 35 cfm (991 lpm) of airflow at 40 psig (2.76 bars) to 80 psig (5.52 bars).

Table 1 Welding Current Types for Different Metals

BASE METAL	ELECTRODE TYPE	CURRENT TYPE AND POLARITY	ADDITIONAL RECOMMENDATIONS
Carbon steel and low alloyed steel	DC	DCEP*	Use DC electrodes with DCEP.
	AC	AC or DCEN**	AC electrodes with a (C) transformer can be used, but AC will only be half as efficient as DC.
Stainless steel	AC	DCEP	Use DC electrodes with DCEP.
	DC	AC or DCEN	AC electrodes with a (C) transformer can be used, but AC will only be half as efficient as DC.
Iron (cast, ductile, or malleable)	AC	AC or DCEN	Best practice to use 1/2" (12.7 mm) or if possible, larger electrodes. These must be used at highest rated amperage.
	DC	DCEP (Max Amps Rated)	
Copper alloys (content 60% and above)	DC	DCEN (Max Amps Rated)	Use at the electrode's highest rated amperage.
	AC	AC	Use AC electrodes with AC.
Copper alloys (content 60% or below)	DC	DCEN (Max Amps Rated)	Use at the electrode's highest rated amperage.
Aluminum bronze and aluminum nickel bronze	DC	DCEN	For special navy propeller alloy.
Nickel alloys (content is over 80% of mass)	AC	AC	
Nickel alloys (content is less than 80% of mass)	DC	DCEP	
Aluminum	DC	DCEP	Use stainless steel wire brush on base metal before welding. Note: Electrode length between the torch and base metal must not exceed 3" (76.2 mm).
	AC	AC or DCEN	

*Direct Current Electrode Positive
**Direct Current Electrode Negative

Compressor Sizes

A 1 horsepower (hp) air compressor can typically supply sufficient air pressure and volume for smaller electrodes. Larger electrodes may require an air compressor of 10hp or greater to supply sufficient air pressure and volume. The torch manufacturer's recommendations for air pressure and flow rates should be followed.

Heavy-duty cutting can require up to 50 cfm (1,416 lpm) of airflow at 80 psig (5.52 bars) to 100 psig (6.89 bars). The regulator should be set for the lowest pressure that will reliably do the job.

1.3.0 Electrodes

A-CAC electrodes used for cutting and gouging are composed of a blend of carbon and graphite with a binder material, which is baked to produce a uniform structure. They may be plain or coated with copper, also referred to as copper-clad. Electrodes designed for A-CAC use are copper-coated and contain a rare-earth, arc-stabilizing metal. Note that rare-earth metals are not necessarily rare, but they are usually widely dispersed rather than concentrated, and are challenging to separate from each other. Coated and uncoated electrodes are available in various types, styles, and sizes. *Figure 6* shows the two types of coated electrodes that are most commonly used.

> **WARNING!**
>
> Do not use wet carbon arc electrodes, as they can explode. The electrodes must be kept dry.

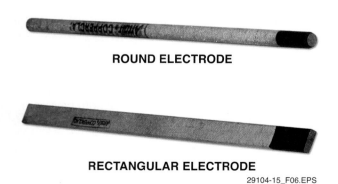

ROUND ELECTRODE

RECTANGULAR ELECTRODE

29104-15_F06.EPS

Figure 6 Copper-coated electrodes.

1.3.1 Electrode Types

A-CAC electrodes are of the following three basic types:

- *Plain for direct current (DC)* – Plain carbon electrodes for DC use do not have a copper coating. This type of electrode runs at hotter surface temperatures, oxidizes along its length, and loses diameter (reduces in thickness) between the torch and the arc tip. This electrode carries less current and burns more easily.
- *Copper-coated for direct current (DC)* – Copper-coated electrodes for DC use are the most common. These electrodes are superior for most applications because they can carry more current, can operate at cooler temperatures, and do not lose diameter through oxidation. However, they cannot be used where copper contamination could be a problem.
- *Copper-coated for alternating current (AC)* – Copper-coated AC electrodes differ from copper-coated DC electrodes because they contain rare-earth materials that help stabilize the arc; copper-coated DC electrodes do not.

1.3.2 Electrode Styles

Electrodes are manufactured in several styles, each with a range of sizes:

- *Round electrodes* – Round electrodes are the most common type used for manual cutting or gouging. They are manufactured with an uncoated tip that tapers as they begin to be consumed.
- *Round-jointed electrodes* – Round-jointed electrodes are designed for use with continuous-feed automatic and semi-automatic torches. Each electrode has a slightly conical tenon machined on one end and a matching socket machined into the opposite end. These ends allow the electrodes to be joined end-to-end to allow continuous cutting or gouging without interruptions for electrode changes. This system also eliminates stub waste.
- *Special shapes* – Special-shape electrodes are designed to make special-shaped grooves or cuts. Special shapes include flat (rectangular) and half-round. The flat electrodes are designed for close tolerance metal removal. Their uses include removing weld buildups, shaping dies, removing welded lugs, and beveling edges. Half-round electrodes are designed for cutting wide grooves with torches designed for smaller electrodes. *Figure 7* shows the cross-section shapes of special electrodes.

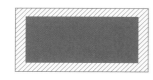

RECTANGULAR ELECTRODE

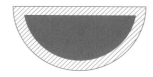

HALF-ROUND ELECTRODE

29104-14_F07.EPS

Figure 7 Cross-section shapes of special electrodes.

1.3.3 Electrode Sizes and Amperages

Round electrode sizes are specified by diameter; flat electrodes are specified by the widths of their widest side.

Plain and copper-coated AC electrodes range in size from ⁵⁄₃₂" (4 mm) to ⁵⁄₈" (16 mm). They are 12" (300 mm) long.

DC copper-coated electrodes range in size from ⅛" (3.2 mm) to 1" (25 mm). The smaller-sized round and flat electrodes are 12" (300 mm) long. Some of the larger-sized jointed electrodes are 17" (432 mm) long. AC copper-coated electrodes range in size from ³⁄₁₆" (4.8 mm) to ½" (12.7 mm).

Electrode current ratings increase with size. Plain electrodes and AC copper-coated electrodes operate at lower amperages than the same size of DC copper-coated electrodes.

Table 2 shows some typical current ratings for plain DC and copper-covered AC electrodes.

These ratings will vary with different manufacturers. For DC copper-covered electrodes, add 10 percent to the maximum amperage values shown in the table.

1.3.4 Electrode Selection

When performing A-CAC cutting or gouging, the minimum groove width is controlled by the size of the electrode. Groove depth and contour are controlled by electrode angle, travel speed, electrode side-to-side movement, current setting, and air pressure.

With copper-coated electrodes up to ½" (13 mm) in size on carbon steel, the minimum groove width will run about ⅛" (≈3 mm) wider than the electrode size.

Table 3 lists copper-coated electrode sizes and travel speeds for various groove sizes in carbon steel.

Table 2 Electrode Current Ratings

ELECTRODE TYPE	ELECTRODE SIZE		MINIMUM CURRENT	MAXIMUM CURRENT*
	IN	MM		
PLAIN DC OR COPPER-CLAD AC	⅛"	3.2	30 AMP	60 AMP
	⁵⁄₃₂"	4.0	90 AMP	150 AMP
	³⁄₁₆"	4.8	150 AMP	200 AMP
	¼"	6.4	200 AMP	400 AMP
	⁵⁄₁₆"	8.0	250 AMP	450 AMP
	⅜"	9.5	350 AMP	600 AMP
	½"	13	600 AMP	1000 AMP
	⅝"	15	800 AMP	1200 AMP
	¾"	19	1200 AMP	1600 AMP
	1"	25	1800 AMP	2200 AMP

*(For copper-clad DC, add 10% to maximum current value)

29104-15_T02.EPS

Table 3 Groove Size, Electrode Size, Amperes, and Travel Speed

GROOVE WIDTH (IN, MM)	GROOVE DEPTH (IN, MM)	ELECTRODE SIZE (DIA.) (IN, MM)	DIRECT CURRENT (AMPERES)	TRAVEL SPEED (IN / MIN)
¼, 6.4	¹⁄₁₆, 1.6	³⁄₁₆, 4.8	200	82
⁵⁄₁₆, 7.9	⅛, 3.2	¼, 6.4	300	51
⁵⁄₁₆, 7.9	¼, 6.4	¼, 6.4	320	29
⁵⁄₁₆, 7.9	⅜, 9.5	¼, 6.4	300	15
⅜, 9.5	⅛, 3.2	⁵⁄₁₆, 7.9	320	65
⅜, 9.5	¼, 6.4	⁵⁄₁₆, 7.9	420	31
⅜, 9.5	½, 12.7	⁵⁄₁₆, 7.9	540	27
⁷⁄₁₆, 11.1	⅛, 3.2	⅜, 9.5	560	82
⁷⁄₁₆, 11.1	¼, 6.4	⅜, 9.5	560	75
⁷⁄₁₆, 11.1	½, 12.7	⅜, 9.5	560	15
⁷⁄₁₆, 11.1	¹¹⁄₁₆, 17.5	⅜, 9.5	560	12
⁹⁄₁₆, 14.3	¼, 6.4	½, 12.7	1200	22
⁹⁄₁₆, 14.3	⅜, 9.5	½, 12.7	1200	21
⁹⁄₁₆, 14.3	½, 12.7	½, 12.7	1200	18
⁹⁄₁₆, 14.3	¾, 19.1	½, 12.7	1200	12
¹³⁄₁₆, 20.6	¼, 6.4	⅝, 15.9	1300	29
¹³⁄₁₆, 20.6	½, 12.7	⅝, 15.9	1300	14
¹³⁄₁₆, 20.6	¾, 19.1	⅝, 15.9	1300	11

29104-15_T03.EPS

To select an electrode, follow these steps:

Step 1 Identify the base metal type or alloy.

> **WARNING!**
> Never cut or burn cadmium material without proper ventilation and respiratory protection. The process produces poisonous gases that can cause severe personal injury.

> **CAUTION**
> Most manufacturers do not recommend using inverters or electronically controlled welding machines for carbon arc cutting. Peak currents are produced that are not usually filtered by the machine, which can damage the internal circuitry.

Step 2 Select a welding machine with the recommended current type (AC or DC), adequate ampere rating, and duty cycle for the base metal type and cut or gouge size.

Step 3 Select an electrode type compatible with the weld current type.

Step 4 Select the electrode size to produce the desired groove width.

1.3.5 Electrode Installation

An A-CAC electrode is usually gripped about 6"(≈150 mm) from the arc tip, as shown in *Figure 8*. For aluminum, a 3" (≈76 mm) stick-out is recommended. As the electrode is consumed, it is periodically advanced in the torch handle to maintain the proper projection.

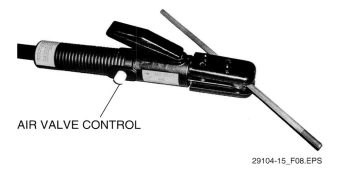

AIR VALVE CONTROL

29104-15_F08.EPS

Figure 8 Properly gripped electrode.

1.4.0 Safety Practices

The following is a summary of safety procedures and practices that must be observed while cutting or welding. Keep in mind that this is just a summary. Complete safety coverage is provided in the *Welding Safety* module. If you have not completed that module, do so before continuing. Above all, be sure to wear appropriate protective clothing and equipment when welding or cutting.

1.4.1 Protective Clothing and Equipment

Welding activities can cause injuries unless you wear all of the protective clothing and equipment that is designed specifically for the welding industry. The following information includes important safety guidelines about protective clothing and equipment:

- Always use safety glasses with a full face shield or a helmet. The glasses, face shield, or helmet lens must have the proper light-reducing tint for the type of welding or cutting being performed. Use corrective glasses if needed; note that some companies will not allow contact lenses. Never directly or indirectly view an electric arc without using a properly tinted lens.
- Wear proper protective leather and/or flame-retardant clothing along with welding gloves that will protect you from flying sparks and molten metal, as well as heat.

- Wear high-top safety shoes or boots. Make sure that the tongue and lace area of the footwear will be covered by a pant leg. If the tongue and lace area is exposed, or if the footwear must be protected from burn marks, wear leather spats under your pants or chaps and over the top front area of the footwear.
- Wear a 100-percent cotton cap with no mesh material included in its construction. The bill of the cap points to the rear. If a hard hat is required for the environment, use one that allows the attachment of rear deflector material and a face shield. A hard hat with a rear deflector is generally preferred when working overhead, and may be required by some employers and job sites.

> **WARNING!**
> Do not wear a cap with a button in the middle. The conductive metal button beneath the fabric represents a safety hazard.

- Wear a face shield over snug-fitting cutting goggles or safety glasses for gas welding or cutting. Either the face shield or the lenses of the welding goggles must be an approved shade for the application. A welding hood equipped with a properly tinted lens is also acceptable. See *Table 4* for a list of lens shades based on arc current.

Table 4 Guide for Shade Numbers

From *AWS F2.2 Lens Shade Selector*. Shade numbers are given as a guide only and may be varied to suit individual needs.

PROCESS	ELECTRODE SIZE IN (MM)	ARC CURRENT (AMPERES)	MINIMUM PROTECTIVE SHADE	SUGGESTED* SHADE NO. (COMFORT)
Shielded Metal Arc Welding (SMAW)	Less than 3/32 (2.4)	Less than 60	7	–
	3/32–5/32 (2.4–4.0)	60–160	8	10
	5/32–1/4 (4.0–6.4)	160–250	10	12
	More than 1/4 (6.4)	250–550	11	14
Gas Metal Arc Welding (GMAW) and Flux Cored Arc Welding (FCAW)		Less than 60	7	–
		60–160	10	11
		160–250	10	12
		250–500	10	14
Gas Tungsten Arc Welding (GTAW)		Less than 50	8	10
		50–150	8	12
		150–500	10	14
Air Carbon Arc Cutting (CAC-A)	(Light)	Less than 500	10	12
	(Heavy)	500–1000	11	14

*As a rule of thumb, start with a shade that is too dark to see the weld zone. Then go to a lighter shade which gives sufficient view of the weld zone without going below the minimum. In oxyfuel gas welding, cutting, or brazing where the torch and/or the flux produces a high yellow light, it is desirable to use a filter lens that absorbs the yellow or sodium line of the visible light spectrum.

29104-15_T04.EPS

Because of the high temperature and the intensity of the arc, choose a shade of helmet lens that is darker than the normal shade you would use for welding on the same thickness of metal. A number 10 to 14 lens shade is recommended for air-carbon arc cutting and gouging, depending on the current level.

- Wear earplugs to protect ear canals from sparks. Wear hearing protection to protect against the consistent sound of the torch.

Ear protection is essential to protect ears from the noise of the torch. Other personal protective equipment (PPE) must be worn to protect the operator from hot metal and slag.

1.4.2 Fire/Explosion Prevention

Welding activities usually involve the use of fire or extreme heat to melt metal. Whenever fire is used, it must be controlled and contained. Welding activities are often performed on vessels that may once have contained flammable or explosive materials. Residues from those materials can catch fire or explode when a welder begins work on such a vessel. The following are fire and explosion prevention guidelines associated with welding:

- Never carry matches or gas-filled lighters in your pockets. Sparks can cause the matches to ignite or the lighter to explode, causing serious injury.
- Always comply with all site and/or employer requirements for a hot-work permit and a fire watch.
- Never use oxygen to blow dirt or dust off clothing. The oxygen can remain trapped in the fabric for a time. If a spark hits the oxygen in the fabric, the clothing can burn rapidly and violently.
- Make sure that any flammable material in the work area is moved or shielded by a fire-resistant covering.
- Approved fire extinguishers must be available before attempting any heating, welding, or cutting operations. Make sure the extinguisher is

DCEP Electrode Consumption

When making a gouge that equals the electrode diameter, you will get approximately 8" (≈200 mm) of groove for every inch of electrode consumed. However, this will vary widely depending on electrode angle, speed, current, and the material being cut.

charged, the inspection tag is valid, and any individual that may be required to operate it knows how to do so.

- Never release a large amount of oxygen or use oxygen in place of compressed air. The presence of oxygen around flammable materials or sparks can cause rapid and uncontrolled combustion. Keep oxygen away from oil, grease, and other petroleum products.
- Never release a large amount of fuel gas, especially acetylene. Methane and propane are heavier than air and tend to migrate to and concentrate in low areas. As a result, they can ignite at a considerable distance from the release point. Acetylene is lighter than air but is even more dangerous than methane; when mixed with air or oxygen, it will explode at much lower concentrations than any other common fuel gas.
- To prevent fires, maintain a neat and clean work area, and make sure that any metal scrap or slag is cold before disposal.

Before cutting or welding containers such as tanks or barrels, find out if they contained any explosive, hazardous, or flammable materials, including petroleum products, citrus products, or chemicals that decompose into toxic fumes when heated. Proper procedures for cutting or welding hazardous containers are described in the *American Welding Society (AWS) F4.1, Safe Practices for the Preparation of Containers and Piping for Welding and Cutting, and ANSI Z49.1*. As a standard practice, always clean and then fill any tanks or barrels with water, or purge them with a flow of inert gas such as nitrogen to displace any oxygen.

Containers must be cleaned by steam cleaning, flushing with water, or washing with detergent until all traces of the material have been removed.

After cleaning the container, fill it with water or a purging gas, such as carbon dioxide, argon, or nitrogen to displace the explosive fumes. Air, which contains oxygen, is displaced from inside the container by the water or inert gas. Without oxygen, combustion cannot take place.

A water-filled vessel is the best alternative. When using water, position the container to minimize the air space. When using an inert gas, provide a vent hole so the inert gas can push the air and other vapors out to the atmosphere.

Keep in mind, though, that even these precautions do not guarantee the absence of flammable materials inside. For that reason, these types of activities should not be done without proper supervision and the use of proper testing methods.

1.4.3 Work Area Ventilation

Vapors and fumes tend to rise in the air from their sources. Welders often have to work above the welding area where the fumes are being created. Welding fumes can cause personal injuries. Good work area ventilation helps to remove the vapors and to protect the welder. The following is a list of work area ventilation guidelines to consider before and during welding activities:

- Make sure confined space procedures are followed before conducting any welding or cutting in the confined space.
- Always perform cutting or welding operations in a well-ventilated area. Cutting or welding operations involving materials such as coatings containing cadmium, lead, zinc, and chromium will result in toxic fumes. For cutting or welding of such materials, always wear an approved respirator as directed by your employer with a filter that may be used on a standard respirator.
- Make sure confined spaces are ventilated properly for cutting or welding purposes.
- Never use pure oxygen from cylinders for the purpose of ventilation or breathing air.

Because the molten metal cut during most A-CAC cutting or gouging activities tends to splatter, make sure that fire-resistant shields or screens are set up to protect any equipment or personnel near the work area. Also, make sure that a fire watch is posted.

Additional Resources

ANSI C5.3, Recommended Practices for Air Carbon Arc Gouging and Cutting. Latest Edition. Miami, FL: American Welding Society.

Air Carbon-Arc Guide, Form Number: 89-250-008. Denton, Texas: Victor Technologies, Inc.

2014 Technical Training Guide. The Lincoln Electric Company, Cleveland, OH. USA. **www.lincolnelectric. com**.

1.0.0 Section Review

1. One of the primary uses for air-carbon arc cutting (A-CAC) is _____.
   a. to gouge weld grooves, to back-gouge welds, and to gouge out defective welds
   b. together with compressed air, to control base metal temperature during cutting
   c. to introduce additional metal alloys into the base metal
   d. to reduce the amperage used in the cutting and gouging process

2. When the electrode is gripped in the torch handle, the electrode tip should be no farther from the heads than _____.
   a. 3" (7.6 cm)
   b. 6" (15 cm)
   c. 8" (20 cm)
   d. 12" (31 cm)

3. Electrodes designed for A-CAC use are copper-coated and contain an arc-stabilizing ingredient referred to as _____.
   a. graphite
   b. copper
   c. carbon
   d. a rare-earth metal

4. One of the procedures to prevent fires or explosions during A-CAC work is to make sure _____.
   a. that a certified welder is a member of the fire watch posted
   b. that the gouge equals the electrode diameter
   c. that the carbon arc electrodes to be used are completely dry
   d. to not use inverters or electronically controlled welding machines

2.0.0 A-CAC SETUP, OPERATION, AND CARE

Objective

Describe how to set up, safely operate, and care for air-carbon arc cutting equipment.

a. Describe how to prepare the equipment and work area for air-carbon arc cutting.
b. Describe how to wash and gouge metals.
c. Describe how to care for air-carbon arc cutting equipment.

Performance Tasks

2. Prepare the work area and air-carbon arc cutting equipment for safe operation.
3. Use air-carbon arc cutting equipment for washing.
4. Use air-carbon arc cutting equipment for gouging.
5. Perform storage and housekeeping activities for air-carbon arc cutting equipment.

Trade Terms

Concentric cable system: A-CAC configuration in which a unique combination fitting is used to connect the torch cable to welding power in order to enable compressed air passage through the power conductor.

Polarity: The condition of a system in which it has opposing physical properties at different points, such as an electric charge.

Now that basic A-CAC equipment has been described, this section will focus on setting up the equipment and the techniques to successfully gouge and wash.

2.1.0 Preparing to Cut

A-CAC equipment includes the torch and its attached cable. The torch cable must be connected to a welding machine and a compressed air source. Some torch cables contain a separate air hose and electrical cable within a common jacket. Others use a special concentric cable system that routes the compressed air through the center of a hollow electrical cable. *Figure 9* shows an inside view of both a concentric cable and a cable with separate air hose and electrical cable within a common jacket.

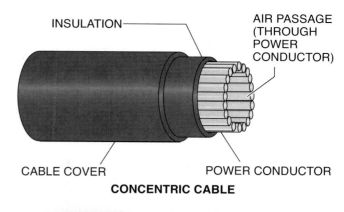

CONCENTRIC CABLE

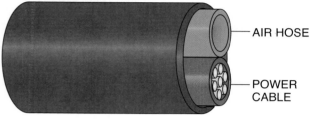

COMMON COVER CABLE

29104-15_F09.EPS

Figure 9 Inside view of concentric cable and common cover cable.

With the concentric system, a unique combination fitting is used to connect the torch cable to welding power and compressed air, as shown in *Figure 10*. It consists of a special female connector assembly and a hose nut with a pierced tab extending from one side. Heavy-duty, high-current foundry torches may have two tabs on the nut. The special fitting is electrically connected to the hollow electrical conductor in the torch cable. A standard threaded air hose fitting attaches to the connector, while the welding machine's electrode cable bolts to the lug on the side of the fitting. A protective rubber boot on the torch cable covers the connections.

2.1.1 Preparing the Work Area for A-CAC

The A-CAC equipment should be set up at a convenient distance from the welding machine. The area should be well ventilated and clear of all combustible material. The work to be cut or gouged should be placed at a comfortable working position.

A-CAC work will blow molten metal considerable distances. It is important that everything combustible be removed from the range of the blown molten metal. Flame-resistant shields or curtains must be erected to protect any workers or equipment within the range of the molten metal. It is also important to station a fire watch in the area. Make sure a fully charged fire extinguisher is at hand.

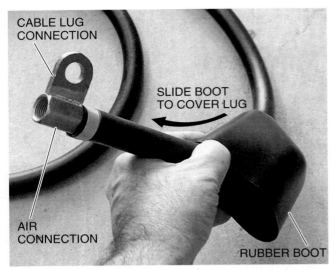

CABLE LUG CONNECTION

SLIDE BOOT TO COVER LUG

AIR CONNECTION

RUBBER BOOT

29104-15_F10.EPS

Figure 10 A-CAC torch combination power and air fitting

A-CAC work also generates loud noise and heavy smoke and fumes.

> **WARNING!**
>
> Ear protection is essential to protect ears from the noise. Other personal protective equipment must be worn to protect the operator from hot metal and slag.

If the site ventilation cannot keep the smoke and fumes away from the operator, an approved respirator must be worn.

2.1.2 Test Operating A-CAC Equipment

A-CAC equipment is simple to operate. To operate the equipment, follow these steps:

Step 1 Verify that the work area is properly shielded.

Step 2 Set the compressed air regulator to the desired pressure and open all valves supplying the A-CAC torch. Operate the air valve to ensure proper flow.

Step 3 Adjust the welding machine for the desired polarity and amperage.

Step 4 Put on a welding shield.

> **WARNING!**
>
> Always use proper hearing protection and appropriate PPE.

Step 5 Put on heat-insulating gauntlet-type welding gloves.

Step 6 Attach the welding machine workpiece clamp to the workpiece.

> **CAUTION**
>
> Be sure to position the workpiece clamp where the welding current will not pass through seals, bearings, or other contacting surfaces that could be damaged from arcing.

Step 7 Open the air jet spool valve. Position the torch at a 55-degree push angle with the air jets below the electrode and between the workpiece and the electrode.

Step 8 Strike an arc by touching the electrode to the workpiece. Do not draw the electrode back after the arc is established. Maintain a short arc and begin washing or gouging as described in the next section.

> **WARNING!**
>
> Do not use wet carbon-arc electrodes, as they can explode. The electrodes must be kept dry.

2.2.0 Washing and Gouging with A-CAC Equipment

Before using A-CAC equipment for washing or gouging activities, always evaluate the work to be done and plan the best way to achieve the desired results. Always be sure you understand how to position the A-CAC electrode to be most effective. It is also important to inspect the surfaces that need to be cut or gouged. Planning the work saves time, effort, and materials.

2.2.1 A-CAC Washing

Washing is the process of cutting objects or projections, such as bolt heads, old welds, and other unwanted materials from the surface of a base metal to leave a flush surface. Typically, the electrode is used at a push angle of about 55 degrees to the surface that contains the projection. The air orifices must blow air between the electrode and the base metal. The washing is performed by weaving the electrode from side to side across the projection. On some torches with available special washing heads, the air jet orifices are located on one side of the electrode. This arrangement reduces the tendency to gouge, but limits the torch to cutting from only one direction. *Figure 11* shows the results that can be achieved with A-CAC washing.

29104-15_F11.EPS

Figure 11 The results of A-CAC washing.

To perform washing with an A-CAC torch, follow these steps:

29104-15_F12.EPS

Figure 12 A-CAC gouging.

> **WARNING!**
>
> Wear eye and ear protection and protective clothing during this procedure.

Step 1 Select the proper electrode and set the correct **polarity** and amperage at the welding machine.

Step 2 Turn on the welding machine and start the air flowing through the jet air orifices.

Step 3 Holding the electrode at about a 55-degree push angle, strike an arc at the edge of the projection to be removed. As the arc melts the metal, swing it gradually across the projection. Continue the process until the projection is removed flush with the surface.

2.2.2 A-CAC Gouging

Gouging is a process of cutting a groove into a surface, usually to prepare it for welding. With A-CAC gouging, the electrode size determines the minimum groove width. Special-shaped electrodes can be used to cut special-shaped grooves. Groove width can be increased by weaving the electrode tip slightly from side to side. *Figure 12* shows the result of an A-CAC gouging process.

Groove depth and contour are controlled by electrode travel angle, tip side movement, and travel speed. For a shallow gouge, the electrode is advanced at a rapid rate. To increase the groove depth, decrease the advance speed or increase the current setting. To increase the width of the groove, move the electrode tip from side to side or use a larger electrode. Practicing A-CAC skills, like all welding skills, will help you to see the effect of very small and different hand motions.

Vertical Washing and Gouging

Positioning the workpiece in a vertical position simplifies metal ejection by using gravity in addition to the air jets. Make sure that adequate shoes and shoe top coverings are worn if vertical washing and gouging positions are used.

To gouge with an A-CAC torch, follow these steps:

Step 1   Select the proper size and shape electrode and set the correct polarity and amperage at the welding machine.

Step 2   Turn on the welding machine and start the air flowing through the air jet orifices.

Step 3   Holding the electrode at about a 55-degree push angle, strike an arc where the groove is to begin. As the arc melts the metal, advance it steadily along the base metal. Keep the electrode at a 0-degree work angle with respect to the groove direction.

2.2.3 Inspecting A-CAC Cut Surfaces

Inspect the A-CAC cut surface for the following conditions:

- Smooth edges that are relatively free of notches
- No dross
- No carbon deposits

Notches left in a base metal have a tendency to trap slag during welding. Dross or carbon deposits left on a surface to be welded will combine with the base metal to form hard, brittle zones and porosity in the weld zone. The topic of weld quality will be covered in depth in another module. Check your site or employer's quality standards for the base metal being cut. Cleaning is always necessary after A-CAC operations to ensure that all carbon and/or dross left by the electrode is removed prior to welding. A-CAC cutting is often used to quickly remove material, shortening the time required for grinding and other preparation tasks.

2.3.0 A-CAC Equipment Care

Proper equipment storage and housekeeping are essential for work efficiency and safety. Good housekeeping requires the following steps:

Step 1   Disconnect the A-CAC torch from the welding machine cable and air hose and store the torch in its assigned location.

Step 2   Neatly store the air hose, welding cables, and any related tools used in their assigned locations.

Step 3   Clean off the welding area and sweep up the slag and debris that were blown around the area.

Step 4   Make sure any remaining electrodes are stored in the proper conditions.

2.3.1 Repair of A-CAC Equipment

Following the proper equipment storage and housekeeping procedures usually results in trouble-free cutting and minimal repairs. However, cables and air hoses may become damaged and require replacement, especially if they have been run over by forklifts or other mobile equipment, been dragged across hot metal, or burned by molten-metal cutting sparks.

Before replacing a cable, always make sure the electrical disconnect and any other sources of energy such as an air supply is in the Off position. Air pressure should be vented from the hose. Remove the cable from the machine using the correct size wrench. If an electrode holder becomes cracked or worn, it may be possible to replace only the damaged portion.

Automatic Gouging System

Manual A-CAC is not usually thought of as a precise way to groove or cut. However, when the process is automated, it produces faster gouging speeds and the accuracy is both surprising and exceptional. A video that demonstrates precision gouging using equipment from Arcair can be viewed at **www.victortechnologies.com/arcair**. From the Home page, look for Media and Downloads under the Featured Products tab.

Additional Resources

ANSI C5.3, Recommended Practices for Air Carbon Arc Gouging and Cutting. Latest Edition. Miami, FL: American Welding Society.

2014 Technical Training Guide. The Lincoln Electric Company, Cleveland, OH. USA. **www.lincolnelectric.com**.

Air Carbon-Arc Guide, Form Number: 89-250-008. Denton, Texas: Victor Technologies, Inc.

2.0.0 Section Review

1. An A-CAC concentric system has compressed air passing through _____.

   a. an air hose that runs next to the power conductor
   b. a special male connector assembly and a hose nut with a pierced fitting
   c. the power conductor
   d. a separate filter to ensure clean compressed air

2. Washing is a process used to prepare the surface of the base metal _____.

   a. and introduce additional metal alloys such as copper to the surface
   b. by creating a narrow groove
   c. by heating it up and then allowing even cooling of the surface
   d. by removing objects and projections from the surface

3. If an electrode holder becomes cracked or worn, _____.

   a. the cable and torch assembly must be replaced
   b. it may be possible to replace only the damaged portion
   c. it can be exchanged with one designed for SMAW welding
   d. it can be exchanged with one designed for GTAW welding

SUMMARY

The A-CAC process is very useful for washing surfaces flush or for gouging out cracks to prepare them for welding repairs. However, because the torch air jets are continuously blasting molten metal from the arc, good site fire-prevention practices and use of appropriate PPE are essential elements. Ear protection is especially important to prevent noise damage from the air blast and to prevent sparks or slag from entering the ears.

1. The A-CAC process uses a high-current electric arc between the base metal and the end of a(n) _____.

   a. carbon-graphite electrode to melt the base metal
   b. high-frequency pilot arc to melt the base metal
   c. plasma jet to melt the base metal
   d. idle arc to melt the base metal

2. An A-CAC torch requires compressed air to blow the _____.

   a. carbon dioxide away from the cut
   b. molten metal away from the cut
   c. nitrogen away from the cut
   d. carbon arc away from the cut

3. When cutting and gouging with A-CAC equipment, the jet orifices of the torch head(s) should be located _____.

   a. beside the electrode
   b. above the electrode
   c. ahead of the electrode
   d. between the electrode and the base metal surface

4. To prevent overheating, the electrode tip to the V-groove head should be no farther than _____

   a. 1" (2 cm)
   b. 2" (5 cm)
   c. 4" (10 cm)
   d. 6" (15 cm)

5. Light-duty gouging or cutting is typically done with a _____.

   a. 10A welding machine
   b. 200A welding machine
   c. 600A welding machine
   d. 1,200A welding machine

6. Heavy-duty cutting can require a torch air-flow rate up to _____.

   a. 20 cfm (566 lpm)
   b. 35 cfm (991 lpm)
   c. 40 cfm (1,133 lpm)
   d. 50 cfm (1,416 lpm)

7. Compressed air is supplied to the torch through a hose attached to a fitting on the _____.

   a. welding machine
   b. end of the torch cable
   c. air jet orifices
   d. workpiece

8. Electrodes designed for A-CAC contain a rare-earth, _____.

   a. magnet metal
   b. undercurrent ingredient
   c. amplifier ingredient
   d. arc-stabilizing metal

9. The most common style of A-CAC electrode used for manual cutting or gouging is the _____.

   a. flat electrode
   b. round electrode
   c. half-round electrode
   d. round-jointed electrode

10. When performing A-CAC cutting or gouging, the minimum groove width is controlled by the _____.

    a. size of the torch
    b. size of the electrode
    c. speed of travel
    d. angle of the electrode holder

11. Because A-CAC activities tend to blow molten metal considerable distances, it is important to station a(n) _____.

    a. observer in the A-CAC work area
    b. slag trap in the A-CAC work area
    c. oxygen monitor in the A-CAC work area
    d. fire watch in the A-CAC work area

12. To wash a projection from a metal surface using A-CAC equipment, the electrode is typically used at a push angle of about _____.

    a. 30 degrees to the base metal surface
    b. 45 degrees to the base metal surface
    c. 55 degrees to the base metal surface
    d. 90 degrees to the base metal surface

13. Torches with special washing heads in which the air jet orifices are located on only one side of the electrode are limited to cutting from _____.

    a. one direction
    b. two directions
    c. three directions
    d. four directions

14. In A-CAC gouging, groove depth is influenced by the electrode _____.

    a. travel angle and speed
    b. size and length
    c. weave and angle
    d. length and weight

15. Dross or carbon deposits left on a surface to be welded will combine with the base metal to form _____.

    a. hard, brittle zones and porosity
    b. oxidation and/or corrosion
    c. soft areas and poor adhesion
    d. an excellent, strong weld

Trade Terms Quiz

Fill in the blank with the correct term that you learned from your study of this module.

1. To combine with oxygen, such as in burning or rusting (corrosion), is to _____.

2. A mixture of soft amorphous carbon and hard graphite carbon, possibly coated in copper, are the ingredients for a(n) _____.

3. A unique combination fitting used to connect the torch cable to welding power that also enables compressed air passage through the power conductor is part of a(n) _____.

4. The condition of opposing physical properties at different points, such as an electric charge, is referred to as _____.

Trade Terms

Carbon-graphite electrode 2
Concentric cable system 3

Oxidize 1
Polarity 4

Appendix

PERFORMANCE ACCREDITATION TASKS

The American Welding Society (AWS) School Excelling through National Skills Standards Education (SENSE) program is a comprehensive set of minimum Standards and Guidelines for Welding Education programs. The following performance accreditation tasks are aligned with and designed around the SENSE program.

The Performance Accreditation Tasks (PATs) correspond to and support the learning objectives in *AWS EG2.0, Guide for the Training and Qualification of Welding Personnel: Entry-Level Welder*.

Note that in order to satisfy all learning objectives in *AWS EG2.0*, the instructor must also use the PATs contained in the second level of the NCCER Welding curriculum.

PAT 1 and 2 correspond to *AWS EG2.0, Module 8 – Thermal Cutting Processes, Unit 4 – Air-Carbon Arc Cutting*, Key Indicators 4, and 5.

PATs provide specific acceptable criteria for performance and help to ensure a true competency-based welding program for students.

The following tasks are designed to develop your competency in preparing base metal using A-CAC processes. Practice each task until you are thoroughly familiar with the procedure.

As you complete each task, take it to your instructor for evaluation. Do not proceed to the next task until instructed to do so by your instructor.

A-CAC WASHING AND GOUGING

Perform A-CAC Washing

Using any of the materials identified below, perform A-CAC washing to remove the portion identified by the instructor. Materials that can be used for this task include:

- Steel backing strip on a butt weld
- Excess buildup on the face of a weld
- Rivets or bolts in a plate
- Blocks, angles, clips, eyes, D-rings, or items welded to a plate

Criteria for Acceptance

- Material removed flush with the base metal surface   _____
- No notching in the surface of the base metal   _____

Perform A-CAC gouging

Using mild steel plate ½" (12.7 mm) thick or thicker, gouge a U-groove at least 8" (20 cm) long, as shown in the figure, in the 1G and 2G positions.

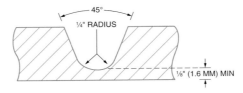

NOTE: GOUGE A WELDING PASS AT LEAST 8" LONG, AS SHOWN IN THE FIGURE BELOW, IN THE 1F AND 2F POSTIONS.

NOTE: BASE METAL = CARBON STEEL PLATE AT LEAST ¼" (6.4 MM) THICK

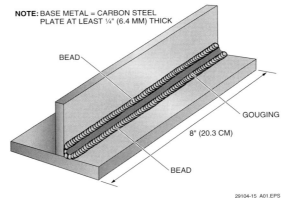

29104-15_A01.EPS

Criteria for Acceptance

- Groove width and depth are uniform   _____
- Groove walls are smooth and uniform   _____
- No dross within the groove   _____

Trade Terms Introduced in This Module

Carbon-graphite electrode: An electrode composed of a mixture of soft amorphous carbon and hard graphite carbon that may be coated with copper.

Concentric cable system: A-CAC configuration in which a unique combination fitting is used to connect the torch cable to welding power in order to enable compressed air passage through the power conductor.

Oxidize: To combine with oxygen, such as in burning (rapid oxidation) or rusting (slow oxidation).

Polarity: The condition of a system in which it has opposing physical properties at different points, such as an electric charge.

Additional Resources

This module presents thorough resources for task training. The following resource material is suggested for further study.

ANSI C5.3, Recommended Practices for Air Carbon Arc Gouging and Cutting. Latest Edition. Miami, FL: American Welding Society.

2014 Technical Training Guide. The Lincoln Electric Company, Cleveland, OH. USA. **www.lincolnelectric. com**.

Air Carbon-Arc Guide, Form Number: 89-250-008. Denton, Texas: Victor Technologies, Inc.

Figure Credits

The Lincoln Electric Company, Cleveland, OH, USA, Module Opener, Figures 3 (photo), 11, 12

Topaz Publications, Inc., Figures 2, 6, 8, 10

Victor Technologies, SA01

AWS F2.2:2001, Lens Shade Selector, reproduced with permission from the American Welding Society (AWS), Miami, FL, USA, Table 4

Section Review Answer Key

Answer	Section Reference	Objective
Section One		
1. a	1.0.0	1a
2. b	1.2.1	1b
3. d	1.3.0	1c
4. c	1.3.0	1d
Section Two		
1. c	2.1.0	2a
2. d	2.2.1	2b
3. b	2.3.1	2c

NCCER CURRICULA — USER UPDATE

NCCER makes every effort to keep its textbooks up-to-date and free of technical errors. We appreciate your help in this process. If you find an error, a typographical mistake, or an inaccuracy in NCCER's curricula, please fill out this form (or a photocopy), or complete the online form at **www.nccer.org/olf**. Be sure to include the exact module ID number, page number, a detailed description, and your recommended correction. Your input will be brought to the attention of the Authoring Team. Thank you for your assistance.

Instructors – If you have an idea for improving this textbook, or have found that additional materials were necessary to teach this module effectively, please let us know so that we may present your suggestions to the Authoring Team.

NCCER Product Development and Revision

13614 Progress Blvd., Alachua, FL 32615

Email: curriculum@nccer.org
Online: www.nccer.org/olf

❏ Trainee Guide ❏ Lesson Plans ❏ Exam ❏ PowerPoints Other _____

Craft / Level: _____ Copyright Date: _____

Module ID Number / Title: _____

Section Number(s): _____

Description: _____

_____

_____

_____

Recommended Correction: _____

_____

_____

_____

Your Name: _____

Address: _____

_____

Email: _____ Phone: _____

29105-15

Base Metal Preparation

OVERVIEW

This module introduces information related to the preparation of metals for welding, including the identification of common base metals, the physical characteristics and mechanical properties of different metals, and cleaning techniques. Welding joint preparation, codes, and specifications related to metal preparation are also described.

Module Five

Trainees with successful module completions may be eligible for credentialing through the NCCER Registry. To learn more, go to **www.nccer.org** or contact us at **1.888.622.3720**. Our website has information on the latest product releases and training, as well as online versions of our *Cornerstone* magazine and Pearson's product catalog.

Your feedback is welcome. You may email your comments to **curriculum@nccer.org**, send general comments and inquiries to **info@nccer.org**, or fill in the User Update form at the back of this module.

This information is general in nature and intended for training purposes only. Actual performance of activities described in this manual requires compliance with all applicable operating, service, maintenance, and safety procedures under the direction of qualified personnel. References in this manual to patented or proprietary devices do not constitute a recommendation of their use.

Objectives

When you have completed this module, you will be able to do the following:

1. Identify safety practices related to preparing base metals and describe basic cleaning procedures.
   a. Identify safety practices related to preparing base metals.
   b. Describe the basic properties and types of carbon and stainless steel.
   c. Describe basic metal cleaning procedures and concerns.
2. Identify and describe basic weld joint design and types of welds.
   a. Identify and describe the loads that are routinely placed on weld joints.
   b. Identify and describe the various types of weld joints.
   c. Describe a welding procedure specification (WPS) and the information it provides.
3. Describe how to prepare joints for welding.
   a. Describe how to mechanically prepare joints for welding.
   b. Describe how to thermally prepare joints for welding.

Performance Tasks

Under the supervision of your instructor, you should be able to do the following:

1. Mechanically or hand grind a bevel on the edge of a ¼"- to ¾"-thick mild steel plate (6 to 20 mm metric plate) at 22½ degrees.

2. Thermally bevel the edge of a ¼"- to ¾"-thick mild steel plate (6 to 20 mm metric plate) at 22½ degrees.

Trade Terms

Annealed	Ferritic	Oxide	Tempering
Austenitic	Ferrous	Piping porosity	Tensile strength
Backing	Load	Porosity	Weathering steel
Base metal	Malleable	Quench	Welding procedure
Castings	Martensitic	Root face	specification (WPS)
Distortion	Melt-through	Root opening	Weldment
Ductile	Nonferrous metal	Surfacing	Wrought

Industry Recognized Credentials

If you are training through an NCCER-accredited sponsor, you may be eligible for credentials from NCCER's Registry. The ID number for this module is 29105-15. Note that this module may have been used in other NCCER curricula and may apply to other level completions. Contact NCCER's Registry at 888.622.3720 or go to **www.nccer.org** for more information.

Contents

Topics to be presented in this module include:

Figures and Tables

SECTION ONE

1.0.0 BASIC WELDING SAFETY AND CLEANING

Objective

Identify safety practices related to preparing base metals and describe basic cleaning procedures.

a. Identify safety practices related to preparing base metals.
b. Describe the basic properties and types of carbon and stainless steel.
c. Describe basic metal cleaning procedures and concerns.

Trade Terms

Annealed: To free from internal stress by heating and gradually cooling.

Austenitic: Containing austenite (a solid solution of carbon, or of carbon and other elements in a ferrous alloy) added through heating.

Base metal: As used in this module, metal to be welded, cut, or brazed.

Castings: Something cast; any article that has been cast in a mold.

Distortion: The expansion and contraction of welded parts caused by the heating and subsequent cooling of the weld joint.

Ductile: Able to undergo change of form without breaking.

Ferritic: Steel containing less than 0.10 percent carbon and is magnetic. This steel can't be hardened via heat treatment.

Ferrous: Containing iron.

Malleable: Capable of being extended or shaped by hammering or by pressure from rollers.

Martensitic: Steel that shares some characteristics with ferritic, but has a higher level of carbon, up to a full 1 percent. It can be tempered and hardened and is used where strength is more important than a resistance to oxidation.

Nonferrous metal: A metal, such as aluminum, copper, or brass, lacking sufficient quantities of iron to have any effect on its properties.

Oxide: The scale that forms on metal surfaces when they are exposed to oxygen or air containing oxygen.

Piping porosity: A form of porosity having a length greater than its width and that is approximately perpendicular to the weld face.

Porosity: Gas pockets, or voids in the weld metal.

Quench: To cool suddenly by plunging into a liquid.

Tempering: To impart strength or toughness to (steel or cast iron) by heating and cooling.

Weathering steel: Steel alloy that, under specific conditions, is designed to form a very dense oxide layer on its outer surfaces, which retards further oxidation.

Weldment: An assembly that is fastened together by welded joints.

Wrought: Produced or shaped by beating with a hammer, as with iron.

To ensure that the safest and highest quality welds are produced, and to comply with welding codes, base metals must be properly prepared prior to welding. Safety must be priority one, and proper safety precautions add to the welder's organizational skills during the preparation process. The type of preparation required depends on the governing code requirements, the base metal type, the condition of the base metal, the welding process to be used, and the equipment available.

1.1.0 Basic Welding Safety

Complete safety coverage is provided in the *Welding Safety* module. If you have not completed that module, do so before continuing.

Welders routinely use manual and powered tools to clean and shape metals. *Figures 1* and *2* are examples of metal being prepared for welding. Everyone entering the welding trade starts by doing basic tasks, which includes preparing the metals for welding.

When using tools to clean metals prior to welding, wires from rotating wire brushes and particles from abrasive discs can become flying objects. Metal particles ground from the workpieces are usually hot and can cause burns, severe eye injuries, or fires. Welding screens or similar protective devices are used to keep the hot particles from igniting combustible materials and harming other workers nearby. Most companies require a fire extinguisher and a fire watch posted near the welding and cutting activities. To prevent injury from the flying particles and debris, always wear the proper personal protective equipment (PPE).

Before working on welding equipment and power tools, ensure that they are disconnected from the power supply. When the welding machine is energized, the open circuit between the electrode holder and the workpiece clamp represents a significant electrical hazard.

While much of the preparation work is performed in the shop where parts can be secured in clamping devices, some of it must be performed in the field on structures or equipment. Welders must sometimes perform at heights where fall prevention and fall arrest equipment is required. Always make sure that fall protection devices are undamaged and the inspections are current before using them.

The following is a summary of safety procedures and practices that must be observed while cutting, cleaning, shaping, and welding. Keep in mind that this is a summary. Complete safety coverage is provided in the *Welding Safety* module. If you have not completed that module, do so before continuing.

1.1.1 Protective Clothing and Equipment

Welding activities can cause injuries unless you wear all of the protective clothing and equipment that is designed specifically for the welding industry. The following information includes important safety guidelines about protective clothing and equipment:

Figure 2 Grinding metal.

- Always use safety glasses with a full face shield or a helmet. The glasses, face shield, or helmet lens must have the proper light-reducing tint for the type of welding or cutting being performed. Use corrective lenses if needed; some companies will not allow contact lenses to be worn during cutting or welding activities. Never view an electric arc directly or indirectly without using a properly tinted lens.
- Wear proper protective leather and or flame-retardant clothing along with welding gloves that will protect you from flying sparks and molten metal (*Figure 3*).

Figure 1 Cutting metal with an oxyfuel torch.

Figure 3 Appropriate personal protective equipment (PPE).

- Wear high-top safety shoes or boots. Make sure that the tongue and lace area of the footwear will be covered by a pant leg or other protective material. Boots with no laces at all are generally preferred for both welding and cutting activities. If the tongue and lace area is exposed, or if the footwear must be protected from burn marks, wear leather spats under your pants or chaps and over the top front area of the footwear.
- Wear a 100-percent cotton cap with no ventilation mesh material included in its construction. The bill of the cap points to the rear. If a hard hat is required for the environment, use one that allows the attachment of rear deflector and a face shield. A hard hat with a rear deflector is generally preferred when working overhead, and may be required by some employers and job sites.

> **WARNING!**
> Do not wear a cap with a button in the middle. The conductive metal button beneath the fabric represents a safety hazard.

- Wear a face shield over snug-fitting cutting goggles or safety glasses for gas welding or cutting. Either the face shield or the lenses of the welding goggles must be an approved shade for the application. A welding hood with a properly tinted lens is also acceptable. See *Table 1* for a list of lens shades based on arc current.
- Wear earplugs to protect ear canals from sparks. Wear hearing protection to protect against the consistent sound of the torch.

> **WARNING!**
> Ear protection is essential to protect ears from the noise of the torch. Other personal protective equipment (PPE) must be worn to protect the operator from hot metal and slag.

1.1.2 Fire and Explosion Prevention

Welding activities usually involve the use of fire or extreme heat to melt metal. Although most work site fires are caused by cutting torches and the hot material they produce, welding sparks and spatter can also start fires easily. Whenever fire is used, it must be controlled and contained.

Welding or cutting activities are often performed on vessels that may once have contained flammable or explosive materials. Residues from those materials can catch fire or explode when a welder begins work on such a vessel. The following are fire and explosion prevention guidelines associated with welding:

- Never carry matches or gas-filled lighters in your pockets. Sparks can cause the matches to ignite or the lighter to explode, causing serious injury.
- Always comply with any site requirement for a hot-work permit and/or a fire watch.
- Never use oxygen to blow dirt or dust off clothing. The oxygen can remain trapped in the fabric for a time. If a spark hits the oxygen in the fabric, the clothing can burn rapidly and violently.
- Never release a large amount of oxygen or use oxygen in place of compressed air. The presence of oxygen around flammable materials or sparks can cause rapid and uncontrolled combustion. Keep oxygen away from oil, grease, and other petroleum products.
- Make sure that any flammable material in the work area is moved or shielded by a fire-resistant covering.
- Approved fire extinguishers must be available before attempting any heating, welding, or cutting operations. Make sure the extinguisher is charged, the inspection tag is valid, and any individual that may be required to operate it knows how to do so.
- Never release a large amount of fuel gas, especially acetylene. Propylene and propane are heavier than air and tend to migrate to and concentrate in low areas. As a result, they can ignite at a considerable distance from the release point. Acetylene is lighter than air but is even more dangerous than methane; when mixed with air or oxygen, it will explode at much lower concentrations than any other common fuel gas.
- To prevent fires, maintain a neat and clean work area, and make sure that any metal scrap or slag is cold before disposal.

Before cutting or welding containers such as tanks or barrels, find out if they contained any explosive, hazardous, or flammable materials, including petroleum products, citrus products, or chemicals that decompose into toxic fumes when heated. Proper procedures for cutting or

Table 1 Guide for Shade Numbers

From *AWS F2.2 Lens Shade Selector*. Shade numbers are given as a guide only and may be varied to suit individual needs.

Process	Electrode Size in (mm)	Arc Current (Amperes)	Minimum Protective Shade	Suggested* Shade No. (Comfort)
Shielded Metal Arc Welding (SMAW)	Less than 3/32 (2.4)	Less than 60	7	–
	3/32–5/32 (2.4–4.0)	60–160	8	10
	5/32–1/4 (4.0–6.4)	160–250	10	12
	More than 1/4 (6.4)	250–550	11	14
Gas Metal Arc Welding (GMAW) and Flux Cored Arc Welding (FCAW)		Less than 60	7	–
		60–160	10	11
		160–250	10	12
		250–500	10	14
Gas Tungsten Arc Welding (GTAW)		Less than 50	8	10
		50–150	8	12
		150–500	10	14
Air Carbon Arc Cutting (CAC-A)	(Light)	Less than 500	10	12
	(Heavy)	500–1000	11	14
Plasma Arc Welding (PAW)		Less than 20	6	6 to 8
		20–100	8	10
		100–400	10	12
		400–800	11	14
Plasma Arc Cutting (PAC)		Less than 20	4	4
		20–40	5	5
		40–60	6	6
		60–80	8	8
		80–300	8	9
		300–400	9	12
		400–800	10	14
Torch Brazing (TB)		–	–	3 or 4
Torch Soldering (TS)		–	–	2
Carbon Arc Welding (CAW)		–	–	14

	Plate Thickness		Suggested* Shade No. (Comfort)
	in	mm	
Oxyfuel Gas Welding (OFW)			
Light	Under 1/8	Under 3	4 or 5
Medium	1/8 to 1/2	3 to 13	5 or 6
Heavy	Over 1/2	Over 13	6 or 8
Oxygen Cutting (OC)			
Light	Under 1	Under 25	3 or 4
Medium	1 to 6	25 to 150	4 or 5
Heavy	Over 6	Over 150	5 or 6

*As a rule of thumb, start with a shade that is too dark to see the weld zone. Then go to a lighter shade which gives sufficient view of the weld zone without going below the minimum. In oxyfuel gas welding, cutting, or brazing where the torch and/or the flux produces a high yellow light, it is desirable to use a filter lens that absorbs the yellow or sodium line of the visible light spectrum.

29105-15_T01.EPS

welding hazardous containers are described in the *American Welding Society (AWS) F4.1, Safe Practices for the Preparation of Containers and Piping for Welding and Cutting,* and *ANSI Z49.1.* As a standard practice, always clean and then fill any tanks or barrels with water, or purge them with a flow of inert gas such as nitrogen to displace any oxygen.

> **WARNING!**
>
> Welding or cutting must never be performed on drums, barrels, tanks, vessels, or other containers until they have been emptied and cleaned thoroughly, eliminating all flammable materials and all substances (such as detergents, solvents, greases, tars, or acids) that might produce flammable, toxic, or explosive vapors when heated. Do not assume that a container that has held combustibles is clean and safe until proven so by proper tests. Do not weld in places where dust or other combustible particles are suspended in air or where explosive vapors are present.

Containers must be cleaned by steam cleaning, flushing with water, or washing with detergent until all traces of the material have been removed.

> **WARNING!**
>
> Clean containers only in well-ventilated areas. Vapors can accumulate during cleaning, causing explosions or injury.

After cleaning the container, fill it with water or a purging gas, such as carbon dioxide, argon, or nitrogen to displace the explosive fumes. Air, which contains oxygen, is displaced from inside the container by the water or inert gas. Without oxygen, combustion cannot take place.

A water-filled vessel is the best alternative. When using water, position the container to minimize the air space. When using an inert gas, provide a vent hole so the inert gas can push the air and other vapors out to the atmosphere. Keep in mind, though, that even these precautions do not guarantee the absence of flammable materials inside. For that reason, these types of activities should not be done without proper supervision and the use of proper testing methods.

GOING GREEN — Welding Water

To better protect the environment, send any water used in a welding/cutting process or used to clean tanks to a waste treatment facility instead of simply allowing it to run into a storm drain or onto the ground.

1.1.3 Work Area Ventilation

Vapors and fumes tend to rise in the air from their sources. Welders often have to work above the welding area where the fumes are being created. Welding fumes can cause personal injuries. Good work area ventilation helps to protect the welder by removing the vapors. The following is a list of work area ventilation guidelines to consider before and during preparation activities:

- Make sure confined space procedures are followed before conducting any welding or cutting in a confined space.
- Always perform cutting or welding operations in a well-ventilated area. Cutting or welding operations involving materials such as coatings that contain cadmium, lead, zinc, or chromium will result in toxic fumes. For cutting or welding of such materials, always wear an approved respirator with an appropriate filter as directed by your employer.
- Grinding wheels containing aluminum oxide, silicon carbide and other materials have a composition that requires the use of respiratory protection. Consult the SDS/MSDS for warnings before starting any grinding activities.
- Make sure confined spaces are ventilated properly for cutting or welding purposes.
- Never use oxygen in confined spaces for purposes of ventilation.

GOING GREEN — Welding Fumes

To better protect the environment, perform any welding activities under a vapor extraction system that can filter the welding fumes before they reach the atmosphere.

1.2.0 Properties of Steel

Metals are a collection of chemical elements bound together by natural forces or manmade production processes. The chemical elements in some metals make them stronger and give them the ability to withstand stress or force better than other metals. The composition of common welding base metals varies widely, from metals made of essentially one metallic element, to alloys (mixtures) of metallic and nonmetallic components.

Metals are classified into two basic groups: ferrous metals, which are composed mainly of iron, and nonferrous metals, which contain very little or no iron. Ferrous metals include all steel, cast iron, wrought iron, malleable iron, and ductile (nodular) iron. Nonferrous metals and their alloys include the light metals (aluminum, magnesium, titanium); the heavy metals (copper, nickel, lead, tin, zinc); and the precious metals (platinum, gold, silver). Ferrous metals contain mostly iron and some carbon. Special steels may contain as much as 2 percent carbon. Carbon steel, the largest group of the ferrous-based metals, generally contains less than 1.7 percent carbon.

At low levels of carbon content, steel has some of the properties of cast iron, but is more ductile and more easily formed. Above 1.4 percent carbon, steel becomes stronger and more wear-resistant, although it becomes less ductile (difficult to form) and more difficult to weld. The higher carbon content lowers the melting temperature. The carbon content of a particular carbon steel product is the primary factor in determining its use.

Steel castings usually contain a higher percentage of carbon than rolled plate and other rolled shapes. Casting is done instead of rolling or shaping due to the loss of ductility in high-carbon steels. The casting foundry can usually provide the alloy formula, including the carbon content.

Steels, including carbon steels, low-alloy carbon steels, alloy steels, and stainless steels, are classified by various systems, such as by specification

Carbon and Alloy Steels

Steel is an alloy of iron and carbon typically containing less than 1 percent carbon. All steels also contain varying amounts of other elements such as manganese, silicon, phosphorus, sulfur, and oxygen. In addition, some standard alloy steels can also contain elements such as nickel, chromium, and molybdenum. There are currently about 3,500 different grades of steel with many different properties. Seventy percent of these steels have been developed in the past 25 years. Most of the steels produced today are carbon and alloy steels. Ninety-five percent of the construction and fabrication metals used worldwide consist of these materials.

29105-15_SA01.EPS

number and grade, or by manufacturer's trade name and number. Steel classification systems include those developed by the American Iron and Steel Institute (AISI), the American Society for Testing Materials International (ASTM), and the Unified Numbering System (UNS), as well as manufacturer's trade names and identification numbers. Stainless steels have their own classification systems. They are sometimes referred to by the percentages of their chromium and nickel content, such as 18/8, 25/20, and 18/10, but this system has largely been replaced by the AISI stainless steel classification system. Common classification systems for steels are explained in the following sections.

1.2.1 Carbon Steel Classification

The principal classification system for carbon steels is the AISI numerical designation of Standard Carbon and Alloy Steels. It was originally a four-digit system developed by the Society of Automotive Engineers (SAE) for carbon steels commonly used in structural shapes, plate, strip, sheet, and welded tubing. Five digits are now used for some alloys. The classification system may be referred to as the AISI, SAE, or the AISI and SAE system.

The AISI and SAE designations are essentially the same, except that the AISI system sometimes uses a letter prefix to indicate the manufacturing process that produced the steel. The following are examples of letter designations:

- A – Open-hearth steel
- B – Acid-Bessemer carbon steel
- C – Basic open-hearth carbon steel
- D – Acid open-hearth carbon steel
- E – Electric furnace steel

The absence of a letter prefix indicates basic open-hearth or acid-Bessemer carbon steel. The prefix letter (if any) designates the manufacturing process. The first two numerical digits represent the series (type and class) of steel. The third and fourth (and sometimes fifth) numerical digits specify the approximate percentage of carbon content.

The following examples explain the AISI and SAE numbers for carbon steels:

- *AISI number C1020*:
  C = Indicates basic open-hearth carbon steel
  10 = Carbon steel, nonresulfurized
  20 = Contains approximately 0.20 percent carbon

- *AISI number E2512*:
  E = Indicates electric furnace steel
  25 = Designates steel alloyed with approximately 5 percent nickel
  12 = Designates steel containing approximately 0.12 percent carbon
- *AISI number E52100*:
  E = Indicates electric furnace steel
  5 = Contains either approximately 0.50 percent, 1.00 percent, or 1.45 percent chromium, designated by the next digit: 0, 1, or 2, respectively
  2 = Designates approximately 1.45 percent chromium
  100 = Designates approximately 1 percent carbon

The common group classification of carbon steels is primarily based on carbon content. Plain carbon steels (AISI series 10XX through 11XX) are basically iron-carbon alloys. High-strength low alloy (HSLA) carbon steels (AISI series 13XX through 98XX) have small amounts of alloying elements to improve strength, hardness, and toughness or to increase resistance to oxidation, heat, and environmental damage. The plain carbon steels are commonly grouped as follows:

- *Low-carbon* – 0.10 percent to 0.15 percent carbon, 0.25 percent to 1.50 percent manganese
- *Mild-carbon* – 0.15 percent to 0.30 percent carbon, 0.60 percent to 0.70 percent manganese
- *Medium-carbon* – 0.30 percent to 0.50 percent carbon, 0.60 percent to 1.65 percent manganese
- *High-carbon* – 0.50 percent to 1.00 percent carbon, 0.30 percent to 1.00 percent manganese

The HSLA carbon steels are alloyed with one or more of the elements manganese, nickel, chromium, or molybdenum to provide higher strength, better toughness, weldability, and, in some cases, greater resistance to oxidation. As the percentage of these elements increases, changes in the weldability of the low-alloy steels make electrode selection and welding procedures more critical. HSLA steels include the following:

- Low-nickel steels
- Low-nickel chrome steels
- Low-manganese steels
- Low-alloy chromium steels
- Weathering steels

Another class of steel in the AISI series is sulfur steel. It is not an alloy steel but is included in the series.

1.2.2 Common Grade Stainless Steel Classifications

Stainless steels are typically classified as being a common grade or a specialty grade. Common-grade stainless steels (oxidation-resistant steels) are iron-based alloys that normally contain at least 11 percent chromium. Other alloying elements, including nickel, carbon, manganese, and silicon, may be present in varying quantities, depending on the specific type of stainless steel, to enhance its physical and mechanical properties. The main physical characteristics of all stainless steels are their resistance to oxidation and heat. Some have good low- and high-temperature mechanical properties. When compared with mild steels, stainless steels have the following characteristics:

- Lower coefficients of thermal conductivity that increase the chances of distortion
- Higher coefficients of thermal expansion that increase the chances of distortion
- Higher electrical resistances that increase the tendency to build up heat from welding current

Common-grade stainless steels are classified by their grain structures. The type of grain structure is determined by the specific alloy content of the stainless steel and its heat treatment during manufacture. Based on their microcrystalline structures, the classification of common-grade stainless steels is divided into these three groups:

- Austenitic
- Ferritic
- Martensitic

Austenitic stainless steels are non-magnetic in the annealed condition and not able to be hardened by heat treatment. However, they can be hardened significantly by cold working. Austenitic stainless steels combine excellent oxidation and heat resistance with good mechanical properties over a broad temperature range. Austenitic stainless steels make up the largest of the three common-grade stainless steel groups. For this reason, austenitic stainless steel is encountered most often by welders. They include the AISI 200 and 300 series stainless steels that all contain significant amounts of both chromium and nickel.

Austenitic steels are sometimes further subdivided in two classifications based on their compositions: chromium-nickel-magnesium (AISI 200 series) and chromium-nickel (AISI 300 series).

The range of applications for austenitic stainless steels includes housewares, containers, industrial piping and vessels, and architectural facades. Austenitic stainless steels are also available in low-carbon (L-grade) and high-carbon (H-grade) types. The letter L after a stainless steel type, such as 304L, indicates a low-carbon content of 0.03 percent or under. Similarly, the letter H after a stainless steel type indicates a high-carbon content, ranging between 0.04 and 0.10 percent. L-grades are typically used when annealing after welding is impractical, such as in the field where pipe and fittings are welded. H-grades are used when the steel will be subjected to extreme temperatures; the higher carbon content helps it retain strength.

Ferritic stainless steels are always magnetic. They are hardened to some extent by cold working, not by heat treatment. Ferritic stainless steels combine oxidation and heat resistance with fair mechanical properties over a narrower temperature range than austenitic stainless steels. Ferritic stainless steels are straight chromium stainless steels containing 11.5 to 27 percent chromium, about 1 percent manganese, and little or no nickel. The carbon content is 0.20 percent or less. Examples of ferritic stainless steels include AISI types 405, 409, 430, 442, 444, and 446. They typically are used for decorative trim, sinks, and automotive applications, particularly exhaust systems.

Martensitic stainless steels are also magnetic and can be hardened by quenching and tempering. They are excellent for use in mild environments such as the atmosphere, freshwater, steam, and with weak acids. However, they are not resistant to severely corrosive solutions. Martensitic stainless steels comprise two groups: martensitic and chromium-martensitic stainless steels.

Martensitic stainless steels contain only 4 to 6 percent chromium and no nickel. Chromium-martensitic stainless steels contain from 11.5 to 18 percent chromium, about 1 percent manganese, and in some cases 0 to 2.5 percent nickel. Because of the low chromium and no nickel, martensitic stainless steels are not considered true stainless steels, although their oxidation resistance is much greater than mild carbon steels, even at elevated temperatures.

Examples of chromium-martensitic stainless steels are AISI types 403, 410, 414, 420, 422, 431, and 440. Examples of martensitic stainless steels are AISI types 501 and 502.

1.3.0 Base Metal Cleaning

All base metals must be cleaned before welding. Even new materials that may look clean often are not. They pick up contaminants and surface oxidation during shipping and handling. When performing maintenance welding, a welder often comes across components that have been exposed to surface contamination such as oxidation, paint, oil, and grease. The heat generated by the welding process could cause paints and coatings to give off toxic fumes, endangering the worker performing the welding.

Do not assume a metal's surface is oxidation-free just because it is shiny. Oxidation on stainless steel and aluminum is difficult to see; it may have a white color or it may be transparent. To ensure quality welds and to conform to code requirements, surface contaminants and oxides must be removed prior to welding. In general, the codes state that the surface to be thermally cut or welded must be clean and free from paint, oil, rust, scale, and other material that would be harmful either to the weld or to the base metal when heat is applied. Cleaning is typically performed by mechanical and or chemical means.

Most metal has some surface oxidation. Oxidation occurs when metal is exposed to oxygen in the air or from other sources. Some oxidation exists as a thin, hard film that merely stains the metal and, in fact, acts as a retardant to further oxidation. Examples of this type of oxidation are found on stainless steel, copper, and aluminum. The thin, hard layer of oxidation bonds tightly to the surface and protects the metal.

A more recognizable type of oxidation found on mild steel is called rust (*Figure 4*). Rust is a very coarse type of oxidation that tends to flake easily, exposing more of the base metal to oxidation. Alloys such as chromium and copper are often added to mild steel to protect it from this coarse type of oxidation. These types of steels are called alloy steels. The alloys cause the rust to be very fine, bonding to the surface and protecting the base metal beneath. Weathering steel is an example of a copper-alloyed steel that protects itself from oxidation by forming a hard, tough layer of brown oxidation (rust) on its surface. Weathering steels are used outdoors and require no painting.

29105-15_F04.EPS

Figure 4 Examples of oxidation.

Regardless of the type of contamination, it must be removed prior to welding. Any type of contamination can cause weld defects. The most common defect caused by surface contamination is porosity. Examples of porosity are shown in *Figure 5*. Porosity occurs when gas pockets or voids appear in the weld metal. When the porosity has a length greater than its width and is approximately perpendicular to the weld face, it is called piping porosity. Piping porosity is formed as the gas pocket or bubble floats toward the surface of the weld, leaving a void. The gas pocket is trapped as the weld metal solidifies, but as the next layer of weld is deposited, the bubble often continues to float up. Porosity may not always be visible on the surface of the weld.

Surface contamination can easily result from any of several causes, including the following:

- Use of gloves contaminated with oil or other substances
- Allowing alloys to come into contact with carbon steel objects, including chain falls, jack stands, and forklifts
- The use of chloride-based markers to write on the base metal

Because of the risk of contamination, metals such as carbon steel, stainless steel, and aluminum must be stored separately. Aluminum must be stored indoors.

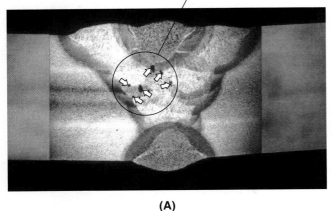

TRAPPED AIR BUBBLES

(A)

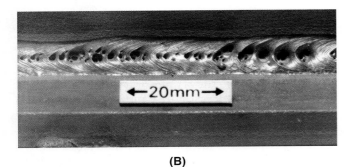

←20mm→

(B)

Reproduced by permission TWI Ltd.

29105-15_F05.EPS

Figure 5 Porosity.

1.3.1 Mechanical Cleaning

Mechanical cleaning is the most common method of removing surface contamination. Tools used for mechanical cleaning include hand tools, power tools, and special sandblasting equipment. When performing mechanical cleaning, be sure to wear safety glasses and a face shield for protection from the flying particles produced during the cleaning operation. In addition, special clothing is required for sandblasting.

Flexible scrapers are used to remove dirt and grease from weldments or weldment components. Rigid scrapers can be used to remove hardened paint or dirt.

Wire brushes are used to remove paint and light-to-medium surface oxidation. Wire brushes are available with bristles made from a variety of metals, including carbon steel, stainless steel, and brass. Use only stainless steel wire brushes on stainless steel and aluminum, and make sure that the stainless steel wire brush used on these metals has never been used with another metal.

Once brushes, grinding wheels, and files have been used on carbon steel and other metals, they should never be used on stainless steel or aluminum. The remaining particles of other metals on the tools will contaminate stainless steel and aluminum, as will a wire brush made of carbon steel. The contamination leads to weld defects.

Files are another way to remove surface oxidation. When filing aluminum or stainless steel, do not use files that have been used on other metals. Before using a file, always be sure it has a handle installed on the file tang. Use a fine file and keep it clean to prevent scratching the base metal surface. Take care not to damage or scratch the base metal outside the weld zone. Clean files by using stainless steel brushes or file cards that either are new or have only been used to brush the same type of metal (*Figure 6*). Even particles from a file card can contaminate a file, which in turn can contaminate metals on which the file is used.

Tight oxidation must be removed by filing, grinding, or sandblasting. For large jobs or jobs where speed is important, power tools are most efficient. Power tools may be electrically or pneumatically powered.

NOTE

Sandblasting is not used on aluminum.

METAL FILE WIRE BRUSH FILE CARD

29105-15_F06.EPS

Figure 6 File and cleaning tools.

Angle grinders (*Figure 7*) are very effective in removing large areas of surface contamination. Die grinders and small angle grinders work very well for small areas such as weld grooves and bevel edges. Grinders have various attachments that can be used for special applications. These attachments include grinding disks and wheels, wire brushes, rotary files, and flapper wheels (*Figure 8*).

Grinding disks and wheels are made for specific types of metals. Be sure you have the correct disk for the type of metal being ground. Always use aluminum oxide disks for grinding aluminum or stainless steel. For stainless steel or aluminum, do not use wheels that have been used on other metals. They will contaminate the surface of aluminum and stainless steel.

When using wire brush attachments on stainless steel or aluminum, brushes with stainless steel bristles must be used. Only use stainless steel brushes that have not been used on other metals. Remember that carbon steel bristles will contaminate stainless steel and aluminum. Over time, the contamination will reveal itself by iron oxide (rust) developing at the contamination sites of the joint.

When grinding, care must be taken to prevent grinding the base metal below the minimum allowable base metal thickness. If this should happen, the base metal will have to be discarded or, if allowable, built up with weld metal. Both of these alternatives are expensive.

Cutoff wheels are used to cut through metals and other materials. They are made for specific types of metals, or for other materials such as stone. Be sure you have the correct cutoff wheel for the type of material being cut.

Weld flux chippers and needle scalers are usually pneumatic rather than electric. These tools are used by welders to clean surfaces and

(A) ELECTRIC ANGLE GRINDER

(B) ELECTRIC DIE GRINDER

(C) PNEUMATIC ANGLE GRINDER

(D) PNEUMATIC DIE GRINDER

29105-15_F07.EPS

Figure 7 Handheld grinders.

remove dross from cuts and welds. Weld flux chippers, and needle scalers are also excellent for removing paint, heavy scale, or hardened dirt, but they are not very effective for removing surface oxidation. Weld flux chippers have a single chisel; needle scalers have about 18 to 20 blunt steel needles about 10 inches (≈25 cm) long. Many weld flux chippers can be converted to needle scalers with an attachment. A weld flux chipper and a needle scaler are shown in *Figure 9.*

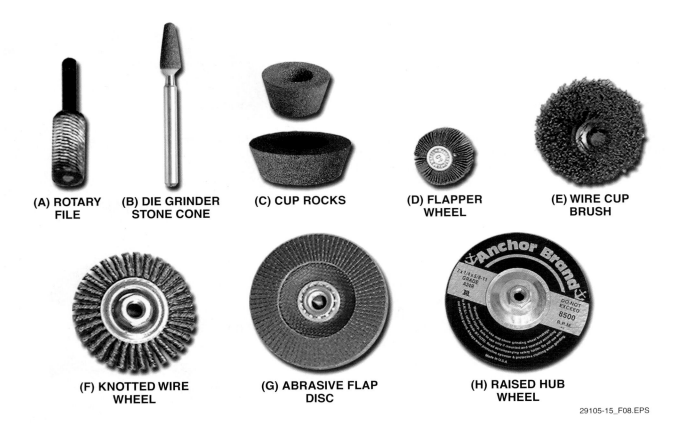

(A) ROTARY FILE

(B) DIE GRINDER STONE CONE

(C) CUP ROCKS

(D) FLAPPER WHEEL

(E) WIRE CUP BRUSH

(F) KNOTTED WIRE WHEEL

(G) ABRASIVE FLAP DISC

(H) RAISED HUB WHEEL

29105-15_F08.EPS

Figure 8 Grinder attachments.

NEEDLE SCALER

WELD FLUX CHIPPER

29105-15_F09.EPS

Figure 9 Weld flux chipper and needle scaler.

1.3.2 *Chemical Cleaning*

There will be times when mechanical cleaning is not enough. Residues left on a metal can contaminate the weld. Even the oil from the skin of the welder handling the metal can contaminate a weld. Welders may have to use some form of chemical cleaner to ensure that the metals being prepared are very clean.

It is essential that only approved cleaners are used. Some chemicals can create hazardous fumes during the welding process. Always follow the prescribed specifications of the approved cleaners and the related SDS/MSDS. The two chemical cleaners most often used by welders are acetone and butanone. Butanone is better known as methyl ethyl ketone, or MEK.

WARNING!

Do not use brake part cleaning fluid to clean surfaces prior to welding. Some brake part cleaning fluids contain chlorinated hydrocarbons. Phosgene gas is formed by the decomposition of chlorinated hydrocarbon solvents by ultraviolet radiation. Phosgene gas reacts with moisture in the lungs to produce hydrogen chloride, which in turn destroys lung tissue. For this reason, any use of chlorinated solvents for any purpose should be kept away from welding operations.

Acetone is a flammable, colorless liquid. Although a manufactured product, it can also be found in some natural environments. It is a primary ingredient in nail polish remover. Anyone who has worked in an auto repair shop has probably used acetone to remove over-sprayed paint from car windows. It will dissolve resins, epoxies, and many plastics. It is a great degreaser that is often used to prepare metals before painting.

From a safety standpoint, acetone is an irritant. Because it can cause permanent eye damage (corneal clouding), workers handling it or working near it must wear safety goggles. Acetone vapors should be avoided, and a suitable respirator is recommended. Prolonged exposure may eventually cause liver damage. Check the product SDS or MSDS closely to ensure that you understand how to safely work with acetone.

MEK is a manufactured organic chemical. It is a colorless liquid that dissolves substances such as gums, resins, vinyl film coatings, and other similar materials that may still be on the metals. MEK is highly flammable, but it presents much less of a health risk than acetone. However, anyone working with MEK must review the SDS or MSDS to make sure all safety precautions are followed.

WARNING!

Acetone and MEK are both extremely flammable. Keep these products away from all potential sources of ignition.

After the metal parts have been mechanically cleaned, the chemical cleaner is applied, usually with a brush or spray. The chemical is allowed to work for a short time before being removed. A clean cloth is used to remove the cleaning agent along with any contamination loosened from the metal. When cleaning metals with chemicals, be aware that the contaminants being removed may give off vapors that could be dangerous. Be sure to use the proper PPE, including respiratory protection in some conditions.

Additional Resources

Steel Metallurgy for the Non-Metallurgist. John D. Verhoeven; Prepared under the direction of the ASM International Technical Book Committee. ASM International® Materials Park, OH 44073-0002, USA. **www.asminternational.org**

American Iron and Steel Institute, 25 Massachusetts Avenue, NW Suite 800, Washington, DC 20001, USA. **www.steel.org**

Association for Iron & Steel Technology, 186 Thorn Hill Road Warrendale, PA, USA. **www.aist.org**

1.0.0 Section Review

1. Grinding wheels containing aluminum oxide, silicon carbide and other materials have a composition that require _____.

   a. a simple dust mask that is comfortable for the wearer
   b. a mask made of 100-percent cotton, such as a handkerchief, or similar type of cover
   c. consultation of the SDS/MSDS for recommended respiratory protection
   d. a safety shield only; a respiratory mask is not necessary

2. Which letter prefix, when used in the AISI numbering system, indicates acid-Bessemer carbon steel?

   a. The letter A
   b. The letter B
   c. The letter C
   d. The letter D

3. When brushing aluminum, use stainless steel brushes that have only been used on _____.

   a. stainless steel
   b. aluminum
   c. bronze
   d. carbon steel

2.0.0 JOINT DESIGN AND WELD TYPES

Objective

Identify and describe basic weld joint design and types of welds.
  a. Identify and describe the loads that are routinely placed on weld joints.
  b. Identify and describe the various types of weld joints.
  c. Describe a welding procedure specification (WPS) and the information it provides.

Trade Terms

Backing: A weldable or non-weldable material used behind a root opening to allow defect-free welding at the open root of a joint.

Load: The amount of force applied to a material or a structure.

Melt-through: Complete joint penetration.

Root face: A small flattened area on the end of a bevel for a groove weld.

Root opening: The space between the base metal pieces at the bottom or root of the joint.

Surfacing: The application by welding, brazing, or thermal spraying of a layer of material to a surface to obtain desired properties or dimensions.

Tensile strength: The measure of the ability of a material to withstand a longitudinal stress without breaking.

Welding procedure specification (WPS): The document containing all the detailed methods and practices required to produce a sound weld.

W elded joints are selected primarily for the safety and strength required for the conditions that will be encountered. When selecting the joint for a particular application, many factors must be taken into account, including load considerations, the environment in which it will be used, materials, processes, and cost.

2.1.0 Load Considerations

The loads in a welded steel structure or component are transferred through the welded joint.

The welded joint must be designed to withstand the stresses caused by the loads.

Any or all of these forces may act on welded parts. When two parts are welded together, in whatever manner, they are intended to become one part. If the surface areas of the two parts being welded are not properly prepared, the welds may fail when one or more of these forces is applied. The following are brief explanations of load terms as they relate to welds. Refer to *Figure 10* for a visual representation of these forces:

- *Tensile* – When metals are pulled in opposite directions by forces (F_1 vs. F_2), they will eventually break. Their strength to remain one is referred to as their tensile strength. When metals are heated, as when they are welded, the molecules in the metals change. When the metals are cooled, their molecules change again. These heating and cooling cycles can weaken the tensile qualities of metals. When two pieces of metal are welded end-to-end, they become one piece. When they are pulled at each end, the tensile strength of the two pieces, and the weld, are tested. If the weld is a good weld, one of the base metal pieces will usually break before the weld breaks.

- *Compression* – The reduction of volume or mass. Metals can be compressed. If you strike a steel nail or bolt with a hammer, you can eventually compress the nail or bolt into a different shape. When two pieces of metal are welded end-to-end and forces (F_1 and F_2) are applied inwardly on each end, the pieces and their weld are compressed.

- *Bending* – To bend is to make curved. For example, you may be able to manually bend a section of straight lightweight flat bar around a 6-inch (DN150) pipe. The pipe in this case is a force (F_2) pushing against the center of the bar as you pull or apply force (F_1) in the opposite direction at each end of the bar. The result of such action is that the flat bar becomes curved. When two pieces of metal are welded end-to-end, they may be tested in what is known as a bend test.

- *Torsion* – Twisting. Torsion is the amount of stress the welded bars can withstand as a counterclockwise force (F_1) is applied at one end and a clockwise force (F_2) is applied at the other end. For example, the driveshaft of a truck is twisted as it transmits the rotational movements of the drive motor and transmission to the drive axle.

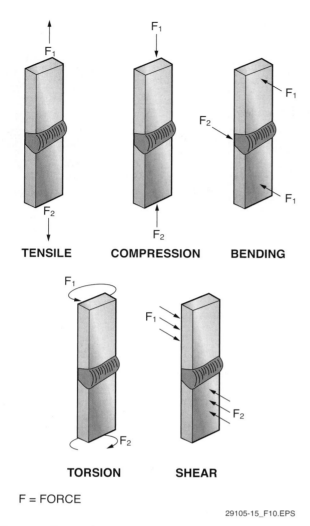

TENSILE   COMPRESSION   BENDING

TORSION   SHEAR

F = FORCE

29105-15_F10.EPS

Figure 10 Forces that act on welded joints.

- *Shear* – To cut off or trim. When a piece of sheet metal is placed into a press brake machine, one side of the sheet metal is resting against a stationary surface of the machine. That stationary surface is applying some force to that side of the sheet metal. That stationary force is shown as F_1. When the brake is activated, a moveable part of the machine applies force to the opposite side of the sheet metal. F_2 represents the forced created by that moveable part. When the moveable force (F_2) is increased, it eventually shears or cuts the metal because the stationary surface (F_1) is applying force to the opposite side of the sheet metal.

2.2.0 Types of Welds

Most types of welds require some degree of base metal edge preparation depending on the type of joint being welded. Common types of welds include the following: surfacing, plug or slot, fillet, square groove, bevel groove, V-groove, J-groove, and U-groove. If a particular type of weld requires a specified root opening, it is typically shown on the welding drawing or welding procedure specification (WPS). Base metals may require that some type of bevel be cut or ground on one or both edges, which form a V-groove when they are mated. Edge preparations will be discussed for each type of weld as it is covered in this section.

There are five basic types of joints (*Figure 11*). They are the butt joint, lap joint, T-joint, edge joint, and corner joint (both inside and outside corners). Each of these joints may be accomplished using various types of welds.

2.2.1 Surfacing Welds

Surfacing is the only type of weld that is applied to just one base metal surface, since it is not used to join two or more components. It is used either to build up a base surface that has become worn below the desired thickness or dimension, or to add material to a base to increase its strength and stability. An example of surfacing is shown in *Figure 12*. When you begin making your first welds, you will build what is referred to as a pad—a section of base metal with weld beads run parallel to each other along its length. This is an example of surface welding.

Did You Know?

Surfacing

Surfacing, as defined by the American Welding Society, is the application by welding, brazing, or thermal spraying of a layer of material to a surface to obtain desired properties or dimensions, as opposed to creating a joint between two or more pieces.

Before applying the first layer of multi-layered welds, the base metal surface must be cleaned to remove contaminants, such as oxidation and dirt. When applying the surfacing weld using an oxyfuel process, the surface should be evenly preheated before surfacing to eliminate warping of the base material.

For proper penetration and layering to take place, each new layer of weld surfacing must be properly chipped, scraped, and thoroughly cleaned before applying additional layers of surfacing.

Surfacing is often applied in the tool and die industry to rebuild relatively expensive tools and dies that have worn down. It is also used in the repair of heavy equipment to build up areas that have experienced excessive wear.

29105-15_F12.EPS

Figure 12 Surfacing welding on an inclined vertical surface.

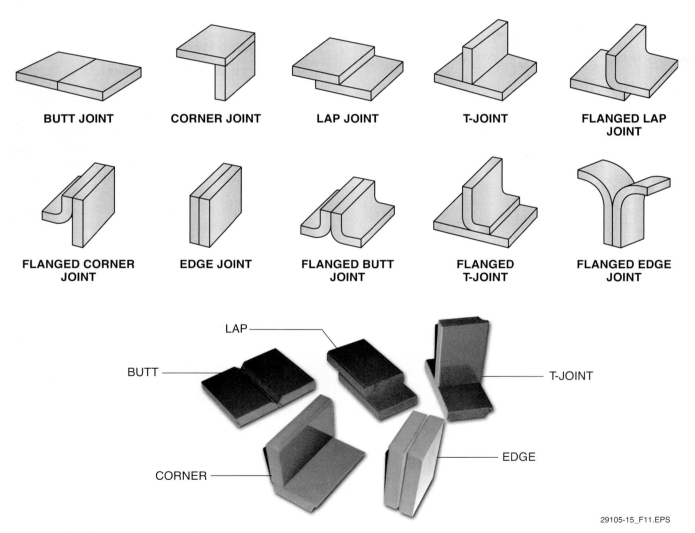

29105-15_F11.EPS

Figure 11 Five basic joint types with variations.

2.2.2 Plug and Slot Welds

Plug and slot welds are used to join metal pieces when the edges cannot be welded. The hole or slot can either be completely or partially filled with weld material when joining it to the other base metal. If the hole is round, the weld is referred to as a plug weld. If the weld is slotted or elongated, it is called a slot weld. *Figure 13* illustrates how a plug or slot weld may be used on a lap joint. A typical application is to overfill the hole or slot, then grind the excess material to a level, finished surface.

Preparation of the base metals for this type of weld requires proper cleaning and removal of any oxides or dirt from both pieces. The hole or slot is then drilled to receive the weld. The walls of the hole or slot must also be cleaned of any oxides, dirt, and oils prior to welding.

Plug and slot welds are commonly used where a finished surface is required, or where overall dimensions are not permitted to exceed those of the plate thickness.

2.2.3 Fillet Welds

Fillet welds may be applied to lap joints, T-joints, or corner joints. The fillet weld does not require any base metal edge preparation other than removing contaminants by cleaning with a wire brush or other appropriate cleaning tools. The two base metals are positioned together with or without a root opening, and one or more passes (welding beads) are applied at the intersection of the two base metal edges (*Figure 14*). Whenever more than one pass is made on a fillet weld—or on any weld—the previous pass must be cleaned of all contaminants such as dross, slag, and oxides before the next pass is applied.

When using a fillet weld on outside corner welds two types of fillet welds may be used. The half-lap joint (*Figure 15*) is easier to assemble, requires less welding material, and is less likely to burn through the corner. The half-lap does require a second weld on the inside of the corner. If a half-lap fit is used, allowances in the plate dimensions must be made for the lap.

The other fillet weld is sometimes called the corner-to-corner joint. The corner-to-corner joint is difficult to assemble because neither plate can support the other, and care must be used when welding to prevent burning through the corner.

PLUG   SLOT

PLUG   SLOT

29105-15_F13.EPS

Figure 13 Plug and slot welds.

Surface Welding

Certain shafts located within gearboxes used in the marine industry turn at relatively high rpms (revolutions per minute) and will corrode and wear rapidly if subjected to salt water. It is not always cost effective to discard and replace each shaft with a new one. Surface welding may be used to bring the shafts back to the desired dimensions by building them up and re-machining them. They are then classified as rebuilt shafts. All rotating parts must be balanced before being installed. Otherwise they will vibrate, which in turn will destroy bearings and may cause the part itself to break.

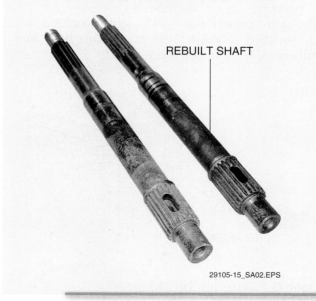

REBUILT SHAFT

29105-15_SA02.EPS

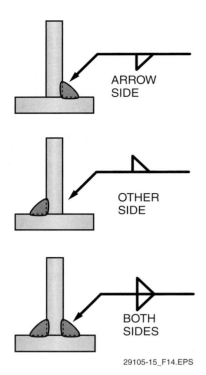

Figure 14 Fillet welds and drawing symbols.

2.2.4 Square-Groove Welds

Square-groove welds can be used with butt joints, corner joints, T-joints, and edge joints. The difference between the fillet and the square groove is that a square groove requires a root opening to be set or prepared between the two base pieces prior to welding, as shown in *Figure 16*. This allows penetration into a greater portion of the surface area of the two base metals. A partial joint-penetration weld will have much less strength than a complete joint-penetration weld.

Welding codes impose restrictions for complete joint-penetration welds. With shielded metal arc welding (SMAW), the maximum base metal thickness is ¼" (roughly equal to 6 mm metric

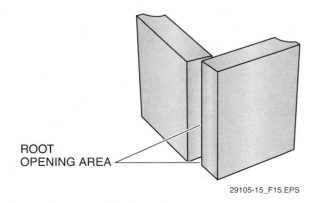

Figure 15 Fit-up of half-lap fillet welds.

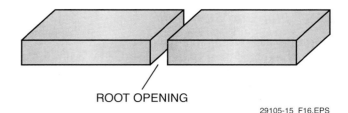

Figure 16 Preparation for a square groove on a butt weld.

steel plate), and welding from both sides is required. In addition, a root opening of half the thickness of the base metal is required, and the root of the first weld must be gouged before the second weld is made. For gas metal arc welding (GMAW) and flux-cored arc welding (FCAW), the maximum base metal thickness is ⅜" (≈10 mm metric plate), with the same requirements for welding from both sides and for back gouging the first pass.

Welding codes and specifications will be discussed in greater detail later in this module.

2.2.5 Single-Bevel Groove Welds

Single-bevel groove welds are commonplace in the welding industry because the base metal preparation is relatively simple and provides greater surface area penetration than the square groove. The single-bevel groove weld can be applied to all five of the basic weld joints. Typically, a single bevel is cut on the edge of one of the base metal pieces. The preparation may include a specified root opening along with the beveled cut, as shown in the butt weld arrangement in *Figure 17*. Although the root opening is small, it can be seen.

2.2.6 V-Groove Welds

V-groove welds are generally applied to inside or outside corner joints, butt joints, or edge joints. This is a very popular joint for both pipe and plate.

V-grooves and bevel grooves are more economical to prepare than other types of welds.

Figure 17 Workpieces for corner and butt bevel-groove welds.

Figure 18 shows a prepared V-groove on a butt joint to be welded. Typically, a 22½- or 30-degree bevel angle must be cut on both workpiece edges to prepare the base metals to receive the weld. In preparing inside or outside corner joints or butt joints for V-groove welds, the angle may not extend from the top of the bevel edge to the bottom of the bevel edge on each piece, but may be cut short of the bottom corner if a root face is required. In this edge preparation, the flat surface from the point where the angle stops to the bottom of the piece is referred to as the root face.

2.2.7 Single Versus Double V-Groove Welds

When possible, a double V-groove weld should be used in place of a single V-groove (*Figure 19*).

Although it may seem unusual, the double V-groove requires half the weld metal compared to the single V-groove. Welding from both sides also reduces distortion, because the forces of distortion work against each other.

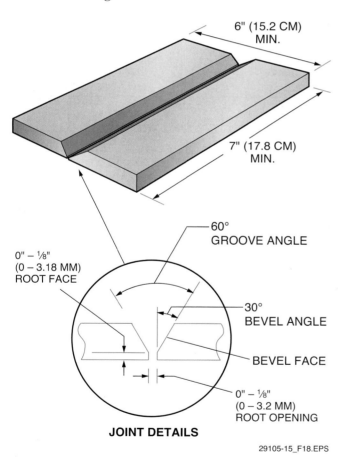

JOINT DETAILS

6" (15.2 CM) MIN.

7" (17.8 CM) MIN.

60° GROOVE ANGLE

0" – 1/8" (0 – 3.18 MM) ROOT FACE

30° BEVEL ANGLE

BEVEL FACE

0" – 1/8" (0 – 3.2 MM) ROOT OPENING

29105-15_F18.EPS

Figure 18 Open root V-groove weld.

2.2.8 J- and U-Groove Welds

J- and U-groove welds (*Figure 20*) are similar. The main difference is that a J-groove weld requires only one base metal to have its edge grooved in the form of a J. A U-groove is formed by preparing two matching J-grooves, one on each base metal edge. Even though both types require much less weld material than bevel or V-groove welds, they require much more preparation time because of the shape of the grooves. A J-groove weld can be applied to all five types of weld joints. The U-groove can only be used on butt, inside and outside corner, and edge joints.

2.2.9 Groove Angles and Root Openings

The purpose of the groove angle is to allow access to the root of the weld. The root preparation is sized to control melt-through.

Increasing or decreasing the root preparation (root opening and root face) will result in excess melt-through or insufficient root penetration. As a general rule, as the groove angle decreases, the root opening increases to compensate (*Figure 21*). Root faces are used with open-root joints, but not when metal backing strips are used.

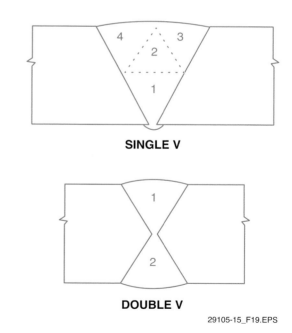

SINGLE V

DOUBLE V

29105-15_F19.EPS

Figure 19 Comparison of double and single V-grooves.

J-GROOVE

TWO MATING
J-GROOVES

U-GROOVE

29105-15_F20.EPS

Figure 20 J- and U-groove welds.

Root Face or Land

The root face is also widely known as the land. The American Welding Society (AWS) defines the land or part of the root face as an extension on the root face. The "table" formed on the extension is considered the land.

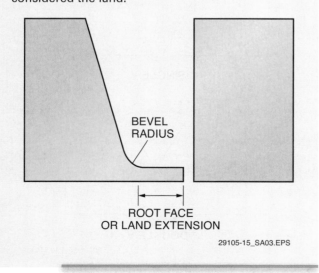

BEVEL
RADIUS

ROOT FACE
OR LAND EXTENSION

29105-15_SA03.EPS

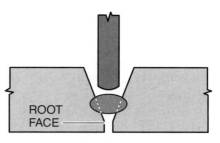

ROOT
FACE

**GROOVE ANGLE AND ROOT
OPENING TOO SMALL**

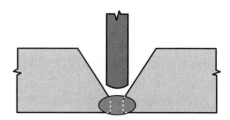

**CORRECT GROOVE ANGLE
AND ROOT OPENING**

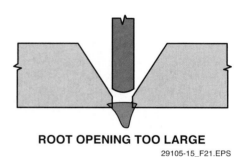

ROOT OPENING TOO LARGE

29105-15_F21.EPS

Figure 21 Groove angles and root openings.

2.2.10 Open-Root Welds

For open-root welds on plate, the groove angle should be 60 degrees, and the maximum size of the root opening and root face is ⅛" (3.2 mm). Remember that the 60-degree groove angle results from a 30-degree bevel angle on each mating workpiece. For open-root welds on pipe, the groove angle is 60 degrees or 75 degrees, depending on the code or specifications used. *Figure 22* shows groove angles for open root welds on plate and pipe.

2.2.11 Welds with Backing on Plate

Backing for plate can be strips made from the same material as the base metal, or flux-coated tape, fiberglass-coated tape, ceramic tape, or gas (*Figure 23*). The WPS will specify when to use backing and what type is to be applied.

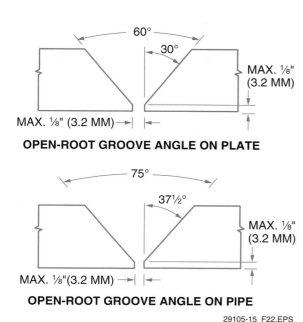

OPEN-ROOT GROOVE ANGLE ON PLATE

OPEN-ROOT GROOVE ANGLE ON PIPE

29105-15_F22.EPS

Figure 22 Open root joint preparation.

When gas is used as a backing, the same groove angles and root preparation used for open-root welds are used. When flux-coated tape, fiberglass-coated tape, or ceramic tape is used as a backing, the groove angle should be 60 degrees, the maximum root opening should be ³⁄₁₆" (4.8 mm), and the maximum root face should also be ³⁄₁₆" (4.8 mm). When applying tape, be sure to clean the surface to ensure that the tape adheres tightly, and make sure that the tape is centered on the weld joint.

When strips made from the same material as the base metal are used, a groove angle of 45 degrees, a root opening of ¼" (6.4 mm), and a feather edge (no root face) are used. The backing strip must be thick enough and wide enough to prevent burn-through at the root and to absorb and dissipate the heat of the root pass.

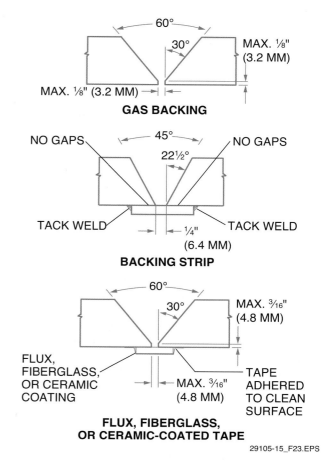

GAS BACKING

BACKING STRIP

FLUX, FIBERGLASS, OR CERAMIC-COATED TAPE

29105-15_F23.EPS

Figure 23 Plate joint preparation for backing.

> **NOTE**
>
> Check the WPS or site procedures for the backing size requirements for your site.

Backing Gas Hazards

Nitrogen and argon are used as backing and purge gases for some welding applications. These gases will displace all the oxygen in the space where they are used. At a refinery in Corpus Christi, Texas, while the welder was at lunch, a helper entered a large section of pipe in which argon had been used as a backing gas. When the welder returned, he found the helper lifeless in the pipe.

Nitrogen is lighter than air and, if allowed, will float upward. Argon, on the other hand, is heavier than air and will settle in low-lying areas. When removing nitrogen from a confined space, open the top to allow it to float out. When removing argon from a confined space, provide some way for it to flow out of the bottom. This is often impractical, however; forced ventilation is by far the most effective method.

Groove Angles

If the groove angle is larger than necessary, it will require additional weld metal to fill. This will increase the time and cost to complete the weld, and also increase distortion. If the groove angle is too small, it may result in weld defects.

The backing strip must be tack welded so that it will be held securely in place during welding. It must also be centered on the groove and be in close contact (no gaps) with the back of the plate being welded.

2.2.12 Welding Position

It is easier and faster to weld groove welds in the flat position and fillets in the flat or horizontal position. Always try to position the weldment so that welding is performed in the flat position. If a weldment is an assembly of parts, look for welds that can be made on the bench before tacking the assembly together. When welding pipe, weld as many fittings as possible before tacking the pipe in position, and try to leave final weld joints that will be the most accessible and the easiest to perform. When preparing weld joints, try to prepare the joint so that welding takes place from the most accessible position.

2.3.0 Codes and Welding Procedure Specifications

A welding code is a detailed listing of the rules and principles that apply to specific welded products. Codes ensure that safe and reliable welded products will be produced and that persons associated with the welding operation will be safe. Clients specify which welding codes should be followed when they place their orders or award project contracts.

Out-of-Position Welding

Out-of-position welding basically includes any welding position other than flat and horizontal. Welding out-of-position generally requires the use of smaller electrodes with more passes to fill the joint. Since it is more difficult, it also increases the likelihood of a weld defect.

If codes are mandated, a WPS must be written for each weld procedure. A WPS (*Figure 24*) is a written set of instructions for producing reliable welds.

The WPS includes the type of joint to be used, as well as the type of weld and any groove preparation that may be required for that particular weld. Each WPS is written and tested in accordance with a particular welding code or specification. All welding requires that acceptable industry standards be followed, but not all welds are governed by codes. The requirement for use of a WPS is often listed on blueprints as a note or in the tail of the welding symbol. If you are unsure whether or not the welding being performed requires a WPS, you should not proceed until you verify this information.

Note that the WPS shown in *Figure 24* uses metric dimensions. Electrodes and filler metals are basically the same size internationally, but they are called out using the metric equivalent of the size in most areas outside of the United States. For example, the 1.6 mm filler metal—wire in this case, for TIG welding—shown on the WPS is actually ⅟₁₆" wire; the 1.2 mm filler is ³⁄₆₄" filler metal. These are the rounded metric equivalents of their actual manufactured Imperial size.

Weld Procedure Number	30 P1 TIG 01 Issue A
Qualifying Welding Procedure (WPAR)	WP T17/A

Manufacturer: National Fabs Ltd 25 Lane End Birkenshaw Leeds Location: Workshop Welding Process: Manual TIG Joint Type: Single Sided Butt Weld	Method Of Preparation and Cleaning: Machine and Degrease Parent Metal Specification: Grade 304L Stainless Steel Parent Metal Thickness 3 to 8mm Wall Pipe Outside Diameter 25 to 100mm Welding Position: All Positions Welding Progression: Upwards

Joint Design

Welding Sequences

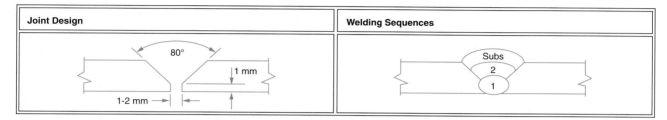

Run	Process	Size of Filler Metal	Current A	Voltage V	Type of Current/Polarity	Wire Feed Speed	Travel Speed	Heat Input
1	TIG	1.2 mm	70 - 90	N/A	DC–	N/A	N/A	N/A
2 And Subs	TIG	1.6 mm	80 - 140		DC–			

Welding Consumables:- Type, Designation Trade Name: Any Special Baking or Drying: Gas Flux: Gas Flow Rate - Shield: - Backing: Tungsten Electrode Type/ Size: Details of Back Gouging/Backing: Preheat Temperature: Interpass temperature: Post Weld Heat Treatment Time, temperature, method: Heating and Cooling Rates*:	BS 2901 Part 2 : 308S92 No Argon 99.99% Purity 8 - 12 LPM 5 LPM 2% Thoriated 2.4mm Dia Gas Backing 5°C Min 200°C Max Not Required	**Production Sequence** 1. Clean weld and 25mm borders to bright metal using approved solvent. 2. Position items to be welded ensuring good fit up and apply purge. 3. Tack weld parts together using TIG, tacks to at least 5mm min length. 4. Deposit root run using 1.2mm dia. wire. 5. Inspect root run internally. 6. Complete weld using 1.6mm dia wire using stringer beads as required. 7. 100% Visual inspection of completed weld

Revision History

Date	Issue	Changes	Authorization
26/11/2000	A	First changes	Jack Straw *Jack Straw*

29105-15_F24.EPS

Figure 24 Example of a WPS.

Additional Resources

2014 Technical Training Guide. Current Edition. The Lincoln Electric Company, Cleveland, OH. USA. **www.lincolnelectric.com**

Welding Handbook. Current Edition. Miami, FL: The American Welding Society.

2.0.0 Section Review

1. When a metal piece is pulled in opposite directions to determine its breaking point, it is a test of its _____ .

   a. tensile strength
   b. torsion strength
   c. shearing strength
   d. compression strength

2. To achieve proper penetration for each layer during multi-layer surface welding, _____ .

   a. the previous weld must be cleaned before each layer application
   b. no cleaning is required since the heat of welding will remove any impurities
   c. post-heating the base metal is needed before layer application
   d. every layer must first be gouged

3. The requirement for use of a WPS is often listed _____ .

   a. within the company's hand book
   b. on blueprints, notes, or on the tail of the welding symbol
   c. in the SDS/MSDS for the electrode
   d. on the base metal

SECTION THREE

3.0.0 WELDING JOINT PREPARATION

Objective

Describe how to prepare joints for welding.
  a. Describe how to mechanically prepare joints for welding.
  b. Describe how to thermally prepare joints for welding.

Performance Tasks

1. Mechanically or hand grind a bevel on the edge of a ¼"- to ¾"-thick mild steel plate (6 to 20 mm metric plate) at 22½ degrees.
2. Thermally bevel the edge of a ¼"- to ¾"-thick mild steel plate (6 to 20 mm metric plate) at 22½ degrees.

There are many different ways to prepare a joint for welding. The joint can be prepared mechanically using nibblers, grinders, or cutters. It can also be prepared thermally using an oxyfuel, plasma arc, or carbon arc cutting torch. The method used to prepare the joint will generally be determined by the base metal type, ease of use, and code or procedure specifications.

Before starting a joint preparation, determine if the joint is covered by a WPS or site quality standard. If this is the case, the joint type and welding parameters will be specified. These specifications must be followed; the guidance provided by a WPS is mandatory. If you are unsure whether or not a joint is covered by a WPS or site quality standard, check with your supervisor before proceeding.

3.1.0 Mechanical Joint Preparation

Mechanical joint preparation is most often used on alloy steel, stainless steel, and nonferrous metal piping or plate. It is slower than thermal methods such as oxyfuel, carbon arc, and plasma arc, but has the advantages of high precision, low heat input, and the absence of oxides commonly left by thermal methods.

3.1.1 *Grinders*

Handheld electric or air-operated grinders are used in welding shops, and even more often in the field, to cut and bevel pipe and plate to prepare them for welding. *Figure 25* shows a handheld angle grinder being used to prepare a stainless steel elbow for welding.

> **WARNING!**
> Cutoff and grinding wheels must be selected based on the material to be cut or beveled. Using an improper wheel may damage the wheel or workpiece, and can create hazardous conditions for the operator should the wheel shatter during operation. Refer to the manufacturer's recommendations and warnings when selecting grinding wheels. Ensure that the maximum revolutions per minute (rpm) rating of the grinding or cutoff wheel is equal to or greater than the speed of the grinder in use.

29105-15_F25.EPS

Figure 25 Preparing a weld joint.

3.1.2 Pipe Beveling Machines

Welded piping is found extensively in all industries. Nearly every piece of pipe that is welded to a fitting, to another piece of pipe, or to any other connection requires that the end be cut square and beveled according to specifications before welding. Welded pipe fittings are typically beveled from the factory and simply need to be cleaned of paint and other contaminants before welding. If a mechanical cutting and beveling process is used (*Figure 26*), the procedure may be done using an electric, hydraulic, or pneumatic beveling machine. Many of these are portable machines that operate much like a lathe, and they are often referred to as pipe-end prep lathes. They have mandrels that hold various cutting tools that cut metal away, as well as numerous adjusting mechanisms that make it easy to set the bevel angle and depth. Various models are available to cut and bevel both large and small pipe. Boiler tube ends are typically prepared using these smaller machines. When specifications call for boiler tubing to be welded, the machine can face, square, and chamfer the ends of the tubes to prepare them.

> **WARNING!**
> Always follow all manufacturers' safety procedures when using grinders and cutoff machines. Failure to follow the manufacturer's safety recommendations could result in serious personal injury.

Special mechanical cutoff machines (*Figure 27*) are also made to be mounted on the outside of the pipe. They can be mounted with a ring or a special chain with rollers. A cutoff blade or tool is used to make the cut.

3.1.3 Nibblers

Nibblers prepare the edge of a plate or pipe with a reciprocal punch that cuts or bites off a chip with each stroke. The bevel angle is set by adjusting the nibbler. A nibbler must have access to an edge on which it will be used (*Figure 28*).

> **WARNING!**
> Always follow all manufacturers' safety procedures when using nibblers and cutters. Failure to follow the manufacturer's safety recommendations could result in serious personal injury.

3.2.0 Thermal Joint Preparation

Thermal joint preparation involves preparing a joint with the oxyfuel or plasma arc processes. It can be done using the air-carbon arc process, but the result is typically poor. Because of carbon deposits and the inaccuracy of the result, the air-carbon arc cutting process is best used for gouging seams or cracks only.

29105-15_F26.EPS

Figure 26 Pipe beveling machine.

29105-15_F27.EPS

Figure 27 Pipe cutoff machine.

Thermal Joint Preparation Precautions

When preparing a joint with the oxyfuel, plasma arc, or air-carbon arc cutting processes, all dross must be removed prior to welding. Any dross remaining on the joint during welding will cause porosity in the weld. Joints prepared with a carbon arc cutting torch must be carefully inspected for carbon deposits. It is common for small carbon deposits to be left behind as the carbon electrode is consumed during the cutting or gouging operation. These carbon deposits will cause defects such as hard spots, loss of ductility, and cracking in the weld. Before welding, use a grinder to clean surfaces prepared with most thermal processes.

The torch used for oxyfuel or plasma arc cutting can be handheld or mounted on a motorized or manual carriage. These are often referred to as track burners. The motorized carriage for plate cutting runs on flat tracks positioned on the surface of the plate to be cut (*Figure 29*). The carriage has an On or Off switch, forward or reverse switches, and a speed adjustment. The torch position can be adjusted as necessary for the proper bevel angle. The carriage can carry a special oxyfuel or plasma arc cutting torch designed to fit into the torch holder.

Other equipment designs are used for cutting pipe (*Figure 30*). A steel ring or special chain with rollers is attached to the outside of the pipe.

29105-15_F28.EPS

Figure 28 Nibbler.

29105-15_F29.EPS

Figure 29 Track burner thermally cutting and beveling plate.

The torch is carried around the pipe on a torch holder that is powered by electricity, air, or a hand crank. The bevel angle is set by pivoting the torch holder, and the torch is adjusted vertically with a handwheel. An out-of-round attachment can be used to compensate for pipe that is out-of-round.

For very large-diameter pipe, special internal cutting and beveling equipment is available (*Figure 31*). The motorized unit shown here uses rare-earth magnets in the wheels, developing 300 pounds (136 kg) of force. No rings, bands, or chains are required. It can also be used for smaller pipe by traveling on the outside surface, and for long cuts on plate.

Automatic Pipe Bevelers

Automatic versions of pipe bevelers and cutters are an improvement over pattern cutters and other similar equipment. This is because there is no need to reset the preheat flame before each cut. Once an automatic pipe beveler has been initially set up, all subsequent cuts can be made with the same settings. These systems also save time and cutting gases, because when the system is switched off, all gases are immediately shut off as well.

GEAR-DRIVEN UNIT (HANDCRANK)

CHAIN-DRIVEN UNIT (HANDCRANK)

29105-15_F30.EPS

Figure 30 Pipe cutting and beveling equipment.

29105-15_F31.EPS

Figure 31 Thermal cutting from inside the pipe.

Additional Resources

2014 Technical Training Guide. Current Edition. The Lincoln Electric Company, Cleveland, OH 44117-1199 USA. **www.lincolnelectric.com**

Welding Handbook. Current Edition. Miami, FL: The American Welding Society.

3.0.0 Section Review

1. Which of the following is an advantage of mechanical joint preparation over the thermal method?

   a. It is more precise with less heat input.
   b. It is faster and less labor intensive.
   c. Mechanically prepared joints require less weld metal.
   d. Mechanically prepared joints require fewer weld passes.

2. Which of the following is a thermal process seldom used due to the deposits left behind which can lead to weld defects?

   a. Air-carbon arc cutting
   b. Oxyfuel cutting
   c. Plasma arc cutting
   d. Pipe-end prep lathe

SUMMARY

The importance of safety cannot be overemphasized within the welding environment. It should be the highest priority at all times.

Performing chemical or mechanical cleaning removes base metal surface contamination. To create reliable weld joints that meet specifications, proper cleaning techniques must always be used.

A great deal of load stress may be placed on a weld joint. Understanding the different types of stress helps engineers to determine what type of welding joint and technique is best for the application. The specifications for a given weld joint are communicated to the welder through the welding procedure specification (WPS).

Weld joints can be prepared mechanically and thermally. Mechanical methods do have advantages, but they can also be labor-intensive. Thermal methods include the use of oxyfuel cutting processes, with the torch mounted on a carriage that allows it to rotate smoothly around the pipe to make a consistent cut and/or bevel.

1. Never directly or indirectly view an electric arc without _____.

   a. having your eyes examined
   b. wearing sunglasses
   c. being at least 10 feet away
   d. using a properly tinted lens

2. Medium-carbon steel contains _____.

   a. 0.10 percent to 0.15 percent carbon
   b. 0.15 percent to 0.30 percent carbon
   c. 0.30 percent to 0.50 percent carbon
   d. 0.50 percent to 1.00 percent carbon

3. A type of copper-alloyed steel that protects itself from damaging oxidation by creating a hard, tough layer of oxidation on its surface is called _____.

   a. spring steel
   b. weathering steel
   c. mild steel
   d. stainless steel

4. A pneumatic tool that uses a number of blunt steel needles to remove surface oxidation is called a _____.

   a. needle scaler
   b. weld flux chipper
   c. needle nose chipper
   d. contamination tattoo scaler

5. Torsion applied to a weld joint refers to a force _____.

   a. pulling in opposite directions
   b. twisting the metal in opposite directions
   c. pushing against a stationary surface
   d. attempting to create a bend

6. The five basic types of joints are butt, lap, corner, T-joint, and _____.

   a. chamfer
   b. square
   c. bevel
   d. edge

7. Lap joints, T-joints, or corner joints can be welded using _____.

   a. surface welds
   b. butt welds
   c. fillet welds
   d. spot welds

8. U-groove welds are formed by preparing two matching J-grooves that require _____.

   a. less preparation time but use more weld material than a V-groove weld
   b. more preparation time but use less weld material than a V-groove weld
   c. more preparation time and use more weld material than a V-groove weld
   d. no preparation time and use the same weld material as a V-groove weld

9. Which of the following materials is *not* a common joint backing material?

   a. The same base metal type
   b. A gas
   c. Ceramic tape
   d. Cast iron

10. When a backing strip of the same material as the base metal is used, the root face should be _____.

    a. a feather edge
    b. ¹⁄₁₆" (1.6 mm)
    c. ⅛" (3.2 mm)
    d. ¼" (6.4 mm)

11. Grinding wheels selected for use must be rated equal to or greater than the grinder's _____.

    a. current
    b. horsepower
    c. rotational speed
    d. air pressure

12. Before a pipe can be butt-welded to a fitting or other connection, its edge must be _____.

    a. burnished
    b. hardened
    c. ground to fit
    d. cut square and beveled

13. Beveling equipment made for pipe is sometimes called _____.

    a. J-prep cutters
    b. ID-prep cutter
    c. pipe-end prep lathes
    d. compound-bevel prep lathes

14. Which of the following processes produces the least desirable result, and is therefore *not* a popular choice for general metal cutting?

    a. Plasma arc cutting
    b. Band saw
    c. Oxyfuel cutting
    d. Air-carbon arc cutting

15. The best approach to gouging seams or cracks is _____.

    a. using a nibbler
    b. carbide-chisel cutting
    c. grinder cutting
    d. air-carbon arc cutting

Trade Terms Quiz

Fill in the blank with the correct term that you learned from your study of this module.

1. An assembly that is fastened together by weld joints is referred to as a(n) _____.

2. A weldable or non-weldable material used behind a root opening to allow defect-free welding at the open root of a joint is a(n) _____.

3. The document containing all the detailed methods and practices required to produce a sound weld is a(n) _____.

4. The primary metal to be welded, cut, or brazed is referred to as the _____.

5. The steel alloy that, under specific conditions, is designed to form a very dense oxide layer on its outer surfaces to retard further oxidation is called _____.

6. The expansion and contraction of welded parts caused by the heating and subsequent cooling of the weld joint is _____.

7. The space between the base metal pieces at the bottom or root of a joint is called a(n) _____.

8. A steel that contains less than 0.10 percent carbon and is magnetic is called _____.

9. The amount of force applied to a material or structure is called a(n) _____.

10. A small flattened area on the end of a bevel for a groove weld is called a(n) _____.

11. Complete joint penetration results in _____.

12. Gas pockets, or voids in the weld metal are called _____.

13. A metal, such as aluminum or brass, lacking sufficient quantities of iron to have any effect on its properties is called a(n) _____.

14. A form of porosity having a length greater than its width and is approximately perpendicular to the weld face is known as a(n) _____.

15. The scale that forms on metal surfaces when they are exposed to the oxygen or air containing oxygen is called a(n) _____.

16. A steel that shares some characteristics with ferritic, but boasts higher levels of carbon, up to a full 1 percent is called _____.

17. To impart strength or toughness to steel or cast iron by heating and cooling is referred to as _____.

18. When a metal is heated and cooled to free it from internal stress, it is said to be _____.

19. Any article that has been removed from a mold is a(n) ____ .

20. A metal that contains austenite (a solid solution of carbon, or of carbon and other elements, in a ferrous alloy), added through heating is called _____.

21. When a metal is _____, it is able to undergo a change of form without breaking.

22. To _____ a piece of metal, you will need a bucket of water.

23. A metal that has been shaped by beating it with a hammer, as is common with iron, is referred to as being _____.

24. Only _____ metals contain any substantial iron.

25. The measure of the ability of a material to withstand a longitudinal stress, without breaking is considered the material's _____.

26. The application by welding, brazing, or thermal spraying of a layer of material to a surface to obtain desired properties or dimensions is _____.

27. Capable of being extended or shaped by hammering or by pressure from rollers, this material is considered to be _____.

Trade Terms

Annealed 18
Austenitic 20
Backing 2
Base metal 4
Castings 19
Distortion 6
Ductile 21

Ferritic 8
Ferrous 24
Load 9
Malleable 27
Martensitic 16
Melt-through 11
Nonferrous metal 13

Oxide 15
Piping porosity 14
Porosity 12
Quench 22
Root face 10
Root opening 7
Surfacing 26

Tempering 17
Tensile strength 25
Weathering steel 5
Welding procedure specification (WPS) 3
Weldment 1
Wrought 23

Appendix

PERFORMANCE ACCREDITATION TASKS

The American Welding Society (AWS) School Excelling through National Skills Standards Education (SENSE) program is a comprehensive set of minimum Standards and Guidelines for Welding Education programs. The following performance accreditation is aligned with and designed around the SENSE program.

The Performance Accreditation Tasks (PATs) correspond to and support the learning objectives in *AWS EG2.0, Guide for the Training and Qualification of Welding Personnel: Entry-Level Welder.*

Note that in order to satisfy all learning objectives in *AWS EG2.0*, the instructor must also use the PATs contained in the second level of the NCCER Welding curriculum.

PAT 1 corresponds to no *AWS EG2.0* reference.

PAT 2 corresponds to *AWS EG2.0, Module 8 – Thermal Cutting Processes, Unit 2 – Manual OFC (e.g., Track Burner)*, Key Indicators 4, 5, and 6.

PATs provide specific acceptable criteria for performance and help to ensure a true competency-based welding program for students.

The following tasks are designed to develop your competency in preparing base metal. Practice each task until you are thoroughly familiar with the procedure.

As you complete each task, take it to your instructor for evaluation. Do not proceed to the next task until instructed to do so by your instructor.

PREPARE PLATE JOINTS MECHANICALLY

Using a nibbler, cutter, or grinder, mechanically or manually prepare the edge of a ¼"- to ¾"-thick carbon steel plate (≈6 to 20 mm metric plate) with a bevel angle of 22½ or 30 degrees, at the discretion of the instructor.

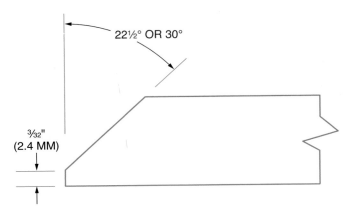

22½° OR 30°

³⁄₃₂"
(2.4 MM)

NOTE: BASE METAL = CARBON STEEL PLATE

29105-15_A01.EPS

Criteria for Acceptance:

• Bevel angle ±2½"   _____

• Bevel face smooth and uniform to ¹⁄₁₆" (1.6 mm)   _____

• Root face ±¹⁄₃₂" (±0.8 mm)   _____

PREPARE PLATE JOINTS THERMALLY

Using oxyfuel or plasma arc cutting equipment, thermally prepare the edge of a ¼"- to ¾"-thick carbon steel plate (≈6 to 20 mm metric plate) with a bevel angle of 22½ or 30 degrees, at the discretion of the instructor.

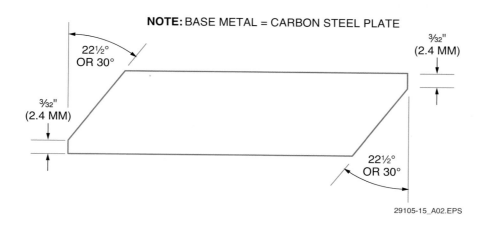

NOTE: BASE METAL = CARBON STEEL PLATE

22½° OR 30°

3/32" (2.4 MM)

3/32" (2.4 MM)

22½° OR 30°

29105-15_A02.EPS

Criteria for Acceptance:

- Bevel angle ±2½" _____

- No dross _____

- Minimal notching not exceeding 1/16" (1.6 mm) deep on the kerf face _____

- Root face ±1/32" (±0.8 mm) _____

Frank Johnson
PBF Energy Toledo Refinery Inc.
Welding Trainer and CWE Weld Procedure Writer

How did you choose a career in the construction industry?
I took welding courses at Penta County Vocational High School. Upon graduation, I passed the state pipe welding test, an accomplishment which was rare for a man my age. I worked as a welder for a year before joining the military. Though I didn't work as a welder during this time, I received increased pay in the military because of my success at passing the pipe test.

Who inspired you to enter the industry?
The reason for choosing this career is directly related to who inspired me to enter the industry. My grandfather owned a weld shop and I began welding when I was a young boy of just 13. My grandfather pushed me to read and learn as much about welding as possible so that I could solve problems as they arose. I was also inspired by my hobby of steam tractors because I had to weld them whenever they were in need of repair.

How important are education and training in construction?
I think training is absolutely essential in order to be a good welder. It is very important to know the codes, variables, and demands of the industry and to be able to perform any type of welding in any type of situation. I feel that the NCCER curriculum is important because I use it to train welders. The NCCER standardized curriculum, developed by expert welders, ensures that employees are trained properly.

What kinds of work have you done in your career?
After my discharge from the service, I started and completed a boilermaking apprenticeship. Six months into my apprenticeship, I was recognized as a skilled welder, which enabled me to be paid as a journeyman. I worked at various refineries and power plants, including the Davis Bessie Nuclear Plant. At that time, the pipefitters at the company needed welders. As a result, I welded at Davis Bessie for about a year, then worked for 2½ years as a pipefitter. I earned journeyman pipefitter and boilermaker journey level cards. Next, I worked at Enrico Fermi Nuclear Power Plant in Monroe, Michigan. While there, I was asked by Detroit Edison to work with their welding engineer because he qualified 95 percent of their welding procedures. I am now a certified welder, journey-level boilermaker, and pipefitter. I have created and qualified 17 welding procedures.

Tell us about your present job.
In my capacity as a trainer, I test and qualify welders, write welding procedures, qualify welding procedures, and serve as a certified welding instructor. The work is challenging and I really enjoy it.

What factors have contributed most to your success?
Factors that have contributed the most to my success are my natural curiosity, as well as my experience working in a nuclear plant where I admired the excellent craftsmanship of seasoned welders. Their professionalism gave me the desire to achieve their level of expertise. In that environment, quality standards are extremely high, so only top-notch welders will succeed there.

Would you suggest construction as a career to others? Why?

I would suggest construction as a career because if you are a hard worker you can attain a well-paying, secure job. I believe that a career in construction is full of challenges that craftworkers can overcome. They can also experience the satisfaction of seeing a project through, which yields them the satisfaction of putting a project into operation and seeing it work as it should. When recommending a career in construction, I've said that it is "Good work, challenging work, and satisfying work that will support you with a good wage."

What advice would you give to those new to the field?

My advice to those new to the field is: "Don't be afraid to try. Even if you don't succeed the first time, you will make it eventually, and you will be successful."

Tell us an interesting career-related fact or accomplishment.

I serve as an advisor to the Historical Boiler Board for the State of Ohio and on the National Board for Historical Boilers, an honor bestowed upon me due to my successful welding career. I was asked to serve as the subject matter expert on the committee that shapes the Welding curriculum for NCCER. Looking back, I never thought I'd be in a position like this. Welding has taken me a long way.

How do you define craftsmanship?

A craftsman is one who takes pride in his or her work.

Trade Terms Introduced in This Module

Annealed: To free from internal stress by heating and gradually cooling.

Austenitic: Containing austenite (a solid solution of carbon or of carbon and other elements in a ferrous alloy), added through heating.

Backing: A weldable or nonweldable material used behind a root opening to allow defect-free welding at the open root of a joint

Base metal: Metal to be welded, cut, or brazed.

Castings: Something cast; any article that has been cast in a mold.

Distortion: The expansion and contraction of welded parts caused by the heating and subsequent cooling of the weld joint.

Ductile: Able to undergo change of form without breaking.

Ferritic: Steel containing less than 0.10 percent carbon and is magnetic. This steel can't be hardened via heat treatment.

Ferrous: Containing iron.

Load : The amount of force applied to a material or a structure.

Malleable: Capable of being extended or shaped by hammering or by pressure from rollers.

Martensitic: Steel that shares some characteristics with ferritic, but has a higher levels of carbon, up to a full 1 percent. It can be tempered and hardened and is used where strength is more important than a resistance to oxidation.

Melt-through: Complete joint penetration.

Nonferrous metal: A metal, such as aluminum, copper, or brass, lacking sufficient quantities of iron to have any effect on its properties.

Oxide: The scale that forms on metal surfaces when they are exposed to oxygen or air containing oxygen.

Piping porosity: A form of porosity having a length greater than its width and that is approximately perpendicular to the weld face.

Porosity: Gas pockets, or voids in the weld metal.

Quench: To cool suddenly by plunging into a liquid.

Root face: A small flattened area on the end of a bevel for a groove weld.

Root opening: The space between the base metal pieces at the bottom or root of the joint.

Surfacing: The application by welding, brazing, or thermal spraying of a layer of material to a surface to obtain desired properties or dimensions.

Tempering: To impart strength or toughness to (steel or cast iron) by heating and cooling.

Tensile strength: The measure of the ability of a material to withstand a longitudinal stress without breaking.

Weathering steel: Steel alloy that, under specific conditions, is designed to form a very dense oxide layer on its outer surfaces, which retards further oxidation.

Welding procedure specification (WPS): The document containing all the detailed methods and practices required to produce a sound weld.

Weldment: An assembly that is fastened together by welded joints.

Wrought: Produced or shaped by beating with a hammer, as with iron.

Additional Resources

This module presents thorough resources for task training. The following resource material is suggested for further study.

2014 Technical Training Guide. Current Edition. The Lincoln Electric Company, Cleveland, OH. USA. **www.lincolnelectric.com**

American Iron and Steel Institute, 25 Massachusetts Avenue, NW Suite 800, Washington, DC 20001, USA. **www.steel.org**

Association for Iron & Steel Technology, 186 Thorn Hill Road Warrendale, PA, USA. **www.aist.org**

Steel Metallurgy for the Non-Metallurgist. John D. Verhoeven; Prepared under the direction of the ASM International Technical Book Committee. ASM International® Materials Park, OH 44073-0002. **www.asminternational.org**

Welding Handbook. Current Edition. Miami, FL: The American Welding Society.

Figure Credits

Topaz Publications, Inc., Module Opener, Figures 4, 6, 8A, 8C–8H, 9, 11 (photos), 12, 13, 17, 20, 25, 29, SA02

The Lincoln Electric Company, Cleveland, OH, USA, Figures 1–3

Reproduced by permission TWI Ltd., Figure 5

Courtesy of Milwaukee Electric Tool Corporation, Figures 7A, 7B

Courtesy of Chicago Pneumatic Tool Co., Figures 7C, 7D

Courtesy of Dremel, Figure 8B

Courtesy of Tri Tool Inc., Figures 26, 27

TRUMPF Inc., Figure 28

Courtesy of Mathey Dearman, Figures 30, 31

AWS F2.2:2001, Lens Shade Selector, reproduced with permission from the American Welding Society (AWS), Miami, FL, USA, Table 1

Photo Courtesy of U.S. Army, SA01

Section Review Answer Key

Answer	Section Reference	Objective
Section One		
1. c	1.1.3	1a
2. b	1.2.1	1b
3. b	1.3.1	1c
Section Two		
1. a	2.1.0	2a
2. a	2.2.1	2b
3. b	2.3.0	2c
Section Three		
1. a	3.1.0	3a
2. a	3.2.0	3b

NCCER CURRICULA — USER UPDATE

NCCER makes every effort to keep its textbooks up-to-date and free of technical errors. We appreciate your help in this process. If you find an error, a typographical mistake, or an inaccuracy in NCCER's curricula, please fill out this form (or a photocopy), or complete the online form at **www.nccer.org/olf**. Be sure to include the exact module ID number, page number, a detailed description, and your recommended correction. Your input will be brought to the attention of the Authoring Team. Thank you for your assistance.

Instructors – If you have an idea for improving this textbook, or have found that additional materials were necessary to teach this module effectively, please let us know so that we may present your suggestions to the Authoring Team.

NCCER Product Development and Revision

13614 Progress Blvd., Alachua, FL 32615

Email: curriculum@nccer.org
Online: www.nccer.org/olf

❏ Trainee Guide   ❏ Lesson Plans   ❏ Exam   ❏ PowerPoints   Other _____

Craft / Level: _____   Copyright Date: _____

Module ID Number / Title: _____

Section Number(s): _____

Description: _____

_____

_____

_____

Recommended Correction: _____

_____

_____

_____

Your Name: _____

Address: _____

_____

Email: _____   Phone: _____

29106-15

Weld Quality

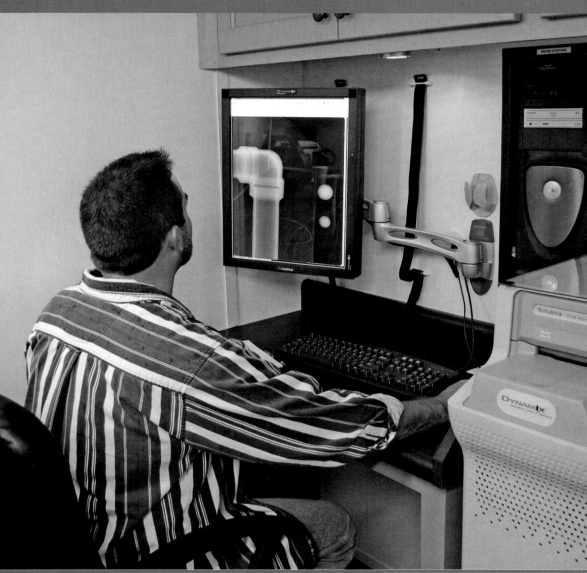

OVERVIEW

The ability to make high-quality welds that will consistently pass inspection is a necessary skill for all welders. An important part of achieving this level of skill is the ability to recognize weld imperfections, as well as their causes.

Module Six

Trainees with successful module completions may be eligible for credentialing through the NCCER Registry. To learn more, go to **www.nccer.org** or contact us at **1.888.622.3720**. Our website has information on the latest product releases and training, as well as online versions of our *Cornerstone* magazine and Pearson's product catalog.

Your feedback is welcome. You may email your comments to **curriculum@nccer.org**, send general comments and inquiries to **info@nccer.org**, or fill in the User Update form at the back of this module.

This information is general in nature and intended for training purposes only. Actual performance of activities described in this manual requires compliance with all applicable operating, service, maintenance, and safety procedures under the direction of qualified personnel. References in this manual to patented or proprietary devices do not constitute a recommendation of their use.

Objectives

When you have completed this module, you will be able to do the following:

1. Identify and describe the various code organizations that apply to welding and their basic elements.
   a. Identify the various welding code organizations and their sponsoring organizations.
   b. Identify and describe the basic provisions of welding codes.
2. Identify and describe weld discontinuities and their causes.
   a. Identify and describe discontinuities related to porosity and inclusions.
   b. Identify and describe discontinuities that result in cracking.
   c. Identify and describe discontinuities related to joint penetration, fusion, and undercutting.
   d. Identify and describe acceptable and unacceptable weld profiles.
3. Describe various nondestructive and destructive weld examination practices.
   a. Describe basic visual inspection methods including measuring devices and liquid penetrants.
   b. Describe magnetic particle and electromagnetic inspection processes.
   c. Describe the radiographic and ultrasonic inspection processes.
   d. Describe destructive testing processes.
4. Describe the welder performance testing process.
   a. Describe the qualification of welders by position.
   b. Describe welder qualification testing to meet American Welding Society (AWS) and American Society of Mechanical Engineers (ASME) requirements.
   c. Describe the process for completing a weld test.

Performance Task

Under the supervision of your instructor, you should be able to do the following:

1. Perform a visual inspection (VT) on a fillet and/or groove weld and complete an inspection report.

Trade Terms

Arc blow	Homogeneity	Specification
Brazing	Inclusion	Standard
Code	Laminations	Supplemental essential variables
Defect	Non-essential variables	Transducer
Discontinuity	Notch toughness	Underbead cracking
Embrittled	Procedure qualification record (PQR)	Welding procedure qualification
Essential variables		
Hardenable materials	Radiographic	

Industry Recognized Credentials

If you are training through an NCCER-accredited sponsor, you may be eligible for credentials from NCCER's Registry. The ID number for this module is 29106-15. Note that this module may have been used in other NCCER curricula and may apply to other level completions. Contact NCCER's Registry at 888.622.3720 or go to **www.nccer.org** for more information.

Contents

Topics to be presented in this module include:

Figures and Tables

Figures and Tables (Continued)

1.0.0 WELDING CODES AND PROVISIONS

Objective

Identify and describe the various code organizations that apply to welding and their basic elements.

   a. Identify the various welding code organizations and their sponsoring organizations.

   b. Identify and describe the basic provisions of welding codes.

Trade Terms

Brazing: A method of joining metal using heat and a filler metal with a melting point above 842°F (450°C). Unlike welding, the base metal is not melted during the brazing process.

Code: A document that establishes the minimum requirements for a product or process. Codes can be, and often are, adopted as laws.

Essential variable: Items in a welding procedure specification (WPS) that cannot be changed without requalifying the WPS.

Non-essential variable: Items in a welding procedure specification (WPS) that can be changed without requalifying the WPS.

Notch toughness: The ability of a material to absorb energy in the presence of a flaw such as a notch.

Procedure qualification record (PQR): The document containing the results of the nondestructive and destructive testing required to qualify a welding procedure specification (WPS).

Specification: A document that defines in detail the work to be performed or the materials to be used in a product or process.

Standard: A document that defines how a code is to be implemented.

Supplemental essential variable:: The variables that must be considered when notch toughness requirements are invoked.

Welding procedure qualification:: A demonstration through testing that welds made following a specific process can meet prescribed standards.

Welding criteria have been established in codes and standards produced by a number of organizations. In addition, each country may have its own welding standards. Welding codes govern welding activities, qualification requirements, and tests that can be performed on weldments to identify imperfections in welds. Standards define how the code requirements are to be achieved. Welders must be familiar with the codes and standards that apply to their work.

1.1.0 Codes and Standards Governing Welding

Welding work is governed by codes, standards, and specifications. A welding code is a set of requirements covering permissible materials, service limitations, fabrication, inspection, testing procedures, and qualification of welders. Welding codes ensure that safe and reliable welded products are produced and that persons associated with the welding operation are safe. Codes are often adopted into law. A standard is a document that defines how a code is to be implemented. Standards are developed by organizations that bring together professionals from all areas of an industry to create standards for that industry. Specifications are detailed instructions for producing a product or performing a particular task. One important difference to note is that standards represent the cooperative effort of people from all parts of an industry, and are published and maintained by organizations established for that purpose. Specifications, on the other hand, can be produced by an individual company or organization without industry consensus and can apply to a single project.

Clients specify in the contract the codes, standards, and specifications that will be used on the project. Since there are a number of codes and each is updated periodically, welders must know which welding codes and code year apply to the project on which they are working. In addition to codes and standards, there are generally specifications that apply to each job. A specification is a document that defines in detail how the work is to be performed, but it is not intended to replace codes and standards.

Codes and standards that apply to welding safety and quality have been developed and are published by a number of nationally recognized agencies. To eliminate the necessity of writing a code for each new job, sections of these existing codes are referenced by the project contract. Agencies and societies that have established widely used welding codes and standards in the United States include the following:

- American Society of Mechanical Engineers (ASME)
- American Welding Society (AWS)
- American Petroleum Institute (API)
- American National Standards Institute (ANSI)

It is beyond the scope of this module to present all the details of the codes and standards organizations mentioned here. However, some of the most important will be discussed along with their impact on the industry. There are also a number of other organizations that have developed codes and standards used internationally. Some of those organizations will also be introduced in this section.

1.1.1 American Society of Mechanical Engineers

The American Society of Mechanical Engineers publishes two codes that welders must be aware of: the *ASME Boiler and Pressure Vessel Code and ASME B31, Code for Pressure Piping*. Both of these codes are endorsed by the American National Standards Institute.

The *ASME Boiler and Pressure Vessel Code (BPVC)* contains eleven sections. The sections most frequently referenced by welders are as follows:

- *Section II, Material Specifications* – This section contains the specifications for acceptable ferrous (Part A) and nonferrous (Part B) base metals and for acceptable welding and brazing filler metals and fluxes (Part C). Many of these specifications are identical to and have the same number designation as AWS specifications for welding consumables. This section is used to match base metals and filler metals.

- *Section V, Nondestructive Examination* – This section covers the methods and standards for nondestructive examination of boilers and pressure vessels.
- *Section IX, Welding and Brazing Qualifications* – This section covers the qualification of welders, welding operators, brazers, and brazing operators. It also covers the welding and brazing procedures that must be used for welding or brazing boilers or pressure vessels. This section of the code is often cited in other codes and standards as the welding qualification standard.

ASME B31, Code for Pressure Piping, consists of eight sections. Each section gives the minimum requirements for the design, materials, fabrication, erection, testing, and inspection of a particular type of piping system. In particular, *B31.1, Power Piping*, covers power and auxiliary service systems for electric generation stations. *B31.3, Process Piping*, covers chemical plant and petroleum refining piping. Other sections include the following:

- *B31.2, Fuel Gas Piping*
- *B31.4, Pipeline Transportation Systems for Liquid Hydrocarbons and Other Liquids*
- *B31.5 Refrigeration Piping and Heat Transfer Components*
- *B31.8 Gas Transmission and Distribution Piping Systems*
- *B31.9 Building Services Piping*
- *B31.11 Slurry Transportation Piping Systems*

All sections of *ASME B31, Code for Pressure Piping*, require qualification of the welding procedures and testing of welders and welding operators. Some sections require these qualifications to be performed in accordance with Section IX of the *ASME Boiler and Pressure Vessel Code*, while in others it is optional.

The Importance of Welding Inspection

It is important that you learn to recognize the various types of weld discontinuities and identify their causes so that you can be the first line of defense against problem welds.

Welding is used to join metals in many critical applications, including buildings, bridges, pipelines, motor vehicles, and heavy machinery. Failure of a weld in any of these applications could mean loss of life or major property damage. For this reason, welding inspectors will frequently check your work to make sure your welds meet the standards established for the project. The greater the potential hazard or cost risk, the more frequent and intense the inspections will be.

Inspection Standards

Among the many documents and standards published by the American Welding Society are documents covering weld inspection. These include the following:

- *AWS B1.10, Guide for the Nondestructive Inspection of Welds*
- *AWS B1.11, Guide for the Visual Inspection of Welds*
- *AWS B5.1, Specification for the Qualification of Welding Inspectors*
- *AWS Welding Inspection Technology*

1.1.2 American Welding Society

The American Welding Society publishes numerous welding-related documents. These documents include codes, standards, specifications, recommended practices, and guides. *AWS D1.1, Structural Welding Code – Steel,* is the code most frequently referenced. It covers welding and qualification requirements for welded structures of carbon and low-alloy steels. It is not intended to apply to pressure vessels, pressure piping, or base metals less than ⅛" (<3.2 mm metric plate) thick. Other commonly used AWS welding codes include the following:

- *AWS D1.2 Structural Welding Code – Aluminum*
- *AWS D1.3 Structural Welding Code – Sheet Steel*
- *AWS D1.6 Structural Welding Code – Stainless Steel*
- *AWS D1.5 Bridge Welding Code*

1.1.3 American Petroleum Institute

The American Petroleum Institute (API) publishes numerous documents in all areas related to petroleum production. *API 1104, Standard for Welding of Pipelines and Related Facilities,* applies to arc and oxyfuel gas welding of piping, pumping, transmission, and distribution systems for petroleum. It presents methods for making acceptable welds by qualified welders using approved welding procedures, materials, and equipment. It also presents methods for ensuring proper analysis of weld quality.

1.1.4 American National Standards Institute

The American National Standards Institute (ANSI) is a private organization that does not actually prepare standards. Instead, it adopts standards that it feels are of value to the public interest. ANSI standards deal with dimensions; ratings; terminology and symbols; test methods; and performance, as well as safety specifications for materials, equipment, components, and products in many fields, including construction. Many codes used today have been adopted as ANSI standards.

1.1.5 Other Standards Organizations

A number of national and international organizations publish standards and specifications governing welding. These organizations include the following:

- *International Standards Organization (ISO)* – This is an independent, non-governmental standards organization with 163 member countries. ISO identifies itself as the world's largest developer of voluntary standards, including nearly 300 welding standards.
- *ASTM International* – Formerly known as the American Society for Testing and Materials, ASTM International develops standards aimed at improving product quality and safety and facilitating market access and trade.
- *American Society for Nondestructive Testing (ASNT)* – ASNT is an organization for professionals involved in nondestructive testing. ASNT provides a forum for the exchange of NDT-related technical information; provides educational materials and programs; and publishes standards related to NDT. The key ASNT standards are as follows:
  - *ANSI/ASNT CP-189, ASNT Standard for Qualification and Certification of Nondestructive Testing Personnel*
  - *ANSI/ASNT CP-105, ASNT Standard Topical Outlines for Qualification of Nondestructive Testing Personnel*

Many countries produce their own standards. Examples include the GOST in Russia; the Canadian Welding Bureau; the European Union (CEN), and many others.

1.1.6 Maritime Welding Guides and Specifications

The AWS website lists the guides and specifications associated with marine welds and provides the following summary statements for each code:

- *AWS D3.5-93R, Guide for Steel Hull Welding* – This guide provides information to users in the marine construction industry about the best and most practical methods to weld steel hulls for ships, barges, mobile offshore drilling units, and other marine vessels. It provides information on steel plates, shapes, castings, and forging, as well as their selection and weldability. It discusses welding processes and proper design for welding. Hull construction is presented in terms of preparation of materials, erection and fitting, and control of distortion. Qualification of procedures and personnel are outlined, and inspection methods are discussed. A common shipyard problem, stray current protection, is discussed as is the health and safety of the workforce. Supplementary non-mandatory appendices are provided for informational purposes.
- *AWS D3.6M, Specification for Underwater Welding* – This specification covers the requirements for welding structures or components under the surface of the water. It includes welding in both dry and wet environments. Sections 1 through 6 constitute the general requirements for underwater welding, while Sections 7 through 10 contain the special requirements applicable to four individual classes of weld: Class A – Comparable to above-water welding; Class B – For less critical applications; Class C – Where load bearing is not a primary consideration; and Class O – To meet the requirements of another designated code or specification.

- *AWS D3.7, Guide for Aluminum Hull Welding* – This guide provides information on the welding of seagoing aluminum hulls and other structures in marine construction. Included are sections on hull materials, construction preparation, welding equipment and processes, qualification requirements, welding techniques, and safety precautions.

Anyone involved in marine welding must be knowledgeable of these welding codes. When in doubt, find a current copy of the applicable code and review it. In addition, welders working on US Navy shipbuilding projects must recognize that the Navy has their own welding specifications, which can differ significantly from those used in commercial work. The main point to remember is to always ensure that the work is being done to the correct codes, standards, and specifications.

1.1.7 Quality Workmanship

The codes and standards discussed in this module were written to ensure that welders consistently make quality welds. Many weldments will be examined and tested. However, due to time and cost, it is not feasible to examine every one, so inspection is done on a sampling basis. Regardless, quality workmanship is expected in every weld a welder makes.

The welder should be able to work with site representatives to ensure that quality work is achieved. To do this, the site organizational structure needs to be understood. If quality problems arise, the welder needs to follow the appropriate chain of command to eliminate any problems. However, there may be instances when the welder should bypass the chain of command. Examples of such instances include the following:

- You have been directed to perform an unsafe act. If you cannot resolve the matter with your immediate supervisor, it is your responsibility to go to the general foreman, superintendent, project manager, or safety engineer.
- You have been directed to perform a weld that requires a specific certification, and you are not certified to perform it. If you cannot resolve the matter with your immediate supervisor, it is your responsibility to go to the general foreman, superintendent, project manager, or quality engineer.

1.2.0 Basic Provisions of Welding Codes

A major function of welding codes is to establish qualification for operators to perform a particular welding operation. All welding codes provide detailed information about qualification in the following general areas:

- Welding procedure qualification
- Welder performance qualification
- Welding operator qualification

Machine welding is covered in some codes but is not common to all codes. Each type of qualification mentioned is different and is subject to different requirements.

> **NOTE**
>
> The information in this module is provided as a general guideline only. Check with your supervisor if you are unsure of the codes and specification requirements for your project.

1.2.1 Welding Procedure Qualification

A welding procedure is a written document that contains materials, methods, processes, electrode types, techniques, and all other necessary and relevant information about the weldment. Welding procedures must be qualified before they can be used. Welding procedure qualification has nothing to do with the skills of the individual welder, but deals only with the process itself.

Welding procedure qualifications are limiting instructions written to explain how a welding operation will be done. These limiting instructions are listed in a document known as a welding procedure specification (WPS). *Figure 1* shows a sample WPS. The purpose of the WPS is to define and document in detail the variables related to project-specific welds. The WPS lists the following information in detail:

- Base metals to be joined by welding
- Filler metal to be used
- Range of preheat and postheat treatment
- Thickness and other variables for each welding process

WPS variables are identified either as essential or nonessential variables. Essential variables are items in the welding procedure specification that cannot be changed without requalifying the welding procedure. Supplemental essential variables are those variables that must be considered when notch toughness requirements are invoked.

Essential variables vary by code or welding process. Refer to the specific code or welding process for the relevant essential variables. The following are some of the essential variables in a welding procedure.

- Filler metal classification
- Material thickness
- Joint design
- Type of base metal
- Welding process
- Current type
- Pre- and post-heat treatment

Nonessential variables are items in the WPS that may be changed within a range identified by the code, but that do not affect the qualification status.

Examples of nonessential variables that may be changed without having to requalify the welding procedure include the following:

- Amperage
- Travel speed
- Shielding gas flow (if applicable)
- Electrode and filler wire size
- Rod travel angle

> **CAUTION**
>
> Do not change any essential or supplemental essential variable without discussing it with your supervisor.

The WPS is qualified for use by welding test coupons as the WPS instructs, and then by testing the coupons in accordance with the applicable code. A test weld is made and test coupons are cut from it. The test coupons are used to make tensile tests, root bends, and face bends as required by the code. The test results are then recorded on a document known as a procedure qualification record (PQR). This documents the tested specimen, testing method, and the results. If the weldment produced by the WPS-guided procedure meets the code requirements, the procedure becomes qualified. Under most codes, each WPS must have a matching PQR to document the quality of the weld produced. *Figure 2* shows a sample PQR.

The methods used to qualify procedures are more detailed and thorough than those used to qualify either welders or welding machine operators. This is because procedures must qualify physical and metallurgical properties.

WELDING PROCEDURE SPECIFICATION (WPS)   Yes ☐
PREQUALIFIED _____   QUALIFIED BY TESTING X
or PROCEDURE QUALIFICATION RECORDS (PQR)   Yes ☐

Identification # **PQR 231** _____
Revision **1** _____   Date **12-1-07**   By **W. Lye**
Authorized by **J. Jones**   Date **1-18-08**
Type—Manual ☐   Semi-Automatic ☐
Machine ☐   Automatic ☐

Company Name **Red Inc.** _____
Welding Process(es) **FCAW** _____
Supporting PQR No.(s) **-** _____

JOINT DESIGN USED
Type: **Butt**
Single **X**   Double Weld ☐
Backing: Yes **X**   No ☐
    Backing Material: **ASTM A131A**
Root Opening **1/4"**   Root Face Dimension **-**
Groove Angle: **35°**   Radius (J–U) **-**
Back Gouging: Yes ☐   No **X**   Method **-**

BASE METALS
Material Spec. **ASTM A131**
Type or (Grade) **A**
Thickness: Groove **1"**   Fillet **-**
Diameter (Pipe) **-**

FILLER METALS
AWS Specification **A5.20**
AWS Classification **E71T-1**

SHIELDING
Flux **-**   Gas **CO_2**
    Composition **-**
Electrode-Flux (Class) _____   Flow Rate **30-40** cfh
    Gas Cup Size **5/8" or 3/4"**

PREHEAT
Preheat Temp., Min **75° Ambient**
Interpass Temp., Min **75°F**   Max **350°F**

POSITION
Position of Groove: **4**G   Fillet: **-**
Vertical Progression: Up ☐   Down ☐

ELECTRICAL CHARACTERISTICS
Transfer Mode (GMAW)   Short-Circuiting ☐
    Globular **X**   Spray ☐
Current: AC ☐   DCEP **X**   DCEN ☐   Pulsed ☐
Other _____
Tungsten Electrode (GTAW)
    Size: _____
    Type: _____

TECHNIQUE
Stringer or Weave Bead: **Stringer**
Multi-pass or Single Pass (per side) **Multipass**
Number of Electrodes **1**
Electrode Spacing   Longitudinal **-**
    Lateral **-**
    Angle **-**

Contact Tube to Work Distance **3/4-1"**
Peening **None**
Interpass Cleaning: **Wire Brush**

POSTWELD HEAT TREATMENT
Temp. **N.A.** _____
Time _____

WELDING PROCEDURE

Pass or Weld Layer(s)	Process	Filler Metals		Current		Volts	Travel Speed	Joint Details
		Class	Diam.	Type & Polarity	Amps or Wire Feed Speed			
1	FCAW	E71T-1	.045"	DC+	180	26	8	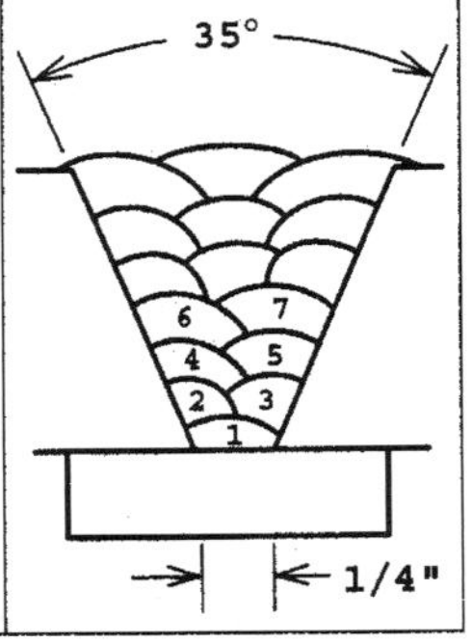
2-8	"	"	"	"	200	27	10	
9-11	"	"	"	"	200	27	11	
12-15	"	"	"	"	200	27	9	
16	"	"	"	"	200	27	11	

Form E-1 (Front)

29106-15_F01.EPS

Figure 1 Example of a welding procedure specification (WPS).

Procedure Qualification Record (PQR) # ___231___
Test Results

TENSILE TEST

Specimen No.	Width	Thickness	Area	Ultimate Tensile Load, lb	Ultimate Unit Stress, psi	Character of Failure and Location
231-1	.75"	1.00"	.75"	52 500	70 000	Ductile
231-3	.75"	1.00"	.75"	52 275	69 700	Ductile

GUIDED BEND TEST

Specimen No.	Type of Bend	Result	Remarks
231-2	Side	Pass	
231-4	Side	Pass	Small (< 1/16") opening acceptable
231-6	Side	Pass	
231-5	Side	Pass	

VISUAL INSPECTION
Appearance _____ acceptable _____
Undercut _____ acceptable _____
Piping porosity _____ none _____
Convexity _____ none _____
Test date _____ 12-3-2002 _____
Witnessed by _____ D. Davis _____

Radiographic-ultrasonic examination
RT report no.: __D231__ Result __passed__
UT report no.: _____ Result _____
FILLET WELD TEST RESULTS
Minimum size multiple pass   Maximum size single pass
Macroetch   Macroetch
1. _____ 3. _____   1. _____ 3. _____
2. _____   2. _____

Other Tests

All-weld-metal tension test

Tensile strength, psi _____ 83,100 _____
Yield point/strength, psi _____ 72,600 _____
Elongation in 2 in., % _____ 28 _____
Laboratory test no. _____ PW 231 _____

Welder's name _____ W. T. Williams _____   Clock no. _____ 261 _____ Stamp no. _____

Tests conducted by _____ RED Inc. & ABC Testing _____ Laboratory

Test number _____ PQR 231 _____

Per _____ D. Miller _____

We, the undersigned, certify that the statements in this record are correct and that the test welds were prepared, welded, and tested in conformance with the requirements of Section 4 of AWS D1.1/D1.1M, _2002_____) Structural Welding Code—Steel.
                                                        (year)

Signed _____ RED Inc. _____
                Manufacturer or Contractor
By _____ R. M. Boncrack _____

Title _____ Q.C. Mgr. _____

Date _____ 12-15-2002 _____

29106-15_F02.EPS

Figure 2 Example of a procedure qualification record (PQR).

1.2.2 Welder Performance Qualification

Once a procedure has been qualified, the welder using it must be qualified to use that procedure by passing a welding performance qualification (WPQ) test. Because no single performance test can qualify welders for all the different types of welding that must be done, a welder may be required to pass a number of performance qualification tests. Performance tests used to qualify welders are covered later in this module.

1.2.3 Welding Machine Operator Qualification

When automatic welding equipment is used, the operators of the equipment must demonstrate their ability to set up and monitor the equipment so that it will produce acceptable welds. The codes also contain qualification tests for these operators.

Additional Resources

2014 Technical Training Guide. Current Edition. Cleveland, OH, USA: The Lincoln Electric Company. **www.lincolnelectric.com**

1.0.0 Section Review

1. Which of the following documents is often adopted into law?

   a. Specifications
   b. Codes
   c. Standards
   d. WPSs

2. A WPQ test is used to qualify a welder to perform a specific procedure.

   a. True
   b. False

Section Two

2.0.0 Weld Discontinuities and Their Causes

Objective

Identify and describe weld discontinuities and their causes.
  a. Identify and describe discontinuities related to porosity and inclusions.
  b. Identify and describe discontinuities that result in cracking.
  c. Identify and describe discontinuities related to joint penetration, fusion, and undercutting.
  d. Identify and describe acceptable and unacceptable weld profiles.

Trade Terms

Arc blow: The deflection of an arc from its intended path by magnetic forces.

Defect: A discontinuity or imperfection that renders a part of the product or the entire product unable to meet minimum acceptable standards or specifications.

Discontinuity: A change or break in the shape or structure of a part that may or may not be considered a defect, depending on the code.

Embrittled: Metal that has been made brittle and that will tend to crack with little bending.

Hardenable materials: Metals that have the ability to be made harder by heating and then cooling.

Homogeneity: The quality or state of having a uniform structure or composition throughout.

Inclusion: Foreign matter introduced into and remaining in a weld.

Underbead cracking: Cracking in the base metal near the weld, but under the surface.

Codes and standards define the quality requirements necessary to achieve the integrity and reliability of a weldment. These quality requirements help ensure that welded joints are capable of serving their intended function for the expected life of the weldment. Weld discontinuities can prevent a weld from meeting the minimum quality requirements.

AWS defines a discontinuity as an interruption of the typical structure of a weldment, such as a lack of homogeneity in the mechanical, metallurgical, or physical characteristics of the material or weldment. A discontinuity is not necessarily a defect. A defect found during inspection will require the weld to be rejected. A single excessive discontinuity or a combination of discontinuities can make the weldment defective (unable to meet minimum quality requirements). However, a weld can have one or more discontinuities and still be acceptable.

The welder should be able to identify discontinuities and understand the effect they have on weld integrity. Some can be determined from visual inspection. Those that are internal to the weldment can only be detected through other testing methods. The most common weld discontinuities are the following:

- Porosity
- Inclusions
- Cracks
- Incomplete joint penetration
- Incomplete fusion
- Undercuts
- Arc strikes
- Spatter
- Unacceptable weld profiles

Ideally, a weld should not have any discontinuities, but most will have one or more. When evaluating a weld, it is important to note the type, size, and location of the discontinuity. Any one of these factors, or all three, can change a discontinuity into a defect, requiring the weld to be rejected during the inspection process. For example, discontinuities located at stress points tend to expand and thus have higher risk than those in other locations. Surface or near-surface discontinuities may be more harmful than similarly shaped internal discontinuities.

2.1.0 Porosity and Inclusions

Porosity and inclusions are weld defects that generally result from improper welding technique. Welders must recognize the causes of these defects and the techniques required to avoid them.

2.1.1 Porosity

Porosity is the presence of voids or empty spots in the weld metal. It is the result of gas pockets being trapped in the weld as it is being made. As the molten metal hardens, the gas pockets form voids.

Unless the gas pockets work up to the surface of the weld and burst open before the metal hardens, porosity cannot be detected through visual inspection.

Porosity can be grouped into the following major types:

- *Linear porosity* – May be aligned along a weld interface, the root of a weld, or a boundary between weld beads (*Figure 3[A]*).
- *Uniformly scattered porosity* – May be located throughout single-pass welds or throughout several passes in multiple-pass welds (*Figure 3[B]*).
- *Clustered porosity* – A localized grouping of pores that result from improperly starting or stopping the welding.
- *Piping porosity* – Normally extends from the root of the weld toward the face. These elongated gas pores are also called wormholes. They do not often extend to the surface, and the porosity cannot be visually detected (*Figure 3[C]*).

(A) LINEAR POROSITY

(B) SCATTERED SURFACE POROSITY

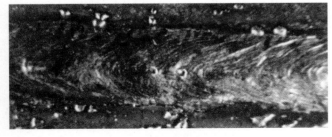

(C) PIPING POROSITY

29106-15_F03.EPS

Figure 3 Examples of porosity.

Most porosity is caused by improper welding technique or contamination. Improper welding techniques may cause an inadequate amount of shielding gas to be formed. As a result, parts of the weld site are left unprotected. Oxygen in the air at the weld site, or moisture in the flux or on the base metal that dissolves in the weld pool, can become trapped and produce porosity.

The intense heat of the weld can decompose paint, dirt, oil, or other contaminants, producing hydrogen. This gas can become trapped in the solidifying weld pool and produce porosity.

Excessive porosity has a serious effect on the mechanical properties of the joint. Although some codes permit a certain amount of porosity in welds, it is best to have as little as possible. This can be accomplished by properly cleaning the base metal, avoiding excessive moisture in the electrode covering, and using proper welding techniques. Any porosity that occurs must be ground out until it is removed.

2.1.2 Inclusions

Inclusions are foreign matter trapped in the weld metal (*Figure 4*), between weld beads, or between the weld metal and the base metal. Inclusions are sometimes jagged and irregularly shaped. Sometimes they form in a continuous line. This concentrates stresses in one area and reduces the structural integrity (strength) of the weld.

Inclusions generally result from faulty welding techniques, improper access to the joint for welding, or both. A typical example of an inclusion is slag, which normally forms over a deposited weld. If the electrode is not manipulated correctly, the force of the arc will cause some of the slag particles to be blown into the molten pool. If the pool solidifies before the inclusions can float to the top, they become lodged in the metal, producing a discontinuity. Sharp notches in joint boundaries or between weld passes also can result in slag entrapment.

Inclusions are more likely to occur in out-of-position welding because the tendency is to keep the molten pool small and allow it to solidify rapidly to prevent it from sagging. Tungsten inclusions may occur in GTAW.

With proper welding technique, along with use of the correct electrode and the proper setting, inclusions can be avoided or kept to a minimum. Other preventive measures include the following:

- Positioning the work to maintain slag control
- Changing the electrode to improve control of molten slag

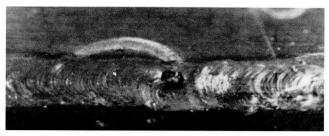

SURFACE SLAG INCLUSIONS

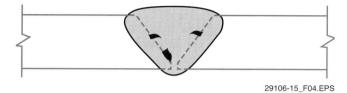

29106-15_F04.EPS

Figure 4 Examples of nonmetallic inclusions.

- Thoroughly removing slag between weld passes
- Grinding or sanding the weld surface if it is rough and likely to entrap slag
- Removing heavy mill scale or rust on weld preparations
- Avoiding the use of electrodes with damaged coverings

2.2.0 Cracks

Cracks are narrow breaks that occur in the weld metal, in the base metal, or in the crater formed at the end of a weld bead (*Figure 5*). Cracks occur when localized stresses exceed the ultimate strength of the metal. Cracks are generally located near other weld or base metal discontinuities.

2.2.1 Weld Metal Cracks

Three basic types of cracks can occur in weld metal: transverse, longitudinal, and crater. As seen in *Figure 5*, weld metal cracks are named to correspond with their location and direction.

Transverse cracks run across the face of the weld and may extend into the base metal. They are more common in joints that have a high degree of restraint.

Longitudinal cracks are usually located in the center of the weld deposit. They may be the continuation of crater cracks or cracks in the first layer of welding. Cracking of the first pass is likely to occur if the bead is thin. If this cracking is not eliminated before the other layers are deposited, the crack will progress through the entire weld deposit.

Crater cracks have a tendency to form in the crater whenever the welding operation is interrupted. These cracks usually proceed to the edge of the crater and may be the starting point for longitudinal weld cracks. Crater cracks can be minimized or prevented by filling craters to a slightly convex shape prior to breaking the welding arc.

Figure 6 shows various kinds of weld metal cracks. Weld metal cracking can usually be reduced by taking one or more of the following actions:

- Improve the contour or composition of the weld deposit by changing the electrode manipulation or electrical conditions
- Increase the thickness of the deposit and provide more weld metal to resist the stresses by decreasing the travel speed
- Reduce thermal stress by preheating
- Use low-hydrogen electrodes
- Balance shrinkage stress by sequencing welds
- Avoid rapid cooling conditions

2.2.2 Base Metal Cracks

Base metal cracking usually occurs within the heat-affected zone of the metal being welded. The possibility of cracking increases when working with **hardenable materials**. These cracks usually occur along the edges of the weld and through the heat-affected zone into the base metal. Types of base metal cracking include **underbead cracking** and toe cracking.

Underbead cracks are limited mainly to steel. They are usually found at regular intervals under the weld metal and usually do not extend to the surface. Because of this, they cannot be detected by visual inspection.

Hot and Cold Cracks

Hot cracks occur while the weld is solidifying. They can be caused by insufficient ductility at high temperature. Cold cracks occur after the weld has solidified. They are often caused by improper welding techniques.

Toe cracks are generally the result of strains caused by thermal shrinkage acting on a heat-affected zone that has been embrittled. They sometimes occur when the base metal cannot handle the shrinkage strains that are imposed by welding.

Base metal cracking can usually be reduced or eliminated by one of the following methods:

- Controlling the cooling rate by preheating
- Controlling heat input
- Using the correct electrode
- Controlling welding materials
- Properly matching the electrode filler metals to the base metals being welded

2.3.0 Other Discontinuities

In addition to those just described, there are several other types of discontinuities. Welders must learn to recognize these discontinuities, understand their causes, and learn the techniques required to avoid them.

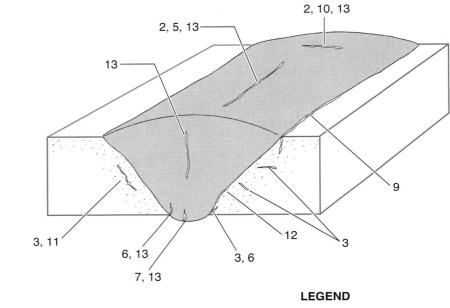

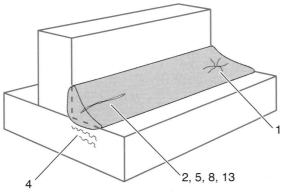

LEGEND

1. CRATER CRACK
2. FACE CRACK
3. HEAT-AFFECTED ZONE CRACK
4. LAMELLAR TEAR
5. LONGITUDINAL CRACK
6. ROOT CRACK
7. ROOT SURFACE CRACK
8. THROAT CRACK
9. TOE CRACK
10. TRANSVERSE CRACK
11. UNDERBEAD CRACK
12. WELD INTERFACE CRACK
13. WELD METAL CRACK

29106-15_F05.EPS

Figure 5 Types of weld metal and base metal cracks.

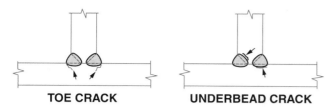

TOE CRACK   UNDERBEAD CRACK

TOE CRACK

LONGITUDINAL CRACK AND LINEAR POROSITY

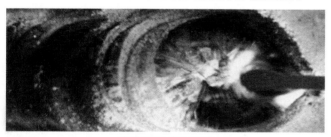

CRATER CRACK

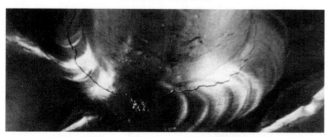

LONGITUDINAL CRACK OUT OF CRATER CRACK

FILLET WELD THROAT CRACK

29106-15_F06.EPS

Figure 6 Examples of weld metal cracks.

2.3.1 *Incomplete Joint Penetration*

Incomplete joint penetration (*Figure 7*) occurs when the filler metal fails to penetrate and fuse with an area of the weld joint. Incomplete penetration will cause weld failure if the weld is subjected to tension or bending stresses.

Insufficient heat at the root of the joint is a frequent cause of incomplete joint penetration. If the metal being joined first reaches the melting point at the surfaces above the root of the joint, molten metal may bridge the gap between these surfaces and screen off the heat source before the metal at the root melts.

Improper joint design is another leading cause of incomplete joint penetration. If the joint is not prepared or fitted accurately, an excessively thick root face or an insufficient root gap may cause incomplete penetration. Incomplete joint penetration is likely to occur under the following conditions:

- If the root face dimension is too big, even though the root opening is adequate
- If the root opening is too small
- If the included angle of a V-groove is too small

Figure 8 is a labeled diagram showing correct and incorrect joint designs.

Even if the welding heat is correct and the joint design is adequate, incomplete penetration can result from poor control of the welding arc. Examples of poor control include the following:

- Using an electrode that is too large
- Excessive travel speed
- Using a welding current that is too low

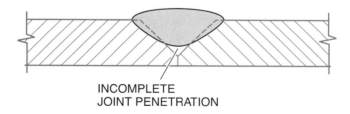

INCOMPLETE
JOINT PENETRATION

|◄——— INCOMPLETE ———►|
JOINT PENETRATION

29106-15_F07.EPS

Figure 7 Incomplete joint penetration.

Incomplete penetration is always undesirable in welds, especially in single-groove welds where the root of the weld is subject either to tension or bending stresses. It can lead directly to weld failure or can cause a crack to start at the unfused area.

2.3.2 Incomplete Fusion

Many welders confuse incomplete joint penetration with incomplete fusion. It is possible to have good penetration without complete root fusion. Incomplete fusion is the failure of a welding process to fuse, or join together, layers of weld metal or weld metal and base metal.

Incomplete fusion may occur at any point in a groove or fillet weld, including the root of the weld. Often, the weld metal simply rolls over onto the plate surface. This is generally referred to as overlap. In many cases, the weld has good fusion at the root and at the plate surface, but because of poor technique and insufficient heat, the toe of the weld does not fuse. *Figure 9* shows incomplete fusion and overlap. Causes for incomplete fusion include the following:

- Insufficient heat as a result of low welding current, high travel speeds, or an arc gap that is too close
- Wrong size or type of electrode
- Failure to remove oxides or slag from groove faces or previously deposited beads
- Improper joint design
- Inadequate gas shielding
- Improper electrode angle
- **Arc blow**

Incomplete fusion discontinuities affect weld joint integrity in much the same way as porosity and slag inclusion.

	CORRECT	INCORRECT
DOUBLE V-GROOVE		ROOT FACE NOT CORRECT
DOUBLE BEVEL GROOVE		ROOT OPENING TOO BIG
U-GROOVE		UNEVEN ROOT FACE PREPARATION
V-GROOVE		MISALIGNED EDGES
BEVEL GROOVE		BEVEL ANGLE TOO SMALL
J-GROOVE		ROOT FACE TOO WIDE

29106-15 _F08.EPS

Figure 8 Correct and incorrect joint designs.

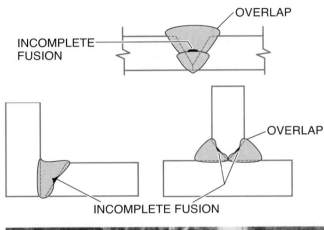
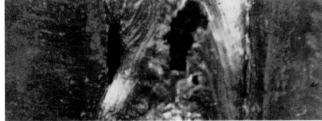
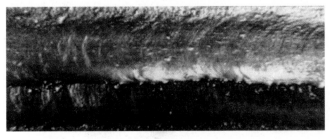

INCOMPLETE FUSION AT WELD FACE

INCOMPLETE FUSION BETWEEN INDIVIDUAL WELD BEADS

29106-15_F09.EPS

Figure 9 Incomplete fusion and overlap.

2.3.3 Undercut

Undercut is a groove melted into the base metal beside the weld. It is the result of the arc removing more metal from the joint face than is replaced by weld metal. On multilayer welds, it may also occur at the point where a layer meets the wall of a groove (*Figure 10*).

Undercutting is usually caused by improper electrode manipulation. Other causes of undercutting include the following:

- Using a current adjustment that is too high
- Having an arc gap that is too long
- Failing to fill up the crater completely with weld metal
- Improper electrode angle
- Incorrect rod travel speed

Most welds have some undercut that can be found upon careful examination. When it is controlled within the limits of the specifications and does not create a sharp or deep notch, undercut is usually not considered a weld defect. However, when it exceeds the limits, undercutting can be a serious defect because it reduces the strength of the joint.

Causes of Incomplete Joint Penetration

Incomplete joint penetration is generally associated with groove welds. It may result from insufficient welding heat, improper joint design (too much metal for the welding arc to penetrate), or poor control of the welding arc.

2.3.4 Arc Strikes

Arc strikes (*Figure 11*) are small, localized points where surface melting occurs away from the joint. These spots may be caused by accidentally striking the arc in the weld zone or by faulty ground connections. The weld zone is the area where the weld bead should be, and/or where the weld metal is to be deposited.

Striking an arc on base metal that will not be fused into the weld metal should be avoided. Arc strikes can cause hardness zones in the base metal and can become the starting point for cracking. Arc strikes can cause a weld to be rejected.

2.3.5 Spatter

Spatter is made up of very fine particles of metal on the plate surface adjoining the weld area. It is usually caused by high current; a long arc; an irregular and unstable arc; or improper shielding. Spatter makes a poor appearance on the weld and base metal and can make it difficult to inspect the weld. Spatter can also cause coating failure.

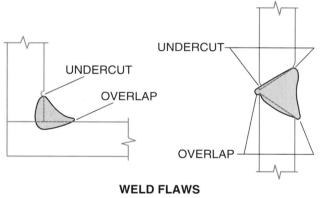

WELD FLAWS

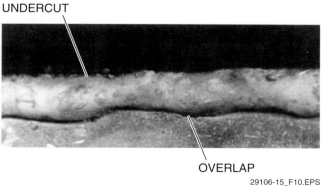

29106-15_F10.EPS

Figure 10 Undercut and overlap.

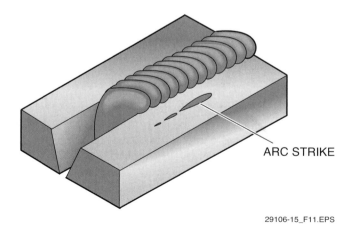

29106-15_F11.EPS

Figure 11 Arc strike.

2.4.0 Acceptable and Unacceptable Weld Profiles

The profile of a finished weld can affect the performance of the joint under load as much as other discontinuities affect it. This applies to the profile of a single-pass weld and to a layer of a multiple-pass weld. An unacceptable profile for a single-pass or multiple-pass weld could lead to the formation of discontinuities such as incomplete fusion or slag inclusions as the other layers are deposited. *Figure 12* shows acceptable and unacceptable weld profiles for both fillet and groove welds.

2.4.1 Fillet Welds

A fillet weld is a weld that is approximately triangular in cross section and is used with T-, lap, and corner joints. The sizes and locations of fillet welds are given as welding symbols. The two types of fillet welds are convex and concave (*Figure 13*). A convex fillet weld has its surface bowed out like the outside surface of a ball. A concave fillet weld has its surface bowed in like the inside surface of a bowl.

The following terms are used to describe a fillet weld:

- *Weld face* – The exposed surface of the weld
- *Leg* – The distance from the root of the joint to the toe of a fillet weld
- *Weld toe* – The junction between the face of a weld and the base metal
- *Weld root* – The point shown in cross section at which the weld metal intersects with the base metal and extends farthest into the weld joint
- *Size* – The leg lengths of the largest right triangle that can be drawn within the cross section of a fillet weld
- *Actual throat* – The shortest distance from the root of the weld to its face
- *Effective throat* – The minimum distance, minus any convexity, from the root of the weld to its face
- *Theoretical throat* – The distance from the beginning of the joint root (with a zero opening) that is perpendicular to the hypotenuse of the largest right triangle that can be inscribed within the cross section of a fillet weld.

Fillet welds may be either equal leg or unequal leg (*Figure 14*). The face may be slightly convex, flat, or slightly concave. Welding codes require that fillet welds have a uniform concave or convex face, although a slightly non-uniform face is acceptable. The maximum convexity of a fillet weld or individual surface bead is dependent on the width of the weld face or individual surface bead, as follows:

- If the weld face or individual surface bead is ≤ 5/16" (8 mm), the maximum convexity is 1/16" (1.6 mm).
- If the weld face or individual surface bead is > 5/16" (8 mm) and < 1" (25 mm), the maximum convexity is 1/8" (3 mm).
- If the weld face or individual surface bead is ≥ 1", the maximum convexity is 3/16" (5 mm).

A fillet weld is unacceptable if the profile has any of the following discontinuities:

- Insufficient throat
- Excessive convexity
- Excessive undercut
- Overlap
- Insufficient leg
- Incomplete fusion

Fillet welds require little base metal preparation except for cleaning the weld area and removing all dross from cut surfaces. Any dross from oxyfuel, plasma arc, or carbon arc cutting will cause porosity in the weld. For this reason, the codes require that all dross be removed prior to welding.

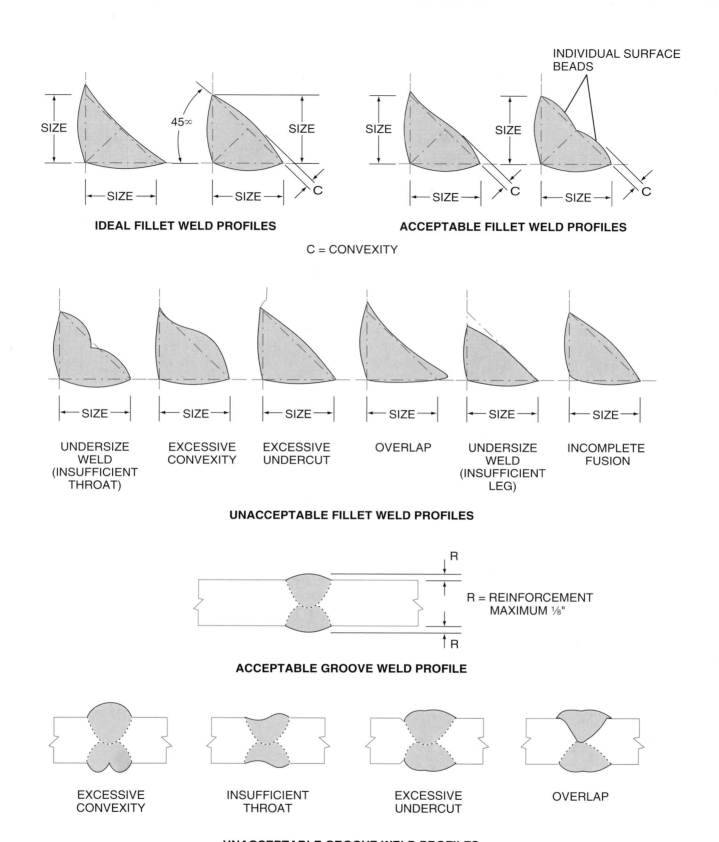

IDEAL FILLET WELD PROFILES

ACCEPTABLE FILLET WELD PROFILES

INDIVIDUAL SURFACE BEADS

45∞

C = CONVEXITY

UNACCEPTABLE FILLET WELD PROFILES

UNDERSIZE WELD (INSUFFICIENT THROAT)

EXCESSIVE CONVEXITY

EXCESSIVE UNDERCUT

OVERLAP

UNDERSIZE WELD (INSUFFICIENT LEG)

INCOMPLETE FUSION

R = REINFORCEMENT MAXIMUM ⅛"

ACCEPTABLE GROOVE WELD PROFILE

EXCESSIVE CONVEXITY

INSUFFICIENT THROAT

EXCESSIVE UNDERCUT

OVERLAP

UNACCEPTABLE GROOVE WELD PROFILES

29106-15_F12.EPS

Figure 12 Acceptable and unacceptable weld profiles.

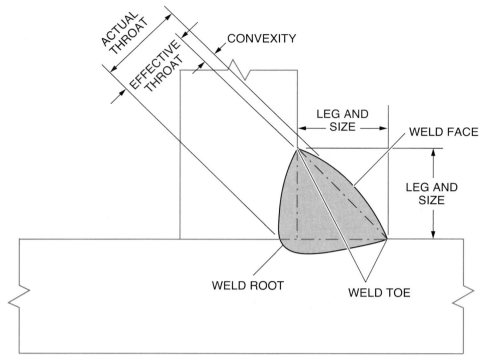

CONVEX FILLET WELD

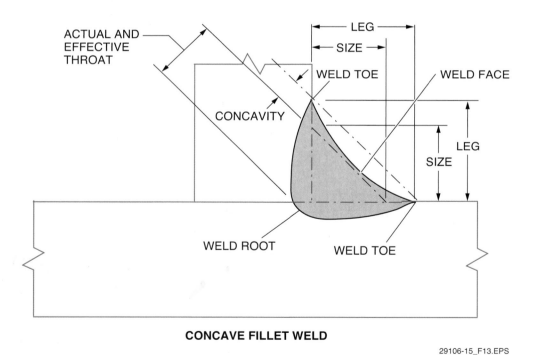

CONCAVE FILLET WELD

29106-15_F13.EPS

Figure 13 Convex and concave fillet welds.

2.4.2 Groove Welds

Groove welds should be made with reinforcement not exceeding ⅛" (3.2 mm) and a gradual transition to the base metal at each toe. Groove welds should not have excess reinforcement, insufficient throat, excessive undercut, or overlap. If a groove weld has any of these defects, it is considered unacceptable. The bead width should not exceed the groove width by more than ⅛" (3.2 mm).

> **NOTE**
>
> Refer to your site's WPS for specific requirements on fillet and butt welds. The information provided here is only a general guideline. The site WPS or quality specifications must be followed for all welds. Check with your supervisor if you are unsure of the specifications for your application.

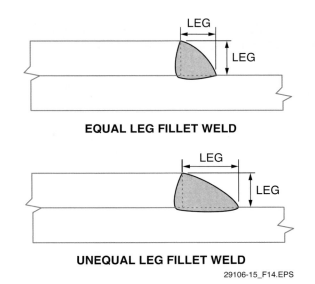

EQUAL LEG FILLET WELD

UNEQUAL LEG FILLET WELD

29106-15_F14.EPS

Figure 14 Equal leg and unequal leg fillet welds.

Additional Resources

Visual Inspection Workshop Reference Manual, AWS VIW-M-2008. Miami, FL: American Welding Society.

Welding Inspection Technology. Miami, FL: American Welding Society.

2.0.0 Section Review

1. The entrapment of gas pockets in a weld will cause _____.

   a. inclusions
   b. cracks
   c. porosity
   d. undercut

2. Which of these discontinuities *cannot* be detected by visual inspection?

   a. Toe cracks
   b. Underbead cracks
   c. Transverse cracks
   d. Face cracks

3. Undercut is usually caused by improper electrode manipulation.

   a. True
   b. False

4. Which of the following describes the leg of a weld?

   a. The shortest distance from the root of the weld to its face
   b. The junction between the face of the weld and the base metal
   c. The minimum distance, minus any convexity, from the root of the weld to its face
   d. The distance from the root of the joint to the toe of a fillet weld

3.0.0 WELD EXAMINATION PRACTICES

Objective

Describe various nondestructive and destructive weld examination practices.

  a. Describe basic visual inspection methods including measuring devices and liquid penetrants.

  b. Describe magnetic particle and electromagnetic inspection processes.

  c. Describe the radiographic and ultrasonic inspection processes.

  d. Describe destructive testing processes.

Performance Task

  1. Perform a visual inspection (VT) on a fillet and/or groove weld and complete an inspection report.

Trade Terms

Laminations: Cracks in the base metal formed when layers separate.

Radiographic: Describes images made by passing X-rays or gamma rays through an object and recording the variations in density on photographic film.

Transducer: A device that converts one form of energy into another.

Nondestructive examination (NDE), sometimes referred to as nondestructive testing (NDT) or nondestructive inspection, is a term used for those inspection methods that allow materials to be examined without changing or destroying them. NDE methods can usually detect the discontinuities and defects previously described.

Nondestructive examination is usually performed by the site quality group as part of the site quality program. Inspectors trained in the proper test methods conduct the examinations.

The welder should be familiar with the following basic nondestructive examination practices:

- Visual inspection (VT)
- Liquid penetrant inspection (PT)
- Magnetic particle inspection (MT)
- Radiographic inspection (RT)
- Ultrasonic inspection (UT)
- Electromagnetic (eddy current) inspection (ET)
- Leak testing (LT)

There is considerable overlap in the application of nondestructive and destructive tests. Destructive tests, which destroy the weld, are frequently used on several sample weldments to supplement, confirm, or establish the limits of nondestructive tests. Destructive testing is also used to provide supporting information. Once this information has been established, nondestructive examinations can be made on similar welds to locate all discontinuities above the critical defect size that was determined by the destructive tests.

3.1.0 Visual Inspection

In visual inspection, the surface of the weld and the base metal are observed for visual imperfections. Certain tools and gauges may be used during the inspection. Visual inspection is the examination method most commonly used by welders and inspectors. It is the fastest and most inexpensive method for examining a weld. However, it is limited to what can be detected by the naked eye or through a magnifying glass.

Properly done before, during, and after welding, visual inspection can detect more than 75 percent of discontinuities before they are found by more expensive and time-consuming nondestructive examination methods.

Prior to welding, the base metal should be examined for conditions that may cause weld defects. The required dimensions of the material, including edge preparations, should also be confirmed by measurements. If problems or potential problems are found, corrections should be made before proceeding any further.

After the parts are assembled for welding, the weld joint should be visually checked for a proper root opening and any other aspects that might affect the quality of the weld. Visually check for the following conditions:

- Proper cleaning
- Joint preparation and dimensions
- Clearance dimensions for backing strips, rings, or consumable inserts
- Alignment and fit-up of the pieces being welded
- Welding procedures and machine settings
- Specified preheat temperature (if applicable)
- Tack-weld quality

Fixed Weld Fillet Gauge

This fixed weld fillet gauge can be used to measure 11 fillet weld sizes in fractions or decimals. Metric versions are also available.

29106-15_SA01.EPS

During the welding process, visual inspection is the primary method for controlling quality. Some of the aspects that should be visually examined include the following:

- Quality of the root pass and the succeeding weld layers
- Sequence of weld passes
- Interpass cleaning
- Root preparation prior to welding a second side
- Conformance to the applicable procedure

After the weld has been completed, the weld surface should be thoroughly cleaned. A complete visual examination may disclose weld surface defects such as cracks, shrinkage cavities, undercuts, incomplete penetration, incomplete fusion, overlap, and crater deficiencies before they are discovered using other nondestructive inspection methods.

An important aspect of visual examination is checking the dimensional accuracy of the weld after it has been completed. Dimensional accuracy is determined by the use of measuring gauges. The purpose of using the gauges is to determine if the weld is within allowable limits as defined by the applicable codes and specifications.

Some of the more common welding gauges are the following:

- Undercut gauge
- Butt weld reinforcement gauge
- Fillet weld blade gauge set

3.1.1 Undercut Gauge

An undercut gauge is used to measure the amount of undercut on the base metal. Typically, codes allow for undercut to be between 0.01" and 0.031" deep. These gauges have a pointed end that is pushed into the undercut. The reverse side of the gauge indicates the measurement in either inches or millimeters. Two types of undercut gauges currently used are the bridge cam gauge and the V-WAC gauge (*Figure 15*). These gauges can be used for measuring undercut and for many other measurements.

3.1.2 Butt Weld Reinforcement Gauge

The butt weld reinforcement gauge has a sliding pointer calibrated to several different scales that are used to measure the size of a fillet weld or the reinforcement of a butt weld. To use the gauge for a fillet weld, position it as shown in *Figure 16*, and then slide the pointer to contact the base metal or weld metal. Be sure to read the correct scale for the measurement being taken. The other end of the gauge is used for butt welds.

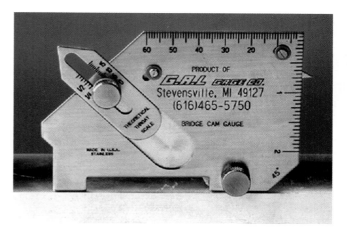

BRIDGE CAM GAUGE

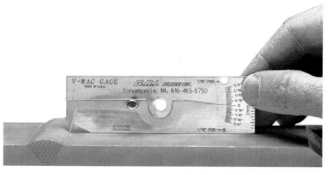

V-WAC GAUGE

29106-15_F15.EPS

Figure 15 Undercut gauges.

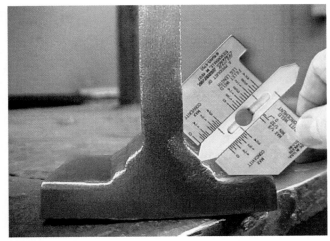

Figure 16 Automatic weld size gauge (AWS).

3.1.3 Fillet Weld Blade Gauge Set

The fillet weld blade gauge set has seven individual blade gauges for measuring convex and concave fillet welds. The individual gauges are held together by a screw secured with a knurled nut. The seven individual blade gauges can measure eleven concave and convex fillet weld sizes: ⅛", ³⁄₁₆", ¼", ⁵⁄₁₆", ⅜", ⁷⁄₁₆", ½", ⅝", ¾", ⅞", and 1" as well as their metric equivalents.

To use the fillet weld blade gauge set, identify the type of fillet weld to be measured (concave or convex) and the size. Select the appropriate blade and position it. Be sure the gauge blade is flush to the base metal with the tip touching the vertical member. *Figure 17* shows an application of a fillet weld blade gauge.

3.1.4 Liquid Penetrant Inspection

Liquid penetrant inspection (PT) examination is a nondestructive method for locating defects that are open to the surface. It cannot detect internal defects. The technique is based on the ability of a penetrating liquid, which is usually red, to wet the surface opening of a discontinuity and be drawn into it. A liquid or dry powder developer, which is usually white, is then applied over the metal. If the flaw is significant, red penetrant bleeds through the white developer to indicate a discontinuity or defect.

The dye, cleaner, and developer are available in spray cans for convenience. Some solvents used in the cleaners and developers contain high amounts of chlorine, a known health hazard, to make the liquids nonflammable.

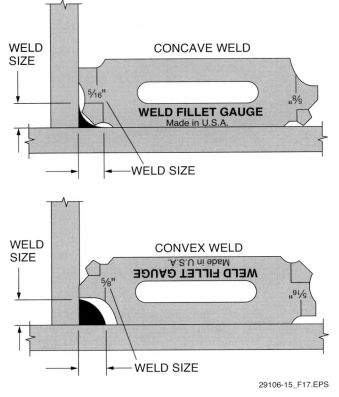

Figure 17 Fillet weld blade gauge.

WARNING!

Refer to the safety data sheet (SDS/MSDS) for hazards associated with the liquid penetrant solvent.

The most common defects found using this process are surface cracks. Most cracks exhibit an irregular shape. The width of the bleed-out (the red dye bleeding through the white developer) is a relative measure of the depth of a crack.

Surface porosity, metallic oxides, and slag will also hold penetrant and cause bleed-out. These indications are usually more circular and have less width than a crack. *Figure 18* shows liquid penetrant materials and an example of the results of liquid penetrant inspection.

The advantages of liquid penetrant inspection are that it can find small defects not visible to the naked eye; it can be used on most types of metals; it is inexpensive; and it is fairly easy to use and interpret. PT is most useful in examining welds that are susceptible to surface cracks. Except for visual inspection, it is perhaps the most commonly used nondestructive examination method for surface inspection.

Figure 18 Liquid penetrant materials and inspection example.

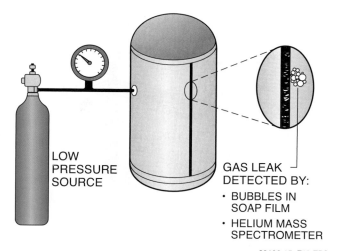

LOW PRESSURE SOURCE

GAS LEAK DETECTED BY:
• BUBBLES IN SOAP FILM
• HELIUM MASS SPECTROMETER

29106-15_F19.EPS

Figure 19 Example of a leak test.

The disadvantages of liquid penetrant inspection are that it takes more time to use than visual inspection, and it can only find surface defects. The presence of weld bead ripples and other irregularities can also hinder the interpretation of indicators. Because chemicals are used, care must be taken when performing the inspection. When testing rough, irregular surfaces, the presence of irrelevant indicators may also make interpretation difficult.

3.1.5 Leak Testing

Leak testing (*Figure 19*) is used to determine the ability of a pipe or vessel to contain a gas or liquid under pressure. Testing methods vary depending on the application of the weldment. In some cases, the vessel is pressurized and tested by immersing it in water or by applying a soap bubble solution to the weld. An open tank can be tested using water that contains fluorescein, which can be detected by ultraviolet light.

A method called the vacuum box test is used to test a vessel when only one side of the weld is accessible. The base of a storage tank is an example. The vacuum box is a transparent box with a soft rubber seal. A vacuum pump is used to extract all the air from the box. A leak is indicated by the presence of bubbles.

In the helium spectrometer leak test, helium is used as a tracer gas. Because of the small size of helium atoms, they can pass through an opening so small that it might not be detectable by other test methods. Sensitive instruments are used to detect the presence of helium.

3.2.0 Magnetic Particle and Electromagnetic Inspection

These non-destructive methods make use of electromagnetism to perform weld examination. Magnetic particle inspection methods are only effective when used with ferrous metals. Electromagnetic inspection can be used with non-ferrous metals as well as ferrous metals.

3.2.1 Magnetic Particle Inspection

Magnetic particle inspection (MT) is a nondestructive examination method that uses electricity to magnetize the weld that will be examined (*Figure 20*). After the metal has been magnetized, metal particles are sprinkled onto the weld surface. If there are defects in the surface or just below it, the metal particles will be grouped into a pattern around the defect. The defect can be identified by the shape, width, and height of the particle pattern.

Magnetic particle inspection is used to test welds for such defects as surface cracks, incomplete fusion, porosity, and slag inclusion. It can also be used to inspect plate edges for surface imperfections prior to welding. Defects can be detected only at or near the surface of the weld. Defects much deeper than this are not likely to be found. Certain discontinuities exhibit characteristic powder patterns that can be identified by a skilled inspector.

For magnetic particle examination, the part to be inspected must be ferromagnetic (made of steel or a steel alloy), smooth, clean, dry, and

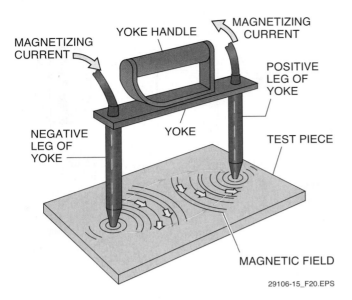

Figure 20 Magnetic particle examination with electromagnetic yokes.

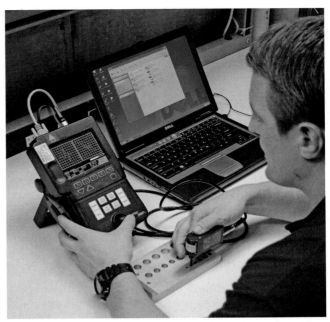

29106-15_F21.EPS

Figure 21 Eddy current test.

free from oil, water, and excess slag. The part is magnetized by using an electric current to set up a magnetic field within the material. The magnetized surface is covered with a thin layer of magnetic powder. If there is a defect, the powder is held to the surface at the defect because of the powerful magnetic field.

When this examination method is used, there is normally a code or standard that governs both the method and the acceptance/rejection criteria of indications.

The advantages of magnetic particle inspection are that it can find small defects not visible to the naked eye, and it is faster than liquid penetrant inspection.

Disadvantages of magnetic particle inspection are that the materials must be capable of being magnetized; the inspector must be skilled in interpreting indications; rough surfaces can interfere with the results; the method requires an electrical power source; and it cannot identify internal discontinuities located deep in the weld.

3.2.2 Electromagnetic (Eddy Current) Inspection

Like magnetic particle testing, electromagnetic, or eddy current inspection (ET) uses electromagnetic energy to detect defects in the joint. A coil, which produces a magnetic field, is placed on or around the part being tested. After being calibrated, the coil is moved over the part to be inspected. The coil produces a current in the metal through induction. The induced current is called

an eddy current. If a discontinuity is present in the test part, it will interrupt the flow of the eddy currents. This change can be observed on the oscilloscope display.

Eddy currents only detect discontinuities near the surface of the part. This method is suitable for both ferrous and nonferrous materials, and it is used in testing welded tubing and pipe. It can determine the physical characteristics of a material and the wall thickness in tubing. It can also check for porosity, pinholes, slag inclusions, internal and external cracks, and incomplete fusion.

The advantage of eddy current inspection is that it can detect surface and near-surface weld defects. It is particularly useful in inspecting circular parts like pipes and tubing.

The disadvantage of eddy current inspection is that eddy currents decrease with depth, so defects farther from the surface may go undetected. The accuracy of the examination depends in large part on the calibration of the instrument and the ability of the inspector. *Figure 21* shows an eddy current test being conducted.

3.3.0 Radiographic and Ultrasonic Inspection

These non-destructive inspection methods rely on technology that is commonly used in medical diagnostics. The ability to apply these methods and to correctly interpret the results requires extensive training and experience.

3.3.1 Radiographic Inspection

Radiography (RT) is a nondestructive examination method that uses radiation (X-rays or gamma rays) to penetrate the weld and produce an image. Like medical X-rays, this weld inspection technology originally worked by exposing treated film to the X-rays. In traditional RT work, X-ray film is placed against the weld and exposed (*Figure 22*). When a joint is radiographed, the radiation source is placed on one side of the weld and the film on the other. The joint is then exposed to the radiation source. The radiation penetrates the metal and produces an image on the film. The film is called a radiograph and provides a permanent record of the weld quality.

Modern RT offers two additional options: digital detector array (DDA) and computed radiography (CR). DDA radiography, also referred to as digital radiography, uses a semiconductor array instead of silver-based film to capture the image. Like film radiography, digital radiography is considered a form of direct radiography. Once the DDA is exposed, the image it captures can be transferred directly to a computer and enhanced for viewing. DDA radiography uses the same type of X-ray source as film radiography.

CR generally uses a cassette that contains a reusable imaging plate. Once the plate is exposed, it is processed through a special laser scanner that captures the image and allows it to be transferred to a computer for enhancement and interpretation. *Figure 23* shows CR radiography being performed.

Radiographic inspection can produce a visible image of weld discontinuities, both surface and subsurface, when they are different in density from the base metal and different in thickness parallel to the radiation. Surface discontinuities are better identified by visual, penetrant, or magnetic particle examination. Radiographic weld inspection requires specialized skill. It should therefore be done and interpreted only by trained, qualified personnel.

The advantages of radiographic inspection are that it provides a permanent record of the weld quality, stored either on film or as a computer image. Through radiography, the entire thickness can be examined, and it can be used on all types of metals.

The disadvantages of radiographic inspection are that it is a slow and expensive method for inspecting welds; some joints are inaccessible to radiography; and excessive exposure to radiation of any type is very hazardous to humans. Cracks can frequently be missed if they are very small or are not aligned with the radiation beam.

3.3.2 Ultrasonic Inspection

Ultrasonic inspection (UT) is a relatively low-cost, nondestructive examination method that uses sound wave vibrations to find surface and subsurface defects in the weld material. Ultrasonic waves are passed through the material being tested and are reflected back by any density change caused by a defect. The reflected signal is shown on the screen display of the instrument (*Figure 24*).

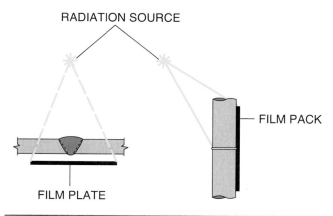

RADIATION SOURCE

FILM PACK

FILM PLATE

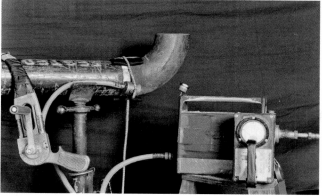

29106-15_F22.EPS

Figure 22 Weld examination using radiography.

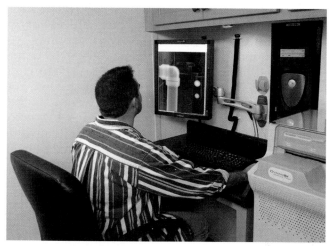

29106-15_F23.EPS

Figure 23 Computed radiography (CR) testing in process.

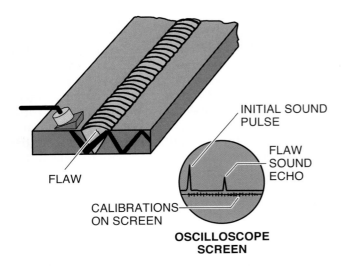

INITIAL SOUND PULSE

FLAW SOUND ECHO

FLAW

CALIBRATIONS ON SCREEN

OSCILLOSCOPE SCREEN

29106-15_F24.EPS

Figure 24 Portable ultrasonic device.

The term *ultrasonic* indicates frequencies above those heard by the human ear. Ultrasonic devices operate very much like depth sounders or fish finders. The key element is a transducer that is passed over the weld to be tested. The transducer converts electrical energy into mechanical energy that is emitted as high-frequency vibrations.

Ultrasonic examination can be used to detect and locate cracks, laminations, shrinkage cavities, pores, slag inclusions, incomplete fusion, and incomplete joint penetration, as well as other discontinuities in the weld. A qualified inspector can interpret the signal on a screen to determine the approximate position, depth, and size of the discontinuity.

Ultrasonic beams can deeply penetrate weld material with high sensitivity and accuracy. The output is readily digitized, and operation of the monitor is entirely nonhazardous. Some of the different types of ultrasonic techniques are the following:

Self-Propelled NDT

This ultrasonic test instrument is often used to inspect pipe and tanks. Its magnetic wheels allow it to cling to the surface as it moves along, controlled by a joystick mechanism.

29106-15_SA02.EPS

- Pulse-echo
- Pulse-transmission
- Ultrasonic attenuation
- Continuous wave resonance
- Ultrasonic spectroscopy
- Phased array

Figure 25 shows the effects of moving an angle-beam transducer back and forth across a weld. *Figure 26* shows the effects of moving an angle-beam transducer in an arc around a weld.

Ultrasonic monitors are very portable, but they require experienced technicians to monitor the equipment. Reading an ultrasonic flaw detector requires some practice because the screen image shows the peaks of back reflection.

In traditional UT, a single-element probe is physically moved around the weldment by the inspector to create a cross-section of the joint being tested. In contrast, the newer phased array (PA) ultrasonic testing method (*Figure 27*) uses a probe containing an array of ultrasonic transducers. The beam created by these transducers is electronically steered. The data from the individual elements is combined electronically to create a cross-section of the weld.

Among the advantages of ultrasonic inspection are that it can find defects throughout the material being examined; it can be used to check materials that cannot be radiographed; it is nonhazardous to personnel and equipment; and it can detect even small defects from one side of the material.

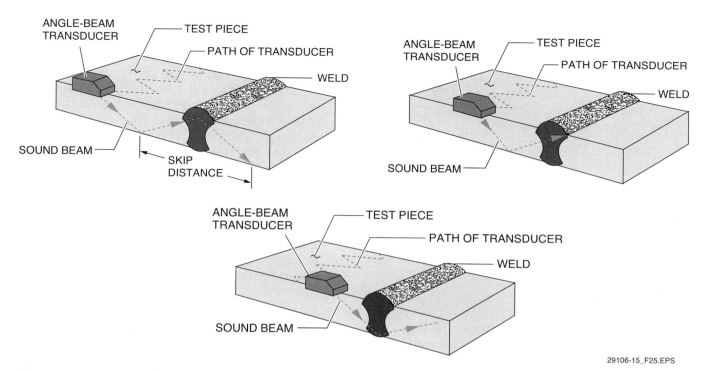

29106-15_F25.EPS

Figure 25 Moving a transducer back and forth.

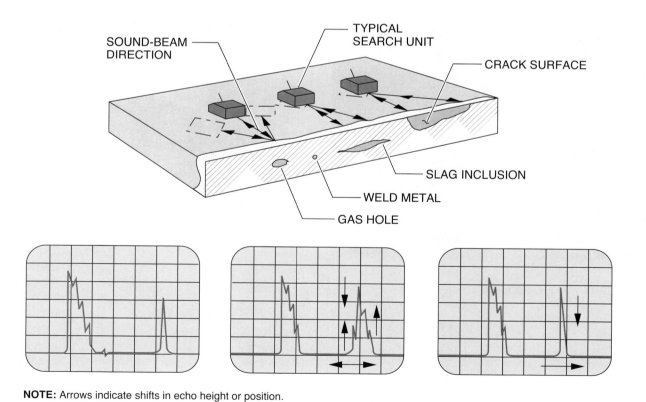

NOTE: Arrows indicate shifts in echo height or position.

29106-15_F26.EPS

Figure 26 Moving a transducer in an arc.

The disadvantages of ultrasonic inspection are that it requires a high degree of skill to properly interpret the patterns. In addition, very small or thin weldments are difficult to inspect using this method.

3.4.0 Destructive Testing

In a destructive test, the test sample is destroyed or damaged in the testing process and is no longer suitable for use. Destructive testing is often done in a laboratory setting using machines designed to apply different stresses to the weld or the base metal, depending upon which is being tested. Examples of destructive tests commonly used to examine welds and base metals include the following:

- *Tensile test* – In this test, a sample is placed in a tensile testing machine and is pulled until it breaks. The tensile test is sometimes used to determine if the weld performs as well as the base metal. In most cases, however, the test is performed to determine specific properties of the weldment, such as strength and ductility (the amount the sample will stretch before breaking).

- *Hardness test* – While strength testing examines the ability of the sample to transmit a load, hardness testing examines its ability to resist penetration. Hardness testing is usually done using a penetrating device that leaves an indentation in the sample. The depth or diameter of the indentation, depending on the type of hardness test being conducted, is then measured to determine the hardness.

- *Impact test* – The ability of a weld to withstand an impact, such as a hammer strike, is measured by this test. The property being tested is referred to as toughness. In a laboratory environment, a pendulum-type machine that simulates a heavy hammer blow is used (*Figure 28*). A notch of a specified size is made in the sample. Then the sample is placed in the jaws of the testing machine and struck with a pendulum blow.

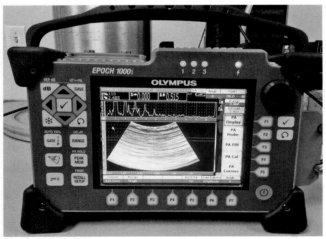

29106-15_F27.EPS

Figure 27 Phased array ultrasonic tester.

29106-15_F28.EPS

Figure 28 Impact test equipment.

- *Soundness test* – The three types of soundness tests include bend, nick-break, and fillet weld break tests.
  - Bend testing is the test most commonly used in determining the qualifications of a welder or welding procedures. Bend tests are performed by placing the sample in a special fixture and applying stress in a way that causes the sample to bend 180 degrees. The bend is then inspected for weld defects. Guided bend tests (*Figure 29*) are used to evaluate groove test welds on both plates and pipes. In this method, specimens are bent into a U-shape with a device called a jig. The bending action places stress on the weld metal and reveals any discontinuities in the weld. The jig has a plunger and die that are dimensioned for the thickness of the specimen being bent (*Figure 30*). Refer to the welding code being used for testing for the required dimensions of the bending jig's plunger and die.

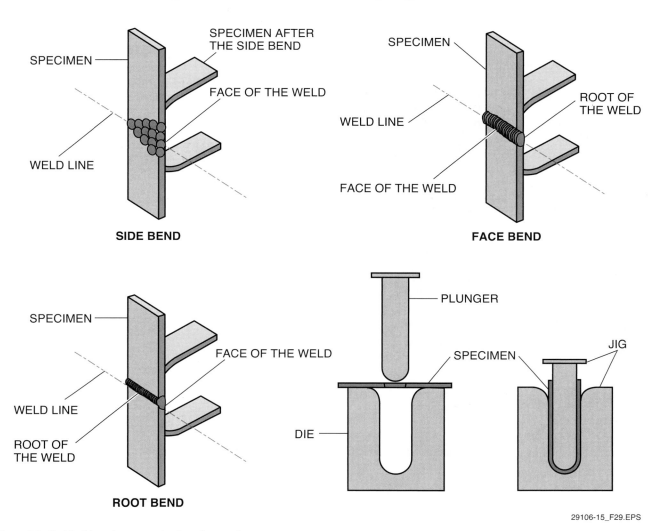

29106-15_F29.EPS

Figure 29 Guided bend test method and samples.

Figure 30 Bend test equipment and specimen.

29106-15_F30.EPS

– The three types of tests performed on the jig are root bends, face bends, and side bends. These bends test for penetration and fusion throughout the root joint. They also test for porosity, inclusions, and other defects, as well as the quality of the fusion to the side walls and the face of the weld joint. Face and root bend tests are used for materials up to $\frac{3}{8}$" (≈10 mm metric plate) thick. For materials thicker than this, side bend testing may be used.

– Nick-break testing is used primarily in the pipeline industry. The specimen is saw-cut so it will break in a specific place. Then it is broken, and the weld zone is examined for defects.

– In a fillet weld break test, a fillet weld is made on one side of a T-joint. Stress is then applied to the T-joint until the weld fractures (*Figure 31*). The weld is then examined for defects and fusion. This is not a weld strength test; it is intended to break the weld so that it can be examined for discontinuities.

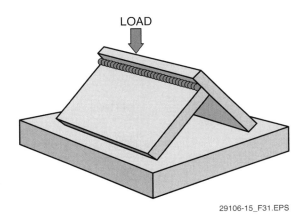

29106-15_F31.EPS

Figure 31 Fillet weld break test.

Additional Resources

Visual Inspection Workshop Reference Manual, AWS VIW-M-2008. Miami, FL: American Welding Society.

Certification Manual for Welding Inspectors. Miami, FL: American Welding Society.

Welding Inspection Technology. Miami, FL: American Welding Society.

B1.10M/B1.10:2009 Guide for the Nondestructive Examination of Welds. Miami, FL: American Welding Society.

3.0.0 Section Review

1. A bridge cam gauge is used to measure _____.
   a. undercut
   b. fillet weld depth
   c. butt weld reinforcement
   d. overlap

2. Electromagnetic inspection can be used to test welds in non-ferrous metals.
   a. True
   b. False

3. Which of these test methods uses an imaging plate placed in a cassette?
   a. Ultrasonic
   b. Phased array UT
   c. Computed radiography
   d. Digital radiography

4. A tensile test is one in which the test coupon is _____.
   a. bent
   b. pulled
   c. struck
   d. compressed

4.0.0 WELDER PERFORMANCE QUALIFICATION TESTS

Objective

Describe the welder performance testing process.

   a. Describe the qualification of welders by position.
   b. Describe welder qualification testing to meet AWS and ASME requirements.
   c. Describe the process for completing a weld test.

The purpose of welder performance qualification is to measure the proficiency of individual welders. As previously discussed, codes require that welders take a test to qualify for performing a welding procedure.

> **NOTE**
>
> Various codes and specifications often require similar methods for qualifying welders. The applicable code or specification should be consulted for specific details and requirements. Ask your supervisor if you are unsure about which codes or specifications apply to your project.

4.1.0 Welding Position Qualification

The welder or welding machine operator is qualified by welding position. Welders may be qualified to perform a welding procedure in only one position, or possibly all positions by passing a welding qualification test. The qualification tests are designed to measure the welder's ability to make groove and fillet welds in different positions on plate, pipe, or both in accordance with the applicable code. Each welding position is designated by a number and a letter, such as 1G. These designations are standard for all codes.

The letter G designates a groove weld. The letter F designates a fillet weld. For plate welding, the positions are designated by the following numbers:

- 1 – Flat position welding
- 2 – Horizontal position welding
- 3 – Vertical position welding
- 4 – Overhead position welding

Figure 32 shows the plate welding positions for both fillet welds and groove welds.

For pipe welding, there are these additional positions: 5F, 5G, 6G, and 6GR. The number indicates that multiple-position welds are required. Also, in the 1G (flat groove) and 1F (flat fillet) positions, the pipe is rotated during welding. *Figure 33* shows the pipe welding positions for both fillet welds and groove welds.

A welder who qualifies in one position does not automatically qualify to weld in all positions. However, in most cases qualification for groove welds qualify the welder for fillet welds; qualification for pipe qualify the welder for plate. Qualification in one code may or may not qualify the welder in other codes. The qualification requirements between codes may not match. Refer to your site requirements/code for qualifying requirements.

4.2.0 Code-Required Testing

AWS, ASME, and API codes require that welders be qualified through testing to perform specific types of welds based on base metal, welding position, joint design, type of electrode, and other variables.

4.2.1 AWS Structural Steel Code

The AWS structural steel code provides information concerning the qualification of welding procedures, welders, and welding operators for the types of welding done by contractors and fabricators in building and bridge construction. Qualification for plate welding also qualifies the welder for rectangular tubing.

The mild steel electrodes used with shielded metal arc welding (SMAW) are classified by F numbers: F1, F2, F3, and F4. Qualification with an electrode in a particular F-number classification will qualify the welder with all electrodes identified in that classification and in lower F-number classifications. *Table 1* shows AWS F-number electrode classifications.

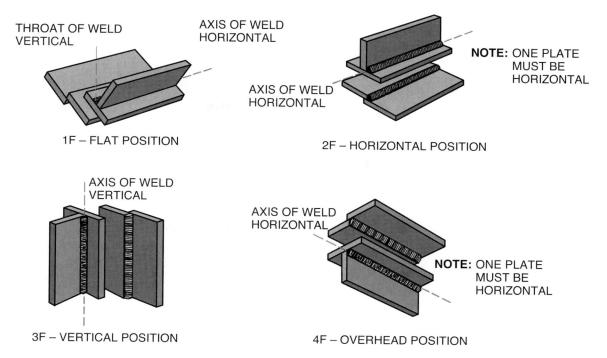

1F – FLAT POSITION

2F – HORIZONTAL POSITION

NOTE: ONE PLATE MUST BE HORIZONTAL

3F – VERTICAL POSITION

4F – OVERHEAD POSITION

NOTE: ONE PLATE MUST BE HORIZONTAL

FILLET WELDS

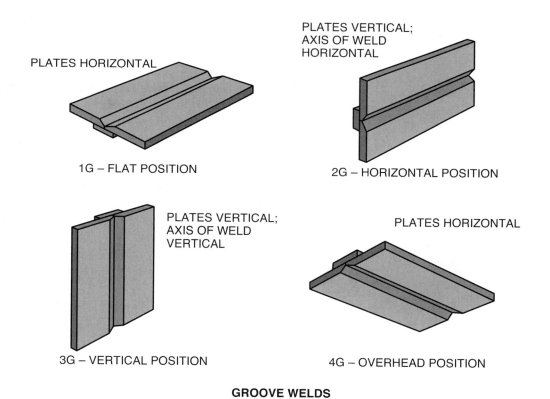

1G – FLAT POSITION

2G – HORIZONTAL POSITION

3G – VERTICAL POSITION

4G – OVERHEAD POSITION

GROOVE WELDS

29106-15_F32.EPS

Figure 32 Welding positions for plate.

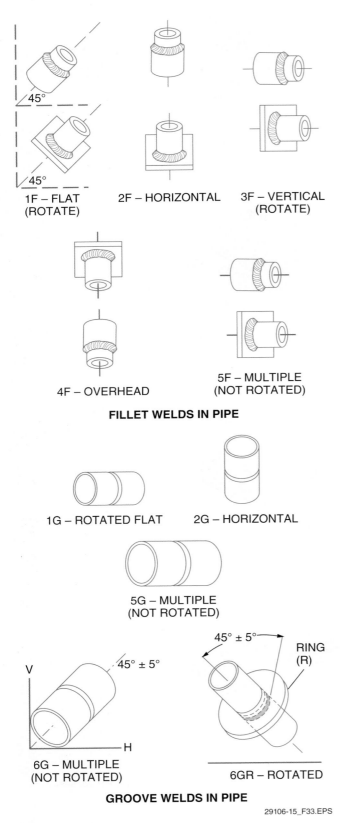

FILLET WELDS IN PIPE

1F – FLAT (ROTATE)
2F – HORIZONTAL
3F – VERTICAL (ROTATE)
4F – OVERHEAD
5F – MULTIPLE (NOT ROTATED)

1G – ROTATED FLAT
2G – HORIZONTAL
5G – MULTIPLE (NOT ROTATED)
6G – MULTIPLE (NOT ROTATED)
6GR – ROTATED
RING (R)

GROOVE WELDS IN PIPE

29106-15_F33.EPS

Figure 33 Welding positions for pipe.

Material thickness is an essential variable in qualification tests under AWS code. Some of the tests in the code qualify the welder only up to twice the thickness of the test piece. Others qualify the welder for unlimited thicknesses.

A typical AWS welder qualification test is a V-groove weld with metal backing in the 3G and 4G positions using an F4 electrode. Passing this test qualifies the welder to weld with F4 or lower electrodes, and to make groove and fillet welds in all positions. *Figure 34* shows an example of a fit-up for the AWS structural test.

4.2.2 ASME Code

Individual welders and welding operators who are required to weld to ASME code must qualify in accordance with Section IX of the *ASME Boiler and Pressure Vessel Code* on either plate or pipe. Qualification on pipe also qualifies the welder to weld plate, but not vice versa. Qualification with groove welds also qualifies the welder for fillet welds, but not vice versa. It is possible under the code to qualify for fillet welds only.

The typical ASME welder qualification test is to weld pipe in the 6G position using an open root. Passing this test qualifies the welder to weld pipe in all positions and plate in all positions (fillet and groove). If F3 electrodes are used all the way out, the welder only qualifies on F3 electrodes on both plate and pipe. If F3 electrodes are used for welding the root and F4 electrodes are used for filler, the welder qualifies to weld pipe or plate with F3 electrodes and pipe or plate with F4 electrodes as long as backing is used. The drawing in *Figure 35* shows a typical ASME pipe test.

4.2.3 API Code

The API requires welders to make butt or fillet welds using qualified procedures. Welds are performed on pipe nipples or segments of pipe nipples. Completed welds are subjected to visual examination and destructive or radiographic testing. Welders may receive a single qualification by performing a single weld in the fixed or rolled position. Multiple-weld qualification can be obtained by first making a butt weld in a fixed position on 6" pipe (DN150) diameter with a minimum thickness of 0.250" (6.4 mm). A second test requires the welder to cut, fit, and weld a pipe of the same size without a backing strip.

Table 1 AWS F-Number Electrode Classification

Group	AWS Electrode Classification				
F4	EXX15	EXX16	EXX18		Low-Hydrogen
F3	EXX10	EXX11			Fast-Freeze
F2	EXX12	EXX13	EXX14		Fill-Freeze
F1	EXX20	EXX24	EXX27	EXX28	Fast-Fill

29106-15_F34.EPS

Figure 34 Typical AWS plate test coupon.

4.3.0 Welder Qualification Tests

The welder becomes qualified by successfully completing a weld made in accordance with the WPS. It is general practice for code welding to qualify welders on groove weld tests. Passing these tests also permits the welder to perform fillet welds.

4.3.1 Making the Test Weld

Although qualification tests are designed to determine the capability of welders, some welders fail for reasons not related to their welding ability. This is due principally to carelessness in the application of the weld and in the preparation of the test specimen. It is important to note prior to welding where the test strips will be cut from the weld coupon. By doing this, you can avoid potential problems such as restarts in the area of the test strips. The following sections explain how to prepare a test specimen.

4.3.2 Removing Test Specimens

After making the qualification test weld, the test specimens are cut from the test pipe or plate by any suitable means. There are specific locations where the test specimen is cut from the pipe or plate.

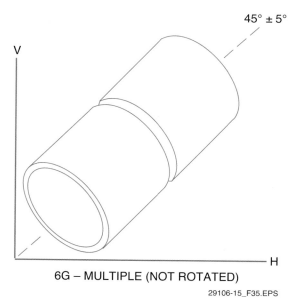

45° ± 5°

6G – MULTIPLE (NOT ROTATED)

29106-15_F35.EPS

Figure 35 Typical ASME pipe test.

Refer to the applicable code for details related to specimen location and quantity.

Typical specimen locations for plate welds are shown in *Figure 36*. For material ⅜" (10 mm metric plate) thick, a face bend and a root bend are required. For material thicker than ⅜" (10 mm metric plate), two side bends are required.

Note that tests are usually given on plate thicknesses of ⅜" (10 mm metric plate) for limited thickness and 1" (25 mm metric plate) for unlimited thickness qualifications.

4.3.3 Preparing the Specimens for Testing

After the specimen has been cut from the test piece, it must be properly prepared for testing (*Figure 37*). Poor specimen preparation can cause a sound weld metal to fail. For example, a slight nick may open up under the severe bending stress of the test, causing the specimen to fail. To properly prepare the test specimen, do the following:

- Grind or machine the surface to a smooth finish. All grinding and machining marks must be lengthwise on the sample. Otherwise, they produce a notch effect, which may cause failure.

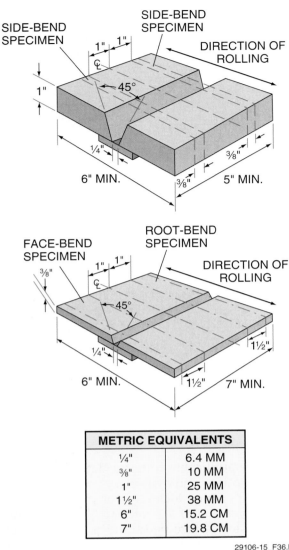

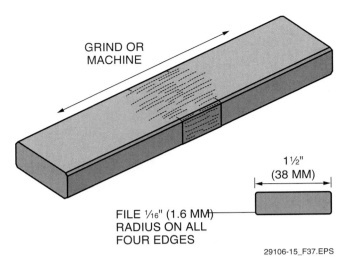

Figure 37 Example of a prepared test specimen.

After bending, the specimen is evaluated by measuring the discontinuities that are exposed. The criteria for acceptance can vary by code or site quality standards. AWS standards require that the surface shall contain no discontinuities exceeding the following dimensions:

- ⅛" (3.2 mm) – Measured in any direction on the surface.
- ⅜" (10 mm) – Sum of the greatest dimensions of all discontinuities exceeding ¹⁄₃₂" (0.8 mm) but less than or equal to ⅛" (3.2 mm).
- ¼" (6.4 mm) – Maximum corner crack, except when the corner crack results from visible slag inclusion or other fusion-type discontinuities, then a ⅛" (3.2 mm) maximum shall apply. A specimen with corner cracks exceeding ¼" (6.4 mm), with no evidence of slag inclusions or other fusion-type discontinuities may be discarded, and a replacement test specimen from the original weldment shall be tested.

METRIC EQUIVALENTS	
¼"	6.4 MM
⅜"	10 MM
1"	25 MM
1½"	38 MM
6"	15.2 CM
7"	19.8 CM

29106-15_F36.EPS

Figure 36 Specimen locations for plate welds.

- Remove any face or root reinforcement from the weldment. This is part of the test requirement and, more important to the welder, failure to do so can cause the failure of a good weld. Grinding of the weld reinforcement may not exceed ¹⁄₃₂" (0.8 mm) or 5 percent of the thickness of the base material, according to AWS standards.
- Round the edges to a smooth ¹⁄₁₆" (1.6 mm) radius. This can be done with a file. Rounded edges help prevent failure caused by cracks starting at a sharp corner.
- Refrain from quenching specimens you are grinding when they are hot. Quenching may create small surface cracks that become larger during the bend test.

> **NOTE**
>
> In some cases, a radiographic inspection will be used instead of the guided bend test. This allows the entire weld to be examined and can detect small discontinuities at any location within the weld.

When the welder passes the qualification tests, the test results and the WPSs that the welder may weld are listed on a record that is kept by the company. This record becomes part of the quality documentation, and the welder becomes qualified to weld to that procedure.

AWS vs. ASME Pipe Welds

AWS and ASME codes have different requirements. The AWS version is shown on the left, while the ASME version is shown on the right.

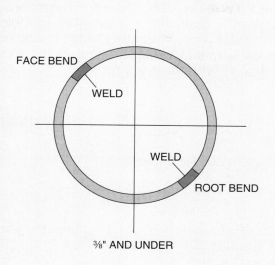

3/8" AND UNDER

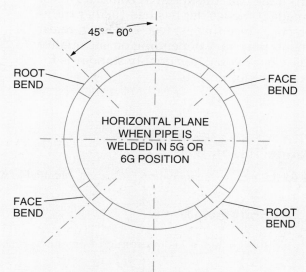

PIPES 1/16" (1.6 MM) UP THROUGH 3/8" (10 MM) THICKNESS

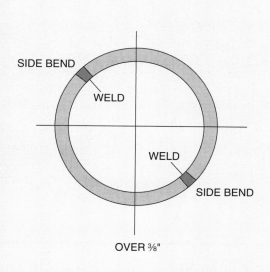

OVER 3/8"

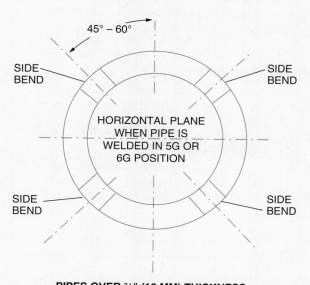

PIPES OVER 3/8" (10 MM) THICKNESS

29106-15_SA03.EPS

4.3.4 Welder Qualification Limits

Welders may retest if they initially fail the test. An immediate retest consists of two test welds of each type of test that was failed. All the test specimens must pass this retest. A complete retest may be made if the welder has had further training or practice since the last test. However, retest requirements may vary depending on site quality standards. Check your site's quality standards for specific retest requirements.

After welders have qualified, they may have to requalify if they have not used the specific process for a certain time period. This time period varies depending on the codes. Welders may also be required to requalify if there is a reason to question their ability to make welds that meet the WPS. Also, since welder performance qualification is limited to the essential variables of a particular procedure, any change in one or more of the essential variables requires the welder to requalify with the new procedure.

Additional Resources

AWS B3.0 Standard Qualification Procedure. Miami, FL: American Welding Society.

4.0.0 Section Review

1. A weld designated 3G is interpreted to mean a _____.

    a. fillet weld in the flat position
    b. groove weld in the flat position
    c. groove weld in the vertical position
    d. fillet weld in the vertical position

2. Which of the following is a *correct* statement regarding qualification to ASME code?

    a. Qualification testing is done only on pipe.
    b. Qualifying on plate qualifies the welder on pipe sizes less than 24".
    c. Qualification on groove welds also qualifies the welder on fillet welds.
    d. Using F3 electrodes all the way out qualifies the welder for all electrodes.

3. Once a welder passes a qualification test, the qualification is permanent.

    a. True
    b. False

SUMMARY

Quality is everyone's responsibility. If the work being done cannot be defined as quality work, it reflects on all those involved in the process. One essential trait of a craftsperson is a sense of quality workmanship. The craftsperson is generally closest to the work and will therefore have a major impact on product quality. Keeping quality in mind as you perform each step of your job will help you identify and correct small problems before they become major ones. This will make everyone's job easier and instill a sense of pride in what has been accomplished.

Safety, quality, and production each has a cost of its own. On the project, each of these factors should have proper guidelines. At the completion of the project, when all records for safety, cost, planning, scheduling, and effectiveness have been evaluated, the papers are usually filed away, but quality remains indefinitely for all eyes to see. Quality is, perhaps, the major reason for repeat business. How well the craftsperson performed will be noticed long after the project has been completed.

1. Which section of the *ASME Boiler and Pressure Vessel Code* contains the specifications for acceptable ferrous (Part A) and non-ferrous (Part B) base metals?
   a. I
   b. II
   c. V
   d. IX

2. The minimum requirements for the design, materials, fabrication, erection, testing, and inspection of various types of piping systems are covered by _____.
   a. *ASME B31*
   b. *AWS D1.1*
   c. *AWS B1.11*
   d. *API Std. 1104*

3. The code that covers the requirements for welding structures or components under the surface of water is _____.
   a. *ASME B31*
   b. *AWS D1.1*
   c. *AWS D3.6M*
   d. *API 1104*

4. If you are directed to perform a weld for which you are not qualified, you should _____.
   a. bypass the chain of command
   b. report directly to the safety engineer as soon as possible
   c. try to resolve the matter with your immediate supervisor
   d. perform the weld to the best of your ability

5. Which of the following is considered a non-essential variable for a welding performance qualification?
   a. Material thickness
   b. Post-heat treatment
   c. Travel speed
   d. Welding process

6. The presence of voids or empty spots in the weld metal is called _____.
   a. porosity
   b. inclusion
   c. defect
   d. discontinuity

7. The type of porosity in which the weld exhibits a localized grouping of pores is known as _____.
   a. uniformly scattered porosity
   b. clustered porosity
   c. linear porosity
   d. piping porosity

8. The type of weld discontinuity referred to as wormholes is _____.
   a. clustered porosity
   b. piping porosity
   c. cold cracks
   d. incomplete fusion

(A)

(B)

(C)

(D)

29106-15_RQ01.EPS

Figure 1

For Questions 9 through 11, match a picture in Review Question Figure 1 to each discontinuity listed by putting the correct letter in the blank.

9. Linear porosity ___A___

10. Surface slag inclusion ___B___

11. Toe crack ___C___

12. One method of reducing weld metal cracking is to use _____.

   a. fast-fill electrodes
   b. fill-freeze electrodes
   c. low-hydrogen electrodes
   d. fast-freeze electrodes

13. Dross from flame cutting, if not removed, is likely to cause _____.

   a. porosity
   b. spatter
   c. undercut
   d. incomplete penetration

14. Dimensional accuracy of a weld is checked by _____.

   a. visual inspection
   b. using a tape measure
   c. radiography
   d. using welding gauges

15. A vacuum box is used to check for _____.

   a. inclusions
   b. porosity
   c. leaks
   d. incomplete fusion

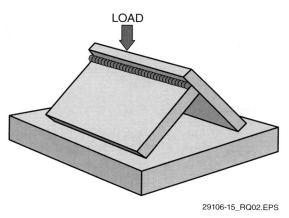

LOAD

29106-15_RQ02.EPS

Figure 2

16. The test method shown in *Review Question Figure 2* is the _____.

    a. nick-break test
    b. fillet weld break test
    c. impact test
    d. hardness test

17. The nick-break test is a type of _____.

    a. tensile test
    b. hardness test
    c. soundness test
    d. impact test

18. A typical AWS welder qualification test is a V-groove weld with metal backing in the 3G and 4G positions using an _____.

    a. F4 electrode
    b. F3 electrode
    c. F2 electrode
    d. F1 electrode

19. An ASME welder qualification that qualifies a welder to weld pipe in all positions and plate in all positions is _____.

    a. 1F position test
    b. 6G position test
    c. 4F position test
    d. 4G position test

20. To properly prepare a test specimen, it is important to _____.

    a. quench the specimen while it is hot
    b. grind it width-wise to produce a notch effect
    c. remove any face or root reinforcement from the weldment
    d. make sure the edges are sharp

Trade Terms Quiz

Fill in the blank with the correct term that you learned from your study of this module.

1. _____ is the demonstration that welds made following a specific procedure can meet prescribed standards.

2. A(n) _____ is a discontinuity or imperfection that renders a part of the product or the entire product unable to meet minimum acceptable standards or specifications.

3. The type of discontinuity that is identified by cracking in the base metal near the weld, but under the surface is _____.

4. Metals that have the ability to be made harder by heating and then cooling are called _____.

5. The document containing the results of the nondestructive and destructive testing required to qualify a welding procedure specification (WPS) is called a(n) _____.

6. _____ is a change or break in the shape or structure of a part that may or may not be considered a defect, depending on the code.

7. The quality or state of having a uniform structure or composition throughout is known as _____.

8. The testing method is which images are made by passing X-rays or gamma rays through an object and recording the variations in density on photographic film is known as _____.

9. Cracks in the base metal formed when layers separate are called _____.

10. _____ metal has been made brittle and will tend to crack with little bending.

11. Foreign matter introduced into and remaining in a weld is called a(n) _____.

12. A document that establishes the minimum requirements for a product or process and can be adopted into law is a(n) _____.

13. A welding parameter that cannot be changed without requalifying a procedure is a(n) _____.

14. The ability of a material to absorb energy in the presence of a flaw is known as _____.

15. A welding parameter that can be changed without requalifying a procedure is a(n) _____.

16. The variables that must be considered when notch toughness requirements are invoked are called _____.

17. A document that defines how a code is to be implemented is a(n) _____.

18. A(n) _____ is a document that defines in detail the work to be performed or the materials to be used in a product or process.

19. A device that converts one form of energy into another is called a(n) _____.

20. _____ is a method of joining metal using heat and a filler metal with a melting point above 842°F (450°C).

21. The deflection of an arc from its intended path by magnetic forces is known as _____.

Trade Terms

Arc blow 21
Brazing 20
Code 12
Defect 2
Discontinuity 6
Embrittled 10
Essential variables 13

Hardenable materials 4
Homogeneity 7
Inclusion 11
Laminations 9
Non-essential variables 15
Notch toughness 14
Procedure qualification record (PQR) 5

Radiographic 8
Specification 18
Standard 17
Supplemental essential variables 16
Transducer 19
Underbead cracking 3
Welding procedure qualification 1

Appendix A

PROCEDURE QUALIFICATION RECORD

FORM QW-483 SUGGESTED FORMAT FOR PROCEDURE QUALIFICATION RECORDS (PQR)
(See QW-200.2, Section IX, ASME Boiler and Pressure Vessel Code)
Record Actual Variables Used to Weld Test Coupon

Organization Name _____

Procedure Qualification Record No. _____ Date _____

WPS No. _____

Welding Process(es) _____

Types (Manual, Automatic, Semi-Automatic) _____

JOINTS (QW-402)

Groove Design of Test Coupon
(For combination qualifications, the deposited weld metal thickness shall be recorded for each filler metal and process used.)

BASE METALS (QW-403)

Material Spec. _____

Type/Grade, or UNS Number _____

P-No. _____ Group No. _____ to P-No. _____ Group No. _____

Thickness of Test Coupon _____

Diameter of Test Coupon _____

Maximum Pass Thickness _____

Other _____

POSTWELD HEAT TREATMENT (QW-407)

Temperature _____

Time _____

Other _____

GAS (QW-408)

	Percent Composition		
	Gas(es)	(Mixture)	Flow Rate
Shielding			
Trailing			
Backing			
Other			

FILLER METALS (QW-404)

	1	2
SFA Specification		
AWS Classification		
Filler Metal F-No.		
Weld Metal Analysis A-No.		
Size of Filler Metal		
Filler Metal Product Form		
Supplemental Filler Metal		
Electrode Flux Classification		
Flux Type		
Flux Trade Name		
Weld Metal Thickness		
Other		

ELECTRICAL CHARACTERISTICS (QW-409)

Current _____

Polarity _____

Amps. _____ Volts _____

Tungsten Electrode Size _____

Mode of Metal Transfer for GMAW (FCAW) _____

Heat Input _____

Other _____

POSITION (QW-405)

Position of Groove _____

Weld Progression (Uphill, Downhill) _____

Other _____

TECHNIQUE (QW-410)

Travel Speed _____

String or Weave Bead _____

Oscillation _____

Multipass or Single Pass (Per Side) _____

Single or Multiple Electrodes _____

Other _____

PREHEAT (QW-406)

Preheat Temperature _____

Interpass Temperature _____

Other _____

(07/13)

29106-15_A01A.EPS

FORM QW-483 (Back)

Tensile Test (QW-150)

Specimen No.	Width	Thickness	Area	Ultimate Total Load	Ultimate Unit Stress, (psi or MPa)	Type of Failure and Location

Guided-Bend Tests (QW-160)

Type and Figure No.	Result

Toughness Tests (QW-170)

Specimen No.	Notch Location	Specimen Size	Test Temperature	Impact Values			Drop Weight Break (Y/N)
				ft-lb or J	% Shear	Mils (in.) or mm	

Comments _____

Fillet-Weld Test (QW-180)

Result — Satisfactory: Yes _____ No _____ Penetration into Parent Metal: Yes _____ No _____

Macro — Results _____

Other Tests

Type of Test _____

Deposit Analysis _____

Other _____

....

Welder's Name _____ Clock No. _____ Stamp No. _____

Tests Conducted by _____ Laboratory Test No. _____

We certify that the statements in this record are correct and that the test welds were prepared, welded, and tested in accordance with the requirements of Section IX of the ASME Boiler and Pressure Vessel Code.

Organization _____

Date _____ Certified by _____

(Detail of record of tests are illustrative only and may be modified to conform to the type and number of tests required by the Code.)

(07/13)

29106-15_A01B.EPS

Appendix B

PERFORMANCE ACCREDITATION TASK – VISUAL WELD TEST INSPECTION

The American Welding Society (AWS) School Excelling through National Skills Standards Education (SENSE) program is a comprehensive set of minimum Standards and Guidelines for Welding Education programs. The following performance accreditation is aligned with and designed around the SENSE program.

The Performance Accreditation Tasks (PATs) correspond to and support the learning objectives in *AWS EG2.0, Guide for the Training and Qualification of Welding Personnel: Entry-Level Welder*.

Note that in order to satisfy all learning objectives in *AWS EG2.0*, the instructor must also use the PATs contained in the second level of the NCCER Welding curriculum.

PAT 1 corresponds to *AWS EG2.0, Module 9 – Welding Inspection and Testing Principles, Unit 4 Welding Inspection and Testing*, Key Indicators 1 and 2.

PATs provide specific acceptable criteria for performance and help to ensure a true competency-based welding program for students.

The following task is designed to develop your competency in performing a visual weld test inspection. Practice this task until you are thoroughly familiar with the procedure.

As you complete the task, take it to your instructor for evaluation. Do not proceed to the next task until instructed to do so by your instructor.

VISUAL WELD TEST INSPECTION

Obtain a completed fillet weld on ¼" plate (3.2 mm metric plate) a minimum of 6" (15.2 cm) long, and a complete joint-penetration groove weld minimum 6" (15.2 cm) long, and then perform a visual test inspection (VT) on each weld using the following guide and acceptance criteria.

Visual Test Inspection Report

Trainee Inspector _____   Instructor _____

Discontinuity Category	Acceptance Criteria	¼" Fillet Weld	CJP Groove Weld
Crack Prohibition	Any crack shall be unacceptable regardless of size or location	Acceptable _____ or Reason for rejection _____	Acceptable _____ or Reason for rejection _____
Weld/Base Metal Fusion	Complete fusion shall exist between adjacent layers of weld metal and base metal	Acceptable _____ or Reason for rejection _____	Acceptable _____ or Reason for rejection _____
Crater Cross Section	All craters shall be filled to provide the specified weld size	Acceptable _____ or Reason for rejection _____	Acceptable _____ or Reason for rejection _____
Weld Profiles	Weld profiles shall be in accordance with page 6.15 figure 14	Acceptable _____ or Reason for rejection _____	Acceptable _____ or Reason for rejection _____
Fillet Weld Size	The specified nominal fillet weld size tolerance is ±¹⁄₁₆ inch	Acceptable _____ or Reason for rejection _____	N/A
Undercut	Undercut shall not exceed ¹⁄₃₂" for any accumulated length up to 2 inches	Acceptable _____ or Reason for rejection _____	Acceptable _____ or Reason for rejection _____
Porosity	The sum of visible porosity ¹⁄₃₂" or greater shall not exceed ⅜" in any linear inch and shall not exceed ¾" in any linear 12 inches of weld	Acceptable _____ or Reason for rejection _____	Acceptable _____ or Reason for rejection _____
Complete Weld	Circle accept or reject for fillet and groove final inspection	Acceptable _____ or Reason for rejection _____	Accept or Reject

Trainee Inspector's Signature _____   Date _____

Instructor's Verification: PASS/FAIL Signature _____

Trade Terms Introduced in This Module

Arc blow: The deflection of an arc from its intended path by magnetic forces.

Brazing: A method of joining metal using heat and a filler metal with a melting point above 842°F (450°C). Unlike welding, the base metal is not melted during the brazing process.

Code: A document that establishes the minimum requirements for a product or process. Codes can be, and often are, adopted as laws.

Defect: A discontinuity or imperfection that renders a part of the product or the entire product unable to meet minimum acceptable standards or specifications.

Discontinuity: A change or break in the shape or structure of a part that may or may not be considered a defect, depending on the code.

Embrittled: Metal that has been made brittle and that will tend to crack with little bending.

Essential variable: Items in a welding procedure specification (WPS) that cannot be changed without requalifying the WPS.

Hardenable materials: Metals that have the ability to be made harder by heating and then cooling.

Homogeneity: The quality or state of having a uniform structure or composition throughout.

Inclusion: Foreign matter introduced into and remaining in a weld.

Laminations: Cracks in the base metal formed when layers separate.

Non-essential variable: Items in a welding procedure specification (WPS) that can be changed without requalifying the WPS.

Notch toughness: The ability of a material to absorb energy in the presence of a flaw such as a notch.

Procedure qualification record (PQR): The document containing the results of the non-destructive and destructive testing required to qualify a WPS.

Radiographic: Describes images made by passing X-rays or gamma rays through an object and recording the variations in density on photographic film.

Specification: A document that defines in detail the work to be performed or the materials to be used in a product or process.

Standard: A document that defines how a code is to be implemented.

Supplemental essential variable: The variables that must be considered when notch toughness requirements are invoked.

Transducer: A device that converts one form of energy into another.

Underbead cracking: Cracking in the base metal near the weld, but under the surface.

Welding procedure qualification: The demonstration that welds made following a specific process can meet prescribed standards.

Additional Resources

This module presents thorough resources for task training. The following resource material is suggested for further study.

2014 Technical Training Guide. Current Edition. Cleveland, OH, USA: The Lincoln Electric Company. **www.lincolnelectric.com**

AWS B3.0 Standard Qualification Procedure. Miami, FL: American Welding Society.

B1.10M/B1.10:2009 Guide for the Nondestructive Examination of Welds. Miami, FL: American Welding Society.

Certification Manual for Welding Inspectors. Miami, FL: American Welding Society.

Visual Inspection Workshop Reference Manual, AWS VIW-M-2008. Miami, FL: American Welding Society.

Welding Inspection Technology. Miami, FL: American Welding Society.

Figure Credits

Section Review Answer Key

Answer	Section Reference	Objective
Section One		
1. b	1.1.0	1a
2. b	1.2.1	1b
Section Two		
1. c	2.1.1	2a
2. b	2.2.2	2b
3. a	2.3.3	2c
4. d	2.4.1	2d
Section Three		
1. a	3.1.1	3a
2. a	3.2.0	3b
3. c	3.3.1	3c
4. b	3.4.0	3d
Section Four		
1. c	4.1.0	4a
2. c	4.2.2	4b
3. b	4.3.4	4c

NCCER CURRICULA — USER UPDATE

NCCER makes every effort to keep its textbooks up-to-date and free of technical errors. We appreciate your help in this process. If you find an error, a typographical mistake, or an inaccuracy in NCCER's curricula, please fill out this form (or a photocopy), or complete the online form at **www.nccer.org/olf**. Be sure to include the exact module ID number, page number, a detailed description, and your recommended correction. Your input will be brought to the attention of the Authoring Team. Thank you for your assistance.

Instructors – If you have an idea for improving this textbook, or have found that additional materials were necessary to teach this module effectively, please let us know so that we may present your suggestions to the Authoring Team.

NCCER Product Development and Revision

13614 Progress Blvd., Alachua, FL 32615

Email: curriculum@nccer.org
Online: www.nccer.org/olf

❑ Trainee Guide   ❑ Lesson Plans   ❑ Exam   ❑ PowerPoints   Other _____

Craft / Level: _____   Copyright Date: _____

Module ID Number / Title: _____

Section Number(s): _____

Description: _____

_____

_____

_____

_____

Recommended Correction: _____

_____

_____

_____

_____

Your Name: _____

Address: _____

_____

Email: _____   Phone: _____

29107-15

SMAW – Equipment and Setup

Overview

Good welding results from a welder's skill, knowledge, and the proper use of technology. Practice develops skill and study increases knowledge. This module presents the tools and equipment used for shielded metal arc welding (SMAW) and how to prepare them for use.

Module Seven

Trainees with successful module completions may be eligible for credentialing through the NCCER Registry. To learn more, go to **www.nccer.org** or contact us at **1.888.622.3720**. Our website has information on the latest product releases and training, as well as online versions of our *Cornerstone* magazine and Pearson's product catalog.

Your feedback is welcome. You may email your comments to **curriculum@nccer.org**, send general comments and inquiries to **info@nccer.org**, or fill in the User Update form at the back of this module.

This information is general in nature and intended for training purposes only. Actual performance of activities described in this manual requires compliance with all applicable operating, service, maintenance, and safety procedures under the direction of qualified personnel. References in this manual to patented or proprietary devices do not constitute a recommendation of their use.

Objectives

When you have completed this module, you will be able to do the following:

1. Identify SMAW-related safety practices and explain how electrical characteristics apply to SMAW.
   a. Define SMAW and identify related safety practices.
   b. Explain how various electrical characteristics apply to SMAW.
2. Identify and describe SMAW equipment.
   a. Identify and describe various types of SMAW machines.
   b. Identify and describe SMAW welding cable and connectors.
   c. Identify common tools used to clean a weld.
3. Explain how to set up and start SMAW equipment.
   a. Explain how to set up SMAW equipment.
   b. Explain how to start, stop, and maintain SMAW equipment.

Performance Task

Under the supervision of your instructor, you should be able to do the following:

1. Set up a machine for SMAW.

Trade Terms

Conductor
Duty cycle
Governor
Motor-generator

Polarity
Rectifier
Step-down transformer

Industry Recognized Credentials

If you are training through an NCCER-accredited sponsor, you may be eligible for credentials from NCCER's Registry. The ID number for this module is 29107-15. Note that this module may have been used in other NCCER curricula and may apply to other level completions. Contact NCCER's Registry at 888.622.3720 or go to **www.nccer.org** for more information.

Contents ──────────

Topics to be presented in this module include:

Figures and Tables

1.0.0 SMAW

Objective

Identify SMAW-related safety practices and explain how electrical characteristics apply to SMAW.

   a. Define SMAW and identify related safety practices.
   b. Explain how various current characteristics apply to SMAW.

Trade Terms

Conductor: A material that will support the flow of an electrical current. Copper wire is the most common conductor.

Polarity: Refers to the direction of electrical current flow in a DC welding circuit.

Rectifier: An electronic device that converts AC voltage to DC voltage.

Step-down transformer: An electrical device that uses two wire coils of different sizes, referred to as the primary and secondary windings, to convert a higher voltage to a lower voltage through induction. Primary windings are connected to the primary power source; the secondary windings provide the reduced voltage to the process.

Shielded metal arc welding (SMAW) is often referred to as stick welding. This welding process can use either alternating current (AC) or direct current (DC) power, with adjustments made on the welding machine to regulate the level of current. The electrodes used, also referred to as welding rods, are typically coated with a clay-like mixture of fluorides, carbonates, oxides, metal alloys, and binding materials to provide a gas shield for the weld puddle. This coating hardens and covers the weld during its cooling cycle. The coating protects the weld from the atmosphere, which can cause weld degradation; then it is removed to expose the weld beneath.

1.1.0 SMAW Safety

Welding can be a dangerous profession, not only because of the welding itself, but also due to the environments in which the work is often performed.

Any welder must consider a number of safety issues associated with welding. Complete safety coverage is provided in *Welding Safety*. If you have not completed that module, do so before continuing. Primary safety concerns involve the following:

- Moving welding equipment
- Electrical hazards
- Lifting hazards
- Working at heights
- Welding hazards

1.1.1 Moving Welding Equipment

Welding in the shop usually means that you are using a welding machine that is either set up on a stationary platform in or near a welding booth, or is on a portable cart that can be rolled around the shop floor. When welding is performed in the field, the welding machines may be mounted on the back of a truck, on a skid, or on a mobile wagon or trailer. *Figure 1* shows a welder using a mobile SMAW welding machine outdoors.

There will be times when the welding machines must be lifted to some elevated location. Whenever a welding machine is moved, it becomes a potential hazard. Machines can fall off their carts or be dropped. Even a 40-pound welding machine can seriously injure a foot or hand when dropped from its cart. Consider the damage a mobile welding machine could do if it detached from its tow vehicle, if the brakes or chocks allowed it to roll down an incline, or if it fell due to poor rigging. Be cautious and observant as welding equipment is being moved by any means.

29107-15_F01.EPS

Figure 1 A mobile SMAW machine being used outdoors.

1.1.2 Electrical Hazards

Electric welding machines typically require 110VAC, 120VAC, 208/240VAC, or 480VAC power sources. Regardless of the voltage level, be cautious of the power supply. Electrical shock or electrocution can result from contact with the electrical power supplied to a welding machine. Always know where the main circuit breaker is located for the outlet supplying power to your machine. Whenever possible, turn that breaker off before connecting or disconnecting the welding machine's corded plug. Always make sure the machine itself is switched off before connecting and disconnecting power cords.

Remember that an arc welder operates on the principle of creating an electrical arc. The potential for the arc occurs between the two welding machine leads. Perhaps more importantly, that potential exists for any other conductor of electricity that bridges the gap between the leads. Respect the welding leads whenever the machine is on; and always be aware of where they are and what is nearby when laying them down.

Another concern with electricity is moisture. Electricity, water, and the human body are a hazardous mix. Make sure that all the welding cables are in good condition, without nicks in the insulation that could allow water in.

1.1.3 Lifting Hazards

Welders sometimes forget the basics of lifting as they transport their welding machines, cables, and even the parts to be welded. Use rigging or lifting devices whenever needed. Get help if you must lift a heavy object by hand. Always remember to use tag lines when hoisting heavy objects that tend to spin or swing.

1.1.4 Working at Heights

Welders may have to be suspended from the side of a structure to weld something. Remember to wear approved fall arrest and fall protection devices when performing such work. Make sure to inspect your fall protection equipment items each time you use them, and be certain that they are current and in the best possible condition.

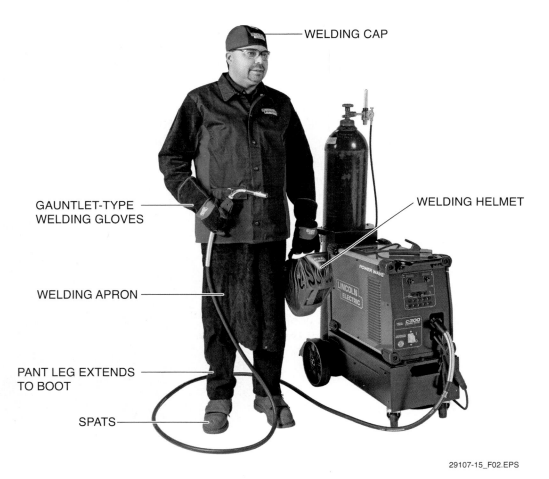

WELDING CAP

GAUNTLET-TYPE WELDING GLOVES

WELDING HELMET

WELDING APRON

PANT LEG EXTENDS TO BOOT

SPATS

29107-15_F02.EPS

Figure 2 Appropriate personal protective equipment (PPE).

1.1.5 Other Welding Hazards

Although this module addresses the preparation of SMAW equipment, there may be a time when you must test-operate a welding machine before using it. When that situation occurs, make sure that you are wearing all the required personal protective equipment (PPE) and that the work area meets all safety requirements. Do not compromise safety, even for a test run.

Many welding machines are coupled with gas- or diesel-fired combustion engines. The engines turn a generator to create the electrical power needed to operate the machine. Remember that combustion engines are extremely hot and have their own set of safety concerns. For example, operating a diesel-fired welding machine in an enclosed or stagnant area could lead to breathing trauma. Flammable fuels must also be handled carefully. Be familiar with all of the relevant safety precautions for the specific equipment being used.

The best way to prevent an accident is to avoid putting yourself in a compromising position. PPE (*Figure 2*) is essential; it will help keep you from being injured, or worse. While welding, always use safety glasses with a full face shield or a helmet. Use corrective lenses if needed; some companies will not allow contact lenses to be worn during welding activities. Never view an electric arc directly or indirectly without using a properly tinted lens. Wear high-top safety shoes or boots. Make sure that the tongue and lace area of the footwear will be covered by a pant leg or spats. Wear a 100-percent cotton cap with no synthetic mesh material included in its construction. The bill of the cap should point to the rear to cover the back of the neck.

1.2.0 Welding Current

In electrical terms, current is the flow of electrons or electrical charge from one point to another. A source of stored electrical energy, such as the battery in your car, would be where the flow of electrons starts. Electrical current will not flow unless a complete circuit is made. In the case of a car, for example, the battery has positive and negative connections. This enables the electric charge to complete a path or circuit from pole to pole, while passing through devices on your car, such as loads and control switches.

The unit of measurement for current is the ampere (abbreviated amp or A). The number of amps produced by the welding machine determines the amount of heat available to melt the workpiece and the electrode. The current is increased or decreased according to the size of electrode being used and the position in which the welding is being performed.

For arc welding to work, an arc must first be created. It is the intense heat of this arc that melts the base metal. This arc is created by bringing the positive and negative electrical connections of the power source (the welding machine) close enough together that the circuit can be completed. To arc weld, an electrical current flow between the two welding machine leads must be established. The base metal being welded and the electrodes used as conductors complete the circuit.

The arc that is produced during SMAW is the result of an electrical current that arcs across a gap between the tip of the electrode and the surface of the workpiece, as long as the workpiece clamp from the SMAW lead is attached to the workpiece (*Figure 3*). The current originates in the transformer or generator of the welding machine, is routed through one welding lead, then through the workpiece, and finally returns to the transformer or generator through the other welding lead. If the electrode tip is touched and held against the workpiece instead of allowing for a gap between the two, the electrode will pass the electricity directly to the workpiece without establishing an arc. As long as the electrode is in physical contact with the base metal, it has the same effect as connecting two wires together. The maximum available current from the welding machine will be passed through this completed circuit. However, no welding will take place in this condition as no continuous arc has been established. The intense heat of an arc crossing the gap is needed to heat the workpiece and the electrode to a molten state. Therefore, the electrode must remain close to the metal, but without making direct contact.

Figure 3 Striking an arc.

1.2.1 Types of Welding Current

SMAW power supplies are designed to produce an output that has a nearly constant current. The current may be either AC or DC. The type of current used depends on the type of welding to be done and the welding equipment available. Voltage is the measure of the electromotive force or pressure that causes current to flow in a circuit.

Alternating current (AC) in welding machines is derived from either a transformer-type welding machine or an alternator-type machine. The transformer-type machine changes high-voltage, low-current commercial AC power to the low voltage, high-current power required for welding. An alternator-type machine uses an electric or fuel-driven motor to turn a generator to produce the current needed for welding.

An AC voltage is developed by rotating conductors in a magnetic field, as shown in *Figure 4(A)*, which symbolizes a generator in its simplest form and the alternating current waveform it produces. AC power alternates constantly from positive to negative, as shown. The waveform is known as a sine wave. The top half of the sine wave is considered positive. The current follows the voltage. In the bottom half of the cycle, the voltage is considered negative, with the current continuing to follow the voltage. In one complete

Striking an Arc

Two general methods of striking the initial arc are scratching the electrode on the workpiece and tapping the electrode on the workpiece. The scratching method is generally used by most beginners and even by many experienced welders. However, it is usually not acceptable for code or finished-product welding if arc strikes are within the area covered by the weld. Some of the newer inverter power sources do not activate the open-circuit voltage until after an electrode touches a surface. With these machines, a scratching motion must be accomplished in order to activate the required open-circuit voltage and start the arc.

cycle, the waveform crosses the zero point twice. The peaks or extreme high or low points on the wave indicate the maximum level of the voltage, either positive (on the top) or negative (on the bottom). The number of cycles completed in one second is called the frequency. Frequency varies depending on the number of poles (magnetic field coils) in the generator and on the rotational speed of the rotor. Frequency, or cycles per second, is stated in Hertz (Hz). The standard frequency in the United States is 60 Hz. In many other regions of the world, the standard power frequency is 50 Hz.

In a power plant, huge turbine generators, like the one shown in *Figure 4(B)* create the AC voltage that is distributed to homes and businesses.

Direct current (DC) is electrical current that has no frequency. It does not travel from positive to negative and negative to positive within the same operation. Rather, the waveform for DC power resembles a straight line. Direct current in welding is derived either from a transformer-**rectifier** type of system or from a generator. In a transformer-rectifier type of system, AC voltage first is reduced (transformed) to a lower voltage level by a transformer, then it is converted (rectified) into DC voltage. Most engine-powered SMAW welding machines produce DC current through a generator. SMAW units that plug into 60 Hz AC power generally produce DC current using a transformer and a rectifier.

1.2.2 Polarity

Polarity only applies to DC current. Polarity in welding is determined by the way the welding leads are connected either to the welding machine or to the workpiece. DC current travels from the negative pole to the positive pole. The minus-marked terminal on the welding machine indicates the negative pole; the plus-marked terminal indicates the positive pole. In welding or in any other DC circuit, one cannot change that process. You can, however, change the path the current takes in going from the negative pole to the positive pole. In welding, this can be done simply by changing the welding lead connections at the machine or, in newer machines, by changing a switch position on the machine.

Most welding leads are configured with one end of one lead equipped with an electrode holder and the other lead equipped with a clamping device to attach to the workpiece. Both of the other ends of these two cables typically are equipped with ring-type or plug-in connectors for attaching the leads to the welding machine.

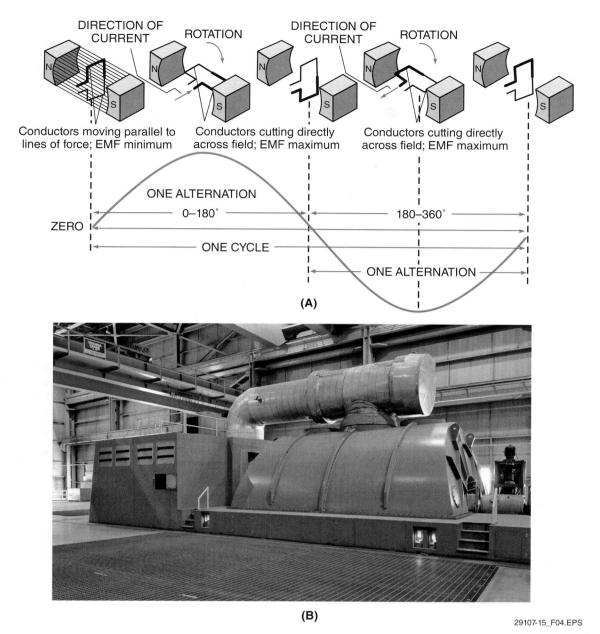

ONE ALTERNATION
0–180°

ONE ALTERNATION
180–360°

ZERO

ONE CYCLE

(A)

(B)

29107-15_F04.EPS

Figure 4 AC development of an AC voltage waveform.

When the electrode-holder lead is connected to the plus (positive) terminal on the machine and the lead with the clamping device is connected to the minus (negative) terminal on the welding machine, direct current electrode positive (DCEP) polarity is established. If these leads are manually switched at the welding machine terminals or through the operation of the manual polarity switch located on the machine, the electrode-holder lead becomes the negative lead and the workpiece clamping device lead becomes the positive lead. This is referred to as direct current electrode negative (DCEN) polarity. *Figure 5* shows labeled diagrams of DCEP and DCEN hookups.

> CAUTION
>
> On machines that have polarity switches, do not switch the polarity during operation. This may damage the machine.

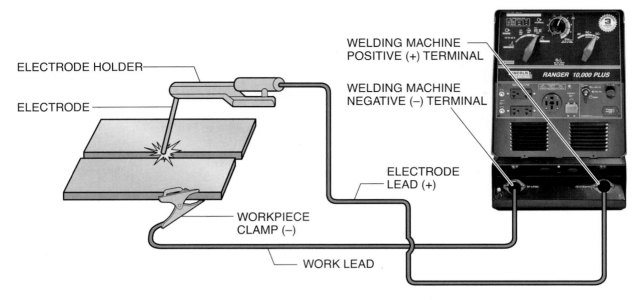

ELECTRODE HOLDER

ELECTRODE

WELDING MACHINE POSITIVE (+) TERMINAL

WELDING MACHINE NEGATIVE (−) TERMINAL

ELECTRODE LEAD (+)

WORKPIECE CLAMP (−)

WORK LEAD

DIRECT CURRENT ELECTRODE POSITIVE (DCEP) HOOKUP

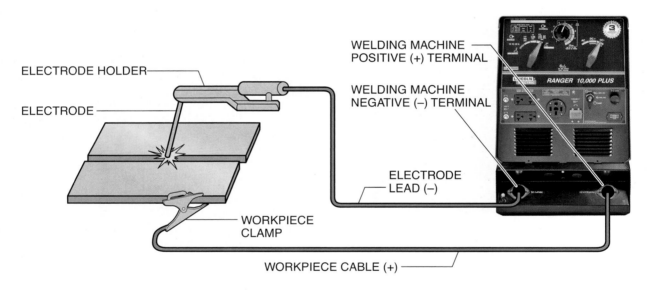

ELECTRODE HOLDER

ELECTRODE

WELDING MACHINE POSITIVE (+) TERMINAL

WELDING MACHINE NEGATIVE (−) TERMINAL

ELECTRODE LEAD (−)

WORKPIECE CLAMP

WORKPIECE CABLE (+)

DIRECT CURRENT ELECTRODE NEGATIVE (DCEN) HOOKUP

29107-15_F05.EPS

Figure 5 DCEP and DCEN hookups.

1.2.3 Characteristics of Welding Current

The voltage supplied to a welding machine is reduced by a **step-down transformer**. The step-down transformer reduces the voltage level and raises the current level. The higher current allows the arc to occur at a lower voltage potential that reduces the risk of electrocution. However, electrical shock in welding remains a potential hazard that can be reduced or eliminated only by using proper PPE and applying safe welding practices.

Figure 6(A) shows a representation of a step-down transformer. *Figure 6(B)* shows the physical appearance of a transformer. It consists of coils of wire wound around an iron coil.

Voltage is the measure of the electromotive force or pressure that causes current to flow in a circuit. Two types of voltage are associated with welding current: open-circuit voltage and operating voltage. Open-circuit voltage is the voltage present when the welding machine is on but no arc is present. For SMAW, the open-circuit voltage

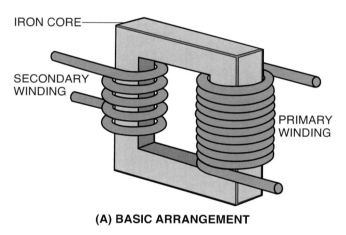

(A) BASIC ARRANGEMENT

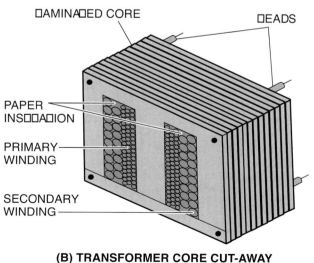

(B) TRANSFORMER CORE CUT-AWAY

29107-15_F06.EPS

Figure 6 A step-down transformer.

Properties of an Arc

An arc produces a temperature in excess of 6,000°F (3,316°C) at the tip of an electrode. This extreme heat is what melts the base metal and electrode, producing a pool of molten metal often referred to as a weld puddle. As the weld puddle solidifies behind the weld, the bond is formed.

potential between the two leads is usually between 50 and 80 volts.

Operating voltage, or arc voltage, is the voltage measured after the arc has been struck. This voltage is lower than open-circuit voltage and usually measures between 18 and 45 volts.

A higher, open-circuit voltage is required to establish the arc because the air gap between the electrode and the work has high resistance to current flow. Once the arc has been established, less voltage is needed, and the welding machine automatically compensates by lowering the voltage. This characteristic of an SMAW machine is called variable-voltage, constant-current power and is shown as a volt-ampere output curve in *Figure 7*. Operating voltage is dependent on the current range selected. However, SMAW voltage can be altered somewhat by a change in the arc gap controlled by the welder.

Amperage is a measurement of the electric current flowing in a circuit. The unit of measurement for current is the ampere (amp or A). The number of amps produced by the welding machine determines the amount of heat available to melt the workpiece and the electrode. The current is increased or decreased according to the size of electrode being used and the position in which the welding is being performed.

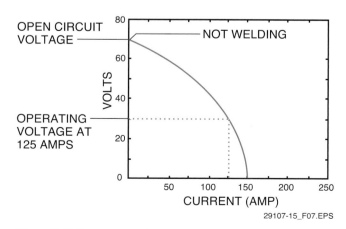

29107-15_F07.EPS

Figure 7 Volt-ampere output curve.

Additional Resources

2014 Technical Training Guide. Current Edition. Cleveland, OH, USA: The Lincoln Electric Company. **www.lincolnelectric.com**

Stick Electrode Product Catalog. Current Edition. Cleveland, OH, USA: The Lincoln Electric Company. **www.lincolnelectric.com**

1.0.0 Section Review

1. When hoisting heavy objects that tend to spin or swing, always use a(n) _____.

   a. pickup line
   b. chain bridle
   c. tag line
   d. synthetic web sling

2. When viewing an alternating current (AC) sine wave, the top half of the cycle of the voltage path is considered _____.

   a. a variable-voltage
   b. the positive half
   c. a constant-current
   d. the negative half

SECTION TWO

2.0.0 SMAW EQUIPMENT

Objective

Identify and describe SMAW equipment.
  a. Identify and describe various types of SMAW machines.
  b. Identify and describe SMAW welding cable and connectors.
  c. Identify common tools used to clean a weld.

Trade Terms

Duty cycle: The percentage of a ten-minute period that a welding machine can continuously produce its rated amperage without overheating.

Governor: A device used to limit engine speed.

Motor-generator: A combination device in which the motor turns the shaft of a generator, which in turn produces an AC voltage.

Welding machines, like any other tools, have specific classifications, special types, and rated capacities. This section will explain the use of SMAW welding machines within these specified areas.

2.1.0 SMAW Machines

SMAW welding machines are classified by the type of welding current they produce: AC, DC, or AC/DC.

An SMAW welding machine classified as an AC machine will produce only AC welding voltage and current, and a DC machine will produce only DC welding voltage and current. If a welding machine produces both AC and DC welding voltage and current, it is classified as an AC/DC welding machine.

Welding machines that produce only DC welding voltage and current can be further classified by the characteristics of the welding current they produce. If the welding current can be varied between the higher open-circuit voltage and the lower operating voltage, it is classified as a variable-voltage constant-current power source. This type of welding machine is used for SMAW. If open-circuit and operating voltage ratings are nearly the same, the machine is classified as a constant-voltage or constant-potential DC welding machine. This type of machine is used for GMAW and FCAW. *Figure 8* shows the difference in outputs between constant-current and constant-voltage welding machines.

Several different types of basic welding machines are available to produce the current necessary for SMAW. They include transformers, transformer rectifiers, inverter power sources, and engine-driven generators.

Transformers, transformer-rectifiers, and electric motor-generators all require electrical power from commercial power lines to operate. This high-voltage, relatively low-amperage current from commercial power lines coming into the welding machine is called the primary current. This is related to the transformer in the machine. The windings connected to the incoming power source are called the primary windings.

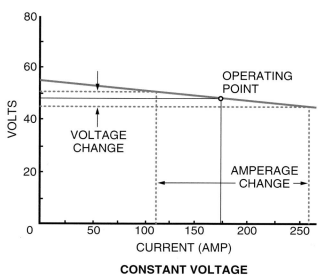

Figure 8 Constant-current and constant-voltage output curves.

29107-15_F08.EPS

The primary current first flows from the power lines through an electrical circuit breaker or fused disconnect switch. From the disconnect switch, the current flows to a receptacle. Alternatively, the welding machine is permanently and directly wired into the power disconnect. The disconnect switch can be used to shut off the flow of power. If necessary, the welding machine power cord is plugged into the receptacle. The primary current required for welding machines can be 120VAC single-phase, 208–240VAC single-phase or three-phase, or 480VAC three-phase. *Figure 9* shows a typical welding machine circuit.

> **WARNING!**
>
> Coming into contact with the primary power source of a welding machine can cause electrocution. Ensure that welding machines are properly grounded.

2.1.1 Transformer Welding Machines

Transformer welding machines without rectifiers produce AC welding current only. They use a step-down transformer, which converts high voltage (line voltage) from commercial power lines to a lower voltage. The construction of the transformer also allows the output current to be significantly higher than the current coming into the machine. This lower-voltage, high-current power supply is needed to arc weld.

The primary power required for a transformer welder may be 240VAC single-phase or three-phase, or 480VAC three-phase. Special light-duty transformer-based welding machines used for sheet metal work are designed to be plugged into a 120VAC outlet. However, most light-duty transformer welding machines operate more effectively on 240VAC primary power. Heavy-duty industrial transformer welders require 480VAC three-phase primary power.

Transformer welders are not as common as other types of welding machines on the job site, but they are used for special jobs. A transformer welder has an On/Off switch, an amperage control, and terminals for connecting the electrode lead and the workpiece lead. *Figure 10* shows the components of a typical transformer welding machine.

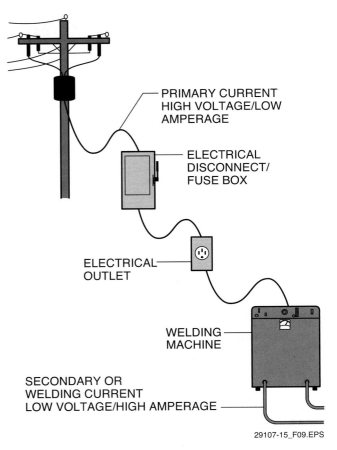

PRIMARY CURRENT HIGH VOLTAGE/LOW AMPERAGE

ELECTRICAL DISCONNECT/ FUSE BOX

ELECTRICAL OUTLET

WELDING MACHINE

SECONDARY OR WELDING CURRENT LOW VOLTAGE/HIGH AMPERAGE

29107-15_F09.EPS

Figure 9 Welding machine primary circuit.

2.1.2 Transformer-Rectifier Welding Machines

A rectifier is a device that converts AC current to DC current. A transformer-rectifier welding machine uses a transformer to reduce the primary voltage to the lower welding voltage. This voltage comes from the secondary winding of the transformer. The rectifier then rectifies the power from AC to DC. Transformer-rectifier welding machines can be designed to produce AC and DC welding current or DC current only. Transformer-rectifier units that produce both AC and DC welding current are usually intended for lighter duty than those that produce DC current only. Depending on their size, transformer-rectifier welding machines may require 208-240V single-phase or three-phase power, or 480V three-phase power.

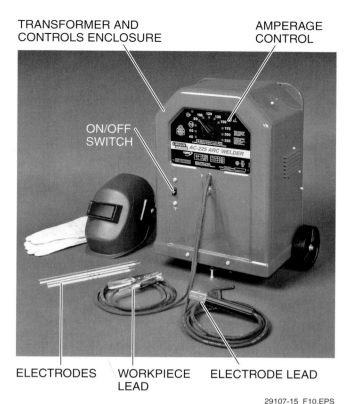

TRANSFORMER AND CONTROLS ENCLOSURE

AMPERAGE CONTROL

ON/OFF SWITCH

ELECTRODES   WORKPIECE LEAD   ELECTRODE LEAD

29107-15_F10.EPS

Figure 10 A basic transformer welding machine.

Transformer-rectifier units can also be designed to produce variable-voltage constant-current or constant-voltage DC welding current. Some are even designed to provide both variable and constant voltage; those are referred to as multi-process power sources since they can be used for many welding processes. Refer to the manufacturer's documentation to determine how the machine is configured.

Transformer-rectifiers have an On/Off switch and an amperage control. The welding cables (electrode lead and workpiece lead) are connected to terminals. They often have selector switches to select DCEP or DCEN, or to select AC welding current if it is available. If there is no selector switch, the cables must be manually changed on the machine terminals to select DCEP or DCEN operation. *Figure 11* shows the various parts of a typical transformer-rectifier welding machine.

Multiple transformer-rectifiers are available grouped into a single cabinet, called a pack.

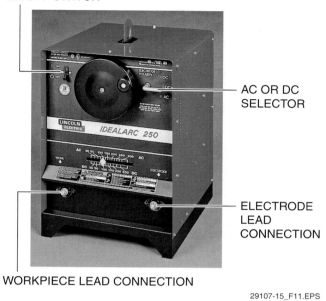

ON/OFF SWITCH

AC OR DC SELECTOR

ELECTRODE LEAD CONNECTION

WORKPIECE LEAD CONNECTION

29107-15_F11.EPS

Figure 11 A transformer-rectifier welding machine.

When the pack contains six welding machines, it is called a six-pack; when it contains eight welding machines, it is called an eight-pack (*Figure 12*).

The pack in *Figure 12* has eight individual welding machines. They receive their electrical power from a power distribution panel at the end of the rack. The panel has one switch that allows all the machines to be turned off at once. Each individual welding machine has its own amperage control and welding cables. In addition, each welding machine has a receptacle for a remote current control. When the remote current control is plugged in, it can be used to adjust the amperage from a remote location without going back to the pack. The pack unit also has 120V receptacles for power tools and lights.

Packs are mounted on skids and have attachments for lifting so they can be moved easily. They require only one primary power connection, which is normally 480VAC three-phase, since a significant power supply is needed to power all the units at once. Packs are common on construction sites and existing facilities during rebuilds when many welding machines are required.

AC/DC Machines

Transformer-rectifier welding machines that can provide either AC or DC current bypass the rectifier circuit when welding with AC current. In this situation, the welding machine functions like a transformer welding machine.

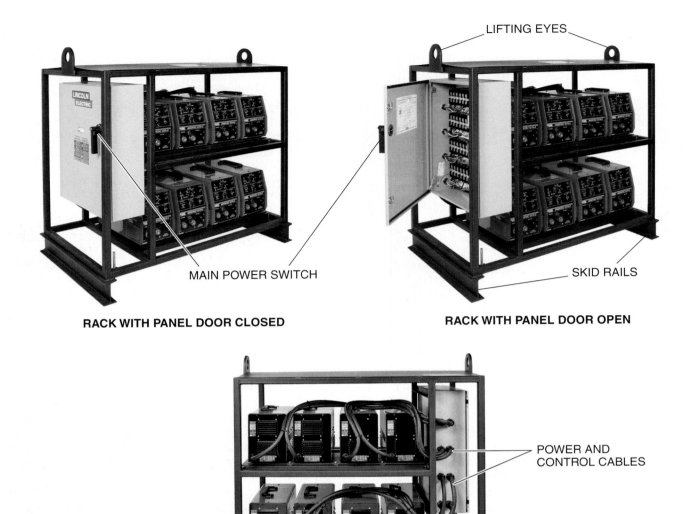

RACK WITH PANEL DOOR CLOSED

RACK WITH PANEL DOOR OPEN

REAR VIEW OF WELDING MACHINE

29107-15_F12.EPS

Figure 12 Eight-pack welding machine.

2.1.3 *Inverter Power Sources*

Inverter-type welding machines provide a smaller, lighter power source with a faster response time and much more waveform control. Inverter power sources are used where limited space and portability are important factors. In contrast to the transformer-types welding machine, an inverter machine first converts the AC input voltage to DC in a rectifier. The resulting DC is then switched on and off at a high rate, resulting in a high-frequency DC voltage, which is then fed into a transformer. The output of the transformer is then converted to DC in a rectifier. A significant advantage of inverter power supplies is that they produce a very stable DC output, without the ripple effect that results at 60 Hz. This results in a very smooth and steady DC welding arc. The controls on these machines vary according to size and application. *Figure 13* shows a close-up view of an inverter power source.

Another advantage lies in size reduction; inverter machines use much smaller transformers than transformer machines, because much of the work is done by lightweight electronic components.

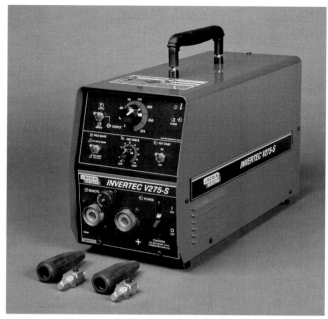

Figure 13 An inverter power source.

29107-15_F13.EPS

The inverter power source shown is designed for outdoor use. It is used either as a single, stand-alone unit or can be mounted in a rack with others. It supports both gas tungsten arc welding (GTAW) and SMAW processes, and can be used for air-carbon arc gouging. With up to 275 amps of output power, it allows the welder to choose from a greater variety of stick electrodes, especially the E6010 and E7018 electrodes.

Another advantage of the inverter power supply is cost of operation. Inverter technology is more energy-efficient than a transformer-based system. An inverter unit may draw roughly 29 amps when operating at a 200 amp output on 230VAC single-phase power. The older transformer-based designs typically draw 50 to 60 amps at the same conditions. Over time, the electrical cost savings is quite significant, helping to return the investment in cost.

2.1.4 Engine-Driven Generator Machines

Welding machines can also be powered by gasoline or diesel engines. The engine is connected to a generator. Engine-driven generators produce DC welding current.

To produce welding current, the generator must turn at a required number of revolutions per minute. Engines that turn generators have governors to control the engine speed. Most governors have a welding speed switch. The switch can be set to idle the engine when no welding is taking place.

When the electrode is touched to the base metal, the governor will automatically increase the speed of the engine to the required rpm for welding. After about 15 seconds of a no-welding condition, the engine will automatically return to idle. This saves fuel and reduces wear on the generator. The switch can also be set to enable the engine to run continuously at welding speed. *Figure 14* shows an example of an engine-driven generator welding machine.

Engine-driven generators often have an auxiliary power unit to produce 120VAC current for lighting, power tools, and other electrical equipment. When 120VAC power is required to operate tools, the engine-driven generator must run continuously at its welding speed.

Engine-driven generators have both engine controls and welding machine controls. The engine controls vary with the type and size but normally include the following items:

- Starter
- Voltage gauge
- Temperature gauge
- Fuel gauge
- Hour meter

Engine-driven generators have an amperage control. To change polarity, there may be a polarity switch, or you may have to manually exchange the welding cables at the terminals.

Many engine-driven generators are mounted on trailers. This makes them portable so they can be used in the field where electricity is not available for other types of welding machines.

29107-15_F14.EPS

Figure 14 Engine-driven welding machine.

The disadvantage of engine-driven generators is that they are costly to purchase, operate, and maintain because of the added cost of the drive unit. However, for field operations where electrical power sources are often very limited, they are essential.

2.1.5 SMAW Machine Ratings

The capacity of a welding machine is determined by the amperage output of the machine at a given **duty cycle**. The duty cycle of a welding machine is based on a ten-minute period of time. It is the percentage of a ten-minute period that the machine can continuously produce its rated amperage without overheating. For example, a machine with a rated output of 300A at a 60-percent duty cycle can deliver 300A of welding current for six minutes out of every ten (60 percent of ten minutes) without overheating.

The duty cycle of a welding machine will generally be 10 percent, 20 percent, 30 percent, 40 percent, 60 percent, or 100 percent. A welding machine having a duty cycle of from 10 to 40 percent is considered a light-duty to medium-duty machine. Most industrial, heavy-duty machines for manual welding have at least a 60-percent duty cycle. Machines designed for automatic welding operations must be capable of a 100-percent duty cycle.

With the exception of 100-percent duty-cycle machines, the maximum amperage that a welding machine will produce is always higher than its rated capacity. A welding machine rated 300A with a 60-percent duty cycle will generally produce a maximum of 375A to 400A. However, since the duty cycle is a function of its rated capacity, the duty cycle will decrease as the amperage is raised over 300A. Welding at 375A with a welding machine rated at 300A will lower the duty cycle from 60 percent to about 30 percent.

Engines

The size and type of engine chosen is based on the capacity and power needs of the welding machine. Single-cylinder engines are used to power small generators. Four- and six-cylinder engines are used to power larger generators.

If welding continues for more than three out of every ten minutes, the machine will probably overheat.

If the amperage is set below the rated amperage, the duty cycle increases. Setting the amperage at 200A for a welding machine rated at 300A with a 60-percent duty cycle will increase the duty cycle to 100 percent. *Figure 15* shows the relationship between amperage and duty cycle.

2.2.0 Welding Cable and Connectors

This section covers the cables and connectors that place welding power into the operator's hands. In addition, this section describes some other common tools used in the welding process.

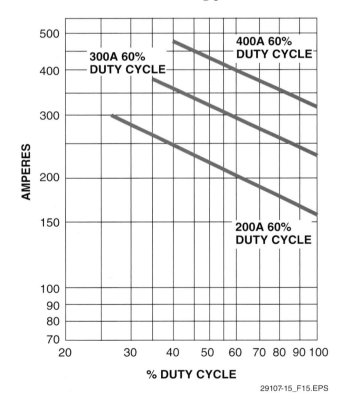

29107-15_F15.EPS

Figure 15 Amperage and duty cycle chart.

Welding Machine Circuit Breakers

Most welding machines have a heat-activated circuit breaker that will shut off the machine automatically when it overheats. The machine cannot be turned on again until it has cooled off and resets.

2.2.1 Welding Cable

Cables that carry welding current (*Figure 16*) are designed for maximum strength and flexibility. The conductors inside the cable are made of fine strands of copper wire. The copper strands are covered with layers of rubber reinforced with nylon or Dacron® cord.

The size of a welding cable is based on the number of copper strands it contains. Large-diameter cable has more copper strands and can carry more welding current. Typically the smallest welding cable size is No. 8 while the largest is No. 4/0 (spoken as *four-aught*).

When selecting welding cable size, the amperage load and the distance the current will travel must be taken into account. The longer the distance the current has to travel, the larger the cable must be to reduce voltage drop and heating caused by the extra electrical resistance of a long cable. This is similar to any other drop cord used on the job site. Longer cords must be a larger wire size to accommodate the load of the connected tools on the other end.

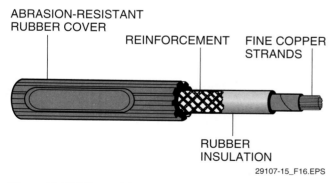

ABRASION-RESISTANT
RUBBER COVER

REINFORCEMENT   FINE COPPER STRANDS

RUBBER INSULATION

29107-15_F16.EPS

Figure 16 Welding cable.

When selecting welding cable, use the rated capacity of the welding machine for the cable amperage requirement. To determine the distance, measure both the electrode and ground leads and add the two lengths together. To identify the welding cable size required, refer to a recommended welding cable size table furnished by most welding cable manufacturers. *Table 1* shows typical welding cable sizes for various welding currents and cable distances.

Welding cables must be equipped with the proper end connections. End connections used on welding cables include lugs, quick disconnects, ground clamps, and electrode holders.

> **CAUTION**
>
> If the end connection is not tightly secured to the cable, the connection will overheat. An overheated connection will cause variations in the welding current and permanent damage to the connector and/or cable. Repair any connectors that overheat.

2.2.2 Lugs and Quick Disconnects

Lugs are used at the end of the welding cable to connect the cable to the welding machine current terminals. Lugs come in various sizes to match the welding cable size and are mechanically crimped onto the welding cable. Lugs, as shown in *Figure 17*, are normally ringed terminals to prevent the welding lead from pulling loose from the stud connection on the machine. Due to the current that passes through these connections, a sound electrical connection must be made.

Quick disconnects are also mechanically connected to the cable ends. They are insulated and serve to splice two lengths of cable together. Quick disconnects are connected or disconnected with a half twist. When using quick disconnects, make sure that they are tightly connected to prevent overheating or arcing in the connector. The DINSE®-style connector offers increased efficiency with a short, strong turn for a positive stop-lock action and maximum conductivity. Like all connectors, they must be kept clean of dirt, sand, mud, and other contaminants to ensure a good connection.

Table 1 Welding Cable Sizes

Machine Size in Amperes	Duty Cycle (%)	Copper Cable Sizes for Combined Lengths of Electrodes Plus Ground Cable				
		Up to 50 Feet	50–100 Feet	150 Feet	150–200 Feet	200–250 Feet
100	20	#8	#4	#3	#2	#1
180	20	#5	#4	#3	#2	#1
180	30	#4	#4	#3	#2	#1
200	50	#3	#3	#2	#1	#1/0
200	60	#2	#2	#2	#1	#1/0
225	20	#4	#3	#2	#1	#1/0
250	30	#3	#3	#2	#1	#1/0
300	60	#1/0	#1/0	#1/0	#2/0	#3/0
400	60	#2/0	#2/0	#2/0	#3/0	#4/0
500	60	#2/0	#2/0	#3/0	#3/0	#4/0
600	60	#3/0	#3/0	#3/0	#4/0	* * *
650	60	#3/0	#3/0	#4/0	* *	* * *

* * Use Double Strand of #2/0
* * * Use Double Strand of #3/0

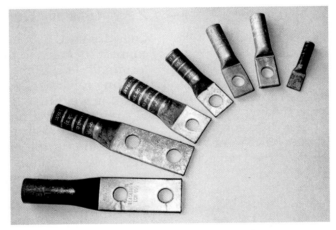

(A) LUGS

MALE

FEMALE

(B) QUICK DISCONNECTS

29107-15_F17.EPS

Figure 17 Lugs and quick disconnects.

2.2.3 Workpiece Clamps

Workpiece clamps (*Figure 18*) establish the connection between the end of the workpiece and the workpiece lead. Workpiece clamps are mechanically connected to the welding cable and come in a variety of shapes and sizes. The capacity or rating of a workpiece clamp is based on the amperage it can carry consistently without overheating.

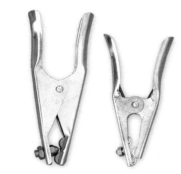

(A) LOW-CURRENT CLAMPS

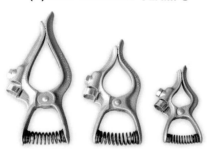

(B) HIGH-CURRENT CLAMPS

(C) ROTARY GROUND CLAMP

(D) MAGNETIC SWIVEL CLAMP

29107-15_F18.EPS

Figure 18 Workpiece clamps.

When selecting a workpiece clamp, be sure that it is rated at least as high as the rated capacity of the welding machine on which it will be used.

2.2.4 Electrode Holders

Electrode holders grasp the electrode and provide the electrical contact between the electrode and the end of the welding cable. Electrode holders are also mechanically connected to the welding cable and come in a variety of styles and sizes. Like the workpiece clamp, the rating of an electrode holder is the rated amperage it will consistently carry without overheating. The physical size of an electrode holder is generally proportional to its current capacity. A smaller 200 amp electrode holder will be light and easy to handle, but will not handle higher currents. When selecting an electrode holder, select the size for the welding current and electrodes being used. *Figure 19* shows two styles of electrode holders. The collet-style typically allows the use of more electrode, as it grips an electrode close to its tip.

2.3.0 Tools for Cleaning Welds

Even though some of the hand tools used to clean welds were presented in another module, it is worthwhile to mention some of them again and explain their functions. Hand tools such as files, chipping hammers, and wire brushes, as well as small power tools such as slag chippers and scalers, are common tools used by welders to prepare surfaces before welding and to clean welds after they are completed.

> **WARNING!**
> Always wear eye and hand protection when preparing and cleaning weldment surfaces. Remember that finished welds remain extremely hot for extended periods and can cause serious burns. Chipping slag and other coatings from finished welds is very hazardous and requires proper eye protection, including face shields.

2.3.1 Hand Tools

It is extremely important that the base metals are thoroughly cleaned and prepared before welding. Dirty or corroded surfaces are very difficult, if not impossible, to weld properly. Also, before any additional weld beads can be applied over previous beads, the previous pass must be cleaned of any slag that was deposited with first bead.

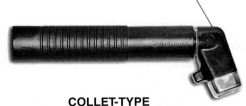

HEAD ROTATES ON HANDLE TO RELEASE
COLLET-TYPE ELECTRODE CLAMP OR TO
ADJUST CLAMP FOR ELECTRODE SIZE

**COLLET-TYPE
ELECTRODE HOLDER**

LEVER-OPERATED
JAW

**CLAMP-TYPE
ELECTRODE HOLDER**

29107-15_F19.EPS

Figure 19 Electrode holders.

Many mechanical methods have been developed to assist in these cleaning and slag-removing processes, including sandblasting and ultrasonic cleaning. However, many welders prefer simple manual procedures, using tools such as files, wire brushes, and chipping hammers. *Figure 20* shows some of the popular hand tools used by welders to prepare base metals and remove slag after a bead has been laid.

2.3.2 Pneumatic Cleaning and Slag Removal Tools

Weld slag chippers and needle scalers (*Figure 21*) are pneumatically powered tools. They are used by welders to clean surfaces and to remove slag from cuts and welds. Weld slag chippers and needle scalers are also excellent for removing paint or hardened dirt, but they are not very effective for removing surface corrosion. Weld slag chippers have a single chisel, while needle scalers have 18 to 20 blunt steel needles approximately 10" (≈25 cm) long. Most weld slag chippers can be converted to needle scalers with an attachment.

29107-15_F20.EPS

Figure 20 Typical hand tools used in welding.

NCCER – *Welding Level One*   29107-15

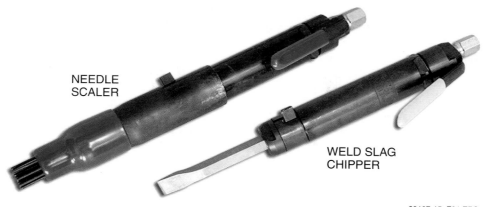

NEEDLE
SCALER

WELD SLAG
CHIPPER

29107-15_F21.EPS

Figure 21 Pneumatic weld slag chipper and needle scaler.

Additional Resources

2014 Technical Training Guide. Current Edition. Cleveland, OH, USA: The Lincoln Electric Company. **www.lincolnelectric.com**

2.0.0 Section Review

1. Heavy-duty industrial transformer welders are most likely to require _____.

    a. 120VAC single-phase power supply
    b. 208VAC single-phase power supply
    c. 240VAC single-phase power supply
    d. 480VAC three-phase power supply

2. The longer the welding leads, the larger the cable should be to _____.

    a. increase conductor resistance
    b. decrease conductor insulation
    c. increase current drop
    d. reduce voltage drop

3. Which of the following tools are more likely to be used for cleaning slag from a weld?

    a. Electric hammer drill, sander, or propane torch
    b. Files, wire brushes, and chipping hammers
    c. A metal pick, steel pry-bar, and drag chain
    d. Cleaning solutions, including alcohol, water, and detergent

3.0.0 SMAW EQUIPMENT SETUP AND STARTUP

Objective

Explain how to set up and start SMAW equipment.
  a. Explain how to set up SMAW equipment.
  b. Explain how to start, stop, and maintain SMAW equipment.

Performance Task

  1. Set up a machine for SMAW.

Because of the nature of electricity in general and specifically as it relates to SMAW, using proper procedure is imperative. SMAW welding equipment is not complicated but there are necessary preparations. This section covers some of the typical setup requirements and the operation of SMAW equipment.

3.1.0 SMAW Equipment Setup

In order to safely and efficiently weld using the SMAW process, the equipment must be properly selected and set up. Accomplishing proper SMAW equipment setup involves the following procedures:

- Selecting the proper SMAW equipment
- Choosing the machine's location
- Moving a welding machine into place
- Stringing welding cable
- Locating the workpiece clamp

3.1.1 Selecting the Proper SMAW Equipment

To select the proper SMAW machine, the following factors must be considered:

- *The welding process* – SMAW requires a variable-voltage power source with a constant current. Other processes may require constant-voltage power sources.
- *The type of welding current (AC or DC)* – Most SMAW processes use DC current.
- *The maximum amperage required* – Consider the composition of the material, its thickness, and other material properties to determine the amperage requirements of the machine.

- *The primary power requirements* – If an AC power source with the proper current rating and receptacle for the welding machine is not available, then an engine-driven unit will be needed.

3.1.2 Welding Machine Location

The welding machine should be located near the work to be performed to minimize lead length and provide convenient access to it as work progresses. Select a site where the machine will not be in the way but will be protected from welding or cutting sparks. There should be good air circulation to keep the machine cool, and the environment should be free from explosive or corrosive fumes. It should also be as free as possible from airborne dust and dirt. Welding machines have internal cooling fans that will pull these materials into the machine if they are present. The site should be free of standing water or water leaks. If an engine-driven generator is used, locate it so it can be easily refueled and serviced.

If the machine is to be plugged into an outlet, be sure the outlet has been properly installed by a licensed electrician. Also, be certain to identify the location of the electrical disconnect or circuit breaker for the outlet before plugging in the welding machine.

> **WARNING!**
> Never run an engine-driven generator indoors, due to the production of hazardous carbon monoxide (CO) and other pollutants.

3.1.3 Moving a Welding Machine

Large, engine-driven generators are mounted on a trailer frame and can be easily moved by a pickup truck or tractor using a trailer hitch. Other types of welding machines may have a skid base or may be mounted on steel or rubber wheels for manual positioning. Be careful when moving welding machines that are mounted on wheels. Some machines are top-heavy and may fall over in a tight turn or if the floor or ground is uneven or soft.

> **WARNING!**
> If a welding machine starts to fall over, do not attempt to hold it. Welding machines are very heavy and can cause severe injury if they roll over.

Many welding machines have lifting eyes. Lifting eyes are usually provided to move machines mounted on skids. Before lifting any welding machine, refer to the equipment specifications for the weight. Be sure the lifting device and tackle are rated at more than the weight of the machine. When lifting a welding machine, always use a proper shackle to connect rigging to the lifting eyes. Never attempt to lift a machine by placing the lifting hook directly in the machine eye. Before lifting or moving a welding machine, make sure the welding cables and other accessories are coiled and secure. *Figure 22* shows how to lift a welding machine. Lifts should be prepared and executed by qualified riggers.

3.1.4 *Stringing Welding Cable*

Before stringing welding cable, inspect the cable for damage. Cuts or breaks in the insulation must be repaired according to designated procedures. Breaks in the welding cable insulation could arc to equipment or to any metal surface the welding contacts. This will damage the surface and cause additional damage to the welding cable. If the welding cable cannot be repaired, it should be recycled.

Welding cables should be long enough to reach the work area but not so long that they must always be coiled. Welding cables must be strung to

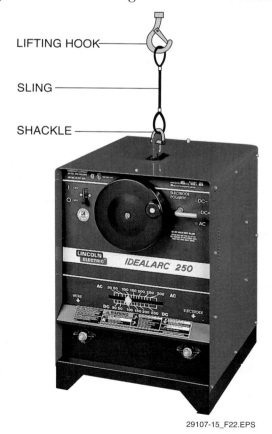

LIFTING HOOK

SLING

SHACKLE

29107-15_F22.EPS

Figure 22 Lifting a welding machine.

prevent tripping hazards or damage from traffic. Keep welding cables out of walkways and aisles. If welding cables must cross a walkway or aisle, string the cables overhead. If the cables cannot be strung overhead, use boards or ramps to prevent workers from tripping and to protect the cable from foot and vehicular traffic. *Figure 23* shows how to protect welding cables with a ramp.

When stringing welding cables, take care not to damage site equipment. Do not string welding cable over instrumentation wires and tubing. These can be easily damaged by the weight or movement of the heavy cables. If there are moving parts, keep the welding cables well away from pinch points and moving parts. If the equipment suddenly starts, the welding cables could be pulled into the equipment. When there is new construction or rebuild work taking place, equipment such as cable trays, conduit, piping, and pipe hangers may not be permanently attached. The weight of welding cables could cause these items to collapse. Check such equipment before stringing welding cables.

When stringing welding cable overhead or between floors, use rope to tie off the cables. The weight of the welding cables could damage connections or pull an individual holding onto the electrode holder off balance. Rope is relatively nonconductive (when dry) and non-abrasive, and will support the weight of the welding cable without damaging it. No significant weight should be placed on the connectors.

WELDING CABLES

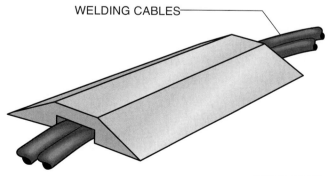

29107-15_F23.EPS

Figure 23 Protecting welding cables.

Recycling

Welding cables contain a tremendous amount of copper. The value of copper has been very high for some time now, even encouraging its consistent theft. They should never be disposed of. Instead, the cable should be turned over to a facility so that the copper can be recovered and recycled.

3.1.5 Locating the Workpiece Clamp

The workpiece clamp must be properly located to prevent damage to surrounding equipment. If the welding current travels through any type of bearing, seal, valve, or contacting surface, it could cause severe damage from arcing at these points, which would require replacement of the item. Carefully check the area to be welded and position the workpiece clamp so the welding current will not pass through any materials or components unnecessarily. If in doubt, ask your supervisor for assistance before proceeding.

> **CAUTION**
>
> Welding current can severely damage bearings, seals, valves, and contacting surfaces. Position the workpiece clamp to prevent welding current from passing through them.

Welding current passing through electrical or electronic equipment will also cause severe damage. Disconnect any other power sources or batteries to protect the electrical system before welding on any type of equipment. If welding will be performed near a battery, remove the battery completely from the equipment and/or work area. Batteries produce hydrogen gas, which is extremely explosive. A welding spark could cause the battery to explode.

> **WARNING!**
>
> Do not weld near batteries. A welding spark could cause a battery to explode, showering the area with battery acid.

Before welding on or near electronic or electrical equipment or cabinets, contact an electrician. The electrician will isolate the system to protect it.

> **CAUTION**
>
> The slightest spark of welding current can destroy sensitive electronic or electrical equipment. Have an electrician check the equipment and, if necessary, isolate the system before welding.

Workpiece clamps must never be connected to pipes carrying flammable or corrosive materials. The welding current could cause overheating or sparks, resulting in an explosion or fire.

The workpiece clamp must make good electrical contact when it is connected. Dirt and paint will inhibit the connection and cause arcing, resulting in overheating of the workpiece clamp. Dirt and paint also affect the welding current and can cause defects in the weld. Clean the surface before connecting the workpiece clamp. If the workpiece clamp is damaged and does not close securely onto the surface, repair or replace it. Many workpiece clamps have replaceable contact assemblies.

3.2.0 Starting SMAW Welding Machines

Starting an electrically powered welder is a simple matter of flipping a switch and making adjustments. The source of energy is constantly available. Starting a motor-driven welding machine is a bit more complicated than starting an electrically powered machine.

Electrically powered welding machines are plugged into an electrical outlet or hard-wired to a disconnect switch. The electrical requirements vary and will be on the equipment data plate displayed prominently on the machine. Machines requiring single-phase power will have a three-prong plug. Machines requiring three-phase power will have a four-prong plug. *Figure 24* shows examples of some grounded electrical plugs for welding machines. Each plug pattern relates to a particular current capacity for the power supply. This prevents connecting a welding machine to an insufficient power source.

> **WARNING!**
>
> Never use a welding machine until you identify the electrical disconnect location. In the event of an emergency, you must be able to quickly turn off the power to the welding machine at the disconnect to prevent injury or electrocution.

(A) SINGLE PHASE AND THREE PHASE PLUGS

(B) VARIATION OF A THREE PHASE PLUG

29107-15_F24.EPS

Figure 24 Grounded electrical plugs.

If a welding machine does not have a power plug, an electrician must connect the machine. The electrician will add a plug or will hardwire the machine directly into a disconnect switch.

Before welding can take place with an engine-driven welding machine, the engine must be checked and then started. As with a car engine, the engine powering the welding machine must also have routine maintenance performed.

3.2.1 Prestart Checks

Many facilities have prestart checklists that must be completed and signed prior to starting and operating an engine-driven welding machine. If your site has such a checklist, you must complete and sign it. If your site does not have a prestart checklist, perform the following basic checks before starting the engine:

- *Engine oil* – Check the oil when the engine has been shut off for at least 10 minutes. This allows the engine oil to drain back into the crankcase. Check the oil level using the engine oil dipstick. If the oil is low, add the proper amount of appropriate grade oil. Do not overfill.
- *Engine coolant* – Do not remove the radiator cap on a hot engine. Check the coolant level in the radiator if the engine is liquid-cooled. The best time to check the engine coolant is before the engine has been used, such as in the morning when the engine is cold. If the coolant level is low, add coolant. Make sure the radiator cap is seated properly.

Case History

Trailer Mud Flaps

A welder was given the job of reattaching a mud flap to the back of a refrigerated trailer. The loaded trailer was parked right next to the welding machine, and it looked to be a very simple job. Because it was so close, the employee reached over and connected the workpiece clamp to the front of the trailer and then pulled the cable with the electrode holder to the back end. After cleaning and positioning the bracket, the mud flap was welded to the trailer. The entire job took less than 15 minutes to complete.

When the customer returned to pick up the trailer, he noticed that the refrigeration unit was off and there was no power at the display panel. A service technician determined that the microprocessor on the refrigeration unit had failed and needed to be replaced. After further investigation, it was determined that the pressure and temperature sensors on the refrigeration unit had been in the path of the welding current. The current was too high for the microprocessor and burned out some of its components. As a result, the welding shop was required to pay more than $1,000 in parts and labor to replace the microprocessor.

The Bottom Line: If the welder had isolated the electronic components from the welding circuit, or at least had located the workpiece clamp close to the weld point, the damage would not have occurred.

- *Engine fuel* – Check the fuel level. The unit likely has a fuel gauge or sight glass. If the fuel is low, add the correct fuel (diesel or gasoline) to the fuel tank. The type of fuel required should be marked on the fuel tank. If not, contact your supervisor to verify the fuel required, and have the unit marked.

- *Engine battery* – If the battery is lead-acid, check the battery water level unless the battery is sealed. Add water if the battery water level is low.

- *Electrode holder* – Look at the electrode holder and make sure it is not touching the workpiece. If the electrode holder is touching the workpiece, it will arc and overheat the welding system when the welding machine is started.
- *Fuel shutoff valve* – Open the fuel shutoff valve if the equipment has one. If there is a fuel shutoff valve, it will be located in the fuel line between the fuel tank and the carburetor or fuel injector(s).
- *Hour meter* – Record the hours from the hour meter if the equipment has one. An hour meter records the total number of hours the engine runs. This information is used to determine when the engine needs to be serviced. The hours will be displayed on a gauge similar to an odometer.

- *Engine appearance* – Clean the engine and equipment after each use. Potential problems can be discovered before they can become larger issues if this is part of the daily maintenance routine. Refer to the manufacturer's instructions for specific areas of attention.

3.2.2 Starting the Engine

Most engines will have an On/Off ignition switch and a starter. They may be combined into a key switch similar to the ignition switch on a car. To start the engine, turn on the ignition switch and press the starter. Release the starter when the engine starts. The engine speed will be controlled by the governor. If the governor switch is set for idle, the engine will slow to an idle after a few seconds. If the governor is set to welding speed, the engine will continue to run at welding speed.

Small engine-driven welding machines may have an On/Off switch and a pull rope. These machines are started by turning on the ignition switch and pulling the cord, which is similar to starting a lawn mower. These engines do not have a battery.

Engine-driven welding machines should be started about five to ten minutes before they are needed for welding. This will allow the engine to warm up before a welding load is placed on it.

3.2.3 Stopping the Engine

If no welding is required for 30 minutes or longer, stop the engine by turning off the ignition switch. If you are finished with the welding machine for the day, also close the fuel valve if there is one. Allow the machine to cool before storing welding leads inside the engine compartment.

3.2.4 Preventive Maintenance

Engine-driven welding machines require regular preventive maintenance to keep the equipment operating properly. Most sites will have a preventive maintenance schedule based on the hours that the engine operates. In severe conditions, such as a very dusty environment or cold weather, maintenance may have to be performed more frequently. The responsibility for performing preventive maintenance will vary from site to site. Check with your supervisor to determine who is responsible for performing preventive maintenance.

When performing preventive maintenance, follow the manufacturer's guidelines. Typical tasks to be performed as part of a preventive maintenance schedule include the following:

- Changing the oil, gas filter, and air filter
- Checking/changing the antifreeze
- Greasing the undercarriage
- Repacking the wheel bearings
- Replacing worn drive belts
- Maintaining/replacing the spark plugs
- Maintaining the battery
- Cleaning surfaces

CAUTION

To prevent equipment damage, perform preventive maintenance as recommended by the site procedures or the manufacturer's maintenance schedule in the equipment manual.

GOING GREEN

Fuel-Powered Machine Maintenance

Proper maintenance of fuel-powered welding machine engines helps to protect the environment. If a fuel-powered engine is smoking or leaking fluids, have it turned in for proper maintenance or repair.

Additional Resources

Covering the Basics – Engine Driven Welder Maintenance. Al Nystrom, Technical Service Representative. Cleveland, OH, USA: The Lincoln Electric Company. **www.lincolnelectric.com**

3.0.0 Section Review

1. When lifting a welding machine, connect the lifting eye to the rigging using a _____.
   a. cable lug
   b. shackle
   c. quick disconnect
   d. swivel clamp

2. To prevent a battery from exploding when welding, you must avoid _____.
   a. exposing the battery to extremely cold temperatures
   b. locating the workpiece clamp close to the joint being welded
   c. disconnecting it from the circuit
   d. exposing the battery to sparks or arcs of any type

SUMMARY

Shielded metal arc welding is often referred to as stick welding. SMAW remains one of the most widely used forms of welding in the industry because of its simplicity and relatively low cost. However, it is not as foolproof as some of the newer technologies now used in the welding industry. It requires practice to be a successful SMAW welder. Whatever type of welding is being used, proper selection and setup of equipment and accessories are necessary to achieve a reliable weld while maintaining a safe working environment.

Included in this module are the welding currents associated with SMAW, the various types of SMAW machines, and some of the accessories needed to connect SMAW equipment. The steps in selecting and setting up SMAW equipment are also covered. SMAW is the basis upon which all other welding processes have been developed. As a widely used welding technique, it is important for you to become familiar with the equipment and develop the skills needed to become a successful SMAW welder.

---

1. SMAW is often referred to as _____.
   a. AC welding
   b. stick welding
   c. gap welding
   d. resistive welding

2. Make sure that your fall protection equipment _____.
   a. is tagged before each use
   b. is inspected prior to each use
   c. is made of nylon
   d. is rated to 600 lbs

3. In the SMAW process, an electrical path must be completed from the tip of the electrode to the surface of the workpiece through the creation of a(n) _____.
   a. gas
   b. arc
   c. electrode holder
   d. transformer

4. In AC current, the number of cycles completed in one second is called the _____.
   a. polarity
   b. travel
   c. frequency
   d. current value

5. The type of power that has no frequency since its waveform is actually a straight line is _____.
   a. direct current (DC)
   b. alternating current (AC)
   c. single-phase power
   d. three-phase power

6. A device that converts AC power to DC power is known as a(n) _____.
   a. conductor
   b. transformer-rectifier
   c. alternator
   d. generator

7. Which of the following terms only applies to DC current?
   a. Frequency
   b. Polarity
   c. Amperage
   d. Voltage

8. The primary voltage in a welding machine is reduced by a(n) _____.
   a. governor
   b. generator
   c. alternator
   d. step-down transformer

9. The voltage that is present at the leads when the welding machine is on but no arc is present is _____ .
   a. open-circuit voltage
   b. operating voltage
   c. zero voltage
   d. minimum voltage

10. If a SMAW welding machine is equipped with a transformer-rectifier, it can be designed to produce _____.
    a. only DC welding current
    b. only AC welding current
    c. both AC and DC welding current
    d. high-frequency pulses for pulse welding

11. Governors on engine-driven generator machines directly control the _____.
    a. engine temperature
    b. engine speed
    c. rectifier
    d. primary power

12. The rated duty cycle of a welding machine is based on _____.
    a. 10 minutes
    b. 15 minutes
    c. 20 minutes
    d. 30 minutes

13. To prevent the weld lead from pulling loose from the welding machine stud connection, terminal welding cables generally use _____.

    a. slotted terminals
    b. soldered-in-place terminals
    c. ringed terminals
    d. spade terminals

14. Engine-driven welding machines should not be operated in poorly ventilated areas because they can produce hazardous _____.

    a. carbon monoxide (CO)
    b. carbon dioxide (CO_2)
    c. hydrogen peroxide (H_2O_2)
    d. methyl ethyl ketone (MEK)

15. Machines requiring single-phase power will have a _____.

    a. two-prong plug
    b. three-prong plug
    c. four-prong plug
    d. five-prong plug

Trade Terms Quiz

Fill in the blank with the correct term that you learned from your study of this module.

1. The direction of flow of electrical current in a DC welding circuit is referred to as the _____.

2. An electrical device that uses two wire coils of different sizes to convert a higher-voltage power source to a lower voltage is a(n) _____.

3. A combination device in which the motor turns the shaft of a generator, which in turn produces an AC voltage is a(n) _____.

4. A material that will support the flow of an electrical current is a(n) _____.

5. A device used to limit engine speed is known as is a(n) _____.

6. An electronic device that converts AC voltage to DC voltage is a(n) _____.

7. The percentage of a ten-minute period that a welding machine can continuously produce its rated amperage without overheating is its _____.

Trade Terms

Conductor 4
Duty cycle 7
Governor 5
Motor-generator 3

Polarity 1
Rectifier 6
Step-down transformer 2

Jerry Trainor

Anchorage, Alaska
Welding Instructor/College Administrator

Currently semi-retired and working as a part-time welding instructor, Jerry started working as a welder in 1959. Since then, he has developed an impressive list of credentials as a vocational trainer, teacher, and college administrator. Jerry was born in Pablo, Montana, and grew up in Washington's Yakima Valley. He learned on the job to become a journeyman welder/fabricator. He even spent some time as a fourth-grade teacher, where he learned he had a natural talent for teaching. He eventually was appointed assistant professor for welding technology at Lewis-Clark State College in Idaho and has since held several positions as an academic administrator.

How did you choose a career in the construction industry?
I got interested in welding as a result of taking a wood and metal class during my senior year in high school. Then I got my first welding job in Davis, California helping to build a food processing plant for Hunt's Foods. From that time on until I started teaching in college, I welded on many construction and production shop projects.

What kinds of work have you done in your career?
I started as a fabricator/welder working at lumber mills, power plants, wood and pulp plants, and a variety of other facilities over a 24-year period. In 1983, I began my career as an educator, first teaching welding at technical schools and colleges in Idaho and the Pacific Northwest. From 1993 to 2012, I served in administrative roles. My last administrative position was as head of the Construction Trades Technology Academic Program at the University of Alaska-Fairbanks.

What advice would you give to those new to the field?
My advice would be to become as proficient as possible in all phases of welding, from using a cutting torch to welding in SMAW, GMAW, and GTAW. Take an interest in the metallurgy and production of different metal materials, understand complex blueprints, and be open to taking advanced theory courses through a community college or other provider. Always be ready to go the extra mile and put in a full day's work for a full day's pay. Be proud of your skills and abilities and take pride in always doing the best job possible. Be able to walk away from a job knowing that your part of the construction project was done with your best expertise and efforts.

Tell us an interesting career-related fact or accomplishment.
In my position as head of construction trades technology, I analyzed construction projects planned for the Interior and Aleutian regions of Alaska. In these remote areas, contractors usually bring workers with them into the villages whenever a school, home, health clinic, water treatment plant, or power plant needs to be built. The purpose of the Construction Trades Technology (CTT) program is to train local people and to decrease the need for outside workers.

The Aleutians are a chain of 14 large islands and 55 smaller ones that extend about 1,200 miles west from the southwestern tip of Alaska. The program has been very successful, mainly because the training program is delivered in the villages, so the students do not have to leave their environment. Instructors spend from 8 to 10 weeks living in a village and taking part in the local community.

What do you think it takes to be a success in the welding trade?
To be successful in welding, a person has to take pride in doing the best job possible, understand the field of metallurgy, and be able to fabricate simple and complex projects. I enjoy fabricating and designing different implements, working on complex drawings of steel structures, and working with other professionals.

Trade Terms Introduced in This Module

Conductor: A material that will support the flow of an electrical current. Copper wire is the most common conductor.

Duty cycle: The percentage of a ten-minute period that a welding machine can continuously produce its rated amperage without overheating.

Governor: A device that limits engine speed.

Motor-generator: A combination device in which the motor turns the shaft of a generator, which in turn produces an AC voltage.

Polarity: Refers to the direction of electrical current flow in a DC welding circuit.

Rectifier: An electronic device that converts AC voltage to DC voltage.

Step-down transformer: An electrical device that uses two wire coils of different sizes, referred to as the primary and secondary windings, to convert a higher voltage to a lower voltage through induction. Primary windings are connected to the primary power source; the secondary windings provide the reduced voltage to the process.

Additional Resources

This module presents thorough resources for task training. The following resource material is suggested for further study.

2014 Technical Training Guide. Current Edition. Cleveland, OH, USA: The Lincoln Electric Company. **www.lincolnelectric.com**

Covering the Basics – Engine Driven Welder Maintenance. Al Nystrom, Technical Service Representative. Cleveland, OH, USA: The Lincoln Electric Company. **www.lincolnelectric.com**

Stick Electrode Product Catalog. Current Edition. Cleveland, OH, USA: The Lincoln Electric Company. **www.lincolnelectric.com**

Figure Credits

Answer	Section Reference	Objective
Section One		
1 c	1.1.3	1a
2. b	1.2.1	1b
Section Two		
1. d	2.1.1	2a
2. d	2.2.1	2b
3. b	2.3.1	2c
Section Three		
1. b	3.1.3	3a
2. d	3.2.1	3b

NCCER CURRICULA — USER UPDATE

NCCER makes every effort to keep its textbooks up-to-date and free of technical errors. We appreciate your help in this process. If you find an error, a typographical mistake, or an inaccuracy in NCCER's curricula, please fill out this form (or a photocopy), or complete the online form at **www.nccer.org/olf**. Be sure to include the exact module ID number, page number, a detailed description, and your recommended correction. Your input will be brought to the attention of the Authoring Team. Thank you for your assistance.

Instructors – If you have an idea for improving this textbook, or have found that additional materials were necessary to teach this module effectively, please let us know so that we may present your suggestions to the Authoring Team.

NCCER Product Development and Revision
13614 Progress Blvd., Alachua, FL 32615

Email: curriculum@nccer.org
Online: www.nccer.org/olf

❏ Trainee Guide   ❏ Lesson Plans   ❏ Exam   ❏ PowerPoints   Other _____

Craft / Level: _____   Copyright Date: _____

Module ID Number / Title: _____

Section Number(s): _____

Description: _____

_____

_____

_____

_____

Recommended Correction: _____

_____

_____

_____

_____

Your Name: _____

Address: _____

_____

Email: _____   Phone: _____

29108-15
SMAW Electrodes

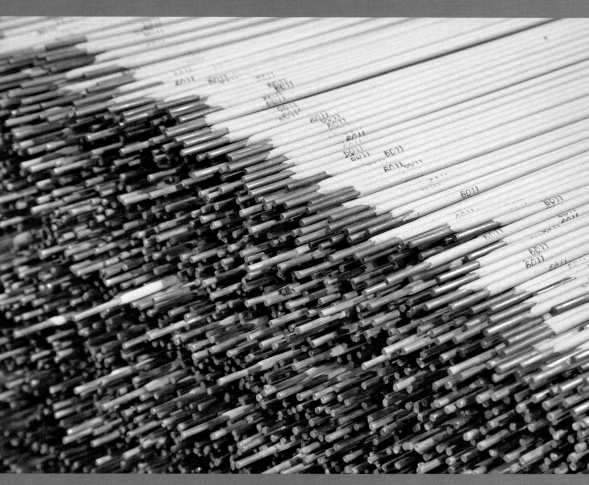

OVERVIEW

Shielded metal arc welding (SMAW) processes depend on the use of stick-like electrodes that provide the filler metal for the weld. However, the welding electrode is designed to contribute more than just filler metal to the welding process. There are many different types of electrodes, each of which has its advantages and disadvantages when coupled with the various base metals, welding positions, and weld types. This module provides information on the classification and selection of SMAW electrodes, and how to identify the various types.

Module Eight

Trainees with successful module completions may be eligible for credentialing through the NCCER Registry. To learn more, go to **www.nccer.org** or contact us at **1.888.622.3720**. Our website has information on the latest product releases and training, as well as online versions of our *Cornerstone* magazine and Pearson's product catalog.

Your feedback is welcome. You may email your comments to **curriculum@nccer.org**, send general comments and inquiries to **info@nccer.org**, or fill in the User Update form at the back of this module.

This information is general in nature and intended for training purposes only. Actual performance of activities described in this manual requires compliance with all applicable operating, service, maintenance, and safety procedures under the direction of qualified personnel. References in this manual to patented or proprietary devices do not constitute a recommendation of their use.

Objectives

When you have completed this module, you will be able to do the following:

1. Describe the SMAW electrode classification system and how to select the proper electrode for the task.
   a. Describe the AWS filler metal specification system and various electrode characteristics.
   b. Describe the characteristics of the four main electrode groups.
2. Explain how to select electrodes and describe their proper care and handling.
   a. Identify various considerations in the selection of the proper electrode.
   b. Describe the proper handling and storage of electrodes.

Performance Tasks

This is a knowledge-based module; there are no Performance Tasks.

Trade Terms

Alloy
Condensation
Ductility
Flux
Heat-affected zone

Hermetically sealed
Low-hydrogen electrode
Notch toughness
Traceability
Vertical welding

Industry Recognized Credentials

If you are training through an NCCER-accredited sponsor, you may be eligible for credentials from NCCER's Registry. The ID number for this module is 29108-15. Note that this module may have been used in other NCCER curricula and may apply to other level completions. Contact NCCER's Registry at 888.622.3720 or go to **www.nccer.org** for more information.

Contents

Topics to be presented in this module include:

Figures and Tables

1.0.0 SMAW ELECTRODES

Objective

Describe the SMAW electrode classification system and how to select the proper electrode for the task.

a. Describe the AWS filler metal specification system and various electrode characteristics.
b. Describe the characteristics of the four main electrode groups.

Trade Terms

Alloy: A metal that has had other elements added to it that substantially change its mechanical properties.

Ductility: Refers to the mechanical property of a material that allows it to be bent or shaped without breaking.

Flux: Material used to prevent, dissolve, or facilitate the removal of oxides and other undesirable substances on a weld or base metal.

Heat-affected zone: The part of the base metal that has been altered by heating, but not melted by the heat.

Hermetically sealed: Having an airtight seal.

Low-hydrogen electrode: An electrode specially manufactured to contain little or no moisture.

Notch toughness: The ability of a material to resist breaking at points where stress is concentrated.

Traceability: The ability to verify that a procedure has been followed by reviewing the documentation step-by-step.

Vertical welding: Welding with an upward or downward progression.

S hielded metal arc welding (SMAW) cannot take place without electrodes. SMAW has often been referred to as stick welding, and the electrodes are referred to as stick electrodes or as filler metal. However, there are many different types of electrodes, including wire rolls used in other welding processes. Stick electrodes are the type used in SMAW and the focus of this module.

The welder must be able to distinguish different types of electrodes and select the correct electrode for the job. In addition, the welder must know how to handle and store electrodes, some of which require special considerations. As simple as they appear, their storage and condition at the time of use is essential to weld quality.

This module explains the electrode classification system, electrodes, and the considerations for selecting the correct electrode for the job. It also explains how to properly store electrodes in compliance with the appropriate welding codes.

1.1.0 Electrode Basics and the AWS Specification System

A SMAW electrode has a wire core coated with **flux**. The wire core transfers the welding current from the electrode holder to the workpiece. The arc at the end of the electrode melts the metal core, the flux coating, and the base metal simultaneously. This occurs at a temperature exceeding 6,000°F (3,316°C). The wire core mixes with the melted base metal, forming and building the weld. The flux coating also has important functions. As the flux coating is melted, it does the following:

- Produces a gaseous shield of carbon dioxide that protects the molten weld by pushing away harmful oxygen and hydrogen present in the atmosphere.
- Acts as all fluxing agents do to clean and deoxidize the molten metal.
- Stabilizes the arc and reduces spatter.
- Forms a slag covering over the weld to further protect it from the atmosphere and slow the cooling rate.

Figure 1 shows an electrode in use.

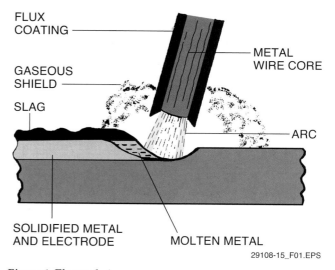

Figure 1 Electrode in use.

29108-15_F01.EPS

Some flux coatings also have powdered metal and alloying elements in them. The powdered metal adds filler metal and helps improve the appearance of the weld. The alloying elements change the chemical composition and strength of the deposited weld metal. **Alloys** are normally added to electrodes so the deposited weld metal will match the base metal chemical composition and strength.

1.1.1 The AWS Filler Metal Specification System

The American Welding Society (AWS) writes specifications for welding consumables, such as electrodes, filler rods, and fluxes, used with all the various welding processes. There are a series of different specifications identified as *AWS A5.xx*. These specifications are used in all major welding codes and accepted in all industries except where additional approvals are specified, such as the American Bureau of Shipping (ABS), the US Coast Guard, the Federal Highway Administration, Military Specifications (MIL), and Lloyds of London. The American Society of Mechanical Engineers (ASME) Boiler Code and the European Norm (EN) electrode specifications closely follow the AWS specifications.

The purpose of the specifications is to set standards that all manufacturers must follow when manufacturing welding consumables of all types. This ensures consistency for the user regardless of which company manufactured the product. The specifications set standards for the following:

- Classification system, identification, and marking.
- Chemical composition of the deposited weld metal.
- Mechanical properties of the deposited weld metal.

The following are some examples of AWS specifications that pertain to shielded metal arc welding electrodes:

- *A5.1 – Specification for Carbon Steel Electrodes for Shielded Metal Arc Welding*
- *A5.3 – Specification for Aluminum and Aluminum Alloy Electrodes for Shielded Metal Arc Welding*
- *A5.4 – Specification for Stainless Steel Electrodes for Shielded Metal Arc Welding*
- *A5.5 – Specification for Low-Alloy Steel Electrodes for Shielded Metal Arc Welding*
- *A5.6 – Specification for Covered Copper and Copper Alloy Arc Welding Electrodes*

- *A5.11 – Specification for Nickel and Nickel-Alloy Welding Electrodes for Shielded Metal Arc Welding*
- *A5.15 – Specification for Welding Electrodes and Rods for Cast Iron*
- *A5.16 – Specification for Titanium and Titanium Alloy Welding Electrodes and Rods*
- *A5.21 – Specification for Bare Electrodes and Rods for Surfacing*
- *A5.24 – Specification for Zirconium and Zirconium Alloy Welding Electrodes and Rods*

Obviously an in-depth study of each classification would require a significant amount of time. As your training and experience progress, a deeper understanding of the less common electrode types will likely be required. The most common electrodes are discussed in the *A5.1* and *A5.5* specifications. These are the electrodes that are most important at this point.

The electrodes described in the *A5.1* specification are used to weld carbon (mild) steel. The electrodes in the *A5.5* specification are used to weld low-alloy, high-strength steels. These two specifications will be explored further in this module.

The specification number is generally followed by the year of publication, such as *A5.1:2004* or *A5.5:2006*. The year of publication is the year that the specification was last revised. These documents are normally on a five-year revision cycle. Always refer to the latest version when information is needed.

1.1.2 A5.1 and A5.5 Classification Systems

The classification system that identifies *AWS Specification A5.1* carbon steel electrodes and *A5.5* low-alloy electrodes uses the prefix E followed by four or five numbers, such as E6010 or E10018. This is printed on each electrode near the end where the electrode holder grasps the electrode. These markings are shown in *Figure 2*.

Refer to *Figure 3* to review the meaning of the markings:

- *E* stands for electrode. Welding current flows through an electrode.
- The first two digits, or three if 100 or over, multiplied by 1,000 designate the minimum tensile strength in pounds per square inch of the deposited weld metal. For example:
  - E6010 has a minimum tensile strength of 60,000 psi.
  - E7018 has a minimum tensile strength of 70,000 psi.
  - E12018 has a minimum tensile strength of 120,000 psi.

Figure 2 Electrode markings.

29108-15_F02.EPS

Sizing and Packaging of SMAW Electrodes

SMAW electrodes are sized according to the wire core diameter, and are commonly packed in 10- and 50-pound cans, as shown in the photo.

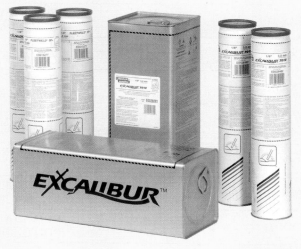

29108-15_SA01.EPS

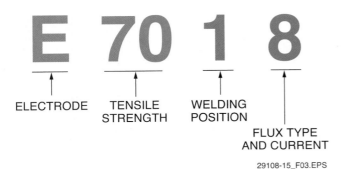

ELECTRODE   TENSILE   WELDING
STRENGTH   POSITION

FLUX TYPE
AND CURRENT

29108-15_F03.EPS

Figure 3 Electrode marking system.

- The third or fourth digit indicates the possible welding positions:
  - *1* can be used to weld in all positions.
  - *2* can be used to weld in flat and horizontal fillet-weld positions only.
  - *4* can be used to weld in flat, horizontal, overhead, and vertical-up positions.
- The fourth or fifth digit indicates special characteristics of the flux coating and the type of welding current required, such as alternating current, direct current electrode negative (DCEN), or direct current electrode positive (DCEP).

Table 1 shows the characteristics of mild steel electrodes covered by the *A5.1* specification. Note that the Xs shown in the AWS Class column refer to the first two digits of the number, which may be 60, 70, or other digits. For example, E6018,

E7018, and E12018 electrodes share the same flux characteristics shown in the bottom row of the table.

The low-alloy steel electrodes that are covered by AWS Specification *A5.5* have a suffix in addition to the standard classification numbers. The suffix has a letter and a number. The letter indicates the chemical composition of the deposited weld metal, and the number indicates the composition of the chemical classification. *Figure 4* shows an electrode classification marking from the *A5.5* specification.

The suffix letters indicate the following:

- *A* – carbon-molybdenum alloy steel
- *B* – chromium-molybdenum alloy steel
- *C* – nickel steel alloy
- *D* – manganese-molybdenum alloy steel
- *G* – other alloys with minimal elements

The meaning of the numbers varies by alloy type, but always indicates the percentages of the following alloying elements:

29108-15_F04.EPS

Figure 4 Markings on a low-alloy, high-strength electrode.

Table 1 Mild Steel Electrode Characteristics

AWS Class	Position	Flux Coating	Current Requirements	Characteristics
EXX 10	ALL	CELLULOSE SODIUM	DCEP	DEEP PENETRATION, FLAT BEADS
EXX 20	FLAT, HOR. FILLET			
EXX 11	ALL	CELLULOSE POTASSIUM	AC, DCEP	DEEP PENETRATION, FLAT BEADS
EXX 12	ALL	TITANIA SODIUM	AC, DCEN	MEDIUM PENETRATION
EXX 13	ALL	TITANIA POTASSIUM	AC, DCEP, DCEN	SHALLOW PENETRATION
EXX 14	ALL	TITANIA IRON POWDER	AC, DCEP, DCEN	MEDIUM PENETRATION, FAST DEPOSIT
EXX 24	FLAT, HOR. FILLET			
EXX 15	All	LOW-HYDROGEN SODIUM	DCEP	MODERATE PENETRATION
EXX 16	All	LOW-HYDROGEN POTASSIUM	AC, DCEP	MODERATE PENETRATION
EXX 27	FLAT, HOR. FILLET	IRON POWDER, IRON OXIDE	AC, DCEP, DCEN	MEDIUM PENETRATION
EXX 18	ALL	IRON POWDER LOW-HYDROGEN	AC, DCEP	SHALLOW TO MEDIUM PENETRATION
EXX 28	FLAT, HOR. FILLET			

- Carbon (C)
- Manganese (Mn)
- Phosphorus (P)
- Sulphur (S)
- Silicon (Si)
- Nickel (Ni)
- Chromium (Cr)
- Molybdenum (Mo)
- Vanadium (V)

Figure 5 shows the electrode marking system for low-alloy, high-strength electrodes.

> **CAUTION**
>
> Be very careful when selecting electrodes. For example, a 7018-B2 cannot be used in place of a 7018 electrode because the additional alloy will cause a significant metallurgical change in the weld. Never assume that electrodes with similar numbers are interchangeable.

Figure 6 shows *A5.5* electrodes listed by their alloy content. Remember that these are subject to change; as mentioned previously, the specifications are periodically updated.

1.1.3 Manufacturers' Classification

All electrodes used for code-governed work must have the AWS classification number printed on them as well as on the container in which they are shipped. In addition, electrode manufacturers generally print their own unique classification name or number on the electrode and on the container in which the electrodes are shipped. Manufacturers may also make more than one electrode within the same AWS classification. For example, Lincoln Electric manufactures two electrodes with the AWS classification E6010. One is called Fleetweld 5P, and the other is called Fleetweld 5P+. Words such as Fleetweld and Pipemaster are trade names and are not related to the AWS specification.

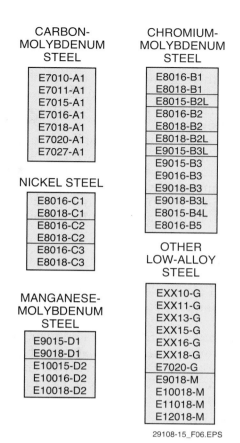

CARBON-MOLYBDENUM STEEL

| E7010-A1 |
| E7011-A1 |
| E7015-A1 |
| E7016-A1 |
| E7018-A1 |
| E7020-A1 |
| E7027-A1 |

NICKEL STEEL

| E8016-C1 |
| E8018-C1 |
| E8016-C2 |
| E8018-C2 |
| E8016-C3 |
| E8018-C3 |

MANGANESE-MOLYBDENUM STEEL

| E9015-D1 |
| E9018-D1 |
| E10015-D2 |
| E10016-D2 |
| E10018-D2 |

CHROMIUM-MOLYBDENUM STEEL

| E8016-B1 |
| E8018-B1 |
| E8015-B2L |
| E8016-B2 |
| E8018-B2 |
| E8018-B2L |
| E9015-B3L |
| E9015-B3 |
| E9016-B3 |
| E9018-B3 |
| E9018-B3L |
| E8015-B4L |
| E8016-B5 |

OTHER LOW-ALLOY STEEL

| EXX10-G |
| EXX11-G |
| EXX13-G |
| EXX15-G |
| EXX16-G |
| EXX18-G |
| E7020-G |
| E9018-M |
| E10018-M |
| E11018-M |
| E12018-M |

29108-15_F06.EPS

Figure 6 Electrodes listed by alloy.

Lincoln Electric recommends Fleetweld 5P for general fabrication and maintenance welding and states that it is good for pipe. However, Lincoln Electric states that Fleetweld 5P+ is the best choice for pipe. The major difference between the two electrodes is the mechanical properties of the deposited weld metal. Even though both fall into the E6010 classification, the 5P+ electrode has improved characteristics to decrease cracking possibilities in **heat-affected zones**. It is also better suited for open root joints. When 5P+ is used for the root pass, its greater elongation results in

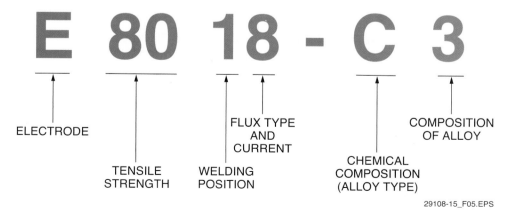

ELECTRODE

TENSILE STRENGTH

WELDING POSITION

FLUX TYPE AND CURRENT

CHEMICAL COMPOSITION (ALLOY TYPE)

COMPOSITION OF ALLOY

29108-15_F05.EPS

Figure 5 Marking system for low-alloy, high strength electrodes.

improved resistance to heat-affected zone cracking. Also, the 5P+ has higher tensile and yield strength.

When there is more than one electrode in an AWS classification and you are unsure about which electrode to use, check with your instructor or supervisor.

Table 2 shows some examples of manufacturers' and AWS classifications for common mild-steel electrodes. Note however, that it is provided as an example only. Trade names and the related product information can change at any time. Only current listings should be consulted on the job.

1.1.4 SMAW Electrode Sizes

Mild steel electrodes for SMAW are available in standard sizes, usually ranging from ³⁄₃₂" to ¼". Note that AWS and ASME specifications were long ago based on fractions of an inch. Due to the precision and required consistency of electrodes, they are not manufactured in metric sizes specifically. A basic metric equivalent, to 0.1 mm of precision, is assigned to each electrode diameter for the convenience of metric system users. However, the specifications are based solely on fractions of an inch.

SMAW electrodes are typically packaged in 10- and 50-pound cans, but also come in smaller containers, as shown in *Figure 7*. The amperage application for which the electrodes are used is proportional to the diameter size of the electrode. The number of electrodes in a 50-pound can varies with the diameter and length of the electrodes. Note that the length of the electrode may vary from manufacturer to manufacturer. *Table 3* is an example of a manufacturer's table listing some mild steel electrodes and the number of electrodes in one pound. Note that there are lengths of 14" (≈36cm) and 18" (≈46cm) listed. Electrodes from other manufacturers will be of very similar weights for a given length.

1.1.5 Filler Metal Traceability Requirements

Welding codes and the Welding Procedure Specification (WPS) on a project or job site normally specify the type of electrode to be used. Once a weld is made, there must be a means to trace and document the type of electrode that was used to make the weld or welds in order to verify that the correct welding codes and WPS were followed.

Traceability requirements vary according to client requirements and job specifications. On many job sites, these traceability requirements may be found in the quality control documentation related to the project. One of the more common requirements followed is found in the *ASME Boiler and Pressure Vessel Code, Section II, Material Specifications*. This section of the ASME code lists acceptable base metal and filler metal combinations.

> **CAUTION**
>
> The traceability requirements in site quality control manuals must be followed. The consequences for violation of the requirements can be severe. Understand traceability requirements before proceeding with a welding job. Verify this information with your supervisor, project engineer, or welding inspector.

1.2.0 Electrode Groups

Because so many different electrodes are available, the welder must be able to distinguish between them and select the correct electrode for the job. The following sections will explain how to select electrodes.

Electrodes are often classified into one of four groups based on their general characteristics:

- Fast-freeze
- Fast-fill
- Fill-freeze
- Low-hydrogen

Table 2 Manufacturers' and AWS Classifications for Common Mild-Steel Electrodes

	AWS Classification				
Manufacturer	**E6010**	**E6011**	**E6013**	**E7014**	**E7018**
Hobart	Pipemaster 60	335A, 335C	447A, 447C	14A	718, 718C
Airco	Pipecraft	6011, 6011C	6013, 6013D	7014	Easy Arc 7018MR
ESAB	AP100, SW-610, SW-10P	SW-14	SW-15	SW-15P	Atom Arc 7018
Lincoln	Fleetweld 5P, 5P1	Fleetweld 35, 35LS, 180	Fleetweld 37, 57	Fleetweld 47	Jetweld LH-70
McKay	6010M	6011	6013	7014	—

50-LB BOX

50-LB CAN

10-LB EASY-OPEN CANS

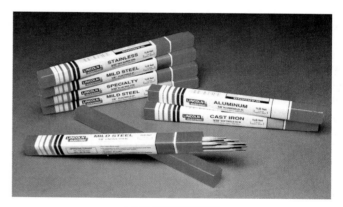

1-LB BOX TUBES

29108-15_F07.EPS

Figure 7 Electrode containers.

Table 3 Number of Electrodes per Pound

Electrode Type	Diameter:	³⁄₃₂" (2.4mm)	⅛" (3.2mm)	⁵⁄₃₂" (4.0mm)	³⁄₁₆" (4.8mm)	⁷⁄₃₂" (5.6mm)	¼" (6.4mm)
	Length:	14"	14"	14"	14"	18"	18"
6010 PM		30	17	12	8	—	—
6011, Soft-Arc 6011		25	15	11	7	—	—
6013		25	15	10	7	—	—
7014		24	13	9	6	—	—
7024		—	10	7	—	4	2
7018 XLM, Soft-Arc 7018-1		32	15	10	7	—	3
7018 AC, Millennia 7018		32	15	10	7	4	3

Material Safety Data Sheet for Electrodes

Every box or can of welding electrodes must contain a copy of the Safety Data Sheet/Material Safety Data Sheet (SDS/MSDS) for that product. The complete MSDS for an Excalibur electrode with an AWS classification of E7018-1H4R is shown in the *Appendix*. As of this writing, many MSDS forms from numerous manufacturers remain valid and in use. Eventually, they will all be replaced by the newer SDS format.

These four groups are identified by F-numbers. Refer to *Table 4* as you progress through the discussion of each group and their characteristics. The mild steel electrodes used with SMAW are classified as F1, F2, F3, and F4. Qualification with an electrode in a particular F-number classification qualifies the welder with all electrodes identified in that classification and in lower F-number classifications. For example, a typical AWS welder qualification test is a V-groove weld with metal backing in the 3G and 4G positions using an F4 electrode. Passing this test qualifies the welder to weld with F4 and all lower F-number electrodes, and to make groove and fillet welds in all positions.

1.2.1 Fast-Fill Electrodes

Fast-fill electrodes (F1) have powdered iron in the flux, which gives them high deposition rates but requires that they only be used for flat welds and horizontal fillet welds. These electrodes have shallow penetration with excellent weld appearance and almost no spatter. The heavy slag covering is easy to remove. Electrodes in the fast-fill group will also have a 2 as the next-to-last number, as shown in *Table 4*. They include electrodes such as the following:

- E6027
- E7024
- E7028 (also in the low-hydrogen F4 group)

Typical applications for fast-fill electrodes are production welds on plate more than ¼" thick and flat and horizontal fillets and lap welds.

Manufacturers' Welding Guides

Most electrode manufacturers publish paperback welding guides that include information on their electrodes, such as recommended amperage settings and uses. These booklets can be obtained through your local distributor for free or at a nominal charge.

1.2.2 Fill-Freeze Electrodes

Fill-freeze electrodes (F2), also called fast-follow electrodes, are all-position electrodes that are commonly used for vertical and flat welding. They have medium deposition rates and penetration and are excellent for sheet metal. The weld bead ranges from smooth and ripple-free to even with distinct ripples. The medium slag covering is easy to remove.

Electrodes in the fill-freeze group include the following:

- E6012
- E6013
- E7014

Typical applications for fill-freeze electrodes are the following:

- Vertical fillet or lap welds
- Sheet metal lap and fillet welds
- Joints having poor fit-up in the flat position
- General purpose welding

Table 4 AWS F-Number Electrode Classification

Group	AWS Electrode Classification				
F4	EXX15	EXX16	EXX18		Low-Hydrogen
F3	EXX10	EXX11			Fast-Freeze
F2	EXX12	EXX13	EXX14		Fill-Freeze
F1	EXX20	EXX24	EXX27	EXX28	Fast-Fill

1.2.3 Fast-Freeze Electrodes

Fast-freeze electrodes (F3) are all-position electrodes that provide deep penetration. However, the arc creates a lot of fine weld spatter. The weld bead is flat with distinct ripples and a light slag coating that can be difficult to remove. The arc is easy to control, making fast-freeze electrodes good for overhead and **vertical welding**. These more challenging positions often make it harder for the welder to maintain a precise arc gap. Electrodes in the fast-freeze group include the following:

- E6010
- E7010
- E6011

Typical applications for fast-freeze electrodes are as follows:

- Vertical and overhead plate welding
- Pipe welding (cross-country and in-plant)
- Welds made on galvanized, plated, or painted surfaces
- Joints requiring deep penetration, such as square butts
- Sheet metal welds

WARNING!

Many surface coatings, such as galvanized coatings, give off very toxic fumes when heated, cut, or welded. Fumes from certain types of electrodes may also be hazardous. Breathing these fumes can cause damage to the lungs and possibly other organs. Identify any hazard for the coatings with which you are working by referring to the SDS/MSDS for that particular coating. Also, read the SDS/MSDS contained in the electrode package. Use proper ventilation and take the necessary precautions to protect yourself and other workers in the immediate area.

1.2.4 Low-Hydrogen Electrodes

Low-hydrogen electrodes (F4) are designed for welding high-sulphur, phosphorus, and medium- to high-carbon steels, which have a tendency to develop porosity and cracks under the weld bead. These problems are caused by hydrogen that is absorbed during welding. Low-hydrogen electrodes are also available with fast-fill and fill-freeze characteristics. They produce dense, X-ray quality welds with excellent **notch toughness** and **ductility**. Electrodes in the low-hydrogen group include the following:

- E7015
- E7016
- E7018
- E7028
- E7048

Typical applications for low-hydrogen electrodes are the following:

- X-ray quality welds
- Welds requiring high mechanical properties
- Crack-resistant welds in medium- to high-carbon steels
- Welds that resist hot-short cracking in phosphorus steels
- Welds that minimize porosity in sulphur steels
- Welds in thick or highly restrained mild or alloy steels to minimize the danger of weld cracking
- Welds in alloy steels requiring a strength of 70,000 psi (≈4,826 bars) or more
- Multiple-pass welds in all positions

If hydrogen is present in the weld zone, it will create defects. The greatest source of hydrogen is airborne water vapor, which is composed of oxygen and hydrogen. Low-hydrogen electrodes are carefully manufactured to eliminate moisture in the flux coating. Low hydrogen is defined as being less than 16 milliliters (ml) per 100 grams of weld metal. The amount of hydrogen in these electrodes varies from 4 to 16 ml per 100 grams of metal weld.

Low-hydrogen electrodes have two unique suffixes. The M suffix is for compliance. Most military codes require an M suffix. When the suffix H and a number are added, the numbers relate to the amount of hydrogen in the electrodes. As a result, the electrodes will have suffixes such as H4, H8, and H16. These three suffixes indicate the dryness of the electrodes, identifying the milliliters of hydrogen per 100 grams of metal weld they contain. An additional suffix may also be added to these dryness suffixes. A suffix such as H4R would be one example. The R indicates that the electrodes are moisture resistant. Without an R suffix, the electrodes may only be exposed to the atmosphere for a maximum of 4 hours. Those with an R suffix have an allowable exposure of up to 8 hours. *Figure 8* shows a 7018 electrode with the H4R suffix.

If low-hydrogen electrodes are exposed to the atmosphere, they will begin to absorb moisture and will no longer be low-hydrogen as a result. This is the reason for their short atmospheric exposure

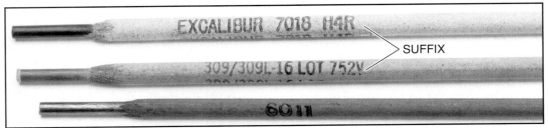

Figure 8 Electrode with suffix.

limits. For this reason, all low-hydrogen electrodes are shipped in **hermetically sealed** metal containers. Once opened, the electrodes must be stored in a heated oven. *Figure 9* shows low-hydrogen electrodes properly stored. The temperature range of this oven is 100°F to 550°F (38°C to 288°C). However, it is important to note that placing the oven in areas below freezing may not allow it to maintain its maximum temperature, if that setting is necessary. Additional guidelines for the storage and handling of low-hydrogen electrodes are covered later in this module.

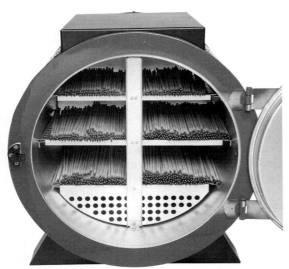

Figure 9 Low-hydrogen electrode storage.

Portable Electrode Storage

Electrodes can also be kept in a portable rod oven for convenience on the job. This ensures that the lowest possible level of hydrogen content is maintained. The model shown has a spring-loaded rod elevator that pushes the rods up for easy grasping when the lid is opened. This is a valuable feature when welding gloves are being worn. It maintains a fixed temperature of 300°F (149°C) when powered. Non-powered dry containers, without electric heating elements, are also available.

Additional Resources

The Welding Handbook for Maritime Welders. Current Edition. Lysaker, Norway: Wilhelmsen Ships Service. **www.wilhelmsen.com**

Welding Consumables Catalog. Current Edition. Cleveland, OH, USA: The Lincoln Electric Company. **www.lincolnelectric.com**

1.0.0 Section Review

1. The slag over a fresh weld primarily forms due to _____ .
   a. impurities in the base metal
   b. impurities in the electrode metal
   c. the flux coating of the electrode
   d. hydrogen and nitrogen in the atmosphere

2. An H8 suffix on the end of an electrode classification number means _____ .
   a. there are 8 grams of hydrogen in each stick
   b. there are 8 ml of hydrogen per 100 grams of weld metal
   c. there is 8 times as much hydrogen in the electrode than normal
   d. the electrode can be exposed to the atmosphere up to 8 minutes

Section Two

2.0.0 Electrode Selection and Care

Objective

Explain how to select electrodes and describe their proper care and handling.
a. Identify various considerations in the selection of the proper electrode.
b. Describe the proper handling and storage of electrodes.

Trade Term

Condensation: The process in which atmospheric water vapor condenses (returns to its liquid state) on a cool surface.

With some basic information concerning electrodes used in SMAW, it will be easier to understand the selection process. The importance of properly caring for these simple consumable items cannot be overstressed. This section provides additional information that will help ensure that electrode quality standards are maintained from the point of manufacture to the job site.

2.1.0 Electrode Selection Considerations

Many factors enter into the proper selection of electrodes for the job. The available welding equipment and its current ranges limit the choice of electrodes, and base metal properties play an important role in the selection as well. Also, the position in which the weld must be made and design factors affect the choice of electrodes for a given application. *Table 5* indicates position, polarity, and common usages for some steel electrodes, listed by their AWS classification.

2.1.1 Welding Procedure Specification

If there is a WPS for the weld to be made, the filler metal will be specified. It will be given as an AWS classification and/or manufacturer's standard and will sometimes even include a manufacturer's name. The filler metal specified in the WPS must be used.

2.1.2 Base Metal Type

The filler metal should match the base metal's chemical composition and mechanical properties. The chemical composition and mechanical properties of the base metal can be obtained by referring to the base metal mill specifications. The mill specifications can then be compared to various tables that list base metal and filler metal compatibility. These tables are available in a number of codes, such as the *AWS Structural Welding Code – Steel D1.1, Table 3.1 Prequalified Base Metal – Filler Metal Combinations for Matching Strength*, and the *ASME Boiler and Pressure Vessel Code, Section II, Parts A, B, and C, Material Specifications*. Tables are also readily available from the electrode and/or base metal manufacturers. Manufacturers acknowledge that an informed user has a great deal to do with the success of their product. An improperly applied filler metal that results in poor weld quality can also reflect on their product, regardless of the circumstances. If the filler metal and base metal are not compatible, the weld produced will be defective.

A weld made on mild steel with a high-tensile electrode such as E12018 may actually create more stress on the joint. The high-tensile weld will be so strong that when stress is placed on the weld, it will not give, transferring all of the stress to the heat-affected zone along the weld. This will cause cracking. A weld made with E7018 will give along with the base metal, distributing the stress across the weld, making a stronger joint. The opposite occurs when mild steel electrodes are used to weld high-strength steel. All the stress occurs in the weld because the surrounding base metal is so much stronger, causing the weld to fail prematurely.

Table 5 Common Steel Electrode Selection Factors

AWS Class	Position	Polarity	Usage
E6011	All	AC, DCEP	Good for dirty, rusty steel
E6013	All	AC, DCEN, DCEP	Good for thin steel and poor fit-up
E7014	All	AC, DCEN, DCEP	Good for thin steel
E7018	All	AC, DCEP	Good for high deposition; good for medium- or high-carbon steels

2.1.3 Base Metal Thickness

Normally, an electrode that is smaller in diameter than the thickness of the metal to be welded is used. For sheet metal, a 3/32" E6013 electrode may be recommended because it has shallow penetration. For very thick sections, always use a low-hydrogen electrode for its ductile qualities. To minimize joint preparation, a deep penetration E6010 or E7010 electrode can be used for the root pass.

2.1.4 Base Metal Surface Condition

The base metal surface should always be cleaned before welding. Surface corrosion, plating, coatings, and paint should be removed. If this is not possible for some reason, use a deep penetrating electrode, such as an E6010.

> **WARNING!**
>
> Many surface coatings, such as galvanized coatings, give off toxic fumes when heated, cut, or welded. Breathing these fumes can cause damage to the lungs and other organs. Identify any hazard for the coatings with which you are working by referring to the MSDS for that particular coating. Use proper ventilation and take the necessary precautions to protect yourself.

2.1.5 Welding Position

The position of welding is important because some electrodes, such as E7024 and E7028, can only be used in the flat and horizontal fillet weld positions. In addition, welding in vertical and overhead positions should never be attempted with electrodes larger than 3/16", except for low-hydrogen electrodes. The limitation for low-hydrogen electrodes and EXX14 electrodes is 5/32". As a general rule, the joint should be positioned for flat welding so you can use the largest electrode possible.

2.1.6 Joint Design

The type of joint to be welded is an important consideration when selecting an electrode. When T- and lap joints can be positioned flat, they are usually welded with E7024 or E7028 electrodes. Butt joints that need deep penetration should be welded with E6010 electrodes. If the joint has poor fit-up, use a shallow penetrating electrode such as an E6012 or E6013. If low-hydrogen electrodes are used for open root joints, back-gouging may be required. To eliminate the need for back-gouging and to make the root pass easier to run, use metal backing or put the root in with an E6010 electrode and then fill the weld out with a low-hydrogen electrode.

2.1.7 Welding Current

The most common type of welding current is DC. Because all electrodes can be run on DC, there is no problem when DC current is available. Welding machines that can produce only AC current cannot use E6010 or E7010 electrodes. Although many other electrodes will run on AC, they will run better using DC current.

The amount of current produced by the welding machine is also a consideration. Running large-diameter electrodes that require high amperages near the maximum output of the welding machine can cause the welding machine to overheat and cut out. Check the welding machine current type, rated output, and duty cycle before selecting the electrode size.

2.2.0 Filler Metal Storage and Control

Electrodes used for SMAW must be dry. The dry mineral flux coatings on electrodes start to absorb moisture as soon as they are exposed to the atmosphere. In as little as 30 minutes, some electrodes can pick up enough moisture from the atmosphere to cause weld defects. Note too, that the local conditions have a lot to do with the rate at which moisture is absorbed from the air. For example, electrodes will absorb moisture faster and in greater volume on a rainy day in Florida than they will on a clear summer day in Arizona. Some of the defects caused, such as under-bead cracking, may not be visible to the naked eye.

Low-hydrogen electrodes pick up moisture faster than any other type of electrode. Also, because of the nature of the base metals welded with low-hydrogen electrodes, a smaller amount of moisture will cause defects. For these reasons, special care must be used when handling and storing low-hydrogen electrodes.

2.2.1 Code Requirements

All welding codes require electrodes that have a low-hydrogen coating to receive special handling. To begin, they must be shipped in hermetically sealed containers. Immediately after opening the sealed containers, low-hydrogen electrodes must be stored in special ovens at a minimum temperature of 250°F (120°C). When low-hydrogen electrodes are removed from the oven, they can be exposed to the atmosphere only for a designated period of time. This time period varies with the classification of the electrode, but generally ranges from one-half hour to ten hours. *AWS D.1.1, Table 5.1 Allowable Atmospheric Exposure of Low-Hydrogen Electrodes* gives allowable exposure times (*Table 6*). The electrodes must then be returned to the oven and generally require a drying process at temperatures above 250°F (>120°C). The exact procedures for handling filler metal at your site are typically written out in a quality standard. These procedures must be followed. If you are unsure of the requirements at your site, check with your supervisor.

> **CAUTION**
> The exposure times shown are general guidelines only. Excessive exposure times are known to cause weld defects. Check your site quality standards for the actual exposure times for your location. To the degree possible, do not allow electrodes to remain exposed any longer than absolutely necessary.

> **CAUTION**
> The consequences of not following your site's standards on electrode handling and storage may be severe. Be sure you understand your site's standards before handling electrodes. Improper handling of electrodes can cause weld defects. The information provided in this module is of a general nature. Your site's particular quality standards must be followed.

2.2.2 Receiving Electrodes

Inspect electrodes as they are received. All low-hydrogen electrodes (EXX15, 16, or 18), stainless steel electrodes, and nickel alloy coated electrodes must be supplied in hermetically sealed containers. If the seal has been broken, reject the electrodes. Mark them "Not to Be Used" and notify your supervisor.

All electrodes other than low-hydrogen electrodes must be shipped in moisture-resistant containers. Inspect these containers for damage. The electrodes can be used if their coatings are not damaged and they have not come into direct contact with water or oil. Mark electrodes that cannot be used and notify your supervisor.

Each container must be marked with at least the following information:

- ASME, AWS, or other electrode specification number and electrode classification designation
- Electrode size
- Weight

Table 6 Allowable Atmospheric Exposure of Low-Hydrogen Electrodes.

Electrode	Column A (hours)	Column B (hours)
A5.1		
E70XX	4 max	
E70XXR	9 max	Over 4 to 10 max
E70XXHXR	9 max	
E7018M	9 max	
A5.5		
E70XX-X	4 max	Over 4 to 10 max
E80XX-X	2 max	Over 2 tp 10 max
E90XX-X	1 max	Over 1 to 5 max
E100XX-X	½ max	Over ½ to 4 max
E110XX-X	½ max	Over ½ to 4 max

Notes:
1. Column A: Electrodes exposed to atmosphere for longer periods than shown shall be redried before use.
2. Column B: Electrodes exposed to atmosphere for longer periods than those established by testing shall be redried before use.
3. Electrodes shall be issued and held in quivers, or other small open containers. Heated containers are not mandatory.
4. The optional supplemental designator, R, designates a low-hydrogen electrode which has been tested for covering moisture content after exposure to a moist environment for 9 hours and has met the maximum level allowed in *AWS A5.1, Specification for Carbon Steel Electrodes for Shielded Metal Arc Welding.*

29108-15_T06.EPS

If the containers are not properly marked, mark them "Not to Be Used" and notify your supervisor.

2.2.3 Storing Filler Metal

All filler metal must be stored in a warm (40°F [4.5°C] minimum), dry storage area. Filler metals are not to be stored directly on the floor. Electrodes that are stored on the floor are more likely to form **condensation**. Use pallets or shelves to store electrodes off the floor. They must also be stored in such a manner that they will not be damaged. Once low-hydrogen, stainless steel, or nickel-alloy electrode containers have been opened, the electrodes must immediately be stored in an oven.

Electrode Mix-Up

Do not mix electrode types in an oven if possible. If electrodes of various types must be stored in the same oven, keep them carefully organized by placing them on separate shelves. Labels on the shelves are another way that mistakes can be avoided.

2.2.4 Storage Ovens

Electrode storage ovens are typically electrical and are controlled by a thermostat. Storage ovens for large shops can hold large quantities of electrodes and can possibly reach temperatures exceeding 800°F (427°C). This maximum temperature is too high for most electrodes. Always follow the electrode manufacturer's guidelines for storing electrodes in ovens. Smaller portable ovens and dry containers that are designed to be used on the job may hold only a few pounds of electrodes. Depending on the WPS or job quality standards, portable electrode ovens may be required at the point of use. The electrodes remaining in the portable ovens after performing the task must be returned to the holding oven. *Figure 10* shows several examples of electrode storage ovens.

Storage ovens should typically be set between 250°F and 300°F (121°C to 149°C). The oven must remain on at all times. Never unplug an oven or plug it into an outlet that could accidentally be shut off.

2.2.5 Exposure and Drying

The time that a low-hydrogen, stainless steel, or nickel alloy electrode can be exposed to the atmosphere is limited and depends on the electrode. It is the welder's responsibility to take from the oven only the number of electrodes that can be used before the maximum exposure time is reached. As a best practice, do not try to push the limits, as the minimal time savings is not worth the risk. This is especially true in wet, humid weather. The exposure time is defined as the period of time that the electrode is not in a heated holding or portable oven.

Samples of electrodes being used at a site are often collected by the quality control group for moisture testing. Depending on the results of this testing, the maximum exposure times may be adjusted higher. However, if high moisture content is found in the electrodes, the welds made with the electrodes may have to be replaced. This can be an extremely costly situation, as well as frustrating to the welder and other affected workers.

Depending on the site quality standards, electrodes that exceed the maximum exposure time must either be destroyed or properly dried. Check your site quality standards for the proper actions. Low-hydrogen electrodes that are exposed to water must be destroyed; drying them to an accepted standard is not practical.

Electrodes that exceed the maximum exposure time can be dried once. They are dried by placing them in a holding oven at an elevated temperature for a period of time. The amount of time and temperature depend on the nature of the moisture pickup and the electrode.

(A) 10-POUND (≈ 4.5 KG) PORTABLE

(B) 50-POUND (≈ 23 KG) PORTABLE

(C) 350-POUND (≈ 159 KG) BENCH MODEL

(D) 1,100-POUND (≈ 499 KG) BENCH MODEL

29108-15_F10.EPS

Figure 10 Electrode storage ovens.

Additional Resources

The Welding Handbook for Maritime Welders. Current Edition. Lysaker, Norway: Wilhelmsen Ships Service. **www.wilhelmsen.com**

Welding Consumables Catalog. Current Edition. Cleveland, OH, USA: The Lincoln Electric Company. **www.lincolnelectric.com**

2.0.0 Section Review

1. Which of the following electrodes would be the best choice for medium- or high-carbon steels?
    a. E6011
    b. E6013
    c. E7014
    d. E7018

2. How many times can electrodes that have exceeded their maximum period of exposure be dried?
    a. Once
    b. Twice
    c. Three times
    d. Four times

SUMMARY

Many different electrodes are produced by many different manufacturers. To ensure quality and uniformity, the American Welding Society has established a classification system for electrodes. Understanding this classification system and why electrodes are placed in a particular classification is essential for the welder to correctly identify and select electrodes for the weld to be made.

Welders must be able to select proper types of filler metals and electrodes to make welds on specific materials. It is also important for welders to understand how to properly store electrodes to prevent damage or deterioration. Properly stored, good quality SMAW electrodes do not guarantee an excellent weld; a great deal of the result is a reflection of the welder's skill. However, electrodes that are poorly maintained and do not meet standards will result in poor, unacceptable welds regardless of the welder's skill level.

1. Welding current is transferred from the electrode holder and into the workpiece through the _____ .
   a. base metal
   b. flux coating
   c. gas shield
   d. wire core of the electrode

2. The harmful oxygen and hydrogen present in the atmosphere surrounding a molten SMAW weld is pushed away by the _____ .
   a. gaseous shield of carbon dioxide
   b. gaseous shield of nitrogen
   c. constant welding spatter
   d. intense heat of the arc

3. The purpose of the specifications written by the American Welding Society (AWS) is to set standards that all manufacturers must follow when manufacturing _____ .
   a. welding consumables
   b. welding machines
   c. chemicals
   d. metals

4. The number placed at the end of a complete AWS specification number indicates the _____ .
   a. type of electrodes discussed in the specification
   b. year the specification was last revised
   c. number of chapters in the document
   d. year the specification was written

5. The traceability requirements of filler metals vary according to _____ .
   a. the electrode trade name
   b. the welder's qualifications
   c. the client requirements and job specifications
   d. state and local laws

6. All-position electrodes that provide deep penetration belong to which electrode group?
   a. F1 fast-fill
   b. F2 fill-freeze
   c. F3 fast-freeze
   d. F4 low-hydrogen

7. Electrodes designed for welding high-sulphur, phosphorus, and medium- to high-carbon steels belong to which electrode group?
   a. F1 fast-fill
   b. F2 fill-freeze
   c. F3 fast-freeze
   d. F4 low-hydrogen

8. Of all electrodes, the ones that collect moisture faster than any other type are the _____ .
   a. F1 fast-fill electrodes
   b. F2 fill-freeze electrodes
   c. F3 fast-freeze electrodes
   d. F4 low-hydrogen electrodes

9. The exposure time for low-hydrogen electrodes, per AWS standards, typically ranges from _____ .
   a. ½ hour to 4 hours
   b. ½ hour to 10 hours
   c. 2 to 5 hours
   d. 3 to 6 hours

10. All filler metal must be stored in a dry storage area at a minimum temperature of _____ .
    a. 30°F (–1°C)
    b. 40°F (4.5°C)
    c. 50°F (10°C)
    d. 60°F (15.5°C)

Trade Terms Quiz

Fill in the blank with the correct term that you learned from your study of this module.

1. Welding upward or downward is called _____ .

2. A(n) _____ is a metal that has had other elements added to it that substantially change its mechanical properties.

3. The part of a base metal that has been altered but not melted by the heat is called the _____ .

4. _____ is the ability of a material to resist breaking at points where stress is concentrated.

5. The process in which water vapor from the atmosphere returns to its liquid state is called _____ .

6. A container that is closed airtight is said to be _____ .

7. _____ is the ability to verify that a procedure has been followed by reviewing the documentation step-by-step.

8. _____ refers to the mechanical property of a material that allows the material to be bent or shaped without breaking.

9. An electrode that is carefully manufactured to contain little or no moisture is called a(n) _____ .

10. A material used to prevent, dissolve, or facilitate the removal of oxides and other undesirable substances from the area of a weld is called _____ .

Trade Terms

Alloy 2
Condensation 5
Ductility 8
Flux 10

Heat-affected zone 3
Hermetically sealed 6
Low-hydrogen electrode 9

Notch toughness 4
Traceability 7
Vertical welding 1

SAMPLE MATERIAL SAFETY DATA SHEET

Date:	1/21/2013	MSDS No.:	**US-M292**
Trade Name:	**Excalibur 7018-1 MR**		
Sizes:	**All**		
Supersedes:	**8/29/2011**		

MATERIAL SAFETY DATA SHEET
For Welding Consumables and Related Products
Conforms to Hazard Communication Standard 29CFR 1910.1200 Rev. October 1988

SECTION I - IDENTIFICATION

Manufacturer/ Supplier:	The Lincoln Electric Company 22801 St. Clair Avenue Cleveland, OH 44117-1199 (216) 481-8100	Product Type: Covered Electrode
		Classification: AWS E7018-1H4R

SECTION II - HAZARDOUS MATERIAL (1)

I M P O R T A N T !
This section covers the materials from which this product is manufactured. The fumes and gases produced during welding with the normal use of this product are covered by Section V; see it for industrial hygiene information.
CAS Number shown is representative for the ingredients listed. All ingredients listed may not be present in all sizes.
(1) The term "hazardous" in "Hazardous Materials" should be interpreted as a term required and defined in the Hazard's Communication Standard and does not necessarily imply the existence of any hazard. All materials are listed on the TSCA inventory.

Ingredients:	CAS No.	Wt.%	TLV mg/m^3	PEL mg/m^3
Iron	7439-89-6	15	10*	15*
Limestone and/or calcium carbonate	1317-65-3	10	10*	15
Titanium dioxides	13463-67-7	< 5	10	15
Fluorides (as F)	7789-75-5	< 5	2.5	2.5
Silicates and other binders	1344-09-8	< 5	10*	15*
Manganese and/or manganese alloys and compounds (as Mn)*****	7439-96-5	< 5	0.2	5 (c)
Mineral silicates	1332-58-7	< 5	5**	5**
Silicon and/or silicon alloys and compounds (as Si)	7440-21-3	1	10*	15*
Zirconium alloys and compounds (as Zr)	12004-83-0	1	5	5
Cellulose and other carbohydrates	65996-61-4	0.5	10*	15*
Quartz	14808-60-7	< 0.5	#0.025**	#0.1**
Molybdenum alloys (as Mo)	7439-98-7	< 0.5	10	10
Lithium compounds (as Li)	554-13-2	< 0.5	10*	15*
Carbon steel core wire	7439-89-6	55	10*	15*

Supplemental Information:

(*) Not listed. The OSHA PEL for nuisance particles is 15 milligrams per cubic meter. The ACGIH guideline for total particulate is 10 milligrams per cubic meter. PEL value for iron oxide is 10 milligrams per cubic meter. TLV value for iron oxides is 5 milligrams per cubic meter.

(**) As respirable dust.

(*****) Subject to the reporting requirements of Sections 311, 312, and 313 of the Emergency Planning and Community Right-to-Know Act of 1986 and of 40CFR 370 and 372.

(c) Value is for manganese fume. Present PEL is 5 milligrams per cubic meter (ceiling value). Values proposed by OSHA in 1989 were 1.0 milligrams per cubic meter TWA and 3.0 milligrams per cubic meter STEL (Short Term Exposure Limit).

(#) Crystalline silica (quartz) is on the IARC (International Agency for Research on Cancer) and NTP (National Toxicology Program) lists as posing a carcinogenic risk to humans.

SECTION III - HAZARD DATA

Non Flammable; Welding arc and sparks can ignite combustibles and flammable products. See Z49.1 referenced in Section VI.
Product is inert, no special handling or spill procedures required. Not regulated by DOT.

Rev 9/07   **(CONTINUED ON SIDE TWO)**

29108-15_A01A.EPS

Product:	Excalibur 7018-1 MR
Date:	1/21/2013

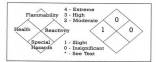

SECTION IV - HEALTH HAZARD DATA

Threshold Limit Value: The ACGIH recommended general limit for Welding Fume NOS- (Not Otherwise Specified) is 5 mg/m³. ACGIH-1999 preface states that the TLV-TWA should be used as guides in the control of health hazards and should not be used as fine lines between safe and dangerous concentrations. See Section V for specific fume constituents which may modify this TLV. Threshold Limit Val ues are figures published by the American Conference of Government Industrial Hygienists. Units are milligrams per cubic meter of air.

Effects of Overexposure: Electric arc welding may create one or more of the following health hazards:
Fumes and Gases can be dangerous to your health. Common entry is by inhalation. Other possible routes are skin contact and ingestion.

Short-term (acute) overexposure to welding fumes may result in discomfort such as metal fume fever, dizziness, nausea, or dryness o r irritation of nose, throat, or eyes. May aggravate pre-existing respiratory problems (e.g. asthma, emphysema).

Long-term (chronic) overexposure to welding fumes can lead to siderosis (iron deposits in lung) and may affect pulmonary function. Manganese overexposure can affect the central nervous system, resulting in impaired speech and movement. Bronchitis and some lung fibrosis have been reported. Repeated exposure to fluorides may cause excessive calcification of the bone and calcificatio n of ligaments of the ribs, pelvis and spinal column. May cause skin rash. Titanium dioxide is listed by the IARC (International Agency for Research on Cancer) as a Group 2B carcinogen (possibly carcinogenic to humans based on animal studies). Respiratory exposure to the crystalline silica present in this welding electrode is not anticipated during normal use. Respiratory overexposure to airborne crystalline silica is known to cause silicosis, a form o f disabling pulmonary fibrosis which can be progress ive and may lead to death. Crystalline silica is on the IARC (International Agency for Research on Cancer) and NTP (National Toxicology Program) lists as posing a cancer risk to humans. WARNING: This product, when used for welding or cutting, produces fumes or gases which contain chemicals known to the State of California to cause birth defects and, in some cases, cancer. (California Health & Safety Code Section 25249.5 et seq.)

Arc Rays can injure eyes and burn skin. *Skin cancer has been reported.*
Electric Shock can kill. If welding must be performed in damp locations or with wet clothing, on metal structures or when in cramped positions such as sitting, kneeling or lying, or if there is a high risk of unavoidable or accidental contact with workpiece, use the following equipment: Semiautomatic DC Welder, DC Manual (Stick) Welder, or AC Welder with Reduced Voltage Control.
Emergency and First Aid Procedures: Call for medical aid. Employ first aid techniques recommended by the American Red Cross
  IF BREATHING IS DIFFICULT give oxygen. IF NOT BREATHING employ CPR (Cardiopulmonary Resuscitation) techniques.
  IN CASE OF ELECTRICAL SHOCK, turn off power and follow recommended treatment. In all cases call a physician.

SECTION V - REACTIVITY DATA

Hazardous Decomposition Products: Welding fumes and gases cannot be classified simply. The composition and quantity of both are dependent upon the metal being welded, the process, procedure and electrodes used.

Other conditions which also influence the composition and quantity of the fumes and gases to which workers may be exposed include: coatings on the metal being welded (such as paint, plating, or galvanizing), the number of welders and the volume of the worker area, the quality and amount of ventil ation, the position of the welder's head with respect to the fume plume, as well as the presence of contaminants in the atmosphere (such as chlorinated hydrocarbon vapors from cleaning and degreasing activities).

When the electrode is consumed, the fume and gas decomposition products generated are different in percent and form from the ingredients listed in Section II. Decomposition products of normal operation include those originating from the volatilizatibn, reaction, or oxidation of the materials shown in Section II, plus those from the base metal and coating, etc., as noted above.

Reasonably expected fume constituents of this product would include: Primarily iron oxide and fluorides; secondarily complex oxides of manganese, potassium, silicon, sodium, and titanium.

Maximum fume exposure guideline for this product (based on manganese content) is 4.0 milligrams per cubic meter.

Gaseous reaction products may include carbon monoxide and carbon dioxide. Ozone and nitrogen oxides may be formed by the radiation from the arc.

Determine the composition and quantity of fumes and gases to which workers are exposed by taking an air sample from inside the welder's helmet if worn or in the worker's breathing zone. Improve ventilation if exposures are not below limits. See ANSI/AWS F1.1, F1.2, F1.3 and F1.5, available from the American Welding Society, 550 N.W. LeJeune Road, Miami, FL 33126.

SECTION VI AND VII
CONTROL MEASURES AND PRECAUTIONS FOR SAFE HANDLING AND USE

Read and understand the manufacturer's instruction and the precautionary label on the product. Request Lincoln Safety Publication E205. See American National Standard Z49.1, "Safety In Welding, Cutting and Allied Processes" published by the American Welding Society, 550 N.W. LeJeune Road, Miami, FL, 33126 (both available for free download at http://www.lincolnelectric.com/community/safety/) and OSHA Publication 2206 (29CFR1910), U.S. Government Printing Office, Superintendent of Documents, P.O. Box 371954, Pittsburgh, PA 15250-7954 for more details on many of the following:
Ventilation: Use enough ventilation, local exhaust at the arc, or both to keep the fumes and gases from the worker's breathing zone and the general area. Train the welder to keep his head out of the fumes. *Keep exposure as low as possible.*
Respiratory Protection: Use respirable fume respirator or air supplied respirator when welding in confined space or general work area when local exhaust or ventilation does not keep exposure below TLV.
Eye Protection: Wear helmet or use face shield with filter lens shade number 12 or darker. Shield others by providing screens and flash goggles.

Protective Clothing: Wear hand, head, and body protection which help to prevent injury from radiation, sparks and electrical shock. See Z49.1. At a minimum this includes welder's gloves and a protective face shield, and may include arm protectors, aprons, hats, shoulder protection, as well as dark substantial clothing. Train the welder not to permit electrically live parts or electrodes to contact skin . . . or clothing or gloves if they are wet. Insulate from work and ground.
Disposal Information: Discard any product, residue, disposable container, or liner as ordinary waste in an environmentally acceptable manner according to Federal, State and Local Regulations unless otherwise noted. No applicable ecological information available.

29108-15_A01B.EPS

Trade Terms Introduced in This Module

Alloy: A metal that has had other elements added to it that substantially change its mechanical properties.

Condensation: The process in which atmospheric water vapor condenses (returns to its liquid state) on a cool surface.

Ductility: Refers to the mechanical property of a material that allows it to be bent or shaped without breaking.

Flux: Material used to prevent, dissolve, or facilitate the removal of oxides and other undesirable substances on a weld or base metal.

Heat-affected zone: The part of the base metal that has been altered by heating, but not melted by the heat.

Hermetically sealed: Having an airtight seal.

Low-hydrogen electrode: An electrode specially manufactured to contain little or no moisture.

Notch toughness: The ability of a material to resist breaking at points where stress is concentrated.

Traceability: The ability to verify that a procedure has been followed by reviewing the documentation step by step.

Vertical welding: Welding with an upward or downward progression.

Additional Resources

This module presents thorough resources for task training. The following resource material is suggested for further study.

The Welding Handbook for Maritime Welders. Current Edition. Lysaker, Norway: Wilhelmsen Ships Service.
www.wilhelmsen.com

Welding Consumables Catalog. Current Edition. Cleveland, OH, USA: The Lincoln Electric Company.
www.lincolnelectric.com

Figure Credits

Section Review Answer Key

Answer	Section Reference	Objective
Section One		
1. c	1.1.0	1a
2. b	1.2.4	1b
Section Two		
1. d	2.1.0	2a
2. a	2.2.5	2b

NCCER CURRICULA — USER UPDATE

NCCER makes every effort to keep its textbooks up-to-date and free of technical errors. We appreciate your help in this process. If you find an error, a typographical mistake, or an inaccuracy in NCCER's curricula, please fill out this form (or a photocopy), or complete the online form at **www.nccer.org/olf**. Be sure to include the exact module ID number, page number, a detailed description, and your recommended correction. Your input will be brought to the attention of the Authoring Team. Thank you for your assistance.

Instructors – If you have an idea for improving this textbook, or have found that additional materials were necessary to teach this module effectively, please let us know so that we may present your suggestions to the Authoring Team.

NCCER Product Development and Revision

13614 Progress Blvd., Alachua, FL 32615

Email: curriculum@nccer.org
Online: www.nccer.org/olf

❏ Trainee Guide   ❏ Lesson Plans   ❏ Exam   ❏ PowerPoints   Other _____

Craft / Level: _____   Copyright Date: _____

Module ID Number / Title: _____

Section Number(s): _____

Description: _____

_____

_____

_____

_____

Recommended Correction: _____

_____

_____

_____

_____

Your Name: _____

Address: _____

_____

Email: _____   Phone: _____

29109-15

SMAW – Beads and Fillet Welds

OVERVIEW

This module provides trainees with their first opportunities to experiment with SMAW processes in hands-on practice. Common preparations for welding are reviewed, along with a brief review of the important safety considerations. The methods of striking an arc are presented, along with the importance of this simple action in weld quality. Both stringer and weave beads are explained, and trainees create many of their own weld beads through a significant amount of practice in the welding booth.

Module Nine

Trainees with successful module completions may be eligible for credentialing through the NCCER Registry. To learn more, go to **www.nccer.org** or contact us at **1.888.622.3720**. Our website has information on the latest product releases and training, as well as online versions of our *Cornerstone* magazine and Pearson's product catalog.

Your feedback is welcome. You may email your comments to **curriculum@nccer.org**, send general comments and inquiries to **info@nccer.org**, or fill in the User Update form at the back of this module.

This information is general in nature and intended for training purposes only. Actual performance of activities described in this manual requires compliance with all applicable operating, service, maintenance, and safety procedures under the direction of qualified personnel. References in this manual to patented or proprietary devices do not constitute a recommendation of their use.

Objectives

When you have completed this module, you will be able to do the following:

1. Explain how to prepare for SMAW welding and how to strike an arc.
   a. Identify safety practices related to SMAW.
   b. Explain how to prepare the area and equipment for welding.
   c. Explain how to strike an arc and respond to arc blow.
2. Explain how to successfully complete various types of beads and welds.
   a. Explain how to properly restart and terminate a weld pass.
   b. Describe the technique required to produce stringer beads.
   c. Describe the technique required to produce weave and overlapping beads.
   d. Describe the techniques required to produce fillet welds in various positions.

Performance Tasks

Under the supervision of your instructor, you should be able to do the following:

1. Set up welding equipment.
2. Strike an arc.
3. Make stringer, weave, and overlapping beads using E6010 and E7018 electrodes.
4. Make corner welds on an angle iron section end welded to a plate coupon.
5. Make fillet welds using E6010 and E7018 electrodes in the specified positions:
   - Flat (1F)
   - Horizontal (2F)
   - Vertical (3F)
   - Overhead (4F)

Trade Terms

Arc blow
Arc strike
Drag angle
Oscillation
Push angle
Quench
Restart

Stringer bead
Undercut
Weave bead
Weld axis
Weld coupon
Work angle

Industry Recognized Credentials

If you are training through an NCCER-accredited sponsor, you may be eligible for credentials from NCCER's Registry. The ID number for this module is 29109-15. Note that this module may have been used in other NCCER curricula and may apply to other level completions. Contact NCCER's Registry at 888.622.3720 or go to **www.nccer.org** for more information.

Contents

Topics to be presented in this module include:

Figures and Tables

1.0.0 SMAW PREPARATIONS

Objective

Explain how to prepare for SMAW welding and how to strike an arc.

a. Identify safety practices related to SMAW.
b. Explain how to prepare the area and equipment for welding.
c. Explain how to strike an arc and respond to arc blow.

Performance Tasks

1. Set up welding equipment.
2. Strike an arc.

Trade Terms

Arc blow: The deflection of the arc from its intended path due to magnetic forces that develop from welding current flow.

Arc strike: A discontinuity consisting of localized melting of the base metal or finished weld caused by the initiation of an arc. It remains visible as a result of striking the arc outside the area to be welded and can lead to rejection of an otherwise good weld.

Weld coupon: Metal pieces to be welded together as a test or practice.

One of the most basic welds is the fillet weld. Before trainees can make a fillet weld, they must be able to set up arc welding equipment, strike an arc, and maintain that arc to run a welding bead. This section reviews the necessary preparations for SMAW, through the establishment of the all-important arc. SMAW is often referred to by the nonstandard name stick welding.

1.1.0 Safety Summary

The following is a summary of safety procedures and practices that must be observed when cutting or welding. Keep in mind that this is a summary only. Complete safety coverage is provided in the *Welding Safety* module. If you have not completed that module, do so before continuing. Above all, be sure to wear the appropriate personal protective equipment (PPE) when welding or cutting.

1.1.1 Protective Clothing and Equipment

To maintain safety and prevent injury, it is essential that you wear the appropriate PPE when cutting or welding metals (*Figure 1*). Be sure to follow these guidelines during all phases of cutting or welding:

- Always use safety glasses with a full face shield or a welding helmet. The glasses, face shield, or helmet lens must have the proper light-reducing tint for the type of welding or cutting being performed. Never view an electric arc, directly or indirectly, without using a properly tinted lens.
- Wear proper protective leather and/or flame retardant clothing along with welding gloves that will protect you from flying sparks and molten metal as well as heat.
- Wear high-top safety shoes or boots. Make sure that the tongue and lace area of the footwear will be covered by a pant leg. If the tongue and lace area is exposed, or if the footwear must be protected from burn marks, wear leather spats under your pants or chaps and over the front of the footwear.
- Wear a 100-percent cotton cap with no mesh material included in its construction. The bill of the cap points to the rear. If a hard hat is required for the environment, use one that allows the attachment of rear deflector material and a face shield. A hard hat with a rear deflector is generally preferred when working overhead, and may be required by some employers and job sites.

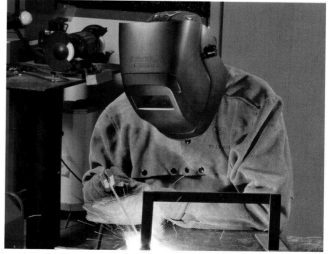

29109-15_F01.EPS

Figure 1 Welder working in full protective equipment.

- Wear a face shield over snug-fitting cutting goggles or safety glasses for gas welding or cutting. Either the face shield or the lenses of the welding goggles must be an approved shade for the application. A welding hood equipped with a properly tinted lens is also acceptable.
- Wear earplugs to protect ear canals from sparks. Wear hearing protection to protect against the consistent sound of the torch.

1.1.2 Fire and Explosion Prevention

Welding activities involve the use of fire or extreme heat to melt metal. Whenever fire is used, it must be controlled and contained. Welding activities are often performed on vessels that may once have contained flammable or explosive materials. Residues from those materials can catch fire or explode when a welder begins work on such a vessel. The following are basic fire and explosion prevention guidelines associated with welding:

- Never carry matches or gas-filled lighters in your pockets. Sparks can cause the matches to ignite or the lighter to explode, causing serious injury.
- Always comply with all site and/or employer requirements for a hot-work permit and a fire watch.
- Never use oxygen to blow dust or dirt off clothing. The oxygen can remain trapped in the fabric for a time. If a spark hits the oxygen in the fabric, the clothing can burn rapidly and violently.
- Make sure that any flammable material in the work area is moved or shielded by a fire-resistant covering.
- Approved fire extinguishers must be available before attempting any heating, welding, or cutting operations. Make sure the extinguisher is charged, the inspection tag is valid, and any individual that may be required to operate it knows how to do so.
- Never release a large amount of oxygen or use oxygen in place of compressed air. The presence of oxygen around flammable materials or sparks can cause rapid and uncontrolled combustion. Keep oxygen away from oil, grease, and other petroleum products.
- Never release a large amount of fuel gas, especially acetylene. Methane and propane are heavier than air and tend to migrate to and concentrate in low areas. As a result, they can ignite at a considerable distance from the release point. Acetylene is lighter than air but is even more dangerous than methane; when mixed with air or oxygen, it will explode at much lower concentrations than any other common fuel gas.
- To prevent fires, maintain a neat and clean work area, and make sure that any metal scrap or slag is cold before disposal.

Before cutting containers such as tanks or barrels, check to see if they have contained any explosive, hazardous, or flammable materials, including petroleum products, citrus products, or chemicals that decompose into toxic fumes when heated. Proper procedures for cutting or welding hazardous containers are described in the *American Welding Society (AWS) F4.1, Safe Practices for the Preparation of Containers and Piping for Welding and Cutting*, and *ANSI Z49.1*. As a standard practice, always clean and then fill any tanks or barrels with water, or purge them with a flow of inert gas such as nitrogen to displace any oxygen.

Containers must be cleaned by steam cleaning, flushing with water, or washing with detergent until all traces of the material have been removed.

After cleaning the container, fill it with water or a purging gas, such as carbon dioxide, argon, or nitrogen to displace the explosive fumes. Air, which contains oxygen, is displaced from inside the container by the water or inert gas. Without oxygen, combustion cannot take place.

A water-filled vessel is the best alternative. When using water, position the container to minimize the air space. When using an inert gas, provide a vent hole so the inert gas can push the air and other vapors out to the atmosphere. Keep in mind, though, that even these precautions do not guarantee the absence of flammable materials inside. For that reason, these types of activities should not be done without proper supervision and the use of proper testing methods.

1.1.3 Work Area Ventilation

Vapors and fumes tend to rise in the air from their sources, and welders often have to work above welding areas where fumes are being created. Welding fumes can be harmful. Good work area ventilation helps to remove the vapors and protect the welder. The following is a list of work area ventilation guidelines to consider before and during welding activities:

- Make sure confined space procedures are followed before conducting any welding or cutting in the confined space.
- Make sure confined spaces are ventilated properly for cutting or welding purposes.
- Never use oxygen from cylinders for the purpose of ventilation or breathing air.
- Always perform cutting or welding operations in a well-ventilated area. Cutting or welding operations involving materials, coatings, or electrodes that contain cadmium, manganese, mercury, lead, zinc, chromium, and beryllium produce toxic fumes. For cutting or welding of such materials, always use proper area ventilation and wear an approved full-face, supplied-air respirator (SAR) that provides breathing air from outside of the work area. For occasional, very short-term exposure to fumes from zinc- or copper-coated materials, a high-efficiency, particulate arresting (HEPA) or metal-fume filters may be used on a standard respirator.

1.2.0 SMAW Equipment Setup

Before welding can take place, the area has to be made ready, the welding equipment must be set up, and the metal to be welded must be prepared. The following sections explain how to set up equipment for welding.

1.2.1 Preparing the Welding Area

To practice welding, a welding table, bench, or stand is needed (*Figure 2*). The welding surface must be steel, and provisions must be made for placing weld coupons out of position.

To set up the area for welding, follow these steps:

Step 1 Ensure that the area is properly ventilated. Make use of doors, windows, and fans.

Step 2 Check the area for fire hazards. Remove any flammable materials before proceeding.

Step 3 Determine the location of the nearest fire extinguisher. Do not proceed unless the extinguisher is charged and you know how to use it.

Step 4 Set up flash shields around the welding area.

1.2.2 Preparing the Weld Coupons

The weld coupons should be carbon steel, ¼" to ¾" (≈6 to 19 mm) thick. Use a wire brush or grinder to remove heavy mill scale or corrosion. Prepare weld coupons to practice the welds indicated as follows:

- *Striking an arc* – The coupons can be any size or shape that is easily handled.

Figure 2 Welding station.

29109-15_F02.EPS

- *Running beads* – The coupons can be any size or shape that is easily handled.
- *Overlapping beads* – The coupons can be any size or shape that is easily handled.
- *Fillet welds* – Cut the metal into 4" × 6" rectangles for the base and 3" × 6" rectangles for the web. To make best use of available materials for welding practice, the instructor may specify different sizes.

Figure 3 shows the weld coupons for fillet welding.

Steel for practice welding is expensive and difficult to obtain. Every effort should be made to conserve and not waste the material that is available. Reuse weld coupons until all surfaces have been welded upon. Weld on both sides of the joint, and then cut the weld coupon apart and reuse the pieces. Use material that cannot be cut into weld coupons to practice striking an arc and running beads.

1.2.3 Electrodes

Obtain a small quantity of the electrodes to be used. Electrodes are sometimes referred to by the nonstandard term welding rods or simply rods. For the welding exercises in this module, $\frac{3}{32}$" (2.4 mm) to $\frac{5}{32}$" (4.0 mm) E6010 and E7018 electrodes will be used. Obtain only the electrodes to be used for a particular welding exercise at one time. The E7018 electrodes should remain in an electrode oven or portable caddy until they are used. Have some type of pouch or rod holder in which to store the electrodes, to prevent them from becoming damaged. Never store electrodes to roll around loose on a table. They may fall on the floor and become damaged or create a tripping hazard. Some type of metal container, such as a pail, must also be available to discard hot electrode stubs.

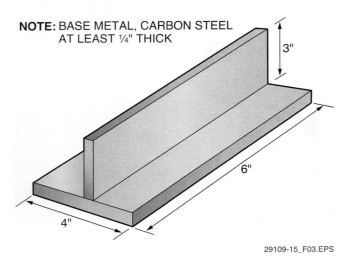

NOTE: BASE METAL, CARBON STEEL AT LEAST ¼" THICK

3"

6"

4"

29109-15_F03.EPS

Figure 3 Fillet weld coupons.

Return unused electrodes to the proper storage location.

> **NOTE**
>
> The electrode sizes indicated are recommendations. Other electrode sizes may be substituted, depending on site conditions. Confirm the electrode size to use with your instructor.

> **WARNING!**
>
> Do not throw electrode stubs on the floor. They easily roll, and could cause someone to slip and fall.

1.2.4 Preparing the Welding Machine

Welders can expect to find different types and makes of welding machines in welding shops and on jobs in the field. Most machines will have no operator manuals with them. As a trainee, and even as a trained welder, you must be able to recognize the different types of welding machines and figure out how to safely set up and operate them. Select a welding machine to use (*Figure 4*) and then follow these steps to set it up for welding:

Step 1 Verify that the welding machine can be used for DC welding.

> **NOTE**
>
> An AC welding machine can be used if a DC machine is not available. If an AC welding machine is used, use E6011 and E7018 electrodes.

Step 2 Verify the location of the primary power disconnect to the machine.

Step 3 Check the area for proper ventilation.

Step 4 Set the polarity to direct current electrode positive (DCEP).

Step 5 Connect the clamp of the workpiece lead to the workpiece.

> **WARNING!**
>
> Even though workpiece leads and clamps are sometimes called ground leads or clamps, they may not actually be grounded. In this case, the full open-circuit voltage of the welding machine may exist between the workpiece lead or clamp and any grounded object.

Work Area Ventilation

There are many different breathing hazards that can be associated with welding processes. Vaporizing a number of base metal and electrode components allows these materials to be inhaled. Chromium, for example, is considered a carcinogenic risk by international organizations. Manganese affects the central nervous system, leading to problems in coordination and walking. The effects are irreversible.

Although employers and organizations generally recognize the potential hazards and provide the correct PPE and ventilation systems to protect welders, you must also take personal responsibility for your safety and well-being. Know the components of base metals and electrodes you are working with, and consider proper ventilation in the work area to be as essential to the task as the welding machine. This welder is using both source capture and a personal air-filtration system to ensure a safe work area. Proper ventilation is also proven to help keep the entire work area cleaner and present a more professional image.

29109-15_SA01.EPS

Step 6 Set the amperage for the electrode type and size to be used. Typical settings are shown in *Table 1*.

> **NOTE**
>
> Amperage recommendations vary by manufacturer, position, current type, and electrode brand. For specific recommendations, refer to the manufacturer's literature for the electrode being used.

Step 7 Ensure that the electrode holder is not touching the workpiece lead clamp, the workpiece, or a grounded object.

Step 8 Turn on the welding machine.

1.3.0 Striking an Arc

The first step in starting a weld is striking the arc. To strike an arc, touch the end of the electrode to the base metal and then quickly raise it to the correct arc length. The general rule is that the arc length should be the diameter of the electrode being used. Thus, if a ⅛" (3.2 mm) electrode is being used, the arc length should be ⅛" (3.2 mm). The arc length is measured from the end of the elec-

GOING GREEN

Recycling Metal

Salvage any coupons or scrap metal that is no longer usable for welding practice and place it in a recycling bin. Almost all metals are recyclable and valuable. One technical school took in donated steel, used it for cutting and welding practice, and then sold all their recyclable metals to fund their school's annual picnic. The bottom line is that metal that is no longer usable for cutting or welding practice can be recycled and kept out of the landfills.

trode core to the base metal. If the electrode has a heavier flux coating, such as E7018 or E6013, the end of the electrode core will be slightly recessed in the flux coating. For these electrodes, the visible arc length (from the end of the coating to the base metal) should be slightly less to compensate for the part of the arc that is recessed. This will ensure that the actual arc length is correct. *Figure 5* shows arc lengths for a ⅛" (3.2 mm) electrode.

There are two ways to strike an arc: the scratching method and the tapping method.

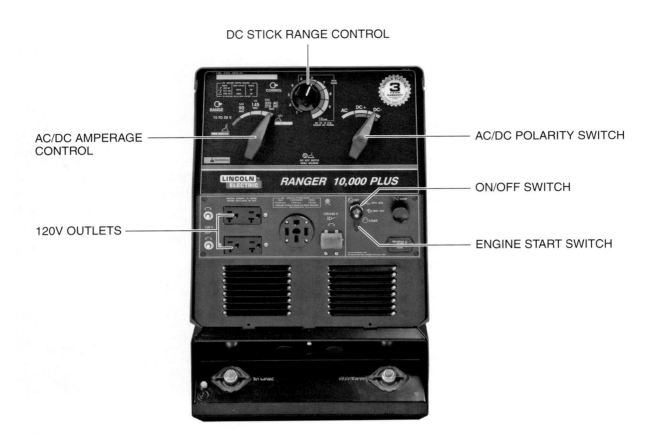

DC STICK RANGE CONTROL

AC/DC AMPERAGE CONTROL

AC/DC POLARITY SWITCH

ON/OFF SWITCH

120V OUTLETS

ENGINE START SWITCH

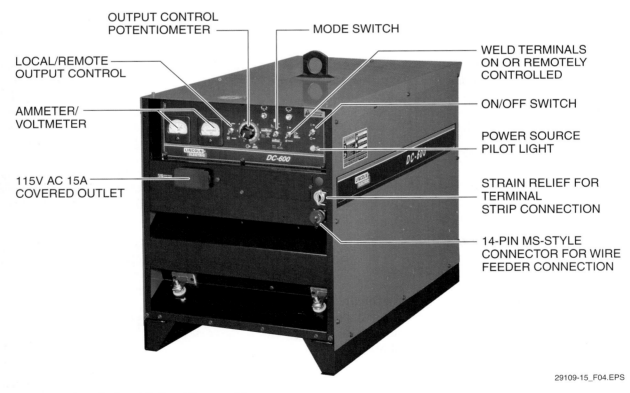

OUTPUT CONTROL POTENTIOMETER

MODE SWITCH

LOCAL/REMOTE OUTPUT CONTROL

WELD TERMINALS ON OR REMOTELY CONTROLLED

AMMETER/ VOLTMETER

ON/OFF SWITCH

POWER SOURCE PILOT LIGHT

115V AC 15A COVERED OUTLET

STRAIN RELIEF FOR TERMINAL STRIP CONNECTION

14-PIN MS-STYLE CONNECTOR FOR WIRE FEEDER CONNECTION

29109-15_F04.EPS

Figure 4 Typical AC/DC and DC welding machines.

Table 1 Amperage Settings for Electrodes

Electrode	Size	Amperage
E6013	1/8"	75A to 130A
E6013	5/32"	90A to 175A
E7018	1/8"	75A to 120A
E7018	5/32"	90A to 160A
E6011	1/8"	90A to 150A
E6011	5/32"	120A to 190A
E6010	1/8"	110A to 150A
E6010	5/32"	150A to 200A

1.3.1 Scratching Method

The scratching method (*Figure 6*) is the easiest way to strike an arc, and it is used by trainees who are learning to weld. It is similar to striking a match. The end of the electrode is simply scratched along the base metal to establish an arc. When the arc has been established, the electrode is raised to establish the correct arc length.

1.3.2 Tapping Method

The tapping method (*Figure 7*) is the best method of establishing an arc when using a transformer DC welding machine. It is more difficult and takes more practice to perfect, but as welders develop their skills, it is the method that should be used. The scratching method leaves **arc strikes** on the base metal that are not allowed by the welding codes unless they occur within the welded area.

Rod Holders

Several different types of rod holders can be purchased. One type is a leather pouch as shown here. Another is a sealable holder that can be used to prevent the electrodes from getting wet. These non-heated rod holders help protect the electrodes from physical damage and at least slow the absorption of moisture.

29109-15_SA02.EPS

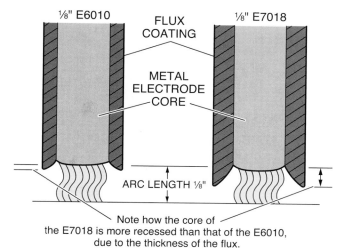

Note how the core of the E7018 is more recessed than that of the E6010, due to the thickness of the flux.

29109-15_F05.EPS

Figure 5 Arc lengths.

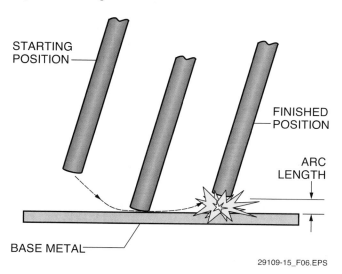

29109-15_F06.EPS

Figure 6 Scratching method of striking an arc.

With the tapping method, the arc is usually established just after the point where the welding should begin (¼" to ½", or ≈6.3 to 12.6 mm). The arc is then moved back to the correct starting point as it stabilizes. To strike an arc with the tapping method, move the electrode quickly to the base metal, lightly tap it, and then raise the electrode to establish the correct arc length.

GOING GREEN

Economizing with Electrodes

Electrodes are expensive, so they should be used economically. One way to conserve electrodes is to save the stubs from a welding project for use in making tack welds.

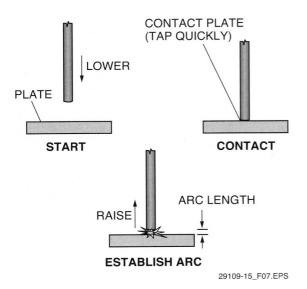

START · **CONTACT** · **ESTABLISH ARC**

29109-15_F07.EPS

Figure 7 Tapping method of striking an arc.

Note that, if you are using an inverter welding machine, the tapping method does not work well. When not actively welding, some inverter units used in the SMAW mode deactivate open-circuit voltage at the electrode. Only a very low sensing voltage exists. For these machines, the scratch method must be used to activate the welding voltage and current and start the arc.

1.3.3 *Practicing Striking and Extinguishing an Arc*

Striking the arc correctly is an important step in learning to weld. Practice both methods using ⅛" (3.2 mm) E6010 or E6011 electrodes until you can strike and maintain an arc consistently. When striking an arc, place the base metal flat on the welding table. To extinguish the arc at the end of a weld, quickly lift the electrode away from the work.

Striking an arc with E7018 (F4 low-hydrogen) electrodes is more difficult than with most other electrodes because they tend to stick to the base metal more easily. Practice striking an arc with E7018 after you become more skilled striking and running beads with E6010 or E6011.

When practicing striking an arc, you will generally experience these two problems:

- The end of the electrode will instantly weld itself to the base metal. This is caused by keeping the electrode in contact with the base metal too long or by trying to maintain too short an arc length. If this condition occurs, free the electrode by quickly moving the electrode holder from side to side. If this fails, release the electrode from the holder and then remove the electrode from the base metal with pliers. Note that when the electrode is released from the holder, arcing generally occurs between the holder jaws and the electrode. Repeated electrode releases can damage the electrode holder jaws. Inspect the holder jaws for any heavy pitting or gouging damage, and replace them if necessary.
- The arc will extinguish. This is caused by raising the end of the electrode too far above the base metal.
- These problems will be eliminated as you gain experience in striking and controlling the arc gap.

1.3.4 *Arc Blow*

When current flows, especially DC current, strong magnetic fields may be created. The magnetic fields tend to concentrate in corners, in deep grooves, or in the ends of the base metal. When the arc approaches these concentrated magnetic fields, it is deflected. This phenomenon is known as **arc blow**. In some cases, ferrous metal welding fixtures or jigs that are part of a consistent DC welding current path can become magnetized and contribute to arc blow. In AC welding, arc blow is rarely a problem because the magnetic field is constantly reversing at twice the frequency of the primary power source. This effectively cancels any strong magnetic field effects. DC arc blow can cause defects such as excessive weld spatter and porosity.

Constant-Current Machines

The AC or AC/DC machines desired for SMAW or gas tungsten arc welding (GTAW) are basically constant-current types of machines, some of which are referred to as droop-current machines or droopers. In a true constant-current machine, the output current varies little over a relatively wide range of circuit voltage. Because of this, raising and lowering the electrode from the workpiece (changing the gap) has little effect on the weld quality. In addition, striking an arc is easier with a true constant-current machine because of the high open-circuit voltage. On the other hand, droopers can be used to control the molten weld-pool temperature more easily. This is because the current and resultant weld heat changes much more with a smaller change in voltage that is caused by raising or lowering the electrode in relation to the workpiece.

Restriking a Cold Electrode

When attempting to strike an arc using a cold, partially consumed electrode, the cup formed by the hardened flux sheath extending beyond the metal core may not allow the core to contact the workpiece to establish an arc. Scratch striking or hard tapping can be used to break off the sheath. Unfortunately, this sometimes causes large pieces of the flux coating to break off the sides, exposing part of the core. Electrodes with missing flux are difficult to strike. If an arc is struck, it will be unstable and will blow or wander until the core is consumed up to where the entire flux sheath cup is reestablished. During this time, poor welds will occur. One method of removing the flux sheath without chipping flux away from the sides of the electrode is to draw the end of it across the face of a medium-coarse file. If code requirement welding is being performed, electrodes with chipped, missing, or cracked flux coatings must be discarded.

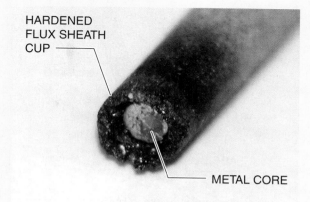

HARDENED
FLUX SHEATH
CUP

METAL CORE

29109-15_SA03.EPS

If arc blow occurs, try one or more of these methods to control it:

- If possible, tack-weld the workpieces together at the ends and in the middle.
- Shorten the arc length. With a shorter arc length, the magnetic field will have less effect on the arc.
- Change the angle of the electrode (*Figure 8*). The normal electrode angle is 10 degrees to 15 degrees in the direction of travel. Raising the electrode angle toward 90 degrees or, in extreme cases, up to 20 degrees in the opposite direction of travel will compensate for the arc blow.
- Change the position of the workpiece lead clamp. This will change the flow of welding current, affecting the way the magnetic fields are created.
- Reducing the weld current when practical may also reduce arc blow.

Accurate Tap-Striking of an Electrode

An accurate method of tap-striking an electrode at an exact spot is to rest the electrode on the finger of your gloved free hand, similar to a pool cue. Angle the rod in the direction of travel. Then move the electrode quickly down and up to strike the arc. This method is best if your helmet can be lowered by nodding your head, or if you are using an auto-darkening lens helmet.

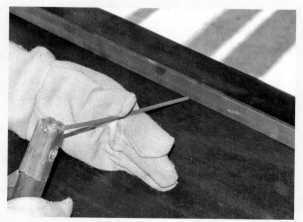

29109-15_SA04.EPS

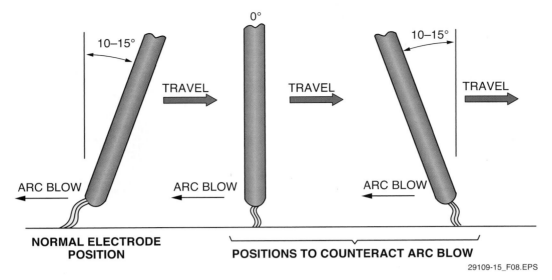

Figure 8 Controlling arc blow.

Reusing a Stuck Electrode

Carefully inspect the end of any electrode that has been stuck to the workpiece and freed by bending. If any of the flux coating is missing, cracked, or badly burned, discard the electrode.

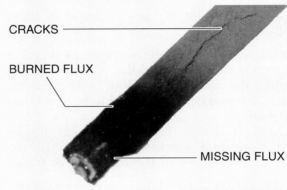

CRACKS

BURNED FLUX

MISSING FLUX

29109-15_SA05.EPS

Arc Shifting or Wandering

When two workpieces are first being joined, the arc may deposit material on one piece only or wander back and forth from one piece to the other. Both of these occurrences are usually the result of too long an arc. If this is the case, the arc will always jump to the closest workpiece. This is especially noticeable with electrodes that are ⅛" (3.2 mm) or smaller when welding gapped-butt or T-joints. If the arc initially deposits material on only one workpiece, it is possible that only one of the workpieces is electrically connected to the workpiece lead from the welder and the other piece is not in the circuit. In this case, weaving the electrode back and forth across the workpiece at the start of the weld will usually bridge the gap between the workpieces and tack the pieces together so that they are both in the circuit. Other solutions are to clamp a piece of metal across both pieces, or to perform the welding on a metal tabletop.

Additional Resources

2014 Technical Training Guide. Current Edition. Cleveland, OH. USA: The Lincoln Electric Company. **www.lincolnelectric.com**

1.0.0 Section Review

1. For electric arc welding, wear safety glasses and a welding hood with a lens tinted to shade _____.

   a. 1 to 4
   b. 4 to 8
   c. 9 to 14
   d. 12 to 16

2. Electrode stubs should be _____.

   a. discarded to the floor for later collection
   b. welded together to form additional electrodes
   c. disposed of in a suitable metal bucket or container
   d. placed in water

3. With the tapping method of striking an arc, the arc is struck _____.

   a. ¼" to ½", or ≈6.3 to 12.6 mm, past where the weld is to begin
   b. ¼" to ½", or ≈6.3 to 12.6 mm, ahead of where the weld is to begin
   c. ½" to 1", or ≈12.6 to 25.3 mm, past where the weld is to begin
   d. ½" to 1", or ≈12.6 to 25.3 mm, ahead of where the weld is to begin

SECTION TWO

2.0.0 SMAW Techniques

Objective

Explain how to successfully complete various types of beads and welds.

a. Explain how to properly restart and terminate a weld pass.
b. Describe the technique required to produce stringer beads.
c. Describe the technique required to produce weave and overlapping beads.
d. Describe the techniques required to produce fillet welds in various positions.

Performance Tasks

3. Make stringer, weave, and overlapping beads using E6010 and E7018 electrodes.
4. Make corner welds on an angle iron section end welded to a plate coupon.
5. Make fillet welds using E6010 and E7018 electrodes in the specified positions:
   - Flat (1F)
   - Horizontal (2F)
   - Vertical (3F)
   - Overhead (4F)

Trade Terms

Drag angle: Describes the travel angle when the electrode is pointing in a direction opposite to the welding bead's progression.

Oscillation: A repetitive and consistent side-to-side motion.

Push angle: Describes the travel angle when the electrode is pointing in the same direction as the welding bead's progression.

Quench: To rapidly cool a hot component such as a freshly welded coupon. Note that only coupons should be quenched; real welds with a function should never be quenched.

Restart: The action of restarting a weld bead that is already in progress that requires the crater left from the termination of the previous arc.

Stringer bead: A type of weld bead made without any significant weaving motion. With SMAW, stringer beads are not more than three times the diameter of the electrode.

Undercut: A defect that occurs when the cross-sectional thickness of the base metal is reduced at the edges of the weld. Excessive weld current can cause this as it causes the base metal at the edges to melt and drain into the weld puddle. A downward sloped edge toward the weld bead is the visual evidence.

Weave bead: A type of weld bead made by transverse oscillation of the electrode in a particular pattern.

Weld axis: A straight line drawn through the center of a weld along its length.

Work angle: An angle less than 90 degrees between a line perpendicular to the major workpiece surface and a plane determined by the electrode axis and the weld axis. 0-degree work angle is common. In a T-joint or corner joint, the line is perpendicular to the non-butting member. The definition of work angle for a pipe weld is covered in a later module, as it differs from that of plate welding.

Now that the work area and equipment have been properly prepared and you understand how to strike an arc and control arc blow, the next step is to begin creating weld beads.

2.1.0 Terminations and Restarts

How a weld bead is terminated is important to the quality of the weld. Since electrodes burn away rather quickly and the welder must stop and replace the electrode repeatedly, the bead must also be restarted with care.

2.1.1 Terminating a Weld

Terminations are made at the end of a weld. A termination leaves a crater. When making a termination, the welding codes require that the crater be filled to the full cross section of the weld. This can be difficult because most terminations are at the edge of a plate where welding heat tends to build up, making filling the crater more difficult.

The technique for making a termination is basically the same for all SMAW. Refer to *Figure 9* as the following steps are reviewed.

Step 1 As you approach the end of the weld, begin to stand the electrode up toward 0 degrees, and slow the forward travel.

NCCER – *Welding Level One* 29109-15

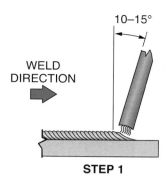

STEP 1

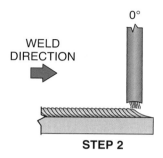

STEP 2

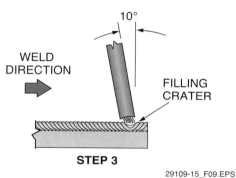

STEP 3

29109-15_F09.EPS

Figure 9 Weld termination.

Step 2 Stop forward movement about ⅛" (≈3.2 mm) from the end of the plate and slowly angle the electrode to about 10 degrees toward the start of the weld.

Step 3 Move about ⅛" (≈3.2 mm) toward the weld, and break the arc when the crater has been filled.

Step 4 Inspect the termination. The crater should be filled to the full cross section of the weld.

Step 5 Remove the electrode from the electrode holder and discard as soon as you stop welding to prevent arcing to nearby surfaces, and place the stub in a proper container.

2.1.2 *Restarting a Weld*

A **restart**, sometimes referred to as a tie-in, is the point where one weld bead stops and another starts (*Figure 10*). Restarts are important because most SMAW welds cannot be made without at least one restart, and an improperly made restart will create a weld defect. A restart must be made so that it blends smoothly into the rest of the weld and does not stand out. The technique for making a restart is the same for both **stringer beads** and **weave beads**. Follow these steps to make a restart:

Step 1 Just before the previous electrode is used up, quickly increase the weld speed to taper the weld for ¼" to ⅜" (≈6.3 to 9.5 mm), and then break the arc.

Step 2 Chip any slag from the tapered section and crater.

Step 3 Remove the electrode from the electrode holder and discard as soon as you stop welding to prevent arcing to nearby surfaces, and place the stub in a proper container.

Step 4 With a new electrode, restrike the arc in front of the crater and in line with the weld. Remember that welding codes do not allow arc strikes outside of the area to be welded.

Step 5 Move the electrode back onto the tapered weld section and develop the weld puddle with a slight circular motion, maintaining the correct arc length and electrode angles.

Step 6 Start the forward motion as the bead achieves the proper width, and continue to weld.

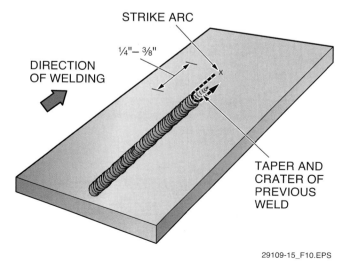

29109-15_F10.EPS

Figure 10 Making a restart.

Step 7 Discard the electrode stub from the holder and inspect the restart. A properly made restart will blend into the previous bead, making it hard to detect.

If the restart has undercut, not enough time was spent on the taper or the crater to fill it. If the undercut is on one side or the other, use more of a side-to-side motion as you move into the taper. If the restart has a hump, it was overfilled; too much time was spent before resuming the forward motion.

Continue to practice restarts until they are correct.

2.2.0 Stringer Beads

A stringer bead is a weld bead that is made with very little or no side-to-side motion of the electrode. The width of a stringer bead will vary with the electrode type. A weave bead is a weld bead that is made with a side-to-side motion of the electrode. Outside of welding joints, stringer and weave beads (*Figure 11*) are used for resurfacing or hard-surfacing. This section will focus on the stringer bead.

The width of weld beads is usually specified in the drawing welding symbols, welding code, or welding procedure specification (WPS) being used for the work. Do not exceed the widths specified for the work.

2.2.1 *Beginning to Weld*

A welder needs to be in a relaxed, comfortable position whenever possible. Because of the limited view of the welding area through a dark helmet lens, a beginning welder may sway a little. This is due to the loss of a visual component in the body's ability to maintain balance. To counteract this, a welder should sit or lean against something to achieve and maintain a stable and relaxed position. This will reduce fatigue and ensure personal safety. With practice and repetition, the body adapts to the visual loss and balance while welding improves.

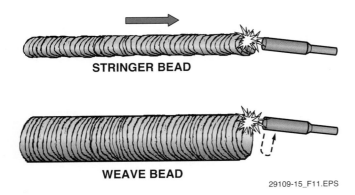

STRINGER BEAD

WEAVE BEAD

29109-15_F11.EPS

Figure 11 Stringer and weave beads.

Starting and Stopping Welds with Weld Tabs

Another method of eliminating welding starting and stopping points is to tack-weld starting (run-on) and stopping (run-off) weld tabs to the workpiece. These tabs allow the arc to stabilize and achieve correct penetration of the workpiece at the start and to terminate with a weld crater off the workpiece at the end. They are especially useful on groove welds requiring multiple passes or when low-hydrogen or fast-fill electrodes are used. Because both ends of a multiple-pass weld may taper down, they help control underfill and back-burn at the start and finish of the welding. After the weld tabs are cut off, the weld is of continuous width and penetration across the entire workpiece.

Beginning welders must learn to view the entire electric arc welding work area through the shaded lens of a helmet and to listen to the sound of the arc. Sound is a vital input. At first, a beginning welder's focus is normally concentrated only on the electric arc, because of the difficulty of striking and keeping an arc established. Once beginners have managed to strike and maintain an arc, however, the tremendous light and heat energy generated within the arc may become fascinating and distracting for a time. With practice, the beginning welder will gradually be able to shift focus and see all sides of a molten weld pool, the weld buildup at the trailing edge of the pool, the cooling slag form over the weld buildup, and even the adjacent welding area on the workpiece. The welder becomes attuned to the correct sound of the arc for the rod and work being welded, and the arc itself may not even be noticed. Once the welder learns to view the entire weld area and listen to the arc, running straight and correct welds is relatively easy.

Before running the first bead, a word of caution is in order. On occasion in this text, you may be instructed to quench your work. This means to dunk the workpiece in water briefly so it can be more easily and safely handled. However, it is very important to understand that this practice relates only to welding coupons used for practice. Real-world welds should never be quenched due to the potential for weld failure due to the rapid temperature change.

2.2.2 Practicing Stringer Beads with E6010

Practice running stringer beads in the flat position using ⅛" (3.2 mm) E6010 or E6011 electrodes. After striking the arc, the electrode axis angle should be a 10-degree to 15-degree **drag angle** in the direction of travel for the weld and at a 0-degree **work angle**. Note that a 0-degree work angle means the electrode is perpendicular (90 degrees) to the base metal. *Figure 12* shows the electrode axis angles.

Travel angles used for SMAW and other processes are categorized as either drag or **push angles**. An electrode drag angle is a travel angle in which the electrode axis points at the weld bead during the running of a weld bead. A push angle is the opposite condition; it occurs when the electrode axis points away from the weld bead and toward the direction of travel when running a weld bead. Except for vertical welds, SMAW is usually accomplished with an electrode drag angle.

The electrode work angle is the side-to-side tilt of a plane containing the electrode axis from a perpendicular line to the major work surface. The plane is formed by the intersection of the **weld axis** and the centerline of the electrode. Work angles typically range from 0 to 25 degrees. In

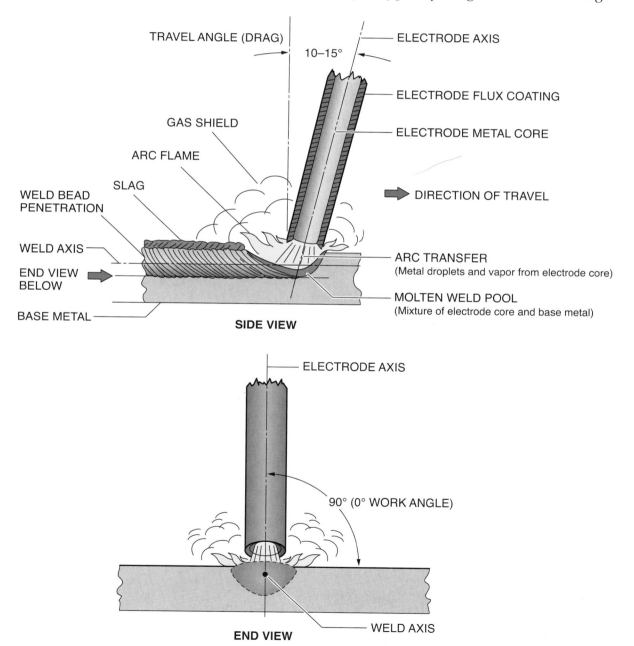

Figure 12 SMAW characteristics with proper travel and work angle.

the case of a single bead being run on a flat plate, the work angle is zero. The work angle for a fillet weld in a T- or corner joint is always taken from a perpendicular line to the non-butting member.

The whipping motion (*Figure 13*), also called a stepping motion, can be used when depositing the stringer bead to control the weld puddle. It is performed by moving the electrode up and forward about ¼" (≈6.3 mm), then down and backward about ³⁄₁₆" (≈4.8 mm). A short pause at the end of the backward travel deposits the weld metal. The exact length of travel can vary. By momentarily lengthening and advancing the arc, the molten weld pool is allowed to cool. Lengthening the arc lowers the arc temperature and reduces metal transfer from the electrode core. Advancing the arc preheats the base metal ahead of the weld and burns off contaminants.

> **NOTE**
>
> This whipping or stepping motion is not used with E7018 electrodes.

Carefully observe and listen to the weld as it is being made. If the arc length, speed, angles, and motions are correct, a distinctive frying sound will be heard. Experiment by making minor changes in these factors and notice the effects.

Follow these steps to run stringer beads:

Step 1 Shield your eyes, strike the arc, and move the electrode back slightly to the correct starting point. Then position the electrode for the travel angle and perpendicular to the base metal.

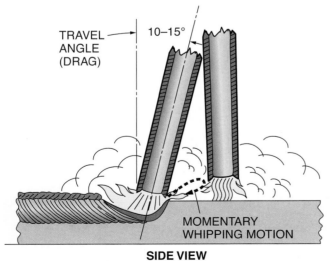

Figure 13 Whipping motion.

Step 2 Hold the arc in place until the weld puddle widens to about two times the diameter of the electrode's metal core.

Step 3 Slowly move the arc forward, keeping a constant arc length. Since the electrode is constantly adding filler metal to the weld, it becomes shorter. The welder, therefore, must be constantly closing in on the workpiece to keep the arc gap (the length of the arc) consistent. As necessary, use the whipping motion and adjust the forward speed to control buildup and preheat the weld zone.

> **WARNING!**
>
> Always wear appropriate face protection and/or a welding hood to prevent hot slag from striking your face.

Step 4 Continue to weld until a bead about 2" to 3" (≈5 to 7.5 cm) long is formed and then break the arc by quickly lifting the electrode straight up. A crater will be left at the point where the arc was broken.

Step 5 Remove the electrode from the electrode holder and discard as soon as you stop welding to prevent arcing to nearby surfaces, and place the stub in a proper container. This should become an unbroken habit.

Step 6 Chip the weld slag that formed with a chipping hammer and clean the bead with a wire brush. This cleaning process should also become habitual, conducted between all weld passes.

Step 7 Have your instructor inspect the bead for the following characteristics:
   - Straightness
   - Uniform appearance on the bead face
   - Smooth, flat transition with complete fusion at the toes (where the edge of the weld meets the base metal) of the weld
   - No porosity
   - No undercut at the toes
   - No inclusions
   - No cracks
   - No overlap

Step 8 Continue welding beads until you can make acceptable welds every time. *Figure 14* shows proper and improper weld beads.

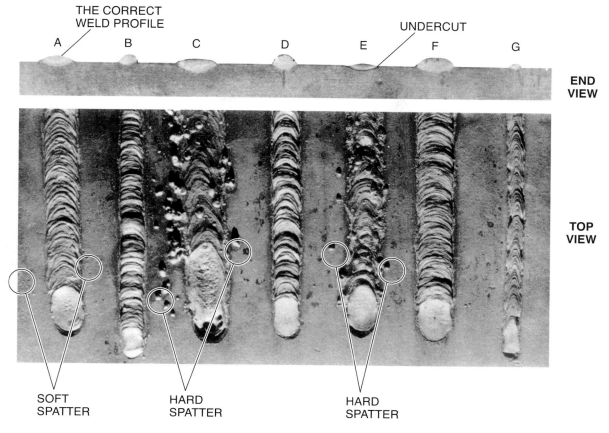

A = Correct current, arc length, and travel speed. Note the easily removed spatter (soft spatter).
B = Current set too low.
C = Current set too high. Note the hard spatter (spatter firmly bonded to base material that must be ground or chiseled off). Note the pointed ends of the bead indicating the weld pool was too hot and cooled too slowly. Impurities are usually trapped in the weld due to the slow cooling.
D = Arc length too short (narrow, high bead caused by arc pressure).
E = Arc length too long. Note the hard spatter and bead undercut of the edges.
F = Travel speed too slow (wide high bead).
G = Travel speed too fast. Note the pointed ends of bead.

© American Welding Society (AWS) *Welding Handbook*

29109-15_F14.EPS

Figure 14 Effect of current, arc length, and travel speed on SMAW beads.

2.2.3 *Practicing Stringer Beads with E7018*

Continue running stringer beads, this time using ⅛" (3.2 mm) E7018 electrodes. Use the same electrode angles, but remember not to whip the E7018 electrode. When running all low-hydrogen (F4) electrodes, the arc should never leave the weld puddle, and the visible arc should be shorter than when using the E6010 electrode. This is due to the thicker, longer sheath of the flux coating. Weld defects, such as porosity or hydrogen embrittlement, can occur if the arc leaves the weld puddle or if the arc is too long. The arc can be moved within the weld puddle to control the bead shape.

2.3.0 Weave Beads and Overlaps

Running weave beads is significantly different from stringer beads. There are a number of different weave patterns that can be used.

2.3.1 *Practicing Weave Beads with E6010*

Practice running weave beads in the flat position using ⅛" (3.2 mm) E6010 or E6011 electrodes. After striking the arc, the electrode angle should be 10 to 15 degrees in the direction of travel and at a 0-degree work angle.

Lash

To ensure good welding results, remember LASH.

L *(length of arc)* – The distance between the electrode and the base metal (usually one times the electrode diameter).

A *(angle)* – Two angles are critical:

- Travel angle – The longitudinal angle of the electrode in relation to the axis of the weld joint
- Work angle – The traverse angle of the electrode in relation to the axis of the weld joint

S *(speed)* – Travel speed is measured in inches per minute (IPM). The width of the weld will determine if the travel speed is correct.

H *(heat)* – Controlled by the amperage setting and dependent upon the electrode diameter, base metal type, base metal thickness, and the welding position.

The weave bead is made by moving the electrode back and forth. Many different patterns can be used to make a weave bead, including zigzags, Js, crescents, boxes, circles, and figure 8s (*Figure 15*). When making a weave bead, use care at the toes to ensure proper tie-in to the base metal. To ensure proper tie-in at the toes, slow down or pause slightly at the edges. Pause longer when using the zigzag motion. The pause at the edges will also flatten out the weld, giving it the proper profile.

Carefully observe and listen to the weld as it is being made. If the arc length, speed, angles, and motions are correct, a distinctive frying sound will be heard. Experiment by making minor changes in these factors and notice the effects.

Follow these steps to run weave beads (*Figure 16*):

Step 1 Strike the arc and position the electrode for a drag angle. Make sure the electrode is perpendicular to the base metal (0-degree work angle).

Step 2 Hold the arc in place until the weld puddle widens to about two times the diameter of the electrode's metal core.

Step 3 Slowly move the arc forward in a weaving motion, keeping a constant arc length.

> **WARNING!**
>
> Always wear the proper PPE to prevent hot slag from hitting your face.

Step 4 Continue to weld until a weave bead about 2" to 3" (≈5 to 7.5 cm) long is formed, and then break the arc by lifting the electrode straight up. A crater will be left at the point where the arc was broken.

Step 5 Remove the electrode from the electrode holder and discard as soon as you stop welding to prevent arcing to nearby surfaces, and place the stub in a proper container.

Step 6 Chip, clean, and wire brush the crater and bead.

Step 7 Make a restart, and continue welding to the end of the plate.

Step 8 Have the instructor inspect the bead, which should have the following characteristics:

- Weld bead straight to within ⅛" (≈3.2 mm)
- Uniform appearance on the bead face
- Smooth, flat transition with complete fusion at the toes of the weld
- Crater and restarts filled to the full cross section of the weld
- No porosity
- No excessive undercut at the toes
- No inclusions
- No cracks
- No overlap

Continue welding weave beads until you can make acceptable welds every time.

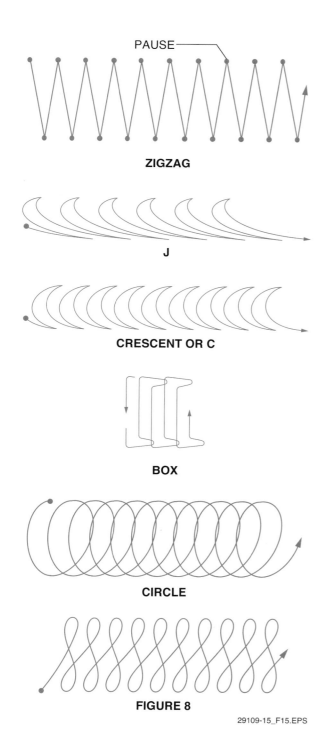

PAUSE

ZIGZAG

J

CRESCENT OR C

BOX

CIRCLE

FIGURE 8

29109-15_F15.EPS

Figure 15 Weave motions.

Weave Bead Width

Although the usual width of a straight bead is two to three times the diameter of the electrode's metal core, the bead width for a weave bead can usually be up to, but should not exceed, eight times the electrode diameter.

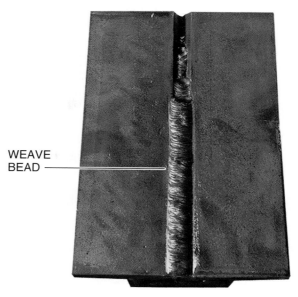

WEAVE BEAD

29109-15_F16.EPS

Figure 16 Weave bead.

2.3.2 Practicing Weave Beads with E7018

Repeat running weave beads using ⅛" (3.2 mm) E7018 electrodes. Use the same electrode angles, but do not whip the E7018 electrode. When running low-hydrogen electrodes, the arc should never leave the weld puddle, and the visible arc should be shorter than when using E6010. Weld defects, such as porosity or hydrogen embrittlement, can occur if the arc is too long or if it leaves the weld puddle. The arc can be moved within the weld puddle to control the bead shape.

It is important to note that there are several weave techniques that should not be used with the E7018 electrode. They are the circle, box, and figure 8 weave techniques.

2.3.3 Practicing Overlapping Beads with E6010

Overlapping beads are made by depositing connective weld beads parallel to one another. The parallel beads overlap, forming a flat surface (*Figure 17*). This technique is also called padding. Overlapping beads are used to build up a surface and to make multi-pass welds. Both stringer and weave beads can be overlapped. Properly overlapped beads, when viewed from the end, will form a relatively flat surface.

Overlapping stringer beads must be cleaned carefully to prevent slag inclusions. The thickness of the overlap must be fairly consistent, so that it will not be necessary to machine the surface. Follow these steps to weld overlapping stringer beads using ⅛" (3.2 mm) E6010 electrodes:

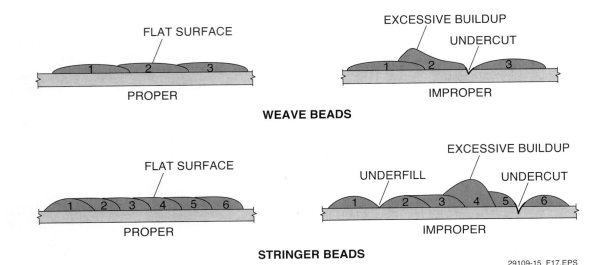

FLAT SURFACE

PROPER

EXCESSIVE BUILDUP

UNDERCUT

IMPROPER

WEAVE BEADS

FLAT SURFACE

PROPER

EXCESSIVE BUILDUP

UNDERFILL

UNDERCUT

IMPROPER

STRINGER BEADS

29109-15_F17.EPS

Figure 17 Proper and improper overlapping beads.

Step 1 Mark out a 4" (102 mm) square on a piece of steel.

Step 2 Weld a stringer bead along one edge.

Step 3 Properly dispose of the electrode stub immediately and chip/clean the weld bead.

Step 4 After striking the arc for the next stringer bead, and with the proper travel angle, position the electrode at a work angle of 10 to 15 degrees to the side of the previous bead to obtain proper tie-in (*Figure 18*).

Step 5 Continue running stringer beads until the marked square is covered.

> **WARNING!**
>
> The base metal will get very hot as it is built up. Using pliers, cool it by plunging it into water. However, be careful of the resulting plume of hot steam.

> **CAUTION**
>
> Only welding coupons for practice should be quenched. Never quench actual welding work.

Step 6 Continue building layers of stringer beads, one on top of the other, until the technique is perfected.

Continue welding overlapping beads using the weaving technique. Remember to angle the electrode toward the previous bead to obtain a good tie-in.

2.3.4 Practicing Overlapping Beads with E7018

Repeat welding overlapping beads using ⅛" (3.2 mm) E7018 electrodes. Build a pad using stringer beads, and then repeat building the pad using weave beads as indicated in *Figure* 18. Keep a short arc and do not whip the electrode.

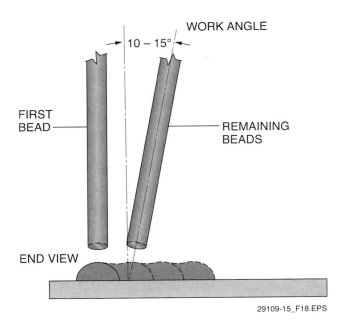

10 – 15°

WORK ANGLE

FIRST BEAD

REMAINING BEADS

END VIEW

29109-15_F18.EPS

Figure 18 Electrode work angle for overlapping beads.

2.4.0 Fillet Welds

A fillet weld is a weld that is approximately triangular in cross section and is used with T-, lap, and corner joints. The sizes and locations of fillet welds are given as welding symbols. The two types of fillet welds are convex and concave (*Figure 19*). A convex fillet weld has its surface bowed out like the outside surface of a ball. A concave fillet weld has its surface bowed in like the inside surface of a bowl. The following terms are used to describe various aspects of fillet welds:

- *Weld face* – The exposed surface of the weld
- *Leg* – The distance from the root of the joint to the toe of a fillet weld
- *Weld toe* – The junction between the face of a weld and the base metal
- *Weld root* – The point shown in cross section in which the weld metal intersects with the base metal and extends farthest into the weld joint
- *Size* – The leg lengths of the largest right triangle that can be drawn within the cross section of a fillet weld
- *Actual throat* – The shortest distance from the root of the weld to its face
- *Effective throat* – The minimum distance, minus any convexity, from the root of the weld to its face
- *Theoretical throat* – The distance from the beginning of the joint root (with a zero opening) that is perpendicular to the hypotenuse of the largest right triangle that can be inscribed within the cross section of a fillet weld

As shown in *Figure 20*, fillet welds may be either equal leg or unequal leg. The face may be slightly convex, flat, or slightly concave. Welding codes require that fillet welds have a uniform concave or convex face, although a slightly non-uniform face is acceptable. The convexity of a fillet weld or individual surface bead should not exceed 0.07 times the actual face width or individual surface bead plus 0.06" (≈0.15 mm).

> **CAUTION**
>
> Refer to your site's WPS for specific requirements on fillet welds. The information in this module is provided as a general guideline only. The site WPS or quality specifications must be followed for all welds. Check with your supervisor if you are unsure of the specifications for your application.

A fillet weld is unacceptable if the profile has insufficient throat, excessive convexity, excessive undercut, overlap, insufficient leg, or incomplete fusion, as shown in *Figure 21*.

Fillet welds require little base metal preparation except for cleaning the weld area and removing all dross from cut surfaces. Any dross from oxyfuel, plasma arc, or carbon arc cutting will cause porosity in the weld. For this reason, the codes require that all dross be removed completely prior to welding.

2.4.1 Fillet Weld Positions

The most common fillet welds are made in lap and T-joints. The weld position for plate is determined by the weld axis and the orientation of the workpiece. The positions for fillet welding on plate are flat, or 1F (the letter F stands for fillet); horizontal, or 2F; vertical, or 3F; and overhead, or 4F (*Figure 22*). In the 1F and 2F positions, the weld axis can be inclined up to 15 degrees. Any weld axis inclination for the other positions varies with the rotational position of the weld face as specified in AWS standards.

2.4.2 Practicing Horizontal Fillet Welds with E6010 (2F Position)

Practice horizontal fillet (2F) welding by placing multiple-pass fillet welds in a T-joint using ⅛" (3.2 mm) E6010 electrodes. When making horizontal fillet welds, pay close attention to the electrode angles and travel speed. For the first bead, the electrode work angle is 45 degrees. The work angle is adjusted for all other welds. Increase or decrease the travel speed to control the amount of weld-metal buildup.

Follow these steps to make a horizontal fillet weld:

Step 1 Tack two plates together to form a T-joint for the fillet weld coupon (*Figure 23*). Clean the tack welds.

Step 2 Position the coupon on the welding table.

Preferred Fillet Weld Contours

In single-pass, weave-bead fillet welds where two workpieces are being joined at an angle (not lap joints), flat or slightly convex faces are usually preferred because weld stresses are more uniformly distributed through the fillet and workpieces.

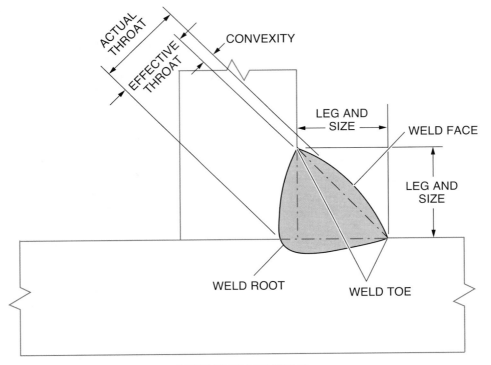

CONVEX FILLET WELD

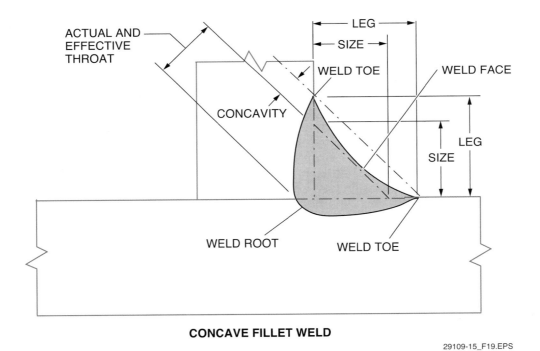

CONCAVE FILLET WELD

29109-15_F19.EPS

Figure 19 Convex and concave fillet welds.

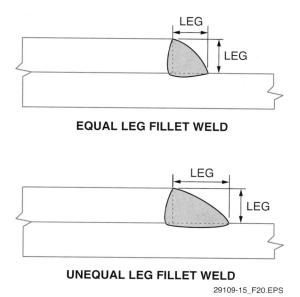

EQUAL LEG FILLET WELD

UNEQUAL LEG FILLET WELD

29109-15_F20.EPS

Figure 20 Equal leg and unequal leg fillet welds.

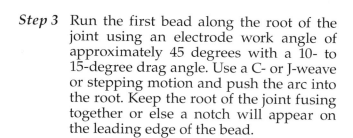

Step 3 Run the first bead along the root of the joint using an electrode work angle of approximately 45 degrees with a 10- to 15-degree drag angle. Use a C- or J-weave or stepping motion and push the arc into the root. Keep the root of the joint fusing together or else a notch will appear on the leading edge of the bead.

Step 4 Properly dispose of the electrode stub immediately and chip/clean the weld bead.

Step 5 Using a slight **oscillation** or stepping motion, run the second bead along the bottom toe of the first weld, overlapping about 75 percent of the first bead. Use the electrode work angle shown in *Figure 24*.

Step 6 Properly dispose of the electrode stub immediately and chip/clean the weld bead.

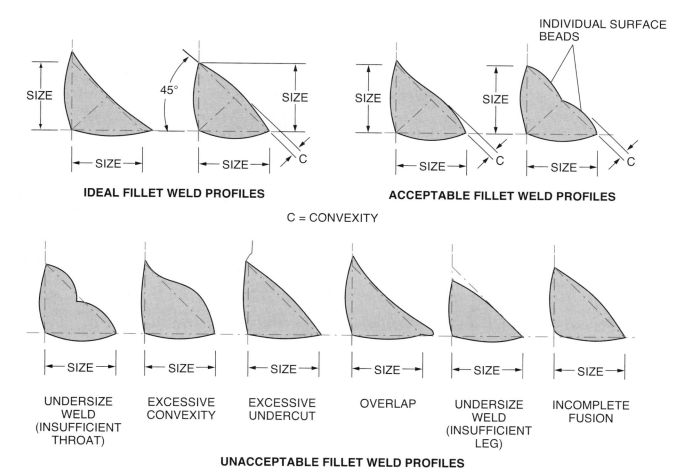

IDEAL FILLET WELD PROFILES

ACCEPTABLE FILLET WELD PROFILES

INDIVIDUAL SURFACE BEADS

C = CONVEXITY

UNDERSIZE WELD (INSUFFICIENT THROAT)

EXCESSIVE CONVEXITY

EXCESSIVE UNDERCUT

OVERLAP

UNDERSIZE WELD (INSUFFICIENT LEG)

INCOMPLETE FUSION

UNACCEPTABLE FILLET WELD PROFILES

29109-15_F21.EPS

Figure 21 Acceptable and unacceptable fillet weld profiles.

Step 7 Repeat Steps 5 and 6 for each of the remaining bead passes. Run the beads along the toes of the underlying beads and overlap them about 50 percent.

Step 8 Have the instructor inspect the weld. The weld is acceptable if it has the following characteristics:

- Uniform appearance on the bead face
- Craters and restarts filled to the full cross section of the weld
- Uniform weld size ±1/16" (±1.6 mm)

- Acceptable weld profile in accordance with the applicable code
- Smooth transition with complete fusion at the toes of the weld
- No porosity
- No undercut
- No overlap
- No inclusions
- No cracks

2.4.3 *Practicing Horizontal Fillet Welds with E7018 (2F Position)*

Repeat horizontal fillet (2F) welding using 1/8" (3.2 mm) E7018 electrodes. Use the same procedure, bead sequence, and electrode angles that were used for the horizontal fillet weld with E6010 electrodes. Use a short arc and remember not to whip the E7018 electrode.

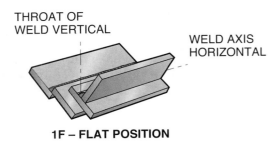

1F – FLAT POSITION

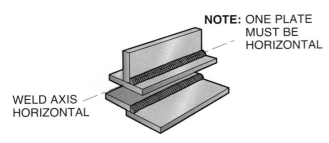

2F – HORIZONTAL POSITION

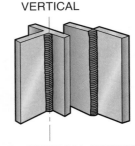

3F – VERTICAL POSITION

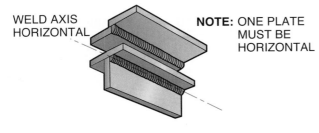

4F – OVERHEAD POSITION

29109-15_F22.EPS

Figure 22 Fillet welding positions for plate.

29109-15_F23.EPS

Figure 23 Fillet weld coupon.

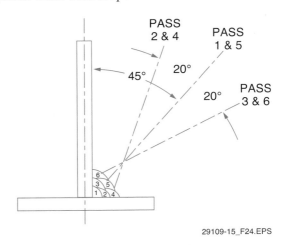

29109-15_F24.EPS

Figure 24 Multiple-pass 2F weld sequences and work angles.

2.4.4 Practicing Vertical Fillet Welds with E6010 (3F Position)

Practice vertical fillet (3F) welding by placing multiple-pass fillet welds in a T-joint using ⅛" (3.2 mm) E6010 electrodes. Normally, vertical welds are accomplished by welding uphill from the bottom to the top using an electrode push angle (up-angle). Because of the uphill welding and push angle, this type of weld is sometimes called vertical-up fillet welding. When vertical welding, either stringer or weave beads can be used. On the job, the site WPS or quality standard will specify which technique to use. Typically, weave beads are used with E6010 electrodes and, for welding carbon steel, with E7018 electrodes. Stringer beads are generally called for when welding alloy steels with low-hydrogen (F4) electrodes.

> **NOTE**
> Check with your instructor to see if you should run stringer beads, run weave beads, or practice both techniques.

Follow these steps to make a vertical-up fillet weld:

Step 1 Tack two plates together to form a T-joint for the fillet weld coupon.

Step 2 Tack-weld the coupon in the vertical position.

Step 3 Run the first bead along the root of the joint (starting at the bottom) using an electrode- (work) angle of approximately 45 degrees with a 0- to 10-degree push angle. Use a whipping motion by quickly raising the electrode about ¼" (≈6 mm) and then dropping it back into the weld puddle. Pause in the weld puddle to fill the crater. For a stringer bead, use the same technique. See *Figure 25* for bead placement.

Step 4 Properly dispose of the electrode stub immediately and chip/clean the weld bead.

Step 5 Run the second bead using a weave technique, such as a C-pattern. Use a slow motion across the face of the weld, pausing at each toe for penetration and to fill the crater. A slight whip can be used to control the puddle when you reach the toe. Adjust the travel speed across the face of the weld to control the buildup. *Figure 26* shows the bead sequence and electrode angles for weave beads (all degrees shown are approximate). In practice, all bead passes run the entire length of the weld.

Step 6 Properly dispose of the electrode stub immediately and chip/clean the weld bead.

Step 7 Continue to run weld beads as shown in *Figure 26*.

Step 8 Properly dispose of the electrode stub immediately and chip/clean the weld bead.

Step 9 Have the instructor inspect the weld. The weld is acceptable if it has the following characteristics:
  - Uniform appearance on the bead face
  - Craters and restarts filled to the full cross section of the weld
  - Uniform weld size ±¹⁄₁₆" (±1.6 mm)
  - Acceptable weld profile in accordance with the applicable code
  - Smooth transition with complete fusion at the toes of the weld
  - No porosity
  - No undercut
  - No overlap
  - No inclusions
  - No cracks

Tacking and Aligning Workpieces

When tacking workpieces together, both sides of the workpieces are usually tacked with welds that are about ½" long. The tacks are used to position the workpieces and minimize distortion as the final welds are made. After the first tack weld, use a hammer or other tool to align the workpieces side-to-side and end-to-end; then tack the opposite side. Tack the far ends of the workpieces in the same manner. Intermediate tack welds can be made every 5" to 6" as necessary, to minimize lengthwise distortion.

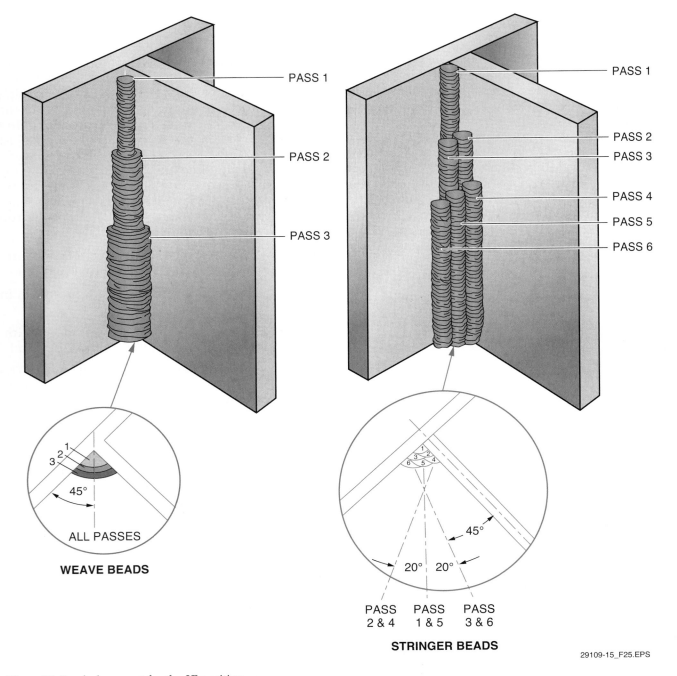

Figure 25 Bead placement for the 3F position.

Test Joint Heat Dissipation

In T-joints, the welding heat dissipates more rapidly in the thicker or the non-butting member. On various bead passes, the arc may have to be concentrated slightly more on the thicker or the non-butting member to compensate for the heat loss.

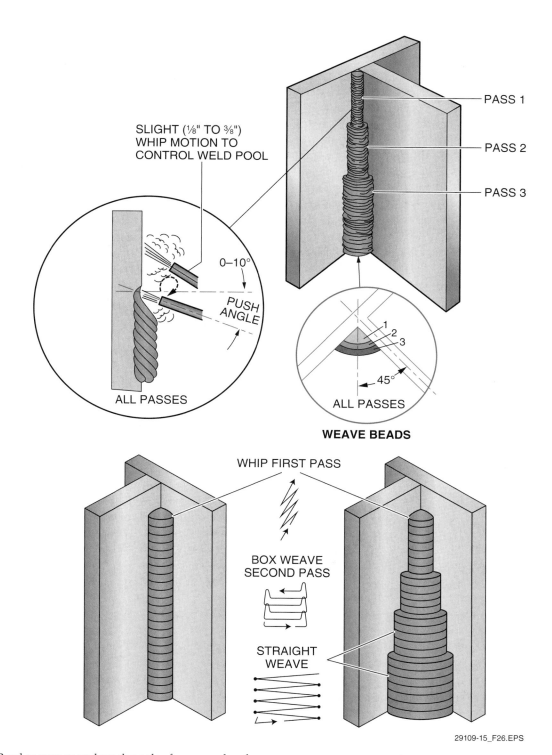

SLIGHT (⅛" TO ⅜")
WHIP MOTION TO
CONTROL WELD POOL

0–10°

PUSH
ANGLE

ALL PASSES

PASS 1

PASS 2

PASS 3

1
2
3

45°

ALL PASSES

WEAVE BEADS

WHIP FIRST PASS

BOX WEAVE
SECOND PASS

STRAIGHT
WEAVE

29109-15_F26.EPS

Figure 26 Bead sequence and work angles for weave beads.

T-Joint Vertical Plate Undercut

The most common defect for T-joints is undercut on the vertical plate of the joint. Use of a J-weave usually eliminates the problem. However, if the problem persists, angling the arc slightly toward the vertical plate and the bead at the top of the weave will force more metal into the bead at the top edge of the weld

2.4.5 Practicing Vertical Fillet Welds with E7018 (3F Position)

Repeat vertical fillet (3F) welding using ⅛" (3.2 mm) E7018 electrodes. For the root pass and stringer fill beads, use a quick side-to-side motion, moving the electrode about ⅛" (3.2 mm) without removing the arc from the weld puddle. Pause slightly at each toe to penetrate and fill the crater to prevent undercut. The electrode work angle should be approximately 45 degrees with a 0- to 10-degree upward push angle. For the remaining stringer beads, use an electrode work angle of approximately ±20 degrees from 45 degrees as required.

If weave beads are to be used, employ the same electrode work angles as the root pass. For the filler beads, move slowly across the face, increasing or decreasing the travel speed to control the buildup. Remember to use a short arc. Do not whip the electrode.

Figure 25 shows the bead pass sequence and electrode angle for both weave and stringer beads. All degrees shown are approximate. Again, all bead passes would extend the entire length of the weld.

2.4.6 Practicing Overhead Fillet Welds with E6010 (4F Position)

Practice overhead fillet (4F) welding by welding multiple-pass convex fillet welds in a T-joint using ⅛" (3.2 mm) E6010 electrodes. When making overhead fillet welds, pay close attention to the electrode angles and travel speed. For the first bead, the electrode work angle is approximately 45 degrees. The work angle is adjusted for all other welds. Increase or decrease the travel speed to control the amount of weld metal buildup.

Vertical Fillet Welds

When making vertical fillet welds, pay close attention to the electrode angles and travel speed. For the first bead, the electrode work angle is approximately 45 degrees. The work angle is adjusted for all other welds. Increase or decrease the travel speed to control the amount of weld metal buildup.

Follow these steps to make an overhead fillet weld:

Step 1   Tack two plates together to form a T-joint for the fillet weld coupon.

Step 2   Tack-weld the coupon so it is in the overhead position.

Step 3   Run the first bead along the root of the joint using an electrode angle of approximately 45 degrees with a 10- to 15-degree drag angle. Use a slight oscillation (circular or side-to-side motion) to tie-in the weld at the toes.

Step 4   Properly dispose of the electrode stub immediately and chip/clean the weld bead.

Step 5   Using a slight oscillation, run the second bead along the bottom toe of the first weld, overlapping about 75 percent of the first bead. Use the electrode work angle shown in *Figure 27*.

Step 6   Properly dispose of the electrode stub immediately and chip/clean the weld bead.

Vertical Weave Bead

To help control undercut, an alternate pattern such as a small triangular weave can be used in vertical welds. By pausing the rod at the edges, the previous undercut can be filled. This action also creates undercut at the existing weld pool, but that will be filled in by the next weave.

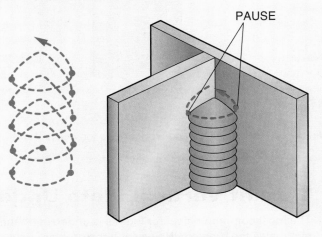

TRIANGULAR WEAVE PATTERN

29109-15_SA06.EPS

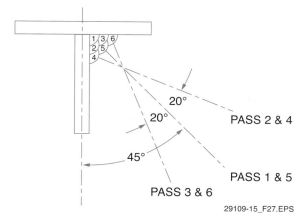

Figure 27 Multiple-pass 4F weld sequences and work angles.

Step 7 Repeat Steps 5 and 6 for each of the remaining bead passes. Run the beads along the weld toes of the underlying beads and overlap them about 75 percent.

Step 8 Have the instructor inspect the weld. The weld is acceptable if it has the following characteristics:

- Uniform appearance on the bead face
- Craters and restarts filled to the full cross section of the weld
- Uniform weld size ±1/16" (±1.6 mm)
- Acceptable weld profile in accordance with the applicable code
- Smooth transition with complete fusion at the toes of the weld
- No porosity
- No undercut
- No overlap
- No inclusions
- No cracks

2.4.7 Practicing Overhead Fillet Welds with E7018 (4F Position)

Repeat overhead fillet (4F) welding using 1/8" (3.2 mm) E7018 electrodes. Use the same procedure, bead sequence, and electrode angles that were used for the overhead fillet weld with E6010 electrodes. Use a short arc and do not whip the electrode.

Additional Resources

2014 Technical Training Guide. Latest Edition. Cleveland, OH. USA: The Lincoln Electric Company. www.lincolnelectric.com

2.0.0 Section Review

1. On a restart, restrike the arc _____.
   a. in front of the crater and in line with the weld
   b. behind the crater and in line with the weld
   c. in front of the crater, just outside the existing bead
   d. behind the crater, just outside the existing bead

2. The whipping motion in SMAW welding helps control the weld puddle by _____.
   a. dramatically increasing the weld current
   b. taking advantage of the E7018 electrode characteristics
   c. depositing additional hydrogen into the puddle
   d. momentarily lengthening and advancing the arc

3. You are working on overlapping stringer bead practice with E6010. After striking the arc for the next stringer bead, and with the proper travel angle, position the electrode at a work angle of _____.
   a. 10 to 15 degrees to the side of the previous bead
   b. 15 to 20 degrees to the side of the previous bead
   c. 25 to 30 degrees to the side of the previous bead
   d. 40 to 45 degrees to the side of the previous bead

4. The position designation for a vertical fillet weld is _____.
   a. 1F
   b. 2F
   c. 3F
   d. 4F

Summary

Striking an arc, running stringer and weave beads, and making fillet welds are basic and essential skills a welder must develop. Even the method used to strike an arc can result in a rejected weld. Developing and honing these skills allows welders to progress to more difficult procedures. It is important to practice these welds until acceptable welds are produced.

Welds can be performed in different positions: flat, vertical, horizontal, and overhead. It is necessary to use various types of electrodes when welding in different positions. The practice provided for in this module provides trainees with their first opportunity to create a weld bead and experience how different electrodes perform in different positions.

Review Questions

1. Which of the following is a gas that would be used to purge a container before welding or cutting?
   a. Oxygen
   b. Methane
   c. Nitrogen
   d. Propane

2. Which of the following compounds or elements are most likely to be encountered in welding and can create toxic fumes?
   a. Cadmium
   b. Brass
   c. Helium
   d. Argon

3. Which of the following electrodes must remain in an oven until they are needed?
   a. E6010
   b. E6011
   c. E7018
   d. E6013

4. The diameter of a SMAW electrode's metal core is also representative of _____.
   a. one-half the proper arc length
   b. twice the proper arc length
   c. four times the arc length
   d. the proper arc length

5. The easiest method of striking an arc is the _____.
   a. scratching method
   b. grounding method
   c. weaving method
   d. tapping method

6. E7018 electrodes are more difficult to strike an arc with because _____.
   a. the flux coating will not break away from the tip
   b. they stick to the base metal more easily
   c. the welding current has to be very low
   d. the welding current has to be very high

7. Arc blow can cause defects such as excessive weld spatter and _____.
   a. porosity
   b. burn through
   c. increased magnetic fields
   d. decreased magnetic fields

8. Which of the following is *not* a common method of controlling or eliminating arc blow?
   a. Tacking the workpieces together.
   b. Changing from an AC welding process to a DC process.
   c. Shortening the length of the arc.
   d. Reducing the weld current.

9. When preparing to terminate a weld bead, _____.
   a. increase the forward travel speed and change from a push angle to a drag angle
   b. begin to lay the electrode down to a more acute work angle
   c. decrease the current setting on the welding machine
   d. slow the forward travel speed and begin to stand the electrode up

10. The point where one weld bead stops and another begins to continue its progression is called a _____.
    a. retrace
    b. restart
    c. weave
    d. blend

11. A weld bead with very little or no side-to-side motion of the electrode required is called a(n) _____.
    a. arc bead
    b. weave bead
    c. stringer bead
    d. whipping bead

12. The width of stringer and weave beads is usually specified in the drawing weld symbols or the _____.
    a. SDS/MSDS for the electrode in use
    b. electrode manufacturer's documentation
    c. welding procedure specification (WPS)
    d. welding machine documentation

13. The whipping motion used to control the weld puddle should not be used with _____.

   a. E6010 electrodes
   b. E6011 electrodes
   c. E7018 electrodes
   d. E6013 electrodes

14. Overlapping beads are made by depositing connective weld beads parallel to one another to form a(n) _____.

   a. flat surface
   b. undercut surface
   c. concave surface
   d. convex surface

15. The part of a fillet weld that can be convex or concave is called the _____.

   a. leg
   b. throat
   c. face
   d. root

Trade Terms Quiz

Fill in the blank with the correct term that you learned from your study of this module.

1. When magnetic forces deflect the arc from its intended path, the action is called _____.

2. A straight line drawn through the center of a weld and along its length is called the _____.

3. _____ is a discontinuity consisting of localized melting of the base metal or finished weld caused by the initiation of an arc.

4. A type of weld bead made by transverse oscillation of the electrode is called a(n) _____.

5. _____ is an angle that is less than 90 degrees between a line perpendicular to the surface of the workpiece and a plane determined by the electrode axis and the weld axis.

6. _____ is the travel angle at which the electrode is pointing in the same direction as the welding bead progression.

7. A consistent and repetitive side-to-side motion is called _____.

8. _____ is the travel angle in which the electrode is pointing in a direction opposite to the welding bead progression.

9. A piece of metal to be welded upon or welded together with another piece as a test or practice is called a(n) _____.

10. _____ is a type of weld bead made without any significant weaving motion.

11. When the cross-sectional thickness of the base metal has been reduced along the edge of a weld bead, this defect is referred to as _____.

12. A _____ is where the welder begins to add to an already existing weld bead.

13. A welder should never _____ a weld that has a function, but it is acceptable for practice coupons.

Trade Terms

Arc blow 1
Arc strike 3
Drag angle 8
Oscillation 7
Push angle 6

Quench 13
Restart 12
Stringer bead 10
Undercut 11
Weave bead 4

Weld axis 2
Weld coupon 9
Work angle 5

Appendix

PERFORMANCE ACCREDITATION TASKS

The American Welding Society (AWS) School Excelling through National Skills Standards Education (SENSE) program is a comprehensive set of minimum Standards and Guidelines for Welding Education programs. The following performance accreditation tasks are aligned with and designed around the SENSE program.

The Performance Accreditation Tasks (PATs) correspond to and support the learning objectives in *AWS EG2.0, Guide for the Training and Qualification of Welding Personnel: Entry-Level Welder.*

Note that in order to satisfy all learning objectives in *AWS EG2.0*, the instructor must also use the PATs contained in the second level of the NCCER Welding curriculum.

PATs 1 and 2 correspond to *AWS EG2.0, Module 4—Shielded Metal Arc Welding*, Key Indicators: 3 and 4.

PATs 3 through 8 correspond to *AWS EG2.0, Module 4—Shielded Metal Arc Welding*, Key Indicators: 3, 4, and 5.

PATs provide specific acceptable criteria for performance and help to ensure a true competency-based welding program for students.

The following tasks are designed to evaluate your ability to run beads and fillet welds with SMAW equipment and techniques. Perform each task when you are instructed to do so by your instructor. As you complete each task, bring it to your instructor for evaluation. Do not proceed to the next task until instructed to do so.

BUILD A PAD WITH E6010 OR E6011 ELECTRODES IN THE FLAT POSITION

Using ³⁄₃₂" to ⁵⁄₃₂" (2.4 to 4.0 mm) E6010 or E6011 electrodes, build up a pad of weld metal on carbon steel plate as indicated.

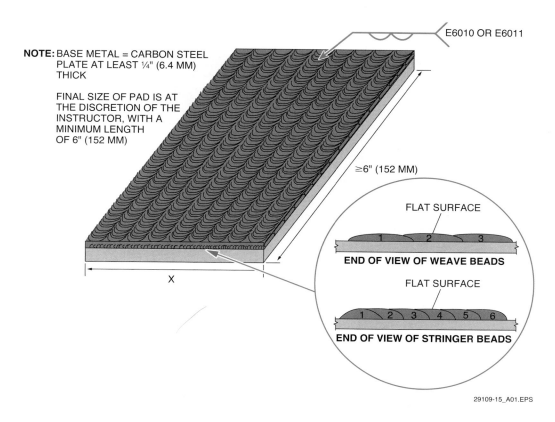

E6010 OR E6011

NOTE: BASE METAL = CARBON STEEL PLATE AT LEAST ¼" (6.4 MM) THICK

FINAL SIZE OF PAD IS AT THE DISCRETION OF THE INSTRUCTOR, WITH A MINIMUM LENGTH OF 6" (152 MM)

≥6" (152 MM)

X

FLAT SURFACE

1 2 3

END OF VIEW OF WEAVE BEADS

FLAT SURFACE

1 2 3 4 5 6

END OF VIEW OF STRINGER BEADS

29109-15_A01.EPS

Criteria for Acceptance:

- Weld beads straight to within ⅛" (3.2 mm)
- Uniform rippled appearance on the bead face
- Craters and restarts filled to the full cross section of the weld
- Face of the pad flat to within ⅛" (3.2 mm)
- Smooth flat transition with complete fusion at the toes of one bead into the face of the previous bead
- No porosity
- No overlap at weld toes
- No excessive undercut
- No inclusions
- No cracks

BUILD A PAD WITH E7018 ELECTRODES IN THE FLAT POSITION

Using ³⁄₃₂" to ⁵⁄₃₂" (2.4 to 4.0 mm) E7018 electrodes, build up a pad of weld metal on carbon steel plate as indicated.

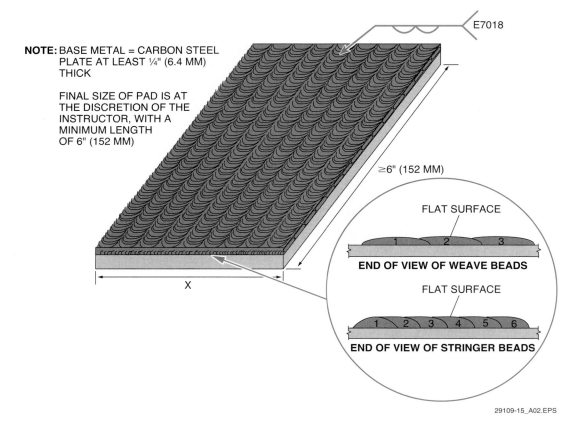

NOTE: BASE METAL = CARBON STEEL PLATE AT LEAST ¼" (6.4 MM) THICK

FINAL SIZE OF PAD IS AT THE DISCRETION OF THE INSTRUCTOR, WITH A MINIMUM LENGTH OF 6" (152 MM)

E7018

≥6" (152 MM)

X

FLAT SURFACE

END OF VIEW OF WEAVE BEADS

FLAT SURFACE

END OF VIEW OF STRINGER BEADS

29109-15_A02.EPS

Criteria for Acceptance:

- Weld beads straight to within ⅛" (3.2 mm)   _____
- Uniform rippled appearance on the bead face   _____
- Craters and restarts filled to the full cross section of the weld   _____
- Face of the pad flat to within ⅛" (3.2 mm)   _____
- Smooth flat transition with complete fusion at the toes of one bead into the face of the previous bead   _____
- No porosity   _____
- No overlap at weld toes   _____
- No excessive undercut   _____
- No inclusions   _____
- No cracks   _____

HORIZONTAL (2F) FILLET WELD WITH E6010 OR E6011 ELECTRODES

Using ³⁄₃₂" to ⁵⁄₃₂" (2.4 to 4.0 mm) E6010 or E6011 electrodes, make a horizontal fillet weld as indicated.

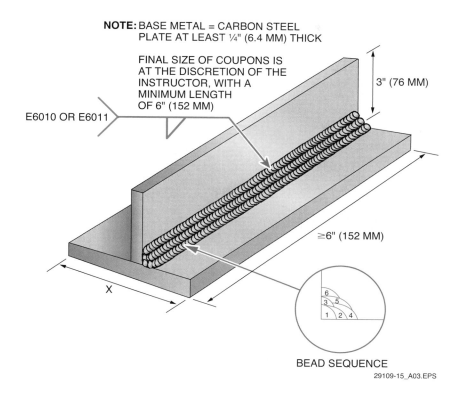

NOTE: BASE METAL = CARBON STEEL PLATE AT LEAST ¼" (6.4 MM) THICK

FINAL SIZE OF COUPONS IS AT THE DISCRETION OF THE INSTRUCTOR, WITH A MINIMUM LENGTH OF 6" (152 MM)

E6010 OR E6011

3" (76 MM)

≥6" (152 MM)

X

BEAD SEQUENCE

29109-15_A03.EPS

Criteria for Acceptance:

- Uniform rippled appearance on the bead face
- Craters and restarts filled to the full cross section of the weld
- Uniform weld size, ±¹⁄₁₆" (1.6 mm)
- Acceptable weld profile in accordance with *AWS D1.1*
- Smooth transition with complete fusion at the toes of the weld
- No porosity
- No excessive undercut
- No inclusions
- No cracks
- No overlap

HORIZONTAL (2F) FILLET WELD WITH E7018 ELECTRODES

Using ³⁄₃₂" to ⁵⁄₃₂" (2.4 to 4.0 mm) E7018 electrodes, make a horizontal fillet weld as indicated.

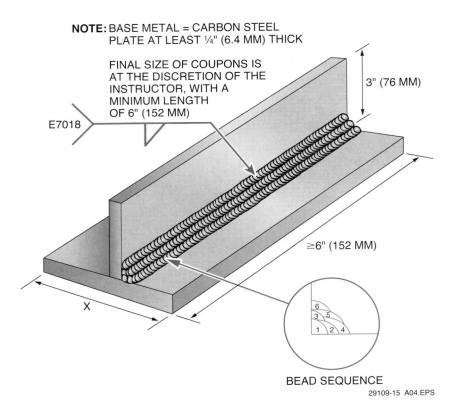

NOTE: BASE METAL = CARBON STEEL PLATE AT LEAST ¼" (6.4 MM) THICK

FINAL SIZE OF COUPONS IS AT THE DISCRETION OF THE INSTRUCTOR, WITH A MINIMUM LENGTH OF 6" (152 MM)

E7018

3" (76 MM)

≥6" (152 MM)

X

BEAD SEQUENCE

29109-15 A04.EPS

Criteria for Acceptance:
- Uniform rippled appearance on the bead face
- Craters and restarts filled to the full cross section of the weld
- Uniform weld size, ±¹⁄₁₆" (1.6 mm)
- Acceptable weld profile in accordance with *AWS D1.1*
- Smooth transition with complete fusion at the toes of the weld
- No porosity
- No excessive undercut
- No inclusions
- No cracks
- No overlap

VERTICAL (3F) FILLET WELD WITH E6010 OR E6011 ELECTRODES

Using ³⁄₃₂" to ⁵⁄₃₂" (2.4 to 4.0 mm) E6010 or E6011 electrodes, make a vertical fillet weld as indicated.

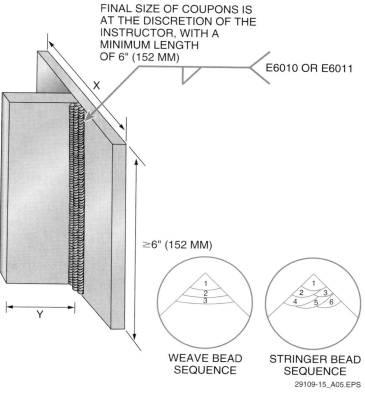

NOTE: BASE METAL = CARBON STEEL PLATE AT LEAST ¼" (6.4 MM) THICK

FINAL SIZE OF COUPONS IS AT THE DISCRETION OF THE INSTRUCTOR, WITH A MINIMUM LENGTH OF 6" (152 MM)

E6010 OR E6011

X

≥6" (152 MM)

Y

WEAVE BEAD SEQUENCE

STRINGER BEAD SEQUENCE

29109-15_A05.EPS

Criteria for Acceptance:

- Uniform rippled appearance on the bead face
- Craters and restarts filled to the full cross section of the weld
- Uniform weld size, ±¹⁄₁₆" (1.6 mm)
- Acceptable weld profile in accordance with *AWS D1.1*
- Smooth transition with complete fusion at the toes of the weld
- No porosity
- No excessive undercut
- No inclusions
- No cracks
- No overlap at weld toes

VERTICAL (3F) FILLET WELD WITH E7018 ELECTRODES

Using ³⁄₃₂" to ⁵⁄₃₂" (2.4 to 4.0 mm) E7018 electrodes, make a vertical fillet weld as indicated.

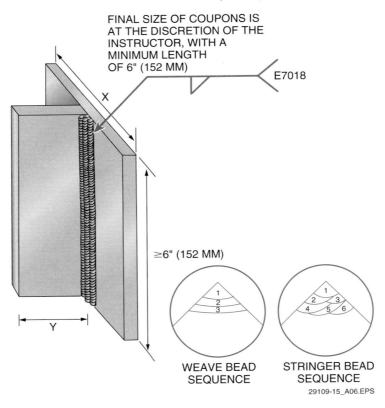

NOTE: BASE METAL = CARBON STEEL
PLATE AT LEAST ¼" (6.4 MM) THICK

FINAL SIZE OF COUPONS IS
AT THE DISCRETION OF THE
INSTRUCTOR, WITH A
MINIMUM LENGTH
OF 6" (152 MM)

E7018

X

≥6" (152 MM)

Y

WEAVE BEAD
SEQUENCE

STRINGER BEAD
SEQUENCE

29109-15_A06.EPS

Criteria for Acceptance:

- Uniform rippled appearance on the bead face
- Craters and restarts filled to the full cross section of the weld
- Uniform weld size, ±¹⁄₁₆" (1.6 mm)
- Acceptable weld profile in accordance with *AWS D1.1*
- Smooth transition with complete fusion at the toes of the weld
- No porosity
- No excessive undercut
- No inclusions
- No cracks
- No overlap at weld toes

OVERHEAD (4F) FILLET WELD WITH E6010 OR E6011 ELECTRODES

Using ³⁄₃₂" to ⁵⁄₃₂" (2.4 to 4.0 mm) E6010 or E6011 electrodes, make an overhead fillet weld as indicated.

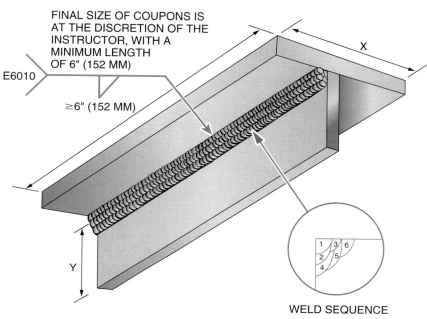

NOTE: BASE METAL = CARBON STEEL PLATE AT LEAST ¼" (6.4 MM) THICK

FINAL SIZE OF COUPONS IS AT THE DISCRETION OF THE INSTRUCTOR, WITH A MINIMUM LENGTH OF 6" (152 MM)

E6010

≥6" (152 MM)

X

Y

WELD SEQUENCE

29109-15_A07.EPS

Criteria for Acceptance:

- Uniform rippled appearance on the bead face
- Craters and restarts filled to the full cross section of the weld
- Uniform weld size, ±¹⁄₁₆" (1.6 mm)
- Acceptable weld profile in accordance with *AWS D1.1*
- Smooth transition with complete fusion at the toes of the weld
- No porosity
- No excessive undercut
- No inclusions
- No cracks
- No overlap

OVERHEAD (4F) FILLET WELD WITH
E7018 ELECTRODES

Using ³⁄₃₂" to ⁵⁄₃₂" (2.4 to 4.0 mm) E7018 electrodes, make an overhead fillet weld as indicated.

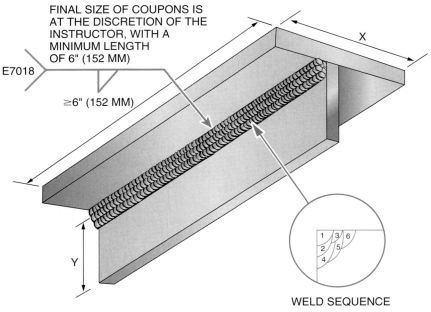

NOTE: BASE METAL = CARBON STEEL
PLATE AT LEAST ¼" (6.4 MM) THICK

FINAL SIZE OF COUPONS IS
AT THE DISCRETION OF THE
INSTRUCTOR, WITH A
MINIMUM LENGTH
OF 6" (152 MM)

E7018

≥6" (152 MM)

X

Y

WELD SEQUENCE

29109-14_A08.EPS

Criteria for Acceptance:

- Uniform rippled appearance on the bead face   _____
- Craters and restarts filled to the full cross section of the weld   _____
- Uniform weld size, ±¹⁄₁₆" (1.6 mm)   _____
- Acceptable weld profile in accordance with AWS D1.1   _____
- Smooth transition with complete fusion at the toes of the weld   _____
- No porosity   _____
- No excessive undercut   _____
- No inclusions   _____
- No cracks   _____
- No overlap   _____

FLAT (1F) FILLET WELD WITH
E6010 OR E6011 ELECTRODES

Using ³⁄₃₂" to ⁵⁄₃₂" (2.4 to 4.0 mm) E6010 or E6011 electrodes, make a flat fillet weld as indicated.

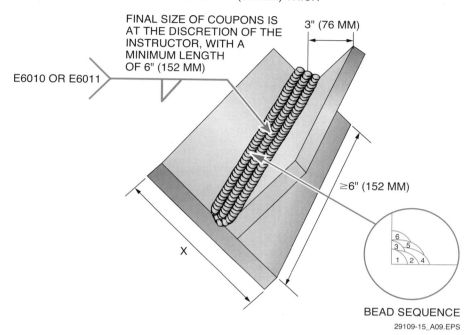

NOTE: BASE METAL = CARBON STEEL
PLATE AT LEAST ¼" (6.4 MM) THICK

FINAL SIZE OF COUPONS IS
AT THE DISCRETION OF THE
INSTRUCTOR, WITH A
MINIMUM LENGTH
OF 6" (152 MM)

E6010 OR E6011

3" (76 MM)

≥6" (152 MM)

X

BEAD SEQUENCE

29109-15_A09.EPS

Criteria for Acceptance:

- Uniform rippled appearance on the bead face
- Craters and restarts filled to the full cross section of the weld
- Uniform weld size, ±¹⁄₁₆" (1.6 mm)
- Acceptable weld profile in accordance with *AWS D1.1*
- Smooth transition with complete fusion at the toes of the weld
- No porosity
- No excessive undercut
- No inclusions
- No cracks
- No overlap

Rod Hellyer
Lee College, Baytown, TX
Department Chair – Industrial Studies

Rod was already working as a mechanic when two of his younger brothers persuaded him to try a career in construction. Rod credits much of his success to his parents, who set very high standards for their children. According to Rod, his dad was a hardworking man who put pride and craftsmanship into everything he did. Inspired by his father, Rod adopted these same values and continues to practice them today as he celebrates 31 years as a college welding instructor. Rod's advice to new welders entering the field is this—take pride in yourself and in your work, do everything to the best of your ability, and be a worker who can be trusted to get the job done.

How did you choose a career in the construction industry?
I owe it all to my two younger brothers who were welders and pipefitters. I was working as a mechanic at the time. My brothers convinced me to switch gears and go into construction.

What kinds of work have you done in your career?
I started out doing structural welding and fabricating. I worked on anything and everything—ship loaders, conveyors, oil storage tanks, towers, pipelines, boilers—you name it. I've performed work in accordance with a wide variety of welding codes.

Tell us about your present job.
I've been a welding instructor at Lee College for more than 30 years. This includes 15 years working at a local high school teaching students interested in welding. I have a deep admiration for welding and cutting, and I try to communicate this to my students and to the instructors I supervise as chair of the industrial studies department. We teach classes in SMAW, GTAW, and GMAW. Since 2008 I've served as department chair of industrial studies, in which I'm responsible for the welding, pipefitting, millwright, and machine shop programs.

What factors have contributed most to your success?
I attribute most of my success to my parents, who had very high moral standards. My dad worked hard all his life. He put a lot of pride and craftsmanship into everything he did, and he instilled the same values in us. Dad believed that anything worth doing was worth doing well. He required that even the least important jobs be done to the best of our ability. Today, I continue to practice the values instilled in me by my father.

What types of training have you been through?
When I first entered construction, I received some on-the-job training. Then I enrolled in a local college and started my formal training in welding. I took classes in SMAW, GTAW, GMAW, and metallurgy, along with enough academic courses to complete my A.A.S. in Welding Technology.

What advice would you give to those new to the field?
I would stress the importance of taking pride in yourself and in your work. Do everything to the best of your ability. You will be judged on your work by your peers and your employers. Don't worry about the work of others. Just focus on the job you are doing. Be one of those workers who will be called upon to tackle the jobs that no one else can get done.

Tell us an interesting career-related fact or accomplishment.
Some of the most interesting work I've done was on pipelines inside the plants making tie-ins on live units. I also enjoyed the manufacturing of products. And what an experience it is to see a new plant built and then started up for operation—knowing that you were a part of building something and transforming it from a bare spot of ground to a large industrial unit.

Trade Terms Introduced in This Module

Arc blow: The deflection of the arc from its intended path due to magnetic forces that develop from welding current flow.

Arc strike: A discontinuity consisting of localized melting of the base metal or finished weld caused by the initiation of an arc. It remains visible as a result of striking the arc outside the area to be welded and can lead to rejection of an otherwise good weld.

Drag angle: Describes the travel angle when the electrode is pointing in a direction opposite to the welding bead's progression.

Oscillation: A repetitive and consistent side-to-side motion.

Push angle: Describes the travel angle when the electrode is pointing in the same direction as the welding bead's progression.

Quench: To rapidly cool a hot component such as a freshly welded coupon. Note that only coupons should be quenched; real welds with a function should never be quenched.

Restart: The action of restarting a weld bead that is already in progress that requires the crater left from the termination of the previous arc.

Stringer bead: A type of weld bead made without any significant weaving motion. With SMAW, stringer beads are not more than three times the diameter of the electrode.

Undercut: A defect that occurs when the cross-sectional thickness of the base metal is reduced at the edges of the weld. Excessive weld current can cause this as it causes the base metal at the edges to melt and drain into the weld puddle. A downward sloped edge toward the weld bead is the visual evidence.

Weave bead: A type of weld bead made by transverse oscillation of the electrode in a particular pattern.

Weld axis: A straight line drawn through the center of a weld along its length.

Weld coupon: Metal pieces to be welded together as a test or practice.

Work angle: An angle less than 90 degrees between a line perpendicular to the major workpiece surface and a plane determined by the electrode axis and the weld axis. 0-degree work angle is common. In a T-joint or corner joint, the line is perpendicular to the non-butting member. The definition of work angle for a pipe weld is covered in a later module, as it differs from that of plate welding.

Additional Resources

This module is intended as a thorough resource for task training. The following reference works are suggested for further study.

2014 Technical Training Guide. Latest Edition. Cleveland, OH. USA: The Lincoln Electric Company. **www.lincolnelectric.com**

Figure Credits

The Lincoln Electric Company, Cleveland, OH, USA, Module Opener, Figures 1, 4, 18 (photo), SA01

Topaz Publications, Inc., Figures 2, 16, SA02–SA05

©American Welding Society (AWS) *Welding Handbook* 1991, Welding Processes Volume No. 2, Edition No. 8, Miami: American Welding Society., Figure 14

Holley Thomas, SME, Figure 23

Section Review Answer Key

Answer	Section Reference	Objective
Section One		
1 c	1.1.1	1a
2. c	1.2.3	1b
3. a	1.3.2	1c
Section Two		
1. a	2.1.2	2a
2. d	2.2.2	2b
3. a	2.3.3	2c
4. c	2.4.1	2d

NCCER CURRICULA — USER UPDATE

NCCER makes every effort to keep its textbooks up-to-date and free of technical errors. We appreciate your help in this process. If you find an error, a typographical mistake, or an inaccuracy in NCCER's curricula, please fill out this form (or a photocopy), or complete the online form at **www.nccer.org/olf**. Be sure to include the exact module ID number, page number, a detailed description, and your recommended correction. Your input will be brought to the attention of the Authoring Team. Thank you for your assistance.

Instructors – If you have an idea for improving this textbook, or have found that additional materials were necessary to teach this module effectively, please let us know so that we may present your suggestions to the Authoring Team.

NCCER Product Development and Revision
13614 Progress Blvd., Alachua, FL 32615

Email: curriculum@nccer.org
Online: www.nccer.org/olf

❏ Trainee Guide   ❏ Lesson Plans   ❏ Exam   ❏ PowerPoints   Other _____

Craft / Level: _____ Copyright Date: _____

Module ID Number / Title: _____

Section Number(s): _____

Description: _____

_____

_____

_____

_____

Recommended Correction: _____

_____

_____

_____

_____

Your Name: _____

Address: _____

_____

Email: _____ Phone: _____

29110-15

Joint Fit-Up and Alignment

OVERVIEW

Regardless of a welder's talent and experience, a joint that is poorly fit or aligned is subject to failure and a liability. All welded joints must be properly fit according to codes, specifications, and industry standards. Due to the intense heat of the welding process, care must also be taken to ensure that a properly fitted joint is not distorted and pulled out of alignment due to expansion and contraction. This module introduces welders to the basic tools and practices related to the proper fit of both plate and pipe joints.

Module Ten

Trainees with successful module completions may be eligible for credentialing through the NCCER Registry. To learn more, go to **www.nccer.org** or contact us at **1.888.622.3720**. Our website has information on the latest product releases and training, as well as online versions of our *Cornerstone* magazine and Pearson's product catalog.

Your feedback is welcome. You may email your comments to **curriculum@nccer.org**, send general comments and inquiries to **info@nccer.org**, or fill in the User Update form at the back of this module.

This information is general in nature and intended for training purposes only. Actual performance of activities described in this manual requires compliance with all applicable operating, service, maintenance, and safety procedures under the direction of qualified personnel. References in this manual to patented or proprietary devices do not constitute a recommendation of their use.

Objectives

When you have completed this module, you will be able to do the following:

1. Identify and describe various types of fit-up and alignment tools.
   a. Identify and describe various fit-up gauges and measuring devices.
   b. Identify and describe common weldment positioning equipment.
   c. Identify and describe various plate alignment tools.
   d. Identify and describe various pipe and flange alignment tools.
2. Describe techniques to avoid weldment distortion and describe the role of codes and specifications.
   a. Describe the causes of weldment distortion.
   b. Describe the techniques and tools used to control weldment distortion.
   c. Describe the role of codes and specifications in welding procedures and techniques.

Performance Tasks

Under the supervision of your instructor, you should be able to do the following:

1. Fit up joints using plate and pipe fit-up tools.
2. Check the joint for proper fit-up and alignment using gauges and measuring devices.

Trade Terms

Addendum
Ambient temperature
Coefficient of thermal expansion
   (linear)

Consumable insert
High-low
Level
Peening

Plumb
Residual stress
Specific heat per unit volume
Tack weld

Industry Recognized Credentials

If you are training through an NCCER-accredited sponsor, you may be eligible for credentials from NCCER's Registry. The ID number for this module is 29110-15. Note that this module may have been used in other NCCER curricula and may apply to other level completions. Contact NCCER's Registry at 888.622.3720 or go to **www.nccer.org** for more information.

Contents

Topics to be presented in this module include:

Figures and Tables

SECTION ONE

1.0.0 JOINT FIT-UP AND ALIGNMENT TOOLS

Objective

Identify and describe various types of fit-up and alignment tools.

a. Identify and describe various fit-up gauges and measuring devices.
b. Identify and describe common weldment positioning equipment.
c. Identify and describe various plate alignment tools.
d. Identify and describe various pipe and flange alignment tools.

Performance Tasks

1. Fit up joints using plate and pipe fit-up tools.
2. Check the joint for proper fit-up and alignment using gauges and measuring devices.

Trade Terms

Consumable insert: Preplaced filler metal that is completely fused into the root of the joint during welding, becoming part of the weld.

High-low: The discrepancy in the internal alignment of two sections of pipe.

Level: A line on a horizontal axis parallel with the earth's surface (horizon). Also refers to the tool used to check if an object is level.

Plumb: A line on a vertical axis perpendicular to the earth's surface (horizon).

Tack weld: A weld made to hold parts of a weldment in proper alignment until the final weld is made.

Joint design and setup affect the safety and quality of a completed weldment. Because joint design and setup are so important, they are covered by written codes and specifications that must be followed. Special tools to measure and support weldment fit-up are also available. This module explains the codes and introduces special tools and measuring devices that aid setup. In addition, joint inspection techniques and procedures to ensure that proper joint setup is maintained during welding will also be explained. This first section of the module focuses on the tools used in fit-up.

1.1.0 Fit-Up Gauges and Measuring Devices

Before making a weld, the joint must be fit up and checked to ensure it conforms to the Welding Procedure Specification (WPS) or site quality standards. The most common tools used to lay out and check joint fit-ups are straightedges, squares, levels, and Hi-Lo gauges.

Before proceeding, it is important to note that fit-up tools and instruments are made from a variety of metals. Of course, weldments can consist of a wide variety of metals as well. The tools and instruments used in the fit-up and preparation of a joint must be compatible with the base metal to avoid contamination. Most combinations are not a large problem, but tools and instruments made from carbon steel must not be used on stainless steel. Stainless steel devices can be used on carbon steel base metals without concern. However, it is important to note that some stainless steel tools, such as wire brushes, should not be used on stainless steel once they have been used with carbon steel. The concern is the contamination of stainless steel by carbon steel, but not the reverse.

1.1.1 Straightedges

Straightedges are used to scribe straight lines and check joint alignment. Many have calibrations along their length for measuring. Straightedges, particularly longer ones, are typically fabricated on the job from small channel or angle iron. When using a straightedge, be careful not to apply heat to it by placing it on hot metal or near the flame of a cutting torch or welding arc. Heat can cause the straightedge to become permanently distorted. Before using a straightedge, check it for straightness by visually sighting down the edge. If it is distorted and cannot be straightened, destroy it.

1.1.2 Squares

Two types of squares are commonly used for layout: pipefitter's squares and combination squares (*Figure 1*). Pipefitter's squares, which are used to measure angles and check squareness, are available in a variety of sizes. Like a carpenter's framing square, the pipefitter's square has a great deal of helpful information stamped into the metal.

Combination squares are smaller than pipefitter's squares, with blades typically 12" or 18" long (≈30 cm or 45 cm). They have replaceable attachments that slide and lock along a groove on the blade. Attachments include a combination 90-degree/45-degree level head, a centering head, and a protractor head. The combination attachment is used to check and lay out 90-degree and

45-degree angles, to check level, and to measure depth. The V-shaped centering head is used to measure round stock and to locate the center of shafts or other round objects. The protractor head is used to lay out and check angles.

When using squares, avoid overheating them as it could cause distortion. Also, protect squares from welding spatter that could stick to measuring surfaces, causing a false reading. To preserve their accuracy, squares should be stored in a dry, protected area when not in use. Carbon steel squares should be kept free of corrosion.

1.1.3 Levels

Levels come in a variety of sizes and shapes (*Figure 2*). Most levels designed for plate and piping work are made from magnesium or aluminum. Some have magnetized bases. Levels are used to check that layouts are level (horizontal) and plumb (vertical). Levels use a bubble in a glass vial to check level and plumb. Centering the bubble between the lines marked on the vial indicates level when the level is in the horizontal position and plumb when the level is in the vertical position.

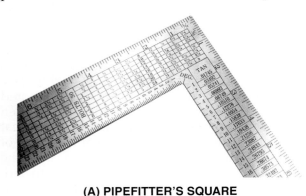

(A) PIPEFITTER'S SQUARE

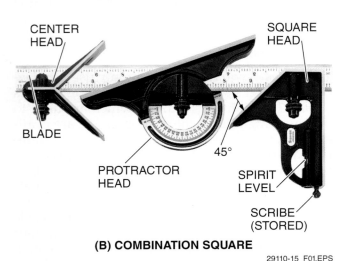

CENTER HEAD

SQUARE HEAD

BLADE

PROTRACTOR HEAD

SPIRIT LEVEL

45°

SCRIBE (STORED)

(B) COMBINATION SQUARE

29110-15_F01.EPS

Figure 1 Pipefitter's square and combination square.

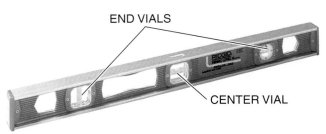

END VIALS

CENTER VIAL

(A) TWO-FOOT LEVEL

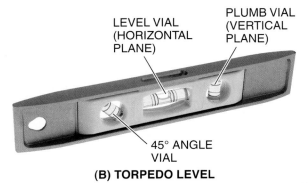

LEVEL VIAL (HORIZONTAL PLANE)

PLUMB VIAL (VERTICAL PLANE)

45° ANGLE VIAL

(B) TORPEDO LEVEL

29110-15_F02.EPS

Figure 2 Common levels.

Some levels also have a vial set at 45 degrees to check 45-degree angles. There are also special levels that have an adjustable protractor scale that can be set to check any angle.

The torpedo level is about 9" (≈23 cm) long and tapered on each end. The torpedo level has three vials: one to check level, one to check plumb, and one to check 45-degree layouts.

Protect levels from heat and weld spatter, which can break the glass tube. Also, be aware that weld spatter can stick to the bottom of the level, giving a false reading. The edge of the level that is designed to be placed against the surface in question must be kept clean and free of significant damage that might alter the reading.

1.1.4 Hi-Lo Gauges

The primary purpose of a Hi-Lo gauge is to check for pipe joint misalignment, although plate joint misalignment can also be checked. The name of the gauge comes from the relationship between the alignment of one pipe or plate to the other, which is called high-low. When either type of weldment is misaligned, one half is considered low and the other half is high.

To check for internal misalignment of pipe, Hi-Lo gauges have two prongs, or alignment stops, that are pulled tightly against the inside diameter of the joint so that one stop is flush with each side of the joint. The variation between the two stops is read on a scale marked on the gauge.

To measure pipe misalignment with a Hi-Lo gauge, insert the prongs of the gauge into the root opening. Pull up on the gauge until the prongs are snug against both inside surfaces, and read the misalignment indicated on the scale. *Figure 3* shows checking internal misalignment with a Hi-Lo gauge.

Many Hi-Lo gauges also have the capability to check the following:

- Root opening width
- Material thickness
- Weld reinforcement
- Scribe lines for socket welds

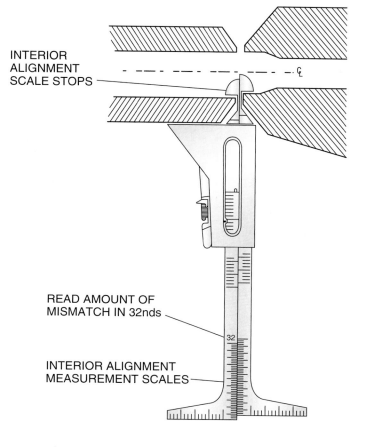

INTERIOR ALIGNMENT SCALE STOPS

READ AMOUNT OF MISMATCH IN 32nds

INTERIOR ALIGNMENT MEASUREMENT SCALES

29110-15_F03.EPS

Figure 3 Checking internal misalignment.

Handy Level and Angle Finder

There might be an angle finder and level closer to you than you might think. The next time you need one, reach instead for your mobile phone. A number of smartphones, including the iPhone, offer a level/angle finder application by default. For the iPhone, access the compass app and, with the compass on screen, swipe left to reveal the level and angle finder tool. For phones not equipped with these apps by default, there are a number of them available for download. Of course, mobile phones are easily scratched and damaged, so you won't want to make a habit of using it on the job. But for leveling a picture or a shelf, it's a handy tool.

When checking for internal misalignment, take two gauge readings: one before tacking together the two pieces of pipe and one after the **tack weld** has been applied. The Hi-Lo gauge is designed to fit into the gap between the two pipes, even after tacking, and are available with both metric and English scales.

The alignment stops are tapered for checking the root opening (*Figure 4*). They are ³⁄₃₂" (2.4 mm) thick and taper to ¹⁄₁₆" (1.6 mm) thick at both ends. Check the root opening by inserting the tapered head and visually checking the opening.

Weld reinforcement is another check that can be performed with the Hi-Lo gauge. It is measured by placing one foot of the gauge on the pipe and the other on the weld face, making sure the foot on the pipe surface is flush against the surface. Weld reinforcement is then read on the material thickness scale (*Figure 5*).

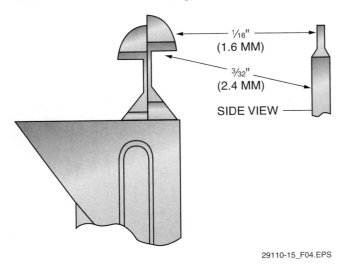

Figure 4 Tapered alignment stops.

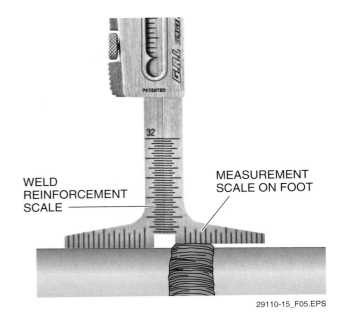

Figure 5 Measuring weld reinforcement.

> **CAUTION**
>
> Actual requirements for socket weld gaps must be verified for your job by checking the WPS or site quality standards. Failure to follow the WPS can cause equipment damage or result in a rejected weld.

When fitting up socket welds, the pipe is scribed so that a measurement can be made to ensure that a gap is left between the end of the pipe and the base of the socket. This gap prevents cracking when the pipe expands during the welding process; the pipe should never be bottomed-out in the socket when it is welded. The scribe-line scale on the feet of the Hi-Lo gauge can be used to measure from the socket to the scribed line to ensure the gap is as specified on the WPS or site quality specifications. Most codes require a minimum ¹⁄₁₆" (1.6 mm) gap, but actual requirements for your job may vary due to pipe size, base metal type, or a unique specification. *Figure 6* shows a Hi-Lo gauge measuring scribe lines on a socket weld.

1.2.0 Common Weldment Positioning Equipment

Many tools are available to aid plate and pipe fit-up. Some are special tools designed specifically for fit-up; others are common hand tools. Before using any tool, be sure to inspect it for damage or missing parts. Never use a tool for a job for which it was not intended. Be sure to clean and return all tools when you have finished using them.

Also remember to ensure the tool material is compatible with the base metal being fitted. Carbon steel tools, as well as some stainless steel tools that have been used with carbon steel, should not be used on stainless steel.

Hydraulic jacks, chain falls, and come-alongs are sometimes used to position parts of a weldment. These types of positioning tools are described in the following sections.

1.2.1 Hydraulic Jacks

When using hydraulic jacks, never weld directly on the jack base or ram. To secure the jack, a ring can be tack welded to act as a socket for the base and/or ram. When welding near hydraulic jacks, protect the ram from weld spatter. If the ram is off the floor, be sure to secure it with rope or chain. Also monitor the jack for oil leaks. Oil will create a fire hazard, and oil in the area to be welded will create weld defects from porosity. Oil leaks must be cleaned up before welding. *Figure 7* shows using a hydraulic jack to aid joint fit-up.

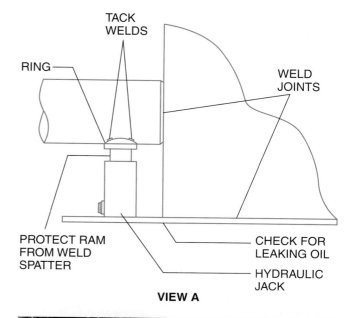

VIEW A

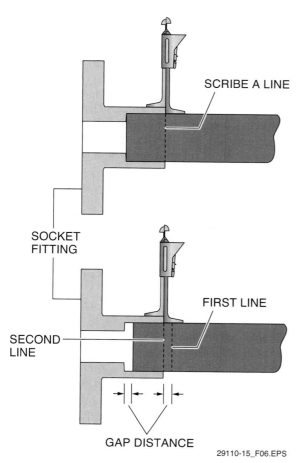

29110-15_F06.EPS

Figure 6 Measuring the distance between scribe lines on a socket weld.

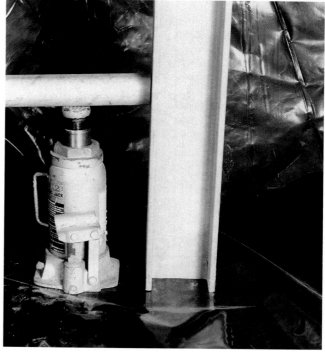

VIEW B

29110-15_F07.EPS

Figure 7 Proper use of a hydraulic jack for joint fit-up.

1.2.2 Chain Hoists, Chain Falls, and Come-Alongs

Chain hoists, also called chain falls, are used to lift or lower weldment parts. Always position the chain hoist directly over the center of gravity of the object being lifted. Secure the chain hoist over the weldment with an approved sling wrapped around a structural member. A suitable structural member may be a large I-beam or other significant building support component.

Never hang chain hoists from piping, ducts, conduit, or raceways. These items are not designed to carry external loads, and damage could result. The ground lead should be attached directly to the item to be welded. The chain fall or hoist should not be part of the welding current path. *Figure 8* shows using a chain hoist to aid joint fit-up.

> **CAUTION**
>
> Never weld a lifting eye to a structural member. If a weld is made on a structural member, it weakens the structural member, and the structural member may have to be replaced.

SLING AROUND STRUCTURAL MEMBER

CHAIN HOIST

WELD JOINT

29110-15_F08.EPS

Figure 8 Chain hoist lifting a pipe into position.

> **WARNING!**
>
> Make sure the current path does not pass through the chain hoist supporting the load you are welding. Extreme heat generated by the current flow can cause the chain hoist to fail, releasing the load support. This can cause serious injury or even death.

Often, more than one come-along or chain fall may be used to precisely position a weldment. Before welding, check the ground lead connection location to be sure the welding current will not pass through the come-along. The ground lead should be attached directly to the item to be welded.

The difference in chain falls, chain hoists, and come-alongs can be a confusing topic. Some manufacturers label their products in ways that adds to the confusion. Some may call an item a chain fall while another item with the same features is referred to as a come-along. As a general rule, chain hoists and chain falls do not have a setting or feature that allows for the free-fall of the load. The device must be operated manually, by pulling the chain in one direction or another, to both lift and descend. Motorized versions also exist. Come-alongs may have chains or cables. Most have a ratcheting handle for operation, and can be switched using a lever or pawl to release the tool's hold on the load completely. If used for vertical lifting, the failure or accidental initiation of this feature could result in serious injury or property damage from a free-falling load. As a result, regardless of what name the device has been given, do not use devices for vertical lifting if they have a feature allowing the free-fall or free-release of the load. They can be safely used for horizontal positioning of a load and similar tasks.

Chain Collectors and Retainers

Many chain hoists are equipped with buckets, canvas bags, or other devices designed to collect the chain loops to prevent them from interfering with the workpiece. If the hoist you are using includes this feature, make sure the collector or retainer is aligned correctly with the drop of the chain. Otherwise, the collector or retainer may tilt, dumping the heavy load of chain on top of you. Wear appropriate PPE, including a hard hat, when using chain hoists or falls.

1.3.0 Plate Fit-Up Tools

The most common method of holding a joint in place after it has been fitted up is to tack weld it in place. This works very well for small weldments and joints that are straight. However, large weldments or long or thick joints often require some type of mechanical means in addition to tack welding to set up and hold them in place for welding. The most common tools for plate fit-up are strongbacks, clips, yokes, and wedges. In addition, special plate alignment tools are available.

1.3.1 Strongbacks

Strongbacks (*Figure 9*) are typically made on the job site from heavy bar stock. They are notched at the weld joint to allow access to the joint so that welds can be made without interference. The strongback can be placed on the face or root side of the weldment. The plates to be joined are clamped or tack welded flush against the strongback, which holds the joint in alignment. If possible, when tack welding strongbacks, place the tack welds on only one side of the strongback. This will make the strongback easier to remove.

A good way to remove a strongback that has been tack welded is to use a grinder to remove or greatly reduce the tack welds. After the tack welds are minimized with a grinder, the tap of

a hammer should break the strongback free of the base metal. After the strongback is removed, the grinder can be used to grind the tacks down flush, but care must be taken to not gouge the base metal during the grinding process.

1.3.2 Clips, Yokes, and Wedges

Clips, yokes, and wedges are also made on the job site. They can be used to align joints and then hold them in place during welding. Clips are welded to the edge of one plate and then wedges are positioned on the other plate and driven under the clips to force the joint into alignment. Yokes work in a similar manner. A yoke is welded to one plate. A slotted plate is then placed over the yoke, and a wedge is driven under the yoke to force the joint into alignment. *Figure 10* shows yokes and wedges being used to align and hold a joint.

1.3.3 Combining Alignment Tools

If the joint needs to be forced into alignment, yokes and wedges or bolts can be used with a strongback (*Figure 11*). When yokes and wedges are used, the strongback is positioned on the weldment straddling the joint. Yokes are positioned over the strongback and welded to the plates. Wedges are then driven under the yokes to force the joint into alignment. When bolts are used, the strongback must be made of channel stock or reinforced angle iron. Holes are cut in the channel or angle iron so that the bolts can pass through. The strongback is then positioned on the weldment straddling the joint. The position of the bolts is marked, and the strongback is removed. Bolts long enough to protrude through the strongback or threaded stock are then welded onto the plates.

The strongback is then positioned with the bolts protruding. Nuts are placed on the bolts and tightened with a ratchet or impact wrench, pulling the joint into position.

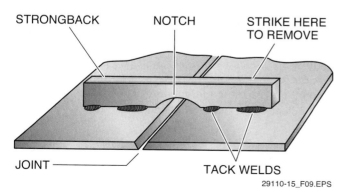

Figure 9 Strongback.

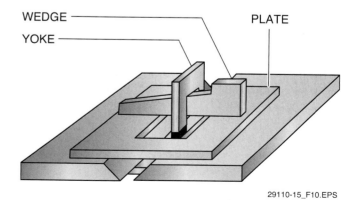

Figure 10 Yokes and wedges.

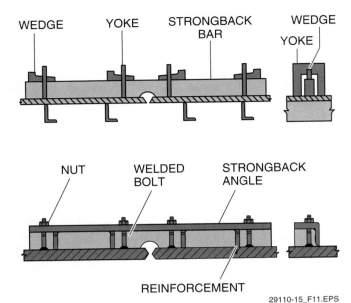

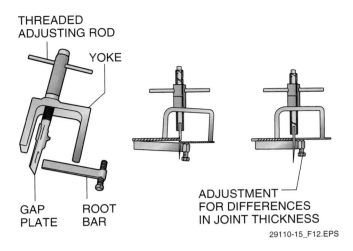

Figure 12 Plate alignment tool.

Figure 11 Aligning a joint using several tools.

1.3.4 *Plate Alignment Tools*

Special tools are manufactured for aligning plate that can make the task easier. A typical configuration of an alignment tool consists of a yoke, threaded adjusting rod, gap plate, and root bar (*Figure 12*). The alignment tool is used by straddling the joint opening with the yoke. The gap plate can be changed to match the specified root opening. The threaded adjusting rod is lowered until the opening in the gap plate is below the joint and the root bar can be inserted. The threaded adjusting rod is then tightened, bringing the joint into alignment. If necessary, the root bar can be adjusted to compensate for uneven joint thicknesses. Remember to properly match the tool materials with the base metal to eliminate contamination. Alternatively, isolating the two surfaces from each other so they don't touch while in use is also an option.

1.4.0 Pipe and Flange Fit-Up Tools

There are many different manufacturers of pipe fit-up tools. The tools they produce are similar, but before using a tool, be sure to review the guidelines provided by its manufacturer.

1.4.1 *Pipe Jacks and Rollers*

Pipe jack stands and rollers (*Figure 13*) are used to support pipe for fit-up and welding. Pipe jack stands typically have either a V-head or roller head and a height adjustment. Rollers, which can be floor-stand or table models, can be adjusted horizontally for various pipe diameters.

Adjustable jack stands and rollers are necessary tools for pipe fabrication, especially in the shop environment. They allow the welder to position various sizes of pipe for alignment and to manually roll the pipe as needed so the weld can be performed from the top, which is the preferred position. When fabricating piping systems in place, however, jack stands and rollers cannot

Using Angle Iron for a Strongback

If angle iron is used for the strongback, the bolts can be welded without removing the angle iron. The angle iron must be reinforced to prevent it from rolling when the bolts are tightened.

Marking Fabricated Alignment Tools

Strongbacks, wedges and similar tools are typically fabricated in the field rather than purchased. Mark these devices in some way so others won't discard them as scrap. In addition, they should be marked to indicate their material of construction.

V-HEAD FOLDING JACK STAND

ADJUSTABLE ROLLER STANDS

HEIGHT
ADJUSTMENT

CASTER
LOCKS

V-HEAD JACK STAND WITH ROLLERS AND CASTERS

29110-15_F13.EPS

Figure 13 Pipe jack stands and rollers.

be used because of positioning limitations. One way to deal with this is to weld straight sections of piping systems using jack stands and rollers. The straight sections can then be lifted and welded in place.

1.4.2 Chain Clamps

Chain clamps are used to align and hold pipe for fit-up and tacking. Some manufacturers use link chain, others use chain similar to bicycle chain. Regardless of the type of chain used, the basic procedure for using chain clamps is the same.

The chain, which is anchored to one side of the clamp, is passed around the pipe and secured. The slack in the chain is then taken up using a screw jack to pull the pipe tightly against the clamp.

Several types of chain clamps can be used to provide precise fit and alignment:

- Single-chain, single-jackscrew clamps
- Single-chain, double-jackscrew clamps
- Double-chain, double-jackscrew clamps
- Rim clamps

The single-chain, single-jackscrew clamp shown in *Figure 14* will align and even re-form out-of-round pipe up to Schedule 40 wall thickness. It can also be used on pipe heavier than Schedule 40 as long as no re-forming is required. Single-chain, double-jackscrew clamps have even more power to reform pipe. These clamps have a one or two jackscrews on each jack bar with optional spacing screws that allow the gap to be set quickly and precisely. A spacer may be used to maintain the required gap in the joint. Both the single- and double-jackscrew clamps are shown in *Figure 14*.

In addition to fitting up straight pipe, attachments for the precision fit-up clamps allow them to be used to fit up elbows, Ts, and flanges. The attachments connect by chain to the main block and to the elbow, T, or flange with a chain or clamp, as shown in *Figure 14*. The attachment has a fine adjustment crank to remove slack from the chain to support the elbow, T, or flange.

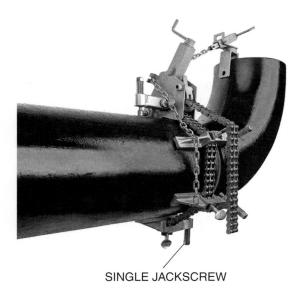

SINGLE JACKSCREW

(A) SINGLE-CHAIN, SINGLE-JACKSCREW PIPE CLAMP

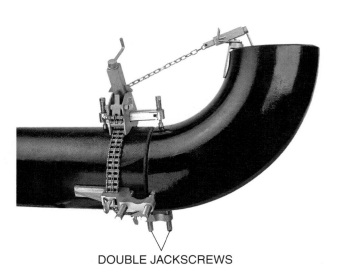

DOUBLE JACKSCREWS

(B) SINGLE-CHAIN, DOUBLE-JACKSCREW PIPE CLAMP

29110-15_F14.EPS

Figure 14 Chain clamps.

1.4.3 Other Pipe Alignment and Clamping Tools

For one section of pipe to be welded end-to-end to another section of pipe, their ends must be cut or ground so that they mate properly. As you have seen in previous sections, jack stands and rollers are used to support sections of pipe to be welded. A chain clamp is one way to secure the two sections of pipe together. Cage clamps are another style of clamp that can also be used for the same purpose. Cage clamps come in a variety of styles and sizes. All must be installed manually, but the ways they are secured to the pipe sections vary. The latches on some of the old-style cage clamps had to be hammered on or off, but hammering tends to damage the clamps. Damaging the clamp can create problems joining the pipe sections.

Newer cage clamps are now available that allow the cages to be custom fitted to the pipes. *Figure 15* shows a cage clamp installed on two pipe sections, a clamp that is manually closed and opened, and a cage clamp that is tightened by a locking screw turned by a ratchet handle.

CAGE CLAMP ON PIPE

MANUAL LOCKING HANDLE

MANUAL CAGE CLAMP

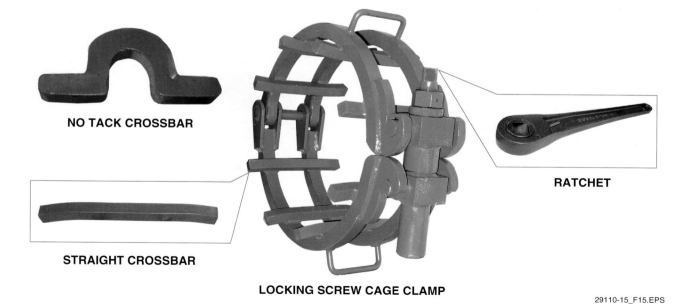

NO TACK CROSSBAR

STRAIGHT CROSSBAR

LOCKING SCREW CAGE CLAMP

RATCHET

29110-15_F15.EPS

Figure 15 Cage clamps.

Nonferrous Alignment Tools

In some industries, fabricated alignment tools such as wedges, strongbacks, and yokes must be constructed of nonferrous materials to prevent sparking. Brass is a popular nonferrous metal for this purpose. This is a common requirement in some segments of the petrochemical industry.

Also shown in *Figure 15* are replaceable bars used with the cage clamps. The straight crossbars have flared ends, which allow the pipe sections to be more easily moved into the clamp. The arched no-tack crossbar allows welds to be made under the clamp or its bars without them being moved.

While cage clamps are good for aligning and holding two sections of straight pipe so that they can be welded, they are of no use when a flange must be welded onto the end of a pipe. One of the devices used when welding flanges onto pipes is a rim clamp (*Figure 16*). A rim clamp is a more versatile, non-chain clamp design. It is also recommended for applications where heavy-duty re-forming is required.

The jackscrews on a rim clamp exert pressure on specific high points of the pipe so that precise alignment can be made. These clamps are ideal for tasks in which 100 percent weld and grind is required before the clamp can be released.

1.4.4 Clamping Devices for Small-Diameter Pipe

Typical chain clamps are often too bulky to use on smaller-diameter pipe. A simple shop-built clamp for small-diameter pipe can be made from a piece of angle iron. The pipe is laid in the angle, and C-clamps or wires are used to secure the pipe. Small blocks are placed on both sides of the angle to hold it in position. When more accuracy and adjustability is required, special clamps for smaller-diameter pipe can be used (*Figure 17*). These clamps have a three-point jaw design to clamp and align the pipe when the clamping screws are turned. Special stainless steel clamps are available for working with stainless steel piping.

1.4.5 Pipe Pullers

Pipe pullers are clamps, often chain-type, that are used to pull together two pipe workpieces so that they can be joined by welding. These types of pullers are often applied in field fabrication or addition projects.

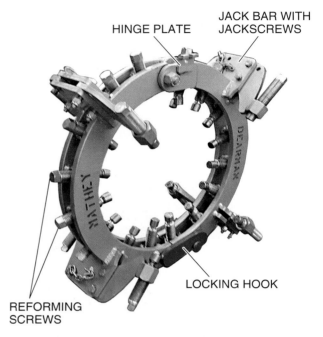

HINGE PLATE

JACK BAR WITH JACKSCREWS

LOCKING HOOK

REFORMING SCREWS

29110-15_F16.EPS

Figure 16 Rim clamp.

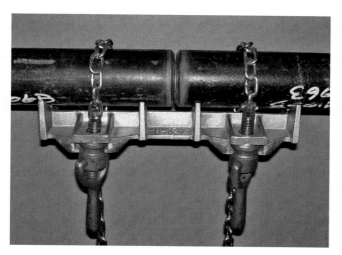

(A) ADJUSTABLE PIPE CLAMP OR JEWEL CLAMP

(B) ULTRACLAMP

29110-15_F17.EPS

Figure 17 Pipe clamps for smaller pipe.

In the petrochemical industry, for example, a new unit may be added to an existing facility. The piping must be joined through the use of manifolds and other piping designs. In order to make the final tie-ins from existing piping to the new piping, pipe pullers are often used to align the joining parts precisely.

> **CAUTION**
>
> When new piping will be coupled to machinery such as valves or pumps, it is extremely important that the flange of the pipe be perfectly aligned to the flange of the valve or pump. Using a come-along, jack, or other such device can put excessive stress on both the piping and the machine. Piping stress on pumps in particular will make them misalign, which in turn will destroy their bearings.
>
> Do not force joints into alignment and then hold them in place with welds. Stress will be exerted at the welded joint in such an alignment, and cracking could result. Precise alignment and fitting without significant stress should be accomplished before finalizing the installation with the weld.

Pipe pullers (*Figure 18*) are frequently used to pull together sections of large-diameter pipe for welding. Fillers or **consumable inserts** may be installed between the faces of the joints to achieve precisely the necessary root opening between the parts. These inserts or fillers may become a permanent part of the finished weld.

> **CAUTION**
>
> Pipe pullers may not be permitted in some applications or projects.

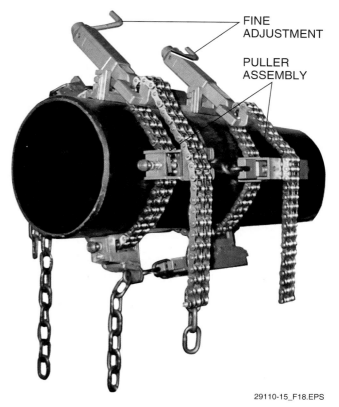

FINE
ADJUSTMENT

PULLER
ASSEMBLY

29110-15_F18.EPS

Figure 18 Pipe puller.

Field-Fabricated Jack Stands

A mechanic was crushed to death when a bus he was working on fell off a set of jack stands. The stands were fabricated of plate steel by a local welding shop. The top plate of each was completely flat; they had no lips, which commercial stands always have. The front tires of the bus were not chocked, and there was nothing to prevent it from falling off the stands.

The investigation of the accident revealed that the jack stands had not been tested or certified for their rated capacity, nor were they marked with such a capacity. They were not fabricated in accordance with commercial jack stand construction. The lips on commercial jack stands cradle the area being supported. Also, commercial jack stands normally have three legs that help them compensate for irregular surfaces.

Sections of large-diameter pipe supported by jack stands may not be as large as buses, but they can cause serious or fatal injuries if they fall from their supports. Always use commercial jack stands that have been tested and certified.

1.4.6 Flange Alignment Tools

Various methods and tools can be used to make sure a flange facing aligns precisely with the pipe before it is permanently welded in place. Flange pins can be installed in the flange bolt holes to provide a reference point from which leveling can be measured for tack welds or complete welding (*Figure 19*). It is important to use a minimum of two flange pins installed in two of the bolt holes to provide a precise reference point from the center of one hole to another. From this flat, level benchmark, other tools can be used to adjust the flange to the pipe for true and precise alignment. The tool shown in *Figure 20* is also used to precisely align a flange for welding.

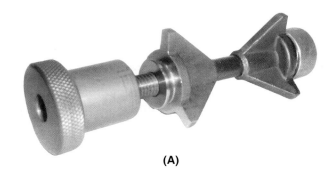

(A)

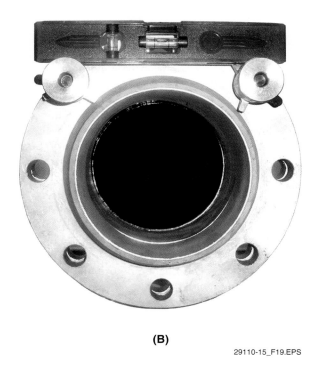

(B)

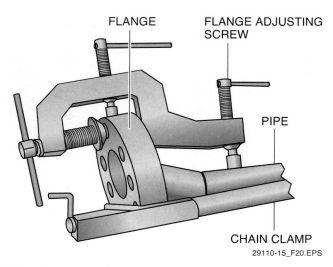

29110-15_F20.EPS

Figure 20 Flange alignment tool.

29110-15_F19.EPS

Figure 19 Using flange pins.

Types of Welded Flanges

Some of the more common flanges generally installed using welded connections are the welding neck flange, the lap joint flange, the socket weld flange, and the slip-on flange. Threaded flanges may also be back-welded, as determined by the process. Welding neck flanges are preferred for use in severe service applications, such as those involving high pressure, subzero temperatures, or elevated temperatures.

Additional Resources

2014 Technical Training Guide. Current Edition. Cleveland, OH. USA: The Lincoln Electric Company. **www.lincolnelectric.com**

Welding Handbook. Current Edition. Miami, FL: The American Welding Society.

Mathey Dearman Product Manuals. The Mathey Dearman Company. **www.matheydearman.com**

Think About It

Root Openings in Aligning and Fitting

Why is it necessary to maintain a root opening between two pieces, especially pipe pieces, when aligning and fitting prior to welding?

1.0.0 Section Review

1. Most levels designed for plate and piping work are made from _____.

   a. plastic
   b. wood
   c. aluminum or magnesium
   d. fiberglass or carbon steel

2. When lifting components into place to prepare for welding, vertical lifts should *not* be conducted using _____.

   a. a device with a free-fall mode or setting
   b. a device without a free-fall mode or setting
   c. any type of chain fall
   d. any type of synthetic lifting strap

3. Clips, yokes, and wedges are generally _____.

   a. purchased by the bag
   b. made on site
   c. shipped with the base metal
   d. made of aluminum

4. A simple shop-built clamp can be made from angle iron to help align _____.

   a. heavy plate
   b. large-diameter pipe
   c. small-diameter pipe
   d. sheet metal

SECTION TWO

2.0.0 WELDMENT DISTORTION

Objective

Describe techniques to avoid weldment distortion and describe the roles of codes and specifications.

   a. Describe the causes of weldment distortion.

   b. Describe the techniques and tools used to control weldment distortion.

   c. Describe the role of codes and specifications in welding procedures and techniques.

Trade Terms

Addendum: Supplementary information typically used to describe corrections or revisions to documents.

Ambient temperature: The temperature of the atmosphere surrounding an object or individual.

Coefficient of thermal expansion (linear): The change in length per unit length of material for a 1°C change in temperature.

Peening: The mechanical working of metal using repeated hammer blows, typically using a ball-shaped hammer head such as that found on a ball peen hammer. Peening may be done for both metallurgical and appearance reasons.

Residual stress: Stress remaining in a weldment as a result of heat.

Specific heat per unit volume: The quantity of heat required to raise one unit mass of a substance by one unit degree.

Besides the obvious cosmetic problems resulting from poor alignment and fit, the proper application of fit-up methods must be applied to deal with distortion and other problems. Distortion is the expansion and contraction of metal as it responds to changes in temperature. Tools such as chain clamps, rim clamps, C-clamps, strongbacks, and other devices can be used to redirect this expansion or contraction. This is referred to as controlling the distortion rather than preventing it. Distortion cannot be prevented as long as the metal experiences changes in temperatures created by the welding process. It can, however, be controlled, resulting in a weldment that is relatively free of damaging stress.

2.1.0 Causes of Distortion

Distortion is caused by the non-uniform expansion and contraction of the weld metal and adjacent base metal during the heating and cooling cycles of welding. When a metal block is heated, it expands uniformly in all directions. As the metal block cools, it contracts uniformly in all directions. If the metal block is restricted on two sides as it is heated, expansion cannot take place in the direction of restriction. Because the metal block must expand the same amount, all expansion occurs in the unrestricted directions. As the metal block cools, it contracts uniformly in all directions, leaving it narrower in the restricted direction and longer in the unrestricted direction. *Figure 21* illustrates the expansion, contraction, and distortion control of a block of steel.

During welding, these same forces act on the weld metal and base metal. As the weld metal is deposited and fuses with the base metal, it is at its maximum expanded state. Upon cooling, it attempts to contract but is restricted by the surrounding base metal. Stresses develop within the weld until the stress reaches the yield point of the base metal, causing the base metal and weld metal to stretch. These same stresses also cause the base metal to move. When the base metal is heated during the welding process, it

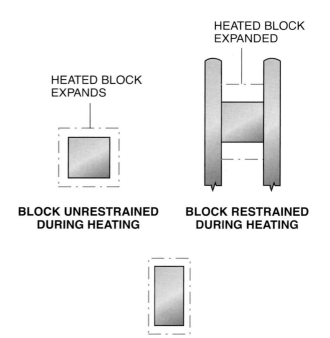

RESTRAINED BLOCK IS DISTORTED AFTER COOLING

29110-15_F21.EPS

Figure 21 Controlling distortion.

will attempt to expand but will be restricted by the surrounding cooler base metal. Some movement will occur, but as the weld cools, the base metal that was welded will contract more due to the stress caused by the restriction of the surrounding base metal. This causes distortion. Even when the weld is at room temperature, stress equal to the strength of the base metal will be locked in the weldment. This is called *residual stress*. *Figure 22* shows how stress and distortion affect a weld.

The degree of distortion is directly related to the stresses generated during welding. Two of the metal properties that figure prominently in these stresses are the *coefficient of thermal expansion (linear)* and the *specific heat per unit volume*. Metals with higher values for these two properties will experience a higher degree of distortion than those with lower values. Stainless steel, for example, has a much higher coefficient of thermal expansion and specific heat per unit volume than carbon steel. It is therefore more likely to be affected by distortion. *Table 1* compares the coefficients of thermal expansion and the specific heats per unit volume for some common metals.

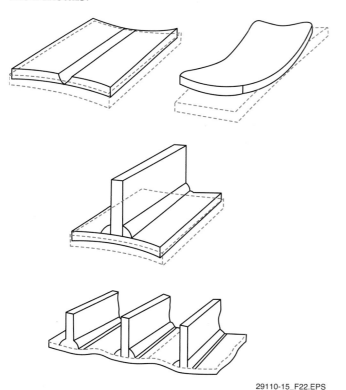

29110-15_F22.EPS

Figure 22 Stress and distortion in welds.

2.2.0 Controlling Distortion

Distortion can be controlled using a variety of techniques and tools. The following sections explain how to control distortion to minimize its effect on weldments.

2.2.1 Clamping and Bracing

Chain clamps, rim clamps, strongbacks, and other field-fabricated alignment devices are often used to hold weldments firmly in place during the welding process in order to prevent warping caused by distortion. These devices must be designed, constructed, and installed such that their overall support can withstand the stresses of welding. They must also be left in place long enough to allow the weldment to cool to *ambient temperature*. Some distortion may still take place once the devices are removed. This is caused by residual stress present in the weldment.

2.2.2 Tack Welding

A tack weld holds parts of a weldment in proper alignment until the finish welds are made. In order for them to be effective, pay attention to the number of tack welds, their length, and the distance between them. If too few tacks are made, the chance of the joint closing up as the welding proceeds is much greater. Clamps and other fit-up and aligning devices must be used to maintain proper alignment and the proper gap during the process.

Latent Distortion

If a weldment is clamped firmly during welding to prevent any movement, it will stay straight as long as the clamps are in place. Even after cooling, when the clamps are removed, residual stress can cause some distortion. When weldments that have residual stress are machined, they distort as the machining process removes metal, allowing the residual stress to overcome the strength of the base metal. This is referred to as latent distortion since it occurs long after the stress was created.

Table 1 Thermal Expansion and Specific Heat per Unit Volume for Common Metals

Metal	Coefficient of Thermal Expansion at 20°C	Specific Heat Per Unit Volume J/g°C
Aluminum	$22–24 \times 10^{-6}/°C$	0.900
Stainless steel (316)	$14–17 \times 10^{-6}/°C$	0.500
Steel	$12–13 \times 10^{-6}/°C$	0.448
Copper	$16–17 \times 10^{-6}/°C$	0.390

If tack welds are to be included with the main weld, they must be applied by qualified welders. They must also follow the qualifying WPS for the project, if one applies. Likewise, the filler metals must meet the same filler metal requirements as the filler metal for the finished welds. For those tack welds that require removal prior to the finish weld, use extreme care to avoid causing defects in the material.

2.2.3 Amount of Weld Material

The more weld metal placed in a joint, the greater the forces of shrinkage. Properly sizing a weld not only minimizes distortion, it also saves weld metal and time. Excess reinforcement on the face of a weld increases the forces of distortion and adds nothing to the strength of the weld. Excess face reinforcement actually reduces the strength of a weld and is therefore prohibited by welding codes. The face of fillet welds may be slightly convex, flat, or slightly concave. Groove welds should have slight reinforcement of no more than 1/8" (3.2 mm) for butt or corner welds. *Figure 23* shows acceptable weld reinforcement profiles.

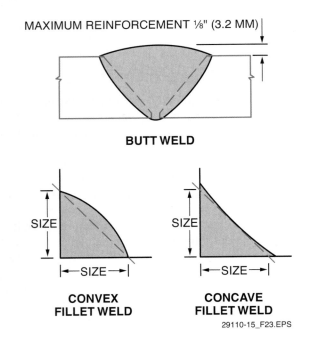

Figure 23 Weld reinforcement.

Proper fit-up and edge preparation also reduce the amount of weld required. Open-root joints should have a root opening from 1/16" to 1/8" (1.6 to 3.2 mm). To control melt-through, a root face of 1/16" to 1/8" (1.6 to 3.2 mm) is used. Each side of the open-root joint is beveled from 30 to 37½ degrees (included angle of 60 to 75 degrees). The greater the angle of bevel, the more weld will be required to fill the joint. The bevel angle must be sufficient to allow access to the root but not so large that it requires excess weld metal to fill. Welds made with GMAW or GTAW generally require larger openings than welds made with SMAW because the nozzles used with GMAW and GTAW are larger in diameter than the electrodes used with SMAW. These nozzles require more room in the joint to reach the root. *Figure 24* shows an open V-groove joint preparation.

When possible, a double V-groove should be used in place of a single V-groove (*Figure 25*). The double V-groove requires half the weld metal as compared to the single V-groove. Also, welding from both sides reduces distortion since the forces of distortion will be working against each other. On thick joints, U- and J-grooves also require less weld metal than V-grooves or bevel grooves, although U- and J-grooves require more preparation time.

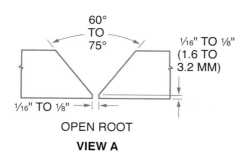

60°
TO
75°

¹⁄₁₆" TO ¹⁄₈"
(1.6 TO
3.2 MM)

¹⁄₁₆" TO ¹⁄₈"

OPEN ROOT

VIEW A

VIEW B

29110-15_F24.EPS

Figure 24 Open V-groove preparation.

2.2.4 Backing Strips on Groove Welds

Various forms of backing materials or strips can be applied to the back side of plate weldments. Temporary strips typically have a groove machined directly into them that controls the shape and size of the penetration material once it cools. Backing strips can be purchased in rolls from which the desired length can be cut and then taped directly to the back of the prepared joint. If the backing strip is designed to become part of the permanent weldment, it must be made of material similar to the alloy or metal being welded. If the backing strip is to be removed, it normally is made of dissimilar metal to prevent it from being welded into the joint. The backing strip can then be peeled from the back side of the finished weldment. Backing strips of copper, for example, are often used on weldments of stainless steel.

WPS

The qualifying WPS that specifies the tack-welding procedures, as well as the consumables to be used, is typically designed as part of the engineering plans. It is not normally a function or responsibility of the welder. However, the welder must be qualified to perform the tasks and follow the procedures specified in the WPS.

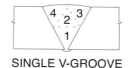

SINGLE V-GROOVE   DOUBLE V-GROOVE

**SINGLE V-GROOVE REQUIRES TWICE
AS MUCH WELD TO FILL JOINT**

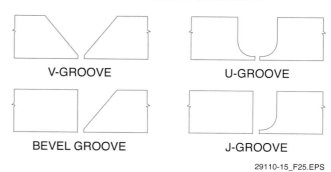

V-GROOVE   U-GROOVE

BEVEL GROOVE   J-GROOVE

29110-15_F25.EPS

Figure 25 Types of groove styles.

When thick metal backing strips are to be used on groove welds, the root opening normally opens to ¼" (6.4 mm). The bevel is reduced to 22½ degrees (a total angle of 45 degrees) as shown in *Figure 26*. A root face is not required because the backing will prevent melt-through. Distortion can be reduced considerably through this control of the size and shape of the penetration.

2.2.5 Open Root Pipe Welds

It is important to note here that, in pipe welding, the alignment of the inside diameter is the most important, as opposed to the alignment of the outside diameter. In some cases, counter-boring of the pipe can be used to achieve alignment. However, there are limitations to counter-boring, since a minimum wall thickness must be maintained.

Molten metal that extends beyond the back or opposite side of the groove is called root reinforcement. Controlling the amount and shape of this root reinforcement is important to the function of the weldment. It is also important because controlling the root reinforcement helps reduce distortion by controlling the weld material uniformly. When welding pipe, the amount of root reinforcement on the inside of the pipe must be controlled. Excessive root reinforcement in the interior of the pipe interferes with the flow within the pipe, causing turbulence, pressure drop, and possibly other problems. Controlling this melt-through is also necessary to reduce stress on joints. However, it is difficult, if not impossible, to gain access to the interior of the pipe. Various methods are used to provide backing and to control penetration in pipe welding.

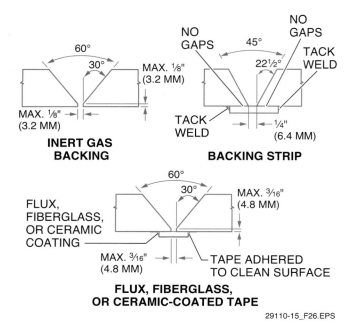

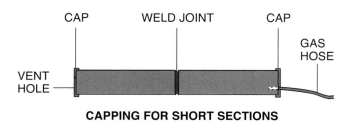

CAPPING FOR SHORT SECTIONS

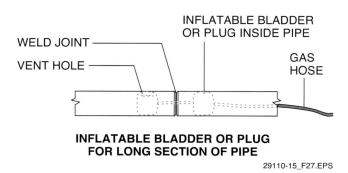

INFLATABLE BLADDER OR PLUG FOR LONG SECTION OF PIPE

29110-15_F27.EPS

Figure 26 Groove welds with backing strip.

29110-15_F26.EPS

Figure 27 Gas backing.

Gas backing using gases such as nitrogen or argon is one method used when welding pipe joints. When gas backing is used, some method of containing the gas inside the pipe must be used (*Figure 27*). For a short section of pipe, the ends of the pipe can be capped by taping or clamping a metal disk over the ends. For larger or longer pipe sections, however, this method may not be feasible due to the amount of gas that would be required to fill the pipe. For these cases, inflatable bladders, water-soluble plugs, or soft plastic bags can be inserted into the pipe near the weld joint. The inflatable bladders can be fished out of the pipe after the weld cools. The water-soluble plugs or soft plastic bags can either be blown or washed out of the pipe after the weld cools.

> **CAUTION**
>
> The pipe end opposite the end into which gas enters must have a vent hole to purge the gas and prevent pressure from building up. What is desired is an inert gas atmosphere in the pipe, rather than gas under pressure.

The type of gas used will depend on the base metal of the pipe being welded. Generally, nitrogen or carbon dioxide is used for carbon steel piping; argon is used for stainless steel, aluminum, and alloy steel piping.

2.2.6 Backing Rings

Backing rings (*Figure 28*) are flat metal strips that have been rolled to fit inside a pipe. They can be ordered in a variety of base metal types to match the base metal being welded. Some of these rings are split, making them easier to insert and adjust to piping that may be slightly out of round. Backing rings are available that have three or more nubs around the outside of the ring. The diameter of the nub is the root opening required for the pipe diameter being welded. Nubs can be pegs, buttons, or indents punched into the ring. Pegs must be removed after the joint is tack welded. They are removed by striking them with a chipping hammer; they snap off rather easily. Buttons can be removed or left in position to be melted during the root pass.

The joint preparation for backing rings (*Figure 29*) is similar to the preparation for open-root joints. The joint angle should be 60 or 75 degrees with a root opening and root face up to ⅛" (3.2 mm), or as specified on the WPS. If backing rings with nubs are used, the root opening will be set automatically by making sure the beveled ends of the pipe are positioned firmly against the nubs.

2.2.7 Consumable Inserts on Pipe Welds

Consumable inserts are similar to backing rings. As the name implies, they are completely consumed during welding and become part of the weld metal.

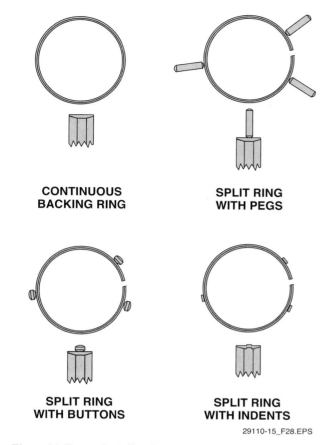

CONTINUOUS BACKING RING

SPLIT RING WITH PEGS

SPLIT RING WITH BUTTONS

SPLIT RING WITH INDENTS

29110-15_F28.EPS

Figure 28 Examples of backing rings.

Because they are part of the finished weld, they must match the filler metal requirements for the weld being made. Gas backing often is required when using consumable inserts. Consumable inserts come in a variety of shapes and are identified by class numbers, as shown in *Figure 30*.

Consumable inserts can be purchased in coils and in preformed split rings. The rectangular Class 3 insert is also available as a solid ring.

Joint preparation for consumable inserts (*Figure 31*) can consist of a V-groove or a J-groove, with the V-groove having a joint angle of 75 degrees and a root face of ¹⁄₁₆" (1.6 mm). The J-groove joint angle is 20 degrees, also with a ¹⁄₁₆" (1.6 mm) root face. The consumable insert usually is tacked against the prepared ends of the pipe. The WPS will specify the tacking requirements.

2.2.8 *Inserts on Socket Joints*

Socket joints are generally used on pipe that is 5" (DN125) or smaller in diameter. A socket joint uses a prefabricated fitting containing sockets on the ends; the pipe slips into the sockets. The fitting and pipe are joined using a fillet weld.

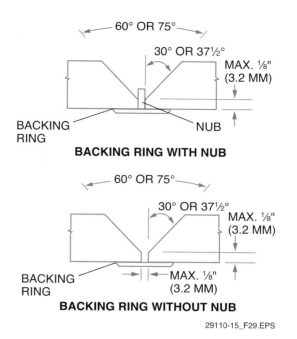

BACKING RING WITH NUB

BACKING RING WITHOUT NUB

29110-15_F29.EPS

Figure 29 Pipe joint with backing ring.

All the common fittings, such as elbows, flanges, couplings, reducers, and even valves, are available as socket fittings. Using socket joints is a quick and easy method for joining pipe evenly.

In order to eliminate stress, distortion, and possible cracking by expansion, the end of the pipe must not touch the bottom of the socket. Depending on the code, socket welds require a ¹⁄₁₆" to ⅛" (1.6 to 3.2 mm) gap between the end of the pipe and the bottom of the socket fitting.

When preparing the joint, there are two ways to ensure that the gap between the end of the pipe and the socket fitting is maintained.

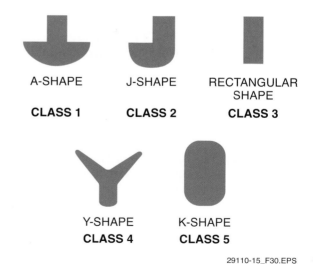

A-SHAPE **CLASS 1**

J-SHAPE **CLASS 2**

RECTANGULAR SHAPE **CLASS 3**

Y-SHAPE **CLASS 4**

K-SHAPE **CLASS 5**

29110-15_F30.EPS

Figure 30 Consumable insert profiles.

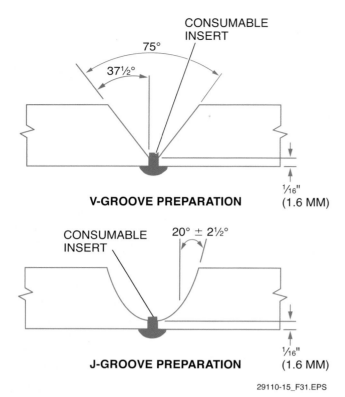

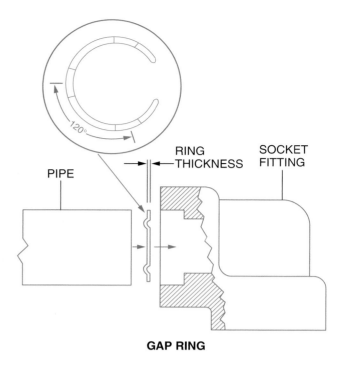

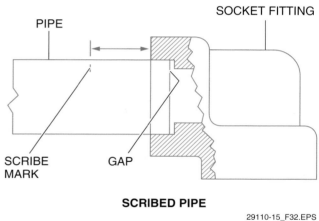

Figure 32 Socket-weld joint preparation.

29110-15_F32.EPS

Figure 31 Joint preparation for consumable inserts.

29110-15_F31.EPS

One way, described earlier, is to scribe the pipe a preset distance from the end. The gap can be checked by measuring from the scribed line to the socket and then adding the socket depth. The scribed lines also can be used to check setup after the joint is tacked up or welded.

The second method of ensuring the correct gap is to use a gap ring. A gap ring is a split ring that is formed to the gap required for the pipe diameter being welded. The gap ring is placed in the bottom of the socket end and becomes a permanent part of the joint. Gap rings are the fastest way to fit up socket welds and are available for various pipe sizes in packages of 20 each or more. *Figure 32* shows both methods of preparing socket-weld pipe joints.

2.2.9 Intermittent Welding

Many joints do not require welds along their full length. Often, stiffeners, brackets, and braces can be intermittently welded rather than continuously welded. Using short intermittent welds instead of continuous welds will reduce distortion.

2.2.10 Backstep Welding

Backstep welding (*Figure 33*) is a welding technique in which the general progression of welding is from left to right, but the weld beads are deposited in short increments from right to left.

This technique reduces distortion by minimizing and interrupting the heat input.

2.2.11 Welding Sequence

A welding sequence involves placing welds at different points on a weldment so that shrinkage forces in one location are counteracted by shrinkage forces in another location. Welding sequences can be quite complicated for large complex weldments or quite simple for less complicated weldments. When possible, use a welding sequence because it is a very effective means of controlling distortion. A simple welding sequence is to make short welds on alternating sides of the joint. This can be used for fillet or groove, intermittent, or continuous welding. *Figure 34* shows examples of intermittent welding sequences.

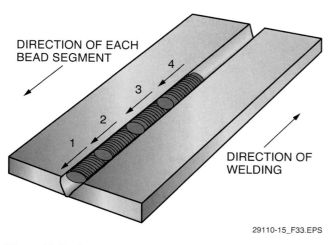

DIRECTION OF EACH
BEAD SEGMENT

4
3
2
1

DIRECTION OF
WELDING

29110-15_F33.EPS

Figure 33 Backstep welding.

Welding sequences are often performed by two individuals welding on opposite sides of a joint at the same time. This type of welding is often called buddy welding and is a very effective way to control distortion.

2.2.12 Heat Treatments

Distortion can be controlled by preheating and postheating. Applying heat before welding is called preheating. Applying heat immediately after welding is called postheating. Preheating and postheating will be covered in detail in another module.

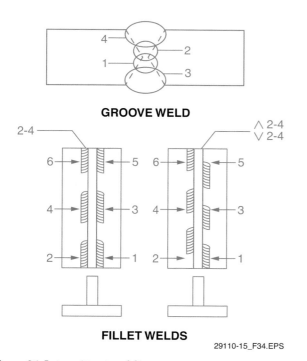

GROOVE WELD

FILLET WELDS

29110-15_F34.EPS

Figure 34 Intermittent welding sequences.

Welding must be stopped periodically to allow inspection of the weld in progress. These hold points should be scheduled strategically. Any work that will be obstructed after welding is complete, and therefore will not be easily inspected, must be inspected before the weld covers it. The following are among the items to be checked:

- Preheat temperature
- Postheat temperature
- Compliance with WPS requirements
- Weld root pass
- Weld layers

2.2.13 Checking Joint Misalignment and Fit-Up

The quality of joint preparation and fit-up directly affects the quality of the completed weld. After a weld is made, it is very costly and time-consuming to repair a defect. By thoroughly checking the joint fit-up, potential problems can be avoided. Follow these steps to check a fit-up:

Step 1 Determine if the weld is covered by a WPS. If the weld has a WPS, obtain a copy and follow it to make the weld.

Step 2 Check that the base metal type and grade are as specified. If either the base metal type or grade is wrong, contact your supervisor.

Step 3 Check, if required, that the welder who will perform the welding is certified/qualified to the WPS.

Alternating Welds

When possible, an alternating weld should be placed directly adjacent to or across from its matching weld on the opposite side. On lengthy welds, however, this sequence may not be advisable because it could leave long runs unwelded in between these welds. In these cases, stagger the welds from side to side.

If welding alternately on either side of the joint is not possible, or if one side must be completed first, a joint preparation may be used that deposits more weld metal on the second side. The greater contraction resulting from depositing more weld metal on the second side will help counteract the distortion on the first side.

Step 4 Check that the joint surfaces are free of contamination such as grease, oil, moisture, and rust. Also, check the surfaces parallel to the root and face of the weld. If there is contamination, clean the joint before continuing.

Step 5 Check the joint surface for cracks or laminations. If you find either, contact your supervisor.

Step 6 Check that the edge preparation is as specified for the joint (*Figure 35*). Check the following:
- Groove type
- Root opening
- Root face
- Included angle
- Bevel angle
- Base metal thickness

Step 7 Check the fit-up of backing strips, backing rings, and consumable inserts if required. Backing strips and backing rings must be tight against the joint along or around its entire length with even root spacing. Consumable inserts must fit tightly at the root of the weld. Correct any fit-up problems before proceeding.

Step 8 Check that the parts of the weldment are square, level, and plumb, and check that any angles are as specified. Correct any problems before proceeding.

Step 9 Check that the tack welds have been cleaned and feathered. Correct any problems before proceeding.

Step 10 Check that the welding process to be used is as specified on the WPS.

Step 11 Check that the consumables (type and size) are as specified on the WPS. Also check that arrangements have been made for the heated storage of consumables if required on the WPS.

Step 12 Check that the welding sequence is available if required.

Step 13 Check that the welding machine settings are as specified on the WPS.

Step 14 Check that arrangements have been made for preheating, interpass temperature control, and postheating if required on the WPS.

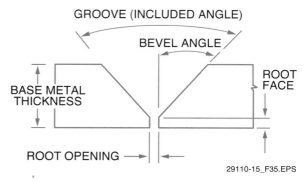

Figure 35 Edge preparation.

2.3.0 Codes and Welding Procedure Specifications

Whenever a bridge, building, ship, or pressure vessel is welded, the manufacturer and the buyer must reach agreement on how each weld will be made. To eliminate the need to write a new code for each job, government agencies, societies, and associations have developed codes. These codes are used universally to ensure safety and quality when welds are made.

2.3.1 Governing Codes and Standards

A welding code is a detailed listing of the rules and principles that apply to specific welded products. Codes ensure that safe and reliable welded products will be produced and that persons associated with the welding operation will be safe. Imagine the potential for disaster if there were no standards to control the quality and technique of welds in a nuclear plant or on an oil pipeline.

All welding should be performed following the guidelines and specifications outlined in various codes, which are specified by clients when they place orders or contract work to be done. In addition, when codes are specified, the use of these codes is mandated with the force of law by one or more government jurisdictions. They are not suggestions; they are requirements. Always check the contract, order, or project specification for the specified code(s). Some of the more common codes are the following:

- *API 1104, Standards for Welding of Pipelines and Related Facilities* – American Petroleum Institute (API); used for pipelines.
- *ASME Boiler and Pressure Vessel Code* – American Society of Mechanical Engineers (ASME); establishes the rules of safety relevant to pressure integrity that govern the design, fabrication, and inspection of boilers and pressure vessels, including nuclear power plant components.

- *ASME B31.1, Power Piping* – American Society of Mechanical Engineers (ASME); used for pressure piping.
- *ASME B31.3, 2006 Process Piping* – American Society of Mechanical Engineers (ASME); used for process piping in refineries, chemical and pharmaceutical, and similar plants.
- *ABS* – American Bureau of Shipping & Coast Guard; sets standards of excellence in Marine and Offshore Classifications. (see: **http://www.eagle.org/company/overview.html**).
- *AWS D1.1, Structural Welding Code* – American Welding Society (AWS); commonly specified for buildings and civil structures of all types for welding standard carbon and low alloy structured steel materials
- *AASHTO/AWS D1.5, 2008 Bridge Welding Code* – American Welding Society (AWS); used for welded bridges.

2.3.2 Code Changes

Periodically, codes are reviewed and changes made. Normally this occurs yearly, but some codes may go more than one year before they are updated. Most codes that are extremely broad in scope may undergo changes one section or category at a time. When a code is updated, the entire code is reissued with an updated year suffix, or addendum sheets for the areas of the code affected by the changes are issued. Changes are typically noted to alert readers to what has changed from the prior edition. The AWS code indicates additions to the code with a double vertical line in the margin alongside the changed area. Editorial changes to the AWS code are indicated by a single vertical line.

The ASME updates the entire code every three years. An addendum is issued annually. The annual addendum is identified by placing the letter A in front of the year on the cover of the code. API updates their entire code every five years.

It is important to recognize code changes. When a client specifies a code year, the code for that year, and that year only, is to be used unless otherwise specified by the client. When referring to a code, be sure the year matches the specifications for the job.

2.3.3 Welding Procedure Specifications

A WPS is a written set of instructions for producing sound welds in a prescribed manner. Each WPS is written and tested in accordance with a particular welding code or specification and must be in accordance with industry practice. Note that all welding requires that acceptable industry standards be followed, but not all welds require a WPS. If a weld does require a WPS, the WPS must be followed. The consequences of not following a required WPS are severe. Consequences include producing an unsafe weldment that could endanger life as well as the rejection of the weldment and lawsuits. Even if an unsafe condition is not known to exist, a structure known to contain numerous welds that were not done according to the WPS may have to be destroyed and redone. The economic impact can be devastating and cost welders their jobs.

When it is required, always follow the WPS. The requirement for the use of a WPS is often listed on job blueprints as a note or in the tail of the welding symbol. If you are unsure whether the welding being performed requires a WPS, do not proceed until you check with your supervisor. As a welder, you have a responsibility to follow the WPS and the applicable codes any time you are on the job. Your employer and fellow workers depend on you to carry that responsibility without deviation.

Each WPS is written by an individual who knows welding codes, specifications, and acceptable industry practices. It then becomes the responsibility of each manufacturer or contractor to test and qualify the WPS before using it. The WPS is tested by welding test coupons. Then, the coupons are tested according to the applicable code. For shielded metal arc welding (SMAW) of complete-penetration groove welds, the testing required includes nondestructive testing (NDT), tensile strength tests, and root, face, or side bend tests. The results of the testing are recorded on a Procedure Qualification Record (PQR). The WPS and PQR must be kept on file.

Many different formats are used for WPSs, but they all contain the same essential information. *Figure 36* shows a portion of a typical WPS.

Information typically found on a WPS includes the following:

- *Scope* – The welding process to which the WPS applies as well as the governing code and/or specification to be used.
- *Base metal* – The chemical composition or specification of the applicable base metal using the industry-standard identification; for example, A36 for carbon steel.
- *Welding process* – The welding process to be used to make the weld. For example, SMAW, gas metal arc welding (GMAW), flux-core arc welding (FCAW), or gas tungsten arc welding (GTAW).
- *Filler metal* – The composition, identifying type, or classification of the filler metal to be used.

 29110-15 Joint Fit-Up and Alignment

Austin Industrial

Austin

An Austin Industries Company

Specification: _____
Date: _____
Revision: _____
Page: _____ Of _____

Welding Procedure Specification

TITLE: _____

PROCESS	APPROX. NUMBER OF PASSES	ROD OR ELECT. SIZE	CURRENT	VOLTAGE	FILLER METALS SFA SPEC. CLASS	TYPE F. NO.	A. NO.

JOINTS (QW-402)
Groove Design _____
Backing: Yes _____ No _____
Backing Material (Type) _____
Other _____
_____

BASE METALS (QW-403)
P. No. _____ Group _____ to P. No. _____ Group _____
Thickness Range _____
Pipe DiameterRange _____
Other _____

FILLER METALS (QW-404)
F. No. _____ Other _____
A. No. _____ Other _____
Spec. No. (SFA) _____
AWS No. (Class) _____
Size of Electrode _____
Size of Filler _____
Electrode-Flux (Class) _____
Consumable Insert _____
Other _____

POSITION (QW-405)
Poistion of Groove _____
Welding Progression _____
Other _____

Preheat Temp. _____
Interpass Temp. _____
Preheat Maintenance _____
Other _____
_____

POSTWELD HEAT TREATMENT (QW-407)
Temperature _____
Time Range _____
Other _____

GAS (QW-408)
Shielding Gas(es) _____
Percent Composition (mixtures) _____
_____
Flow Rate _____
Gas Backing _____
Trailing Shielding Gas Composition _____
_____
Other _____

ELECTRICAL CHARACTERISTICS (QW-409)
Current AC or DC _____ Polarity _____
Amps (range) _____ Volts (range) _____
Other _____

TECHNIQUE (QW-410)
String or Weave bead _____
Orifice or Gas Cup Size _____
Initial & Interpass Cleaning _____
(Brushing, Grinding, etc.) _____

Method of Back Gouging _____
Oscillation _____
Contact Tube to Work Distance _____
Multiple or Single Pass (per side) _____

Multiple or Single Electrodes _____
Travel Speed (Range) _____
Other _____

AND FIT-UP

SUPPORTING PQR NO(S) _____
_____
_____
_____

Written By _____
_____

29110-15_F36.EPS

Figure 36 Portion of typical WPS.

- *Type of current* – Either alternating current (AC) or direct current (DC). If DC is specified, the polarity must also be given.
- *Arc voltage and travel speed* – An arc voltage range. Ranges for travel speed for automatic processes are mandatory and are recommended for semiautomatic processes.
- *Joint design and tolerances* – Joint design details, tolerances, and welding sequences that are given as a cross-sectional drawing or as a reference to drawings or specifications.
- *Joint and surface preparation* – The methods that can be used to prepare joint faces and the degree of surface cleaning required.
- *Tack welding* – Details pertaining to tack welding. Tack welders must use the WPS.

- *Welding details* – The size of electrodes to use for different portions and positions of the joint, the arrangement of welding passes to fill the joint, and the pass width and weave limitations.
- *Positions of welding* – The welding positions that can be used; for example, 1G, 3F, 5G.
- *Peening* – The details of peening and the type of tool to be used if peening is permitted.
- *Heat input* – The details to control heat input.
- *Second-side preparation* – The method used to prepare the second side when joints are welded from two sides.
- *Postheat treatment* – The details about postheat treatment or a reference to a separate document.

Essential and Nonessential Variables on a WPS

A different WPS is required for each change to an essential variable. Some essential variables for welds made with SMAW include the following:

- Base-metal thickness (within range listed)
- Composition of base metal
- Strength and composition of filler metal
- Base-metal preheat temperature

Nonessential variables for welds made with SMAW include the following:

- Type of groove used
- Omission of backing material in butt joints
- Electrode size

Changes in these nonessential variables would not normally require a different WPS.

Additional Resources

2014 Technical Training Guide. Current Edition. Cleveland, OH. USA: The Lincoln Electric Company. **www.lincolnelectric.com**

Welding Handbook. Current Edition. Miami, FL: The American Welding Society.

Mathey Dearman Product Manuals. The Mathey Dearman Company. **www.matheydearman.com**

2.0.0 Section Review

1. Stainless steel is more likely to be affected by distortion than carbon steel because it has a _____.

   a. higher coefficient of thermal expansion and a lower specific heat per unit volume
   b. lower coefficient of thermal expansion and a lower specific heat per unit volume
   c. lower coefficient of thermal expansion and a higher specific heat per unit volume
   d. higher coefficient of thermal expansion and a higher specific heat per unit volume

2. If a WPS is being followed, it also applies to the _____.

   a. tack weld
   b. welding machine model
   c. ground lead
   d. alignment tool

3. Existing welding codes eliminate the need for _____.

   a. welder training and certification
   b. creating standards for each and every job
   c. Welding Procedure Specifications (WPS)
   d. quality control inspections

SUMMARY

It is very important to perform proper joint fit-up and alignment to ensure an acceptable weld. The proper fit-up measuring devices and tools must be used to accomplish this. Welders must know how to compensate for welding distortion as well as how to check for poor fit-up after the weld has been completed. Even though many welders are not required to fit joints to be welded, especially pipe joints, many others are able to fit as well as they weld, adding to their value and versatility. When the welder is not responsible for fitting the joint, he or she must be familiar with the techniques and tools used to create a reliable joint that meets the specifications.

Government agencies, professional societies, and associations have written guidelines for joint fit-up and alignment. These guidelines not only help ensure quality welds, but also help ensure safe welds and welding environments. All fit-up procedures must embrace these guidelines that have been proven over the years through use and repeated testing. When specified, these codes must be followed under penalty of law.

1. Which tool is often fabricated on the job from small channel or angle iron?

   a. Straightedge
   b. Plumb bob
   c. Square
   d. Level

2. Before welding a 45-degree joint, check the angle of the joint with a _____.

   a. straightedge
   b. Hi-Lo gauge
   c. framing square
   d. combination square

3. The primary use of a Hi-Lo gauge is to check for _____.

   a. pipe joint misalignment
   a. I-beam joint alignment
   b. 45-degree joints
   c. 90-degree joints

4. Secure a chain hoist over a weldment to be raised with an approved sling wrapped around _____.

   a. conduit
   b. piping
   c. a structural member
   d. raceways

5. When using a strongback to hold two weldment parts that have been aligned and fitted up, the tack welds used to hold the strongback to the weldment parts should be placed on only _____.

   a. opposite sides of the strongback
   b. the notched side of the strongback
   c. the vertical sides of the strongback
   d. one side of the strongback

6. If a joint needs to be forced into alignment, yokes and wedges, or bolts, can be used with _____.

   a. strongbacks
   b. cage clamps
   c. peening
   d. a come-along

7. When working with pipe over 10" (DN250), OSHA requires _____.

   a. only specific filler metals be used
   b. a WPS
   c. chain hoists with a minimum capacity of 10 tons
   d. four-legged jack stands

8. The arched crossbar available for cage clamps is designed to _____.

   a. strengthen the clamp
   b. allow welds to be made under the clamp
   c. allow the clamp to be used on a many pipe sizes
   d. hold a section of plate to the pipe

9. Which of the following devices may *not* be allowed on some applications or projects?

   a. Strongbacks
   b. Cage clamps
   c. Chain clamps
   d. Pipe pullers

10. The stress locked in a weldment when it returns to the ambient temperature, and equal to the strength of the base metal, is called _____.

    a. base stress
    b. residual stress
    c. thermal conductivity
    d. coefficient of expansion

11. The more weld metal placed into a joint, the greater the forces of _____.

    a. stress
    b. magnetism
    c. shrinkage
    d. thermal conductivity

12. Socket-welded joints are generally found on _____.

    a. lightweight plate
    b. heavy plate
    c. pipe sizes 5" (DN125) or smaller
    d. pipe sizes larger than 6" (DN150)

13. A detailed listing of the rules and principles that apply to specific welded products is called a _____.

    a. work order
    b. procedure
    c. punch list
    d. welding code

14. The written set of instructions for producing sound welds in a prescribed manner is called a _____.

    a. contract
    b. Welding Procedure Specification (WPS)
    c. welding code
    d. PQR

15. Welding test coupons prepared for a job are tested according to the _____.

    a. applicable code
    b. WPS guidelines
    c. manufacturer's guidelines
    d. PQR

Trade Terms Quiz

Fill in the blank with the correct term that you learned from your study of this module.

1. _____ refers to the difference in the internal alignment of two pipe sections.

2. The stress remaining in a weldment as a result of heating and cooling is called _____.

3. The quantity of heat needed to raise one unit mass of a substance by one unit degree is called _____.

4. Preplaced filler metal called a(n) _____ is completely fused into the root of a joint during welding.

5. The term _____ refers to a vertical line that is perpendicular to the earth's horizontal axis.

6. A(n) _____ is extra information used to identify corrections or revisions made to documents and shared with document users.

7. Mechanically working metal with repeated hammer blows is called _____.

8. The _____ refers to the temperature of air surrounding an object or individual.

9. The term _____ refers to a line that is parallel to the earth's horizontal axis.

10. A weld made to hold weldment parts in alignment until a final weld is made is called a(n) _____.

11. _____ is the change in length, per unit length of material, for each 1°C of change in temperature.

Trade Terms

Addendum 6
Ambient temperature 8
Coefficient of thermal expansion 11
   (linear)

Consumable insert 4
High-low 1
Level 9
Peening 7

Plumb 5
Residual stress 2
Specific heat per unit volume 3
Tack weld 10

Trade Terms Introduced in This Module

Addendum: Supplementary information typically used to describe corrections or revisions to documents.

Ambient temperature: The room temperature of the atmosphere completely surrounding an object on all sides.

Coefficient of thermal expansion (linear): The change in length per unit length of material for a 1°C change in temperature.

Consumable insert: Preplaced filler metal that is completely fused into the root of the joint during welding, becoming part of the weld.

High-low: The discrepancy in the internal alignment of two sections of pipe.

Level: A line on a horizontal axis parallel with the earth's surface (horizon). Also refers to the tool used to check if an object is level.

Peening: The mechanical working of metal using repeated hammer blows, typically using a ball-shaped hammer head such as that found on a ball peen hammer. Peening may be done for both metallurgical and appearance reasons.

Plumb: A line on a vertical axis perpendicular to the earth's surface (horizon).

Residual stress: Stress remaining in a weldment as a result of heat.

Specific heat per unit volume: The quantity of heat required to raise one unit mass of a substance by one unit degree.

Tack weld: A weld made to hold parts of a weldment in proper alignment until the final weld is made.

Additional Resources

This module presents thorough resources for task training. The following resource material is suggested for further study.

2014 Technical Training Guide. Current Edition. Cleveland, OH. USA: The Lincoln Electric Company. **www.lincolnelectric.com**

Welding Handbook. Current Edition. Miami, FL: The American Welding Society.

Mathey Dearman Product Manuals. The Mathey Dearman Company. **www.matheydearman.com**

Figure Credits

Section Review Answer Key

Answer	Section Reference	Objective
Section One		
1. c	1.1.3	1a
2. a	1.2.2	1b
3. b	1.3.2	1c
4. c	1.4.4	1d
Section Two		
1. d	2.1.0	2a
2. a	2.2.2	2b
3. b	2.3.0	2c

NCCER CURRICULA — USER UPDATE

NCCER makes every effort to keep its textbooks up-to-date and free of technical errors. We appreciate your help in this process. If you find an error, a typographical mistake, or an inaccuracy in NCCER's curricula, please fill out this form (or a photocopy), or complete the online form at **www.nccer.org/olf**. Be sure to include the exact module ID number, page number, a detailed description, and your recommended correction. Your input will be brought to the attention of the Authoring Team. Thank you for your assistance.

Instructors – If you have an idea for improving this textbook, or have found that additional materials were necessary to teach this module effectively, please let us know so that we may present your suggestions to the Authoring Team.

NCCER Product Development and Revision

13614 Progress Blvd., Alachua, FL 32615

Email: curriculum@nccer.org
Online: www.nccer.org/olf

❑ Trainee Guide   ❑ Lesson Plans   ❑ Exam   ❑ PowerPoints   Other _____

Craft / Level: _____ Copyright Date: _____

Module ID Number / Title: _____

Section Number(s): _____

Description: _____

_____

_____

_____

_____

Recommended Correction: _____

_____

_____

_____

_____

Your Name: _____

Address: _____

_____

Email: _____ Phone: _____

29111-15

SMAW – Groove Welds with Backing

OVERVIEW

There are a number of different groove styles used in welding. The standard V-groove is one of the most popular. Completing groove welds in the various positions requires attention to detail and an understanding of work, drag, and push angles to ensure a quality weld. This module explores the differences in technique used to complete V-groove welds with backing, and provides a significant amount of time to develop these essential skills.

Module Eleven

Trainees with successful module completions may be eligible for credentialing through the NCCER Registry. To learn more, go to **www.nccer.org** or contact us at **1.888.622.3720**. Our website has information on the latest product releases and training, as well as online versions of our *Cornerstone* magazine and Pearson's product catalog.

Your feedback is welcome. You may email your comments to **curriculum@nccer.org**, send general comments and inquiries to **info@nccer.org**, or fill in the User Update form at the back of this module.

This information is general in nature and intended for training purposes only. Actual performance of activities described in this manual requires compliance with all applicable operating, service, maintenance, and safety procedures under the direction of qualified personnel. References in this manual to patented or proprietary devices do not constitute a recommendation of their use.

Objectives

When you have completed this module, you will be able to do the following:

1. Identify various types of groove welds and describe how to prepare for groove welding.
   a. Identify various types of groove welds and define related terms.
   b. Describe how to prepare for groove welding.
2. Describe the technique required to produce various groove welds.
   a. Describe the technique required to produce groove welds in the 1G and 2G positions.
   b. Describe the technique required to produce groove welds in the 3G and 4G positions.

Performance Tasks

Under the supervision of your instructor, you should be able to do the following:

1. Safely set up arc welding equipment for making groove welds.
2. Make flat welds with backing on V-groove joints using E7018 electrodes.
3. Make horizontal welds with backing on V-groove joints using E7018 electrodes.
4. Make vertical welds with backing on V-groove joints using E7018 electrodes.
5. Make overhead welds with backing on V-groove joints using E7018 electrodes.

Industry Recognized Credentials

If you are training through an NCCER-accredited sponsor, you may be eligible for credentials from NCCER's Registry. The ID number for this module is 29111-15. Note that this module may have been used in other NCCER curricula and may apply to other level completions. Contact NCCER's Registry at 888.622.3720 or go to **www.nccer.org** for more information.

Contents

Topics to be presented in this module include:

Figures

1.0.0 GROOVE WELDS

Objective

Identify various types of groove welds and describe how to prepare for groove welding.

a. Identify various types of groove welds and define related terms.
b. Describe how to prepare for groove welding.

Performance Task

1. Safely set up arc welding equipment for making groove welds.

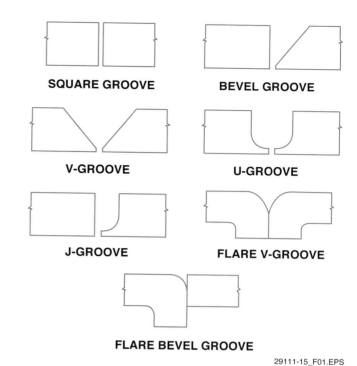

Figure 1 Single groove weld styles.

29111-15_F01.EPS

Different welding applications require differently shaped welds at the point where two metal parts are to be joined. This module explains groove welds and the various techniques and arc welding equipment used to make them.

A groove weld is made in an opening in a part or between two parts. Groove welds can be made in all five of the basic types of joints as well as on the surface of a part. There are several different groove styles. The name of the groove is based on its shape. Some must be prepared, while others appear as a result of fitting together two or more parts to be welded.

If a groove only goes part of the way through the joint, welding in the groove results in what is called a partial joint penetration (PJP) weld. If the groove allows for welding the complete thickness of the base metals, it is called a complete joint penetration (CJP) weld.

1.1.0 Typical Groove Weld Styles

Groove weld styles include the square, V, U, bevel, J, flare V, and flare bevel. *Figure 1* shows these typical single-groove weld styles.

For a part whose shape must be changed to a shape required for a specific groove, you need to start with an object having a squared (90-degree) end or edge, such as the square groove objects in *Figure 1*. Grinders are normally used to bevel metal parts. Beveling the parts means removing enough material to form an incline, such as the bevel groove image in *Figure 1*. Beveling the metal

serves two purposes. First, it cleans all of the areas the weld must contact. Second, it creates a greater or wider space, which requires more weld to fill the joint. The bevel angle must be sufficient to allow access to the root, but not so large that it requires an excessive amount of filler metal to fill it.

1.1.1 Double-Groove Welds

V-, bevel, U-, and J-grooves can be used and welded on one side of a joint or on both sides. If they are used on both sides of a joint, they are called double-groove welds (*Figure 2*).

> **NOTE**
>
> Throughout this module, both metric and Imperial units have been provided for most measurements. Weld symbols are generally provided in one measurement system or another, but not both. For this reason, metric values have not been placed alongside Imperial values in weld symbols.

1.1.2 Combination Groove and Fillet Welds

Groove welds are often used in combination with fillet welds to maintain the strength of the welded joint and use less weld metal. *Figure 3* shows a number of groove and fillet weld combinations.

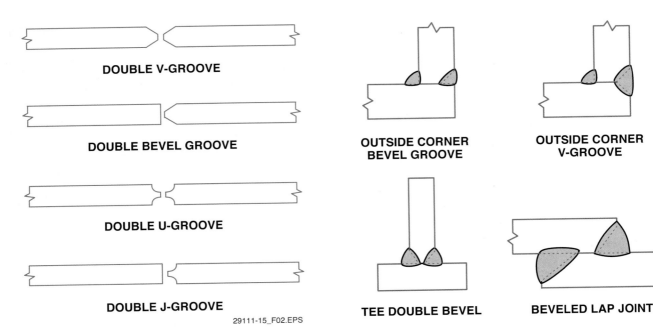

Figure 2 Double groove weld styles.

1.1.3 Groove Weld Terms

Part of a welder's certification includes the proper understanding of related welding terms. The terms associated with a groove are the following:

- *Bevel angle* – The angle formed between the prepared edge of a member and a line perpendicular to the surface of the member.
- *Bevel face* – The prepared surface of a bevel edge shape.
- *Bevel radius* – The radius used to form the shape of a J- or U-groove.
- *Depth of bevel* – The distance the joint preparation extends into the base metal (used for partial joint penetration welds).
- *Groove angle* – The included (total) angle between the groove faces of a weld groove.
- *Groove face* – Any surface in a weld groove prior to welding. See *bevel face*, *root face*, and *bevel radius*.
- *Joint root* – The portion of a joint where the members approach closest to each other.
- *Root face* – The portion of the edge of a part to be joined by a groove weld that has not been beveled or grooved.
- *Root opening* – The separation between the members to be joined at the root of the joint.

Figure 4 shows examples of grooves and their named parts.

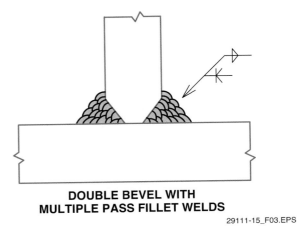

Figure 3 Groove and fillet weld combinations.

The terms used to describe features of a groove weld are as follows:

- *Face reinforcement* – Reinforcement of the weld by excess weld metal on the side from which the weld was made.
- *Groove weld size* – The joint penetration (depth of joint preparation plus root penetration). The size can never be greater than the base metal thickness.
- *Weld face* – The exposed surface of the weld.
- *Weld root* – The point at which the back of the weld extends the farthest into the weld joint.
- *Weld toe* – The points where the face of the weld metal and the base metal meet.

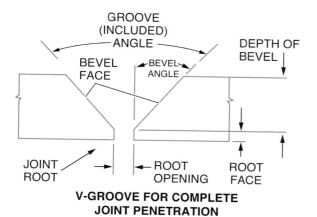

GROOVE
(INCLUDED) ANGLE

BEVEL FACE

BEVEL ANGLE

DEPTH OF BEVEL

JOINT ROOT

ROOT OPENING

ROOT FACE

V-GROOVE FOR COMPLETE JOINT PENETRATION

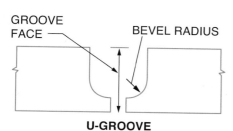

GROOVE FACE

BEVEL RADIUS

U-GROOVE

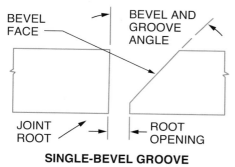

BEVEL FACE

BEVEL AND GROOVE ANGLE

JOINT ROOT

ROOT OPENING

SINGLE-BEVEL GROOVE

29111-15_F04.EPS

Figure 4 Grooves.

1.1.4 Backings

Although single groove welds can be done without backing, single groove welds with backing are used for high-strength, high-quality welds. Backing also makes welding the root easier. Thick metal pieces that are to be joined must be beveled so that the weld will penetrate the full thickness. The simplest form of backing is a strip of metal matching the base metal being welded. Other types of backing include dissimilar metals, ceramic, and coated tapes that can be removed when the welding has been completed.

Figure 5 shows some V-groove welds with backing. Although a weld joint done with an inert gas on the opposite side of the weld has no physical, solid backing material, it is still considered a joint with backing. The other two backings shown have a true, solid backing.

This module will explain the preparation and welding of V-groove welds with metal backing using low-hydrogen electrodes in all positions. Other types of groove welds can be performed in the same manner. Performance demonstrations will require that the welds meet the qualification standards in the applicable code for visual and destructive testing.

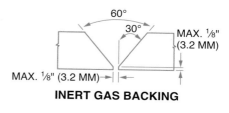

60°
30°
MAX. ¹⁄₈" (3.2 MM)
MAX. ¹⁄₈" (3.2 MM)

INERT GAS BACKING

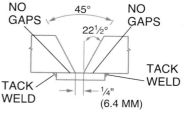

NO GAPS
45°
NO GAPS
22½°
TACK WELD
TACK WELD
¼" (6.4 MM)

BACKING STRIP

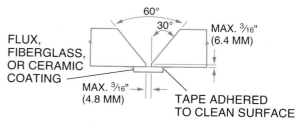

60°
30°
MAX. ³⁄₁₆" (6.4 MM)
FLUX, FIBERGLASS, OR CERAMIC COATING
MAX. ³⁄₁₆" (4.8 MM)
TAPE ADHERED TO CLEAN SURFACE

FLUX, FIBERGLASS, OR CERAMIC-COATED TAPE

29111-15_F05.EPS

Figure 5 V-groove welds with backings.

1.2.0 Welding Preparations

Before welding can take place, the area has to be made ready, the welding equipment must be set up, and the metal to be welded must be prepared. The following sections explain how to set up arc welding equipment for welding.

1.2.1 Safety Practices

The following is a summary of safety procedures and practices that must be observed when cutting or welding. Keep in mind that this is a summary. Complete safety coverage is provided in the *Welding Safety* module. If you have not completed that module, do so before continuing. Above all, be sure to wear appropriate protective clothing and equipment when welding or cutting.

To maintain safety and prevent injury, it is essential that you wear the appropriate protective clothing and equipment when cutting or welding metals. Be sure to follow these guidelines during all phases of cutting or welding:

- Always use safety glasses along with a full face shield or a helmet. The glasses, face shield, or helmet lens must have the proper light-reducing tint for the type of welding or cutting being performed. Never directly or indirectly view an electric arc without using a properly tinted lens.
- Wear proper protective leather and/or flame retardant clothing along with welding gloves that will protect you from flying sparks and molten metal as well as heat.
- Wear high-top safety shoes or boots. Make sure that the tongue and lace area of the footwear will be covered by a pant leg. If the tongue and lace area is exposed, or if the footwear must be protected from burn marks, wear leather spats under your pants or chaps to cover the front top of the footwear.
- Wear a 100-percent cotton cap with no mesh material included in its construction. The bill of the cap points to the rear. If a hard hat is required for the environment, use one that allows the attachment of rear deflector material and a face shield. A hard hat with a rear deflector is generally preferred when working overhead, and may be required by some employers and job sites.

WARNING!
Do not wear a cap with a button in the middle. The conductive metal button beneath the fabric represents a safety hazard.

- Wear a face shield over snug-fitting cutting goggles or safety glasses for gas welding or cutting. Either the face shield or the lenses of the welding goggles must be an approved shade for the application. For electric arc welding or cutting, wear safety gasses and a welding hood with the correct tinted lens.
- Do not open your welding hood lens and expose unprotected eyes when grinding or cleaning a weld. Either use a hood that maintains a protective lens when the shaded lens is opened, a hood with an auto-shading lens, or remove the hood and use a full-face shield.
- Wear earplugs to protect ear canals from sparks. Wear hearing protection to protect against the consistent sound of the torch.

WARNING!
Ear protection is essential to protect ears from the noise of the torch. Other personal protective equipment (PPE) must be worn to protect the operator from hot metal and slag.

Welding activities involve the use of fire or extreme heat to melt metal. Whenever fire is used, it must be controlled and contained. Welding activities are often performed on vessels that may once have contained flammable or explosive materials. Residues from those materials can catch fire or explode when a welder begins work on such a vessel. The following are fire and explosion prevention guidelines associated with welding:

- Never carry matches or gas-filled lighters in your pockets. Sparks can cause the matches to ignite or the lighter to explode, causing serious injury.
- Always comply with all site and/or employer requirements for a hot-work permit and a fire watch.
- Never use oxygen to blow dust or dirt off clothing. The oxygen can remain trapped in the fabric for a time. If a spark hits the oxygen in the fabric, the clothing can burn rapidly and violently out of control.
- Make sure that any flammable material in the work area is moved or shielded by a fire-resistant covering.
- Approved fire extinguishers must be available before any heating, welding, or cutting operations are attempted. Make sure the extinguisher is charged, the inspection tag is valid, and any individual that may be required to operate it knows how to do so.

- Never release a large amount of oxygen or use oxygen in place of compressed air. The presence of oxygen around flammable materials or sparks can cause rapid and uncontrolled combustion. Keep oxygen away from oil, grease and other petroleum products.
- Never release a large amount of fuel gas, especially acetylene. Methane and propane are heavier than air and tend to migrate to and concentrate in low areas. As a result, they can ignite at a considerable distance from the release point. Acetylene is lighter than air but is even more dangerous than methane; when mixed with air or oxygen, it will explode at much lower concentrations than any other common fuel gas.
- To prevent fires, maintain a neat and clean work area, and make sure that any metal scrap or slag is cold before disposal.

Before cutting containers such as tanks or barrels, check to see if they have contained any explosive, hazardous, or flammable materials, including petroleum products, citrus products, or chemicals that decompose into toxic fumes when heated. As a standard practice, always clean and then fill any tanks or barrels with water, or purge them with a flow of inert gas such as nitrogen to displace any oxygen.

> **WARNING!**
>
> Welding or cutting must never be performed on drums, barrels, tanks, vessels, or other containers until they have been emptied and cleaned thoroughly, eliminating all flammable materials and all substances (such as detergents, solvents, greases, tars, or acids) that might produce flammable, toxic, or explosive vapors when heated. Do not assume that a container that has held combustibles is clean and safe until proven so by proper tests. Do not weld in places where dust or other combustible particles are suspended in air or where explosive vapors are present.

Containers must be cleaned by steam cleaning, flushing with water, or washing with detergent until all traces of the material have been removed.

> **WARNING!**
>
> Clean containers only in well-ventilated areas. Vapors can accumulate during cleaning, causing explosions or injury.

Vapors and fumes tend to rise in the air from their sources. Welders often have to work above welding areas where fumes are being created. Welding fumes can be harmful. Good work area ventilation helps to remove the vapors and protect the welder. Always perform cutting or welding operations in a well-ventilated area. Cutting or welding operations involving zinc or cadmium materials or coatings result in toxic fumes. For long-term cutting or welding of such materials, always wear an approved full-face, supplied-air respirator (SAR) that uses breathing air supplied- from outside the work area. For occasional, very short-term exposure, a HEPA-rated or metal-fume filter may be used on a standard respirator.

1.2.2 Preparing the Welding Area

To practice welding, a welding table, bench, or stand is needed (*Figure 6*). The welding surface must be steel, and provisions must be made for placing welding coupons out of position.

To set up the area for welding, follow these steps:

Step 1 Make sure that the area is properly ventilated. Make use of doors, windows, and fans.

Step 2 Check the area for fire hazards. Remove any flammable materials before proceeding.

29111-15_F06.EPS

Figure 6 Welding station with vacuum system.

Step 3 Know the location of the nearest fire extinguisher. Do not proceed unless the extinguisher is charged and you know how to use it.

Step 4 Position a welding table near the welding machine, or vice versa.

Step 5 Set up flash shields around the welding area.

1.2.3 Preparing the Weld Coupons

If possible, the weld coupons should be ⅜" thick carbon steel to conform with AWS limited-thickness test coupon requirements. If this size is not readily available, ¼" to 1" thick steel can be used for practice welds. If you are working with metric metals, a 10 mm thickness is preferred. If not readily available, 6 to 25 mm plate can be used for practice. Clean the steel plate prior to welding by using a wire brush or grinder to remove heavy mill scale or corrosion.

WARNING!

Be extremely careful when using wire brush attachments on powered tools. Always ensure that the rotational speed rating of the wire brush matches or exceeds the rated speed of the powered tool. Loose wires from the brush and debris tend to fly off and embed into anything or anyone near the work. Make sure that all appropriate PPE is worn when wire brushing or grinding with all tools, but be especially cautious with powered tools.

For each weld coupon, you will need two pieces of plate 3" × 7" (7.6 cm × 17.8 cm) and one piece of backing. The backing should be ¼" × 1" × 8" (6 mm × 2.5 cm × 20.3 cm). The backing is 1" (2.5 cm) longer than the coupons, so that it can extend past them ½" (12.7 mm) on each end of the joint. Cut the coupons into rectangles, with the long sides beveled at approximately 22½ degrees. A 37½-degree bevel can also be used, subject to your instructor's choice. The backing is not beveled.

Note that this 3" × 7" size is the minimum size allowed for an AWS test coupon. However, for the purpose of practice and material conservation, your instructor may select a smaller size.

Follow these steps to prepare the V-groove with metal backing weld coupon:

Step 1 Check the bevel. There should be no root face and no dross, and the bevel angle should be approximately 22½ or 37½ degrees.

Step 2 Obtain clamps large enough to hold the coupons to the backing strips.

Step 3 Lay the backing strip on a flat surface and then position the two beveled coupons on it so that their root opening is ¼" (6.4 mm) wide. *Figure 7* shows coupons set up for a V-groove with metal backing.

Step 4 Clamp the beveled strips to the backing strip, making sure to keep the root opening at ¼" (6.4 mm). Also, make sure that the backing strip is held tight against the back of the beveled strips with the clamps.

Step 5 Turn the strips over and place the tack welds on the backside of the joint in the lap formed by the backing strip and the beveled strips. Use a tack in the center and one on each end, applied to each half of the coupon. The result is six tack welds (*Figure 8*).

Step 6 Remove the electrode from the electrode holder and discard as soon as you stop welding to prevent arcing to nearby surfaces, and place the stub in a proper container.

NOTE

At this point, the strips can be positioned into any of the four welding positions as required.

Backing Strip

When the backing strip is ½" to 1" (1.3 to 2.5 cm) longer at each end of the weld groove, it allows the welder to start and stop the bead outside the weld groove.

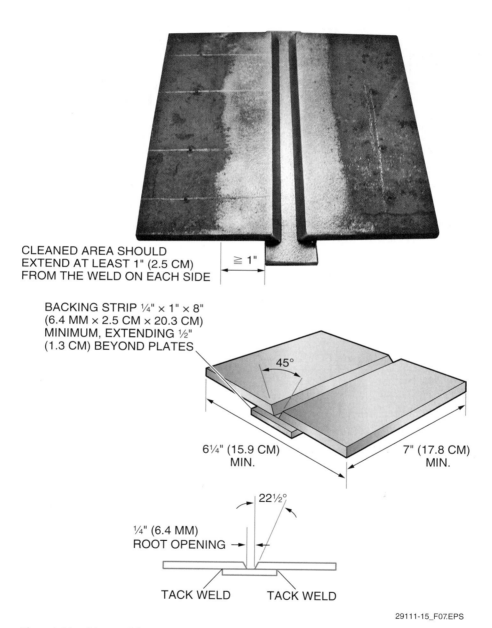

CLEANED AREA SHOULD EXTEND AT LEAST 1" (2.5 CM) FROM THE WELD ON EACH SIDE

≧ 1"

BACKING STRIP ¼" × 1" × 8" (6.4 MM × 2.5 CM × 20.3 CM) MINIMUM, EXTENDING ½" (1.3 CM) BEYOND PLATES

45°

6¼" (15.9 CM) MIN.

7" (17.8 CM) MIN.

22½°

¼" (6.4 MM) ROOT OPENING

TACK WELD   TACK WELD

29111-15_F07.EPS

Figure 7 V-groove with metal backing weld coupon.

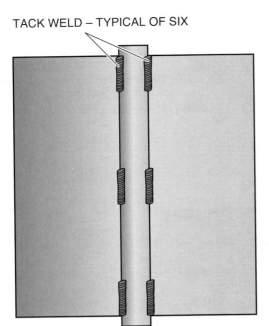

TACK WELD – TYPICAL OF SIX

29111-15_F08.EPS

Figure 8 Coupon tack welds.

When welding in the horizontal position, an alternate joint preparation can be used. The alternate joint has one plate beveled at approximately 45 degrees and the other plate at approximately 90 degrees. The plates are positioned with the beveled plate above the 90-degree plate and a ¼" (6.4 mm) root opening.

Figure 9 is a diagram showing the alternate horizontal weld coupons. Degrees shown are approximate.

> **NOTE**
> Check with your instructor about whether to use the standard V-groove preparation or the alternate preparation for horizontal weld coupons.

1.2.4 Electrodes

Obtain only a small quantity of the electrodes to be used. The welding exercises in this module use ³⁄₃₂" (2.4 mm), ⅛" (3.2 mm), or ⁵⁄₃₂" (4.0 mm) E7018 electrodes.

Keep the following general rules in mind when running low-hydrogen electrodes:

- Remember not to whip low-hydrogen electrodes.
- Maintain a short arc because a low-hydrogen weld relies on its slag to protect the molten metal. It does not have a heavy gaseous shield.

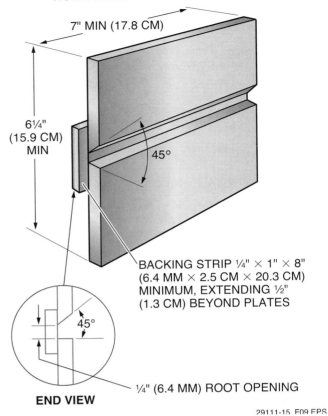

NOTE: BASE METAL MILD STEEL

7" MIN (17.8 CM)

6¼" (15.9 CM) MIN

45°

BACKING STRIP ¼" × 1" × 8" (6.4 MM × 2.5 CM × 20.3 CM) MINIMUM, EXTENDING ½" (1.3 CM) BEYOND PLATES

45°

END VIEW

¼" (6.4 MM) ROOT OPENING

29111-15_F09.EPS

Figure 9 Alternate horizontal weld coupons.

- Remove all slag between passes.
- Remove only a small number of electrodes at a time from the oven because low-hydrogen electrodes absorb moisture quickly.
- Restart by striking the arc ahead of the crater, move quickly back into the crater, and then proceed as usual. This technique welds over the arc strike, eliminating porosity and giving a smoother restart.
- When running stringer beads, use a slight side-to-side motion with a small pause at the weld toe to tie-in and flatten the bead.
- When running weave beads, use a slight pause at the weld toe to tie-in and flatten the bead.
- Do not use low-hydrogen electrodes with chipped or missing flux.
- Set the amperage in the lower portion of the suggested range for better puddle control when welding vertically.

Obtain only the electrodes to be used for a particular welding exercise at one time. Have some type of pouch or rod holder in which to store the electrodes to prevent them from becoming damaged. Never store electrodes loose on a table.

They may fall on the floor where they can become a tripping hazard or become damaged. Some type of metal container or bucket must also be available to discard hot electrode stubs.

> **WARNING!**
>
> Do not throw electrode stubs on the floor. They roll easily and could cause someone to slip and fall.

1.2.5 Preparing the Welding Machine

Welders can expect to find different types and makes of welding machines in welding shops and on jobs in the field. Many machines will not have operator manuals with them. As a trainee, and even as a trained welder, you must be able to recognize the different types of welding machines and to figure out how to safely set up and operate them. Select a welding machine to use and then follow these steps to set it up for welding:

Step 1   Verify that the welding machine can be used for DC welding.

Step 2   Check the area for proper ventilation.

Step 3   Verify the location of the power disconnect for the unit.

Step 4   Set the polarity to direct current electrode positive (DCEP).

Conserving Material

Steel for practice welding is expensive and difficult to obtain. Every effort should be made to conserve and not waste the material that is available. Reuse weld coupons until all surfaces have been welded upon by cutting the weld coupon apart and reusing the pieces. Use material that cannot be cut into weld coupons to practice running beads.

Step 5   Connect the clamp of the workpiece lead to the workpiece.

Step 6   Set the amperage for the electrode type and size to be used. Typical settings are as follows:

Electrode	Size	Amperage
E7018	$^3/_{32}$" (2.4 mm)	70A to 100A
E7018	$^1/_8$" (3.2 mm)	100A to 150A
E7018	$^5/_{32}$" (4.0 mm)	130A to 200A

> **NOTE**
>
> Amperage recommendations vary by manufacturer, welding position, current type, and electrode brand. Refer to the Welding Procedure Specification (WPS) as the primary source of information, or to the manufacturer's recommendations for the electrode being used if a WPS is not in use.

Step 7   Check to be sure the electrode holder is not grounded.

Step 8   Turn on the welding machine.

Figure 10 shows DC and AC/DC welding machines with their various parts labeled.

Electrode Housekeeping

A floor littered with electrode stubs is an accident waiting to happen. Not only does throwing electrode stubs on the floor or ground create a hazard, it can also result in disciplinary action from the employer. In some cases, a worker may even lose his or her job for this infraction.

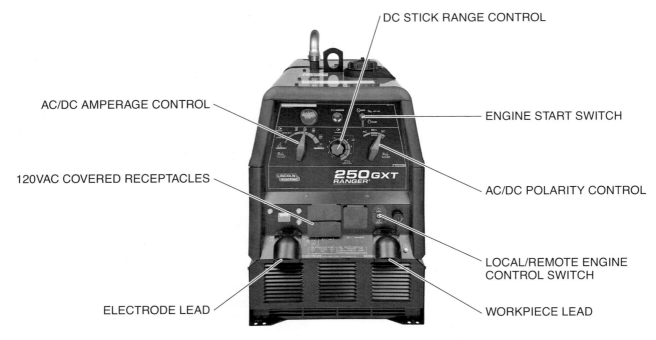

DC STICK RANGE CONTROL

AC/DC AMPERAGE CONTROL

ENGINE START SWITCH

120VAC COVERED RECEPTACLES

AC/DC POLARITY CONTROL

LOCAL/REMOTE ENGINE CONTROL SWITCH

ELECTRODE LEAD

WORKPIECE LEAD

AC/DC ENGINE-DRIVEN WELDING MACHINE

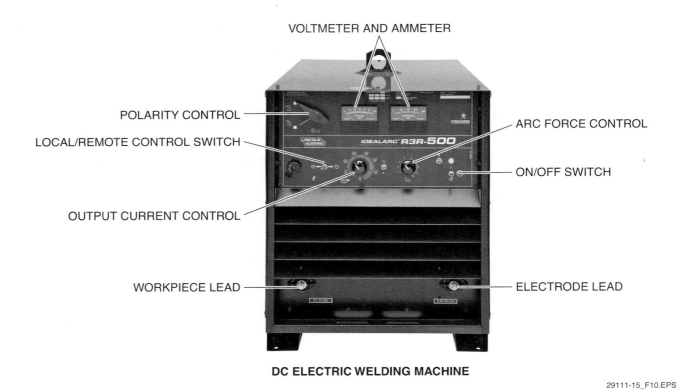

VOLTMETER AND AMMETER

POLARITY CONTROL

ARC FORCE CONTROL

LOCAL/REMOTE CONTROL SWITCH

ON/OFF SWITCH

OUTPUT CURRENT CONTROL

WORKPIECE LEAD

ELECTRODE LEAD

DC ELECTRIC WELDING MACHINE

29111-15_F10.EPS

Figure 10 DC and AC/DC welding machines.

Lash

To ensure good welding results, remember LASH.

L *(length of arc)* – The distance between the electrode and the base metal (usually one times the electrode diameter).

A *(angle)* – Two angles are critical:

- Travel angle – The longitudinal angle of the electrode in relation to the axis of the weld joint
- Work angle – The traverse angle of the electrode in relation to the axis of the weld joint

S *(speed)* – Travel speed is measured in inches per minute (IPM). The width of the weld will determine if the travel speed is correct.

H *(heat)* – Controlled by the amperage setting and dependent upon the electrode diameter, base metal type, base metal thickness, and the welding position.

Additional Resources

2014 Technical Training Guide. Current Edition. Cleveland, OH. USA: The Lincoln Electric Company. **www.lincolnelectric.com**

Welding Handbook. Current Edition. Miami, FL: The American Welding Society.

1.0.0 Section Review

1. The angle formed between the prepared edge of a member and a line perpendicular to the surface of the member is called the _____.
   a. bevel radius
   b. bevel angle
   c. groove angle
   d. groove face

2. Welding operations involving zinc or cadmium create _____.
   a. heavy slag
   b. stronger welds
   c. heavy weld spatter
   d. toxic fumes

2.0.0 V-GROOVE WELDS WITH BACKING

Objective

Describe the technique required to produce various groove welds.

   a. Describe the technique required to produce groove welds in the 1G and 2G positions.

   b. Describe the technique required to produce groove welds in the 3G and 4G positions.

Performance Tasks

1. Make flat welds with backing on V-groove joints using E7018 electrodes.
2. Make horizontal welds with backing on V-groove joints using E7018 electrodes.
3. Make vertical welds with backing on V-groove joints using E7018 electrodes.
4. Make overhead welds with backing on V-groove joints using E7018 electrodes.

The V-groove with backing weld is a common groove weld normally made with low-hydrogen electrodes. The backing can be of dissimilar material, steel, or a backing weld run with E7018. Depending on the code or procedure, low-hydrogen electrodes may be used for the root pass on an open root weld. However, if low-hydrogen electrodes are used on an open root, back gouging and back welding may be required. Using a steel backing eliminates the need for back gouging, and is a fast and effective way to prepare a joint for welding. It also makes the root pass easier to run.

The V-groove with steel backing procedure is also the standard AWS qualification test for plate welding. The most common test is the limited-thickness qualification (up to ¾", or 19 mm), which requires the use of ⅜" (10 mm) carbon steel plate and welding with low-hydrogen electrodes.

Backing

Backing can either be left in place or removed after welding. If the backing is to be removed, an R is inserted in the welding symbol. Backing that is to be removed is usually made of a different material than the workpiece, to keep it from being welded into the joint.

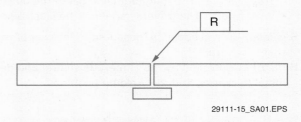

29111-15_SA01.EPS

2.0.1 Groove Weld Positions

Groove welds can be made in all positions. The weld position for plate is determined by the weld and the orientation of the workpiece (*Figure 11*). Groove weld positions for plate are flat, or 1G; horizontal, or 2G; vertical, or 3G; and overhead, or 4G. The G represents a groove weld. In the 1G and 2G positions, the weld axis can be inclined up to 15 degrees. Any weld axis inclination for the other positions varies with the rotational position of the weld face as specified in AWS standards.

2.0.2 Acceptable and Unacceptable Groove Weld Profiles

Groove welds should be made with slight reinforcement and a gradual transition to the base metal at each toe. Groove welds must not have excessive reinforcement, underfill, undercut, or overlap. If a groove weld has any of these defects, it is unacceptable. *Figure 12* shows examples of acceptable and unacceptable groove weld profiles.

> **NOTE**
>
> Refer to your site's WPSs for specific requirements on groove welds. The information in this module is provided as a general guideline only. The site WPS or quality specifications must be followed for all welds. Check with your supervisor if you are unsure of the specifications for your application.

2.1.0 Groove Welding—1G and 2G Positions

The 1G and 2G positions will be practiced first. The following sections describe the types of beads applied, as well as the techniques used to ensure an acceptable weld.

2.1.1 Practicing Flat V-Groove Welds with Backing (1G Position)

Practice flat V-groove welds using ³⁄₃₂", ⅛", or ⁵⁄₃₂" (2.4, 3.2, or 4 mm) E7018 electrodes. Stringer beads or weave beads can be used. When using weave beads, keep the electrode at a 90-degree work angle with a 10- to 15-degree drag angle. When using stringer beads, use a 10- to 15-degree drag angle, but adjust the electrode work angle to tie in the weld on one side or the other as needed. Pay particular attention at the termination of the weld to fill the crater. If the coupon gets too hot between passes, you may have to quench it in water.

CAUTION

Quenching with water is only done on practice coupons. Never cool testing coupons or functional welds with water. Rapid cooling with water can cause weld cracks and affect the mechanical properties of the base metal.

WARNING!

Use pliers and wear gloves to handle hot practice coupons and quench. Steam will rise off the coupon quickly and can burn or scald unprotected hands. Do not cool practice coupons in a sink used for washing hands, or in a water fountain. Use a bucket or bin to avoid clogging the sink with slag and dirt.

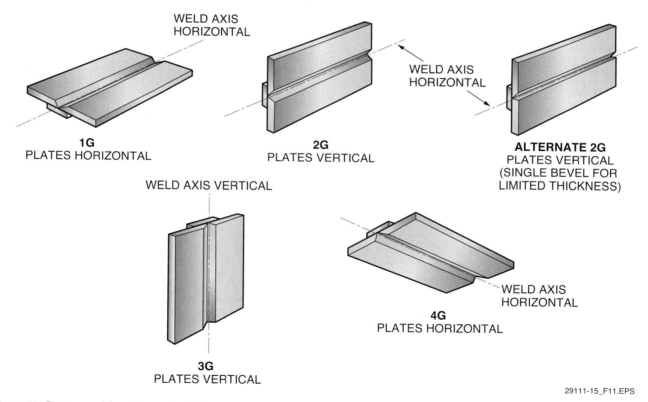

Figure 11 Groove-weld positions for plate.

29111-15_F11.EPS

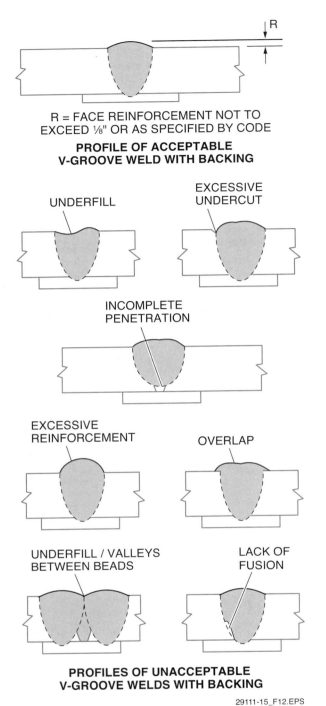

R = FACE REINFORCEMENT NOT TO
EXCEED ⅛" OR AS SPECIFIED BY CODE

**PROFILE OF ACCEPTABLE
V-GROOVE WELD WITH BACKING**

UNDERFILL

EXCESSIVE
UNDERCUT

INCOMPLETE
PENETRATION

EXCESSIVE
REINFORCEMENT

OVERLAP

UNDERFILL / VALLEYS
BETWEEN BEADS

LACK OF
FUSION

**PROFILES OF UNACCEPTABLE
V-GROOVE WELDS WITH BACKING**

29111-15_F12.EPS

Figure 12 Acceptable and unacceptable profiles of V-groove welds with backing.

Follow these steps to practice V-groove welds with metal backing in the flat position:

Step 1 Tack weld the practice coupon together (*Figure 13*).

Step 2 Position the weld coupon flat on the welding table.

Step 3 Run the root pass using E7018 electrodes. *Figure 14* shows the root, filler, and cover pass sequences and work angles. Degrees shown are approximate.

Step 4 Chip, clean, and brush the weld between all passes.

Step 5 Run the fill and cover passes, cleaning between each bead, to complete the weld using E7018 electrodes. Use stringer or weave beads as directed by your instructor.

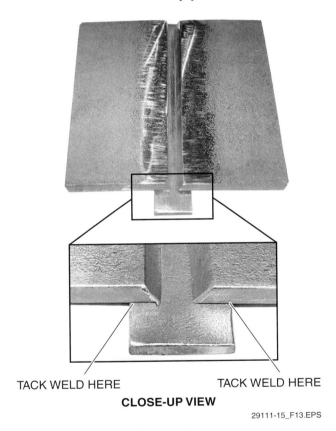

TACK WELD HERE   TACK WELD HERE

CLOSE-UP VIEW

29111-15_F13.EPS

Figure 13 Tack welding the practice coupon.

Step 6 Remove the electrode stub from the electrode holder and discard as soon as you stop welding to prevent arcing to nearby surfaces, and place the stub in a proper container.

Figure 15 shows a flat V-groove weld with typical filler and cover passes.

2.1.2 Horizontal Welds (2G Position)

Horizontal welds can be made with or without backing. This section explains welding horizontal beads and making horizontal V-groove welds with backing.

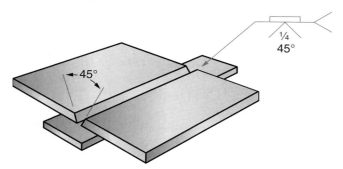

NOTE: The actual number of weld beads will vary depending on the plate thickness and electrode diameter.

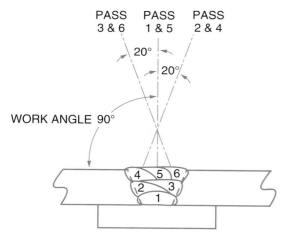

STRINGER BEAD SEQUENCE

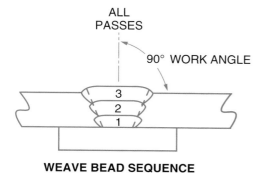

WEAVE BEAD SEQUENCE

29111-15_F14.EPS

Figure 14 Multiple-pass 1G weld sequences and work angles.

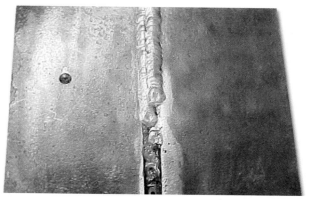

29111-15_F15.EPS

Figure 15 V-groove weld filler and cover pass.

Before welding a V-groove in the horizontal position, practice running horizontal stringer beads by building a pad in the horizontal position. Use ³⁄₃₂", ⅛", or ⁵⁄₃₂" E7018 electrodes. The electrode drag angle should be 10 to 15 degrees, and the work angle should be 0 degrees. To control the weld puddle, the electrode work angle can be dropped slightly, but no more than 10 degrees. Follow these steps to practice welding beads in the horizontal position:

Step 1 Tack weld a flat plate welding coupon in the horizontal position.

Step 2 Run the first pass along the bottom of the coupon using E7018 electrodes.

Step 3 Clean the weld bead with a chipping hammer and wire brush between all passes.

Step 4 Weld the second bead just above the first bead.

Step 5 Continue running beads to complete the pad. *Figures 16* and *17* show the building of a pad in the horizontal position. Degrees shown are approximate.

Step 6 Remove the electrode from the electrode holder and discard as soon as you stop welding to prevent arcing to nearby surfaces, and place the stub in a proper container.

Practice horizontal V-groove welds using ³⁄₃₂", ⅛", or ⁵⁄₃₂" E7018 electrodes and stringer beads. Use a 10- to 15-degree drag angle, but adjust the electrode work angle to tie-in the weld as needed. Pay particular attention at the termination of the weld to fill the crater. The root pass is done using a weave bead.

Follow these steps to practice welding V-groove welds with metal backing in the horizontal position:

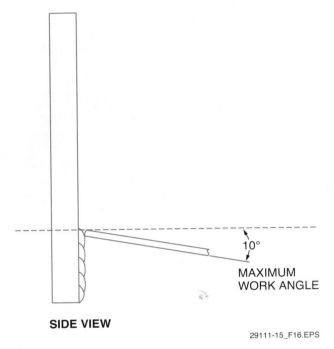

SIDE VIEW

29111-15_F16.EPS

Figure 16 Work angle in the horizontal position.

Step 1 Tack weld the practice coupon together following the example given earlier. Use the standard or alternate horizontal weld coupon as directed by your instructor.

Step 2 Tack weld the weld coupon in the horizontal position.

Step 3 Run the root pass. Note that the first pass, or root pass, is critical because it needs to tie in the backing plate to each of the beveled test coupons (*Figure 18*). A welder's ability to weave in this bead will determine the quality of the root pass.

Step 4 Clean the weld bead with a chipping hammer and wire brush between all passes.

Step 5 Run the fill and cover passes, cleaning the weld between passes, to complete the weld.

Step 6 Remove the electrode from the electrode holder and discard as soon as you stop welding to prevent arcing to nearby surfaces, and place the stub in a proper container.

Figure 19 shows a horizontal V-groove weld with metal backing. Degrees shown are approximate.

29111-15_F17.EPS

Figure 17 A pad built in the horizontal position.

2.2.0 Groove Welding – 3G and 4G Positions

The vertical and overhead welds, positions 3G and 4G respectively, can be more challenging than the 1G and 2G positions.

2.2.1 Vertical Welds (3G Position)

Vertical V-groove welds can be made with or without backing. This section explains welding vertical beads and making vertical V-groove welds with backing.

Before welding a joint in the vertical position, practice running vertical stringer beads by building a pad in the vertical position. Use $^3/_{32}$", $^1/_8$", or $^5/_{32}$" E7018 electrodes. Set the amperage in the lower part of the range for better puddle control. The electrode should be at a 0- to 10-degree push angle with a 90-degree work angle. Use a slight side-to-side motion to control the weld puddle. The width of the stringer beads should be no more than three times the electrode diameter.

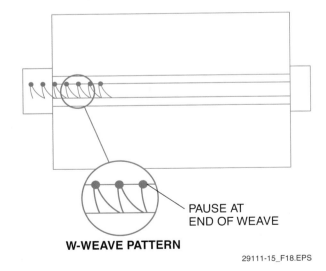

PAUSE AT END OF WEAVE

W-WEAVE PATTERN

29111-15_F18.EPS

Figure 18 Weaving the root pass.

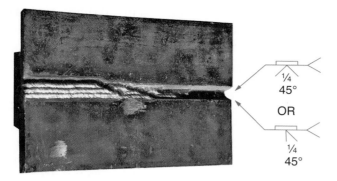

NOTE: The actual number of weld beads will vary depending on the plate thickness and electrode diameter.

Follow these steps to practice welding beads in the vertical position:

Step 1 Tack weld a flat plate welding coupon in the vertical position.

Step 2 Run the first pass along the vertical edge of the coupon using E7018 electrodes.

Step 3 Clean the weld bead with a chipping hammer and wire brush between all passes.

Step 4 Continue running beads to complete the pad. Clean the weld between each pass. *Figure 20* is a diagram showing the travel angle in the vertical position. Degrees shown are approximate.

Step 5 Remove the electrode from the electrode holder and discard as soon as you stop welding to prevent arcing to nearby surfaces, and place the stub in a proper container.

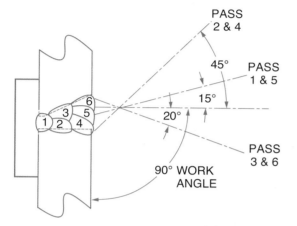

ALTERNATE JOINT REPRESENTATION

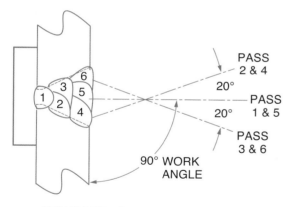

STANDARD JOINT REPRESENTATION

29111-15_F19.EPS

Figure 19 Multiple-pass 2G weld sequences and work angles.

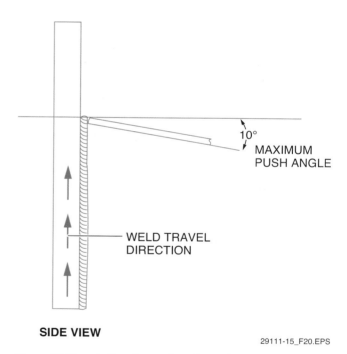

SIDE VIEW

29111-15_F20.EPS

Figure 20 Travel direction and push angles in the vertical (3G) position.

Practice vertical V-groove welds using ³⁄₃₂", ⅛", or ⁵⁄₃₂" E7018 electrodes with stringer beads and/or weave beads. Set the amperage in the lower part of the range for better puddle control. Use a 0- to 10-degree push angle, but adjust the electrode work angle to tie in the weld on one side or the other as needed. If using a weave bead, pause slightly at each toe to penetrate and fill the crater to prevent undercut. Pay particular attention at the termination of the weld to fill the crater.

Follow these steps to practice welding V-groove welds with metal backing in the vertical position:

Step 1 Tack weld the practice coupon together following the example given earlier.

Step 2 Tack weld the weld coupon in the vertical position.

Step 3 Run the root pass with a slight oscillation, pausing at each side.

Step 4 Clean the weld bead with a chipping hammer and wire brush between all passes.

Step 5 Run the fill and cover passes, cleaning each pass, to complete the weld.

Step 6 Remove the electrode from the electrode holder and discard as soon as you stop welding to prevent arcing to nearby surfaces, and place the stub in a proper container.

Figure 21 shows a vertical V-groove weld with metal backing. Degrees shown are approximate.

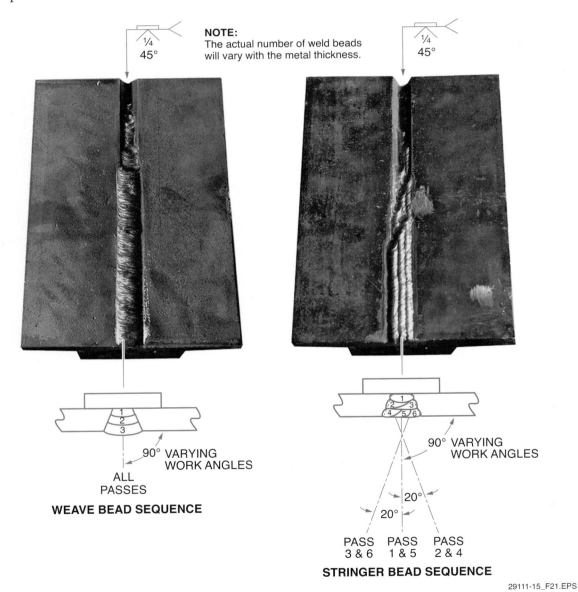

NOTE:
The actual number of weld beads will vary with the metal thickness.

WEAVE BEAD SEQUENCE

STRINGER BEAD SEQUENCE

29111-15_F21.EPS

Figure 21 Multiple-pass 3G weld sequences and work angles.

2.2.2 Overhead Welds (4G Position)

Overhead V-groove welds (4G) can be made with or without backing. This section explains welding overhead beads and making overhead V-groove welds with backing.

Before welding a V-groove in the overhead position, practice running overhead stringer beads by building a pad in the overhead position. Use ³⁄₃₂", ⅛", or ⁵⁄₃₂" E7018 electrodes. The electrode should have a 90-degree work angle and a 10- to 15-degree drag angle.

> **WARNING!**
>
> Try not to stand directly under the coupons while welding because hot slag and molten metal can cause very severe burns. Always wear the proper PPE and any additional clothing required for overhead welding.

Follow these steps to practice welding beads in the overhead position:

Step 1 Tack weld the welding coupon in the overhead position.

Step 2 Run the first pass along the edge of the coupon using E7018 electrodes.

Step 3 Clean the weld bead with a chipping hammer and wire brush between all passes.

Step 4 Continue running and cleaning the beads to complete the pad.

Step 5 Remove the electrode from the electrode holder and discard as soon as you stop welding to prevent arcing to nearby surfaces, and place the stub in a proper container.

Figure 22 is a diagram showing how a pad is built in the overhead position.

Practice making overhead V-groove welds using ³⁄₃₂", ⅛", or ⁵⁄₃₂" E7018 electrodes and stringer beads. Use a 10- to 15-degree drag angle, but adjust the electrode work angle to tie in the weld as needed. Pay particular attention at the termination of the weld to fill the crater.

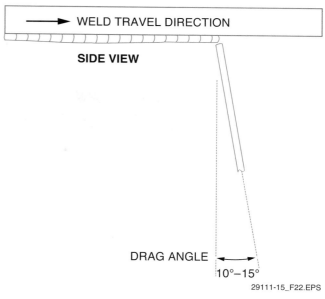

WELD TRAVEL DIRECTION

SIDE VIEW

DRAG ANGLE
10°–15°

29111-15_F22.EPS

Figure 22 Building a pad in the overhead position.

Follow these steps to practice welding V-groove welds with metal backing in the overhead position:

Step 1 Tack weld the practice coupon together following the example given earlier.

Step 2 Tack weld the weld coupon in the overhead position.

Step 3 Run the root pass.

Step 4 Clean the weld bead with a chipping hammer and wire brush between all passes.

Step 5 Run the fill and cover passes, cleaning after each pass, to complete the weld.

Step 6 Remove the electrode from the electrode holder and discard as soon as you stop welding to prevent arcing to nearby surfaces, and place the stub in a proper container.

Figure 23 shows an overhead V-groove weld with metal backing. Degrees shown are approximate.

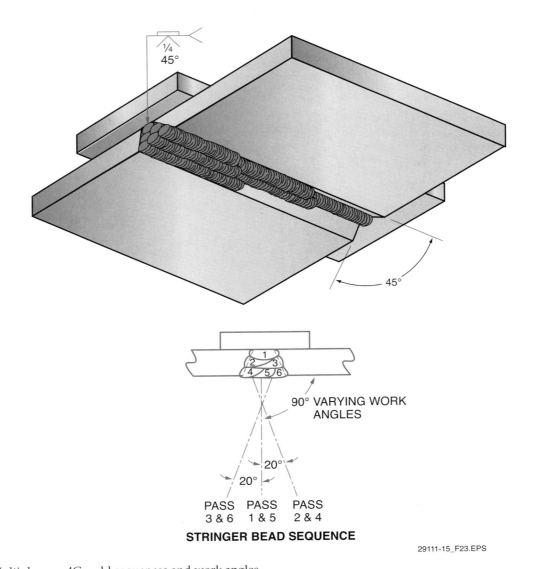

$\frac{1}{4}$
45°

45°

90° VARYING WORK
ANGLES

20°

20°

PASS
3 & 6

PASS
1 & 5

PASS
2 & 4

STRINGER BEAD SEQUENCE

29111-15_F23.EPS

Figure 23 Multiple-pass 4G weld sequences and work angles.

Using Your Bifocals and Trifocals

Many welders have used what are known as cheater lenses in welding hoods for years to help sharpen their vision when the hood is down. A cheater lens is simply a magnification lens. Although many welders wear prescription glasses with bifocal or trifocal lenses, it was impractical to use the glasses due to the design of the hood. Adjustment of the hood's position in front of the face was nonexistent, and the lens size restricted vision.

Newer welding hoods, like the Viking model shown here, allow welders to make use of their bifocal and trifocal prescription glasses. The hood distance from the front of the face can be adjusted in the headgear. This allows the hood to be tilted down farther, even to the chest. The welder can then tilt the head back and forth slightly, using the different focal lengths of the lenses, without losing sight of the task. The significantly larger lens also ensures that the task at hand remains in the welder's field of view.

29111-15_SA02.EPS

Additional Resources

2014 Technical Training Guide. Current Edition. Cleveland, OH. USA: The Lincoln Electric Company. **www.lincolnelectric.com**

Welding Handbook. Current Edition. Miami, FL: The American Welding Society.

2.0.0 Section Review

1. Which of the following can be quenched without concern?

   a. Weld test coupons
   b. Functional welds
   c. Practice coupons
   d. Welds done in the 3G position

2. When V-groove welding overhead, you are welding in the _____.

   a. 1G position
   b. 2F position
   c. 3F position
   d. 4G position

SUMMARY

The ability to make groove welds on plate with metal backing using low-hydrogen electrodes is an essential welding skill. These skills are necessary to progress to more difficult pipe welding procedures in the future. Several groove-weld styles and combination styles, including V-grooves with backing, were covered in this module.

Proper preparation of the equipment, base metal, and backing is important for making quality groove welds. Welds in the vertical and overhead positions are typically required for welder certification on plate.

Groove welds are made in the flat (1G), horizontal (2G), vertical (3G), and overhead (4G) positions. It is important to commit these position designations to memory. Practice these welds until you can consistently produce acceptable welds.

1. If a groove for a weld allows for the entire thickness of the base metals to be welded, it is called a(n) _____.

   a. partial joint penetration (PJP) weld
   b. complete joint penetration (CJP) weld
   c. axial joint penetration (AJP) weld
   d. subsurface joint penetration (SJP) weld

2. The portion of a joint where the two separate components to be welded are the closest to each other is called the _____.

   a. joint root
   b. depth of bevel
   c. groove radius
   d. weld toe

3. When preparing welding coupons for V-groove welds with backing, the backing should extend beyond the end of the coupons _____.

   a. ⅛" (3.2 mm)
   b. ¼" (6.4 mm)
   c. ½" (12.7 mm)
   d. ¾" (19.1 mm)

4. Which of the following is an *incorrect* statement about running low-hydrogen electrodes?

   a. You should maintain a short arc.
   b. You should take only a few electrodes from the oven at a time.
   c. You should always whip the electrode.
   d. You should not use electrodes with missing flux.

5. The V-groove weld with backing is a common weld normally made with _____.

   a. fast-fill electrodes
   b. fill-freeze electrodes
   c. fast-freeze electrodes
   d. low-hydrogen electrodes

6. A way to eliminate the need for back gouging, and to quickly and effectively prepare a joint for welding, is to use _____.

   a. tape backing
   b. steel backing
   c. a backing weld run with E7018 electrodes
   d. a backing weld run with fast-freeze electrodes

7. When using weave beads on flat V-groove welds, keep the electrode at a 10- to 15-degree drag angle and a work angle of _____.

   a. 0 degrees
   b. 45 degrees
   c. 60 degrees
   d. 90 degrees

8. When practicing horizontal (2G) V-groove welding using stringer beads, the work angle should be _____.

   a. 0 degrees
   b. 45 degrees
   c. 90 degrees
   d. adjusted as needed to ensure tie-in

9. When using stringer beads on V-groove welds in the vertical (3G) position, keep the electrode at a _____.

   a. 0-degree push angle at all times
   b. 0- to 10-degree push angle
   c. 45-degree push angle
   d. 90-degree push angle

10. Before welding a V-groove in the overhead position, you should _____.

    e. change to E6010 electrodes
    a. increase the welding current significantly
    b. build a pad with stringer beads in the ~~horizontal~~ overhead position
    c. change to AC welding current

Appendix

PERFORMANCE ACCREDITATION TASKS

The American Welding Society (AWS) School Excelling through National Skills Standards Education (SENSE) program is a comprehensive set of minimum Standards and Guidelines for Welding Education programs. The following performance accreditation tasks are aligned with and designed around the SENSE program.

The Performance Accreditation Tasks (PATs) correspond to and support the learning objectives in *AWS EG2.0, Guide for the Training and Qualification of Welding Personnel: Entry-Level Welder*.

Note that in order to satisfy all learning objectives in *AWS EG2.0*, the instructor must also use the PATs contained in the second level of the NCCER Welding curriculum.

PATs 1 through 4 correspond to *AWS EG2.0, Module 4—Shielded Metal Arc Welding*, Key Indicators 3, 4, and 6.

PATs provide specific acceptable criteria for performance and help to ensure a true competency-based welding program for students.

The following tasks are designed to develop your competency in preparing base metal. Practice each task until you are thoroughly familiar with the procedure.

As you complete each task, take it to your instructor for evaluation. Do not proceed to the next task until instructed to do so by your instructor.

V-GROOVE WELDS WITH BACKING IN THE FLAT (1G) POSITION

Using ³⁄₃₂", ⅛", or ⁵⁄₃₂" (2.4, 3.2, or 4.0 mm) E7018 electrodes, make a V-groove weld with steel backing on carbon steel plate in the flat position as indicated.

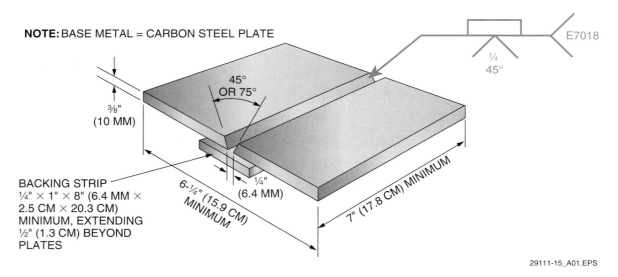

NOTE: BASE METAL = CARBON STEEL PLATE

45° OR 75°

E7018

¼
45°

³⁄₈" (10 MM)

BACKING STRIP
¼" × 1" × 8" (6.4 MM ×
2.5 CM × 20.3 CM)
MINIMUM, EXTENDING
½" (1.3 CM) BEYOND
PLATES

6-¼" (15.9 CM) MINIMUM

¼" (6.4 MM)

7" (17.8 CM) MINIMUM

29111-15_A01.EPS

Inspection Hold Points:

- Fit-up   _____
- Root pass   _____
- Final   _____

Criteria for Acceptance:

- No arc strikes outside the weld area   _____
- Uniform rippled appearance on the bead face with no valley between the beads and acceptable tie-in   _____
- Craters and restarts filled to the full cross section of the weld   _____
- Uniform weld size ±¹⁄₁₆" (1.6 mm)   _____
- Acceptable weld profile in accordance with _AWS D1.1_   _____
- Smooth transition with complete fusion at the toes of the weld   _____
- No porosity   _____
- No overlap   _____
- No excessive undercut   _____
- No inclusions   _____
- No cracks   _____

Note to Instructor: Make sure all trainees complete each pass before starting another.

V-GROOVE WELDS WITH BACKING IN THE HORIZONTAL (2G) POSITION

Using ³⁄₃₂", ⅛", or ⁵⁄₃₂" (2.4, 3.2, or 4.0 mm) E7018 electrodes, make a V-groove weld with steel backing on carbon steel plate in the horizontal position as indicated.

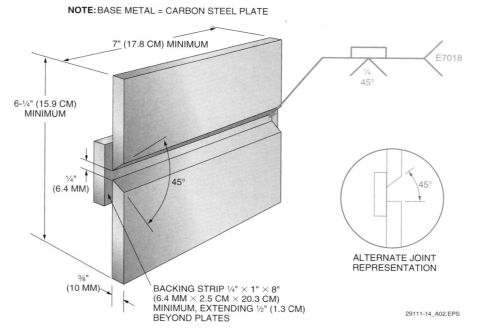

NOTE: BASE METAL = CARBON STEEL PLATE

7" (17.8 CM) MINIMUM

6-¼" (15.9 CM) MINIMUM

¼" (6.4 MM)

45°

³⁄₈" (10 MM)

BACKING STRIP ¼" × 1" × 8" (6.4 MM × 2.5 CM × 20.3 CM) MINIMUM, EXTENDING ½" (1.3 CM) BEYOND PLATES

E7018

¼
45°

ALTERNATE JOINT REPRESENTATION

29111-14_A02.EPS

Inspection Hold Points:

- Fit-up   _____
- Root pass   _____
- Final   _____

Criteria for Acceptance:

- No arc strikes outside the weld area   _____
- Uniform rippled appearance on the bead face with no valley between the beads and acceptable tie-in   _____
- Craters and restarts filled to the full cross section of the weld   _____
- Uniform weld size ±¹⁄₁₆" (1.6 mm)   _____
- Acceptable weld profile in accordance with *AWS D1.1*   _____
- Smooth transition with complete fusion at the toes of the weld   _____
- No porosity   _____
- No excessive undercut   _____
- No overlap   _____
- No inclusions   _____
- No cracks   _____
- Acceptable guided bend test results per *AWS QC-10*   _____

V-GROOVE WELDS WITH BACKING IN THE VERTICAL (3G) POSITION

Using ³⁄₃₂", ¹⁄₈", or ⁵⁄₃₂" (2.4, 3.2, or 4.0 mm) E7018 electrodes, make a V-groove weld with steel backing on carbon steel plate in the vertical position as indicated.

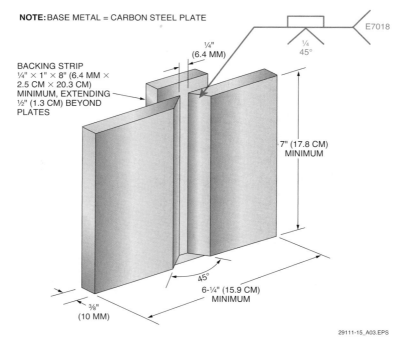

NOTE: BASE METAL = CARBON STEEL PLATE

E7018

¹⁄₄"
(6.4 MM)

¹⁄₄
45°

BACKING STRIP
¹⁄₄" × 1" × 8" (6.4 MM ×
2.5 CM × 20.3 CM)
MINIMUM, EXTENDING
¹⁄₂" (1.3 CM) BEYOND
PLATES

7" (17.8 CM)
MINIMUM

45°

6-¹⁄₄" (15.9 CM)
MINIMUM

³⁄₈"
(10 MM)

29111-15_A03.EPS

Inspection Hold Points:

- Fit-up
- Root pass
- Final

Criteria for Acceptance:

- No arc strikes outside the weld area
- Uniform rippled appearance on the bead face with no valley between the beads and acceptable tie-in
- Craters and restarts filled to the full cross section of the weld
- Uniform weld size ±¹⁄₁₆" (1.6 mm)
- Acceptable weld profile in accordance with *AWS D1.1*
- Smooth transition with complete fusion at the toes of the weld
- No porosity
- No excessive undercut
- No overlap
- No inclusions
- No cracks
- Acceptable guided bend test results per *AWS QC-10*

V-GROOVE WELDS WITH BACKING IN THE OVERHEAD (4G) POSITION

Using ³⁄₃₂", ⅛", or ⁵⁄₃₂" (2.4, 3.2, or 4.0 mm) E7018 electrodes, make a V-groove weld with steel backing on carbon steel plate in the overhead position as indicated.

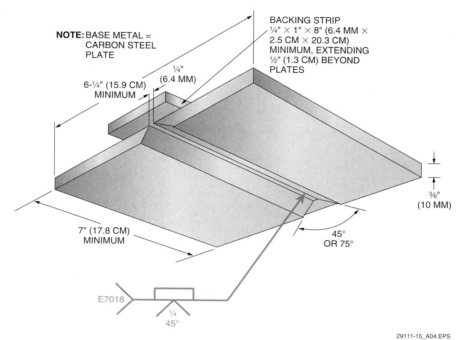

NOTE: BASE METAL = CARBON STEEL PLATE

BACKING STRIP ¼" × 1" × 8" (6.4 MM × 2.5 CM × 20.3 CM) MINIMUM, EXTENDING ½" (1.3 CM) BEYOND PLATES

¼" (6.4 MM)

6-¼" (15.9 CM) MINIMUM

7" (17.8 CM) MINIMUM

45° OR 75°

³⁄₈" (10 MM)

E7018

¼
45°

29111-15_A04.EPS

Inspection Hold Points:

- Fit-up
- Root pass
- Final

Criteria for Acceptance:

- No arc strikes outside the weld area
- Uniform rippled appearance on the bead face with no valley between the beads and acceptable tie-in
- Craters and restarts filled to the full cross section of the weld
- Uniform weld size ±¹⁄₁₆" (1.6 mm)
- Acceptable weld profile in accordance with *AWS D1.1*
- Smooth transition with complete fusion at the toes of the weld
- No porosity
- No excessive undercut
- No overlap
- No inclusions
- No cracks
- Acceptable guided bend test results per *AWS QC-10*

Tom Ashley
Robins and Morton, Birmingham, AL
Quality Assurance/Quality Control Manager

Following his discharge from military service, Tom went to college and trained and worked as a paramedic and firefighter. He subsequently served as a police officer before becoming a welder. That was the end of the career changes. He eventually became a welding inspector and then moved into management as manager of QA/QC.

How did you choose a career in the construction industry?
My father gave me a strong push to learn a trade and suggested that I consider a career in construction.

Who inspired you to enter the industry?
In addition to my father, my grandfather was a strong inspiration.

What types of training have you been through?
I did three years of welding school while serving a pipefitting apprenticeship.

How important are education and training in construction?
Education and training are essential. A person can learn to run beads without formal training, but will not achieve the level of knowledge and skill that lead to a well-paying, secure career. A welder who is well-trained will create a finished product that complies with industry standards.

How important are NCCER credentials to your career?
NCCER credentials provide the means to evaluate a person's knowledge and ability and provide a prescription for additional training. If a craft worker earns a Certified Plus credential, for example, you can be assured that he or she meets the journey-level standards of the craft.

How has training/construction impacted your life?
Training has given me the satisfaction of seeing a project progress from concept to completion. Being involved in multiple projects also provides some very rewarding challenges.

What kinds of work have you done in your career?
Following my discharge form the military, I went to school and became a firefighter and paramedic, then later became a police officer. None of those jobs gave me the satisfaction I needed, so I decided to become a welder.

Tell us about your present job.
I'm QA/QC manager for the Power and Industrial Division of my company. In that position, I supervise inspectors in all disciplines and train and coordinate other QA/QC managers.

What do you enjoy most about your job?
I enjoy working in a variety of disciplines because every day is different and every day presents a new challenge.

What factors have contributed most to your success?
I have to say it's my willingness to learn all disciplines, keeping an open mind, good communication skills, and a strong work ethic.

Would you suggest construction as a career to others? Why?
Absolutely! A construction career can be very rewarding because you are only limited by your desire and level of commitment. Many leaders and business owners in the construction industry started out on the ground floor and worked their way up.

What advice would you give to those new to the welding field?
Stay focused on your goals, get all the training you can, and listen to the advice of those with more experience.

Tell us an interesting career-related fact or accomplishment.
I started out as a public servant with no intention of entering the construction industry. Once I did, I came to love the field and I realized that it's what I'm best at.

How do you define craftsmanship?
Craftsmanship is taking pride in your work and completing it safely. It's creating a product that meets applicable specifications and standards, as well as expectations for its intended use and reliability. lines, boilers—you name it. I've performed work in accordance with a wide variety of welding codes.

Additional Resources

This module presents thorough resources for task training. The following resource material is suggested for further study.

2014 Technical Training Guide. Current Edition. Cleveland, OH. USA: The Lincoln Electric Company. **www.lincolnelectric.com**

Welding Handbook. Current Edition. Miami, FL: The American Welding Society.

Figure Credits

The Lincoln Electric Company, Cleveland, OH, USA, Module Opener, Figures 6, 10, 17, SA02

Topaz Publications, Inc., Figures 7 (photo), 19 (photo), 21 (photo)

Zachry Industrial, Inc., Figures 13, 15

Section Review Answer Key

Answer	Section Reference	Objective
Section One		
1. b	1.1.3	1a
2. d	1.2.1	1b
Section Two		
1. c	2.1.1	2a
2. d	2.2.2	2b

NCCER CURRICULA — USER UPDATE

NCCER makes every effort to keep its textbooks up-to-date and free of technical errors. We appreciate your help in this process. If you find an error, a typographical mistake, or an inaccuracy in NCCER's curricula, please fill out this form (or a photocopy), or complete the online form at **www.nccer.org/olf**. Be sure to include the exact module ID number, page number, a detailed description, and your recommended correction. Your input will be brought to the attention of the Authoring Team. Thank you for your assistance.

Instructors – If you have an idea for improving this textbook, or have found that additional materials were necessary to teach this module effectively, please let us know so that we may present your suggestions to the Authoring Team.

NCCER Product Development and Revision
13614 Progress Blvd., Alachua, FL 32615

Email: curriculum@nccer.org
Online: www.nccer.org/olf

❏ Trainee Guide   ❏ Lesson Plans   ❏ Exam   ❏ PowerPoints   Other _____

Craft / Level: _____   Copyright Date: _____

Module ID Number / Title: _____

Section Number(s): _____

Description: _____

_____

_____

_____

_____

Recommended Correction: _____

_____

_____

_____

_____

Your Name: _____

Address: _____

_____

Email: _____   Phone: _____

SMAW – Open-Root Groove Welds – Plate

OVERVIEW

SMAW groove welds in plate can be done with or without a backing. When a backing is not used, there is no material on the opposite side of the weld to prevent molten metal from passing through the groove. There is also less mass involved in the welded pieces. These differences, together with the small differences in technique and fit-up dimensions, make the process different from welding plate with a backing. This module focuses on those differences and on the practicing of SMAW on open-root grooves in plate.

Module Twelve

Trainees with successful module completions may be eligible for credentialing through the NCCER Registry. To learn more, go to **www.nccer.org** or contact us at **1.888.622.3720**. Our website has information on the latest product releases and training, as well as online versions of our *Cornerstone* magazine and Pearson's product catalog.

Your feedback is welcome. You may email your comments to **curriculum@nccer.org**, send general comments and inquiries to **info@nccer.org**, or fill in the User Update form at the back of this module.

This information is general in nature and intended for training purposes only. Actual performance of activities described in this manual requires compliance with all applicable operating, service, maintenance, and safety procedures under the direction of qualified personnel. References in this manual to patented or proprietary devices do not constitute a recommendation of their use.

Objectives

When you have completed this module, you will be able to do the following:

1. Identify various types of groove welds and describe how to prepare for groove welding.
   a. Identify various types of groove welds and define related terms.
   b. Describe how to prepare the work area and plate for groove welding.
2. Describe the technique required to produce various open V-groove welds.
   a. Describe the technique required to produce open V-groove welds in the 1G and 2G positions.
   b. Describe the technique required to produce open V-groove welds in the 3G and 4G positions.

Performance Tasks

Under the supervision of your instructor, you should be able to do the following:

1. Make open V-groove welds with E6010 and E7018 electrodes in the following positions:
   - Flat (1G) position
   - Horizontal (2G) position
   - Vertical (3G) position
   - Overhead (4G) position

Trade Terms

Feather
Root face

Industry Recognized Credentials

If you are training through an NCCER-accredited sponsor, you may be eligible for credentials from NCCER's Registry. The ID number for this module is 29112-15. Note that this module may have been used in other NCCER curricula and may apply to other level completions. Contact NCCER's Registry at 888.622.3720 or go to **www.nccer.org** for more information.

Contents

Topics to be presented in this module include:

Figures

SECTION ONE

1.0.0 OPEN-GROOVE WELDS

Objective

Identify various types of groove welds and describe how to prepare for groove welding.

   a. Identify various types of groove welds and define related terms.

   b. Describe how to prepare the work area and plate for groove welding.

Trade Term

Root face: The portion of the edge of a part to be joined by a groove weld that has not been beveled or grooved; previously referred to as the land.

This module covers open V-groove welds, which are commonly used for welding plate. The open V-groove weld is easy to prepare because it only requires single-face beveling of the matching members. Without backing, however, welding the root pass requires practice to perfect. This module focuses on V-groove welds without a backing.

This module will explain the joint preparation and performance of open-root V-groove welds on plate using E6010 and E7018 electrodes in all positions. It will also prepare the welder to advance to the more difficult pipe welding modules. Performance demonstrations will require that the welds meet the qualification standards in the *ASME Boiler and Pressure Vessel Code, Section IX – Welding and Brazing Qualifications* for visual and destructive testing. Always remember that quality control begins with the welder.

1.1.0 Typical Groove Styles

A groove weld is made in an opening in a part or between two parts. Groove welds can be made in all five of the basic types of joints as well as on the surface of a part. There are several different groove styles, the name of the groove is derived from its shape. Some must be prepared; others appear as a result of fitting together the parts to be welded together.

Note that, if a groove only goes part of the way through the joint, welding in the groove results in what is called a partial joint penetration (PJP) weld. If the groove allows for welding the complete thickness of the base metals, it is called a complete joint penetration (CJP) weld.

1.1.1 Single- and Double-Groove Welds

Groove weld styles include the square, V, U, bevel, J, flare V, and flare bevel. *Figure 1* shows typical single-groove weld styles.

V-, bevel, U-, and J-grooves can be used and welded on one side of a joint or on both sides. If they are used on both sides of a joint, they are called double-groove welds (*Figure 2*).

> **NOTE**
>
> Throughout this module, both metric and Imperial units have been provided for most measurements. Weld symbols are generally provided in one measurement system or another, but not both. For this reason, metric values have not been placed alongside Imperial values in weld symbols.

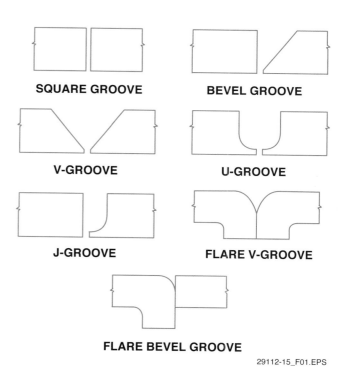

Figure 1 Single-groove weld styles.

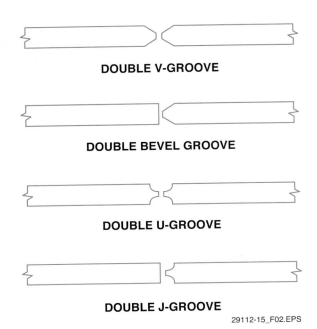

DOUBLE V-GROOVE

DOUBLE BEVEL GROOVE

DOUBLE U-GROOVE

DOUBLE J-GROOVE

29112-15_F02.EPS

Figure 2 Double-groove weld styles.

1.1.2 Combination Groove and Fillet Welds

Groove welds are often used in combination with fillet welds to maintain the strength of the welded joint and use less weld metal. *Figure 3* shows two groove and fillet weld combinations.

1.1.3 Groove Weld Terms

Part of a welder's certification includes the proper understanding of related welding terms. The terms associated with a groove are the following:

- *Bevel angle* – The angle formed between the prepared edge of a member and a line perpendicular to the surface of the member.
- *Bevel face* – The prepared surface of a bevel edge shape.
- *Bevel radius* – The radius used to form the shape of a J- or U-groove.
- *Depth of bevel* – The distance the joint preparation extends into the base metal (used for partial joint penetration welds).
- *Groove angle* – The included (total) angle between the groove faces of a weld groove.
- *Groove face* – Any surface in a weld groove prior to welding. See *bevel face*, root face, and *bevel radius*.
- *Joint root* – The portion of a joint where the members approach closest to each other.
- *Root face* – The portion of the edge of a part to be joined by a groove weld that has not been beveled or grooved; previously referred to as the land.
- *Root opening* – The separation between the members to be joined at the root of the joint.

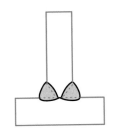

TEE DOUBLE BEVEL

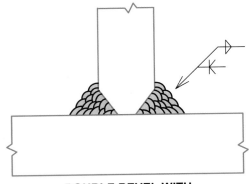

DOUBLE BEVEL WITH MULTIPLE PASS FILLET WELDS

29112-15_F03.EPS

Figure 3 Groove and fillet weld combinations.

Figure 4 shows examples of grooves and their named parts.

The terms used to describe features of a groove weld are as follows:

- *Face reinforcement* – Reinforcement of a joint done from the same side as that from which the weld was made.
- *Groove weld size* – The joint penetration (depth of joint preparation plus root penetration). The size can never be greater than the base metal thickness.
- *Weld face* – The exposed surface of the weld.
- *Weld root* – The point at which the back of the weld extends the farthest into the weld joint.
- *Weld toe* – The points where the face of the weld metal and the base metal meet.

Figure 5 shows the common dimensions of an open-root V-groove weld.

> **NOTE**
>
> The dimensions and specifications in this module are designed to be representative of codes in general and are not specific to any particular code. Always follow the Welding Procedure Specification (WPS) and proper codes for your site.

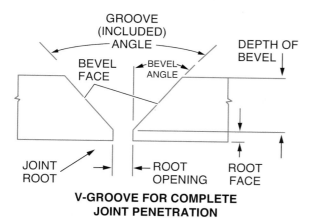

V-GROOVE FOR COMPLETE
JOINT PENETRATION

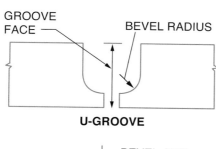

U-GROOVE

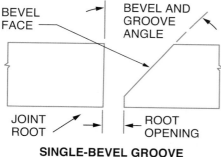

SINGLE-BEVEL GROOVE

29112-15_F04.EPS

Figure 4 Grooves.

1.2.0 Welding Preparations

Before welding can take place, the area has to be made ready, the welding equipment must be set up, and the metal to be welded must be prepared. The following sections explain how to set up equipment for SMAW.

1.2.1 *Safety Practices*

The following is a summary of safety procedures and practices that must be observed when cutting or welding; complete safety coverage is provided in the *Welding Safety* module. If you have not completed that module, do so before continuing. Above all, be sure to wear appropriate protective clothing and equipment when welding or cutting.

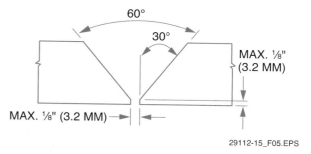

29112-15_F05.EPS

Figure 5 Common open-root V-groove dimensions.

To maintain safety and prevent injury, it is essential that you wear the appropriate protective clothing and equipment when cutting or welding metals (*Figure 6*). Be sure to follow these guidelines during all phases of cutting or welding:

- Always use safety glasses along with a full face shield or a helmet. The glasses, face shield, or helmet lens must have the proper light-reducing tint for the type of welding or cutting being performed. Never directly or indirectly view an electric arc without using a properly tinted lens.
- Wear proper protective leather and/or flame retardant clothing along with welding gloves that will protect you from flying sparks and molten metal as well as heat.
- Wear high-top safety shoes or boots. Make sure that the tongue and lace area of the footwear will be covered by a pant leg. If the tongue and lace area is exposed, or if the footwear must be protected from burn marks, wear leather spats under your pants or chaps to cover the front top of the footwear.

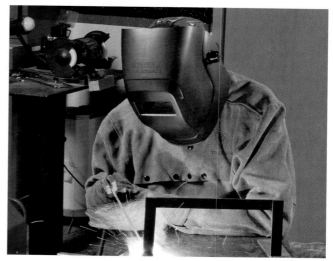

29112-15_F06.EPS

Figure 6 Welder working in full protective equipment.

- Wear a 100-percent cotton cap with no mesh material included in its construction. The bill of the cap points to the rear. If a hard hat is required for the environment, use one that allows the attachment of rear deflector material and a face shield. A hard hat with a rear deflector is generally preferred when working overhead, and may be required by some employers and job sites.

> **WARNING!**
> Do not wear a cap with a button in the middle. The conductive metal button beneath the fabric represents a safety hazard.

- Wear a face shield and snug-fitting safety glasses for gas welding or cutting. Either the face shield or the lenses of the welding glasses must be an approved shade for the application. For electric arc welding or cutting, wear safety glasses and a welding hood with the correct tinted lens.
- Wear earplugs to protect ear canals from sparks. Wear hearing protection to protect against the consistent sound of the torch.

> **WARNING!**
> Ear protection is essential to protect ears from the noise of the torch. Other personal protective equipment (PPE) must be worn to protect the operator from hot metal and slag.

- Do not open your welding hood lens and expose unprotected eyes when grinding or cleaning a weld. Use either a hood that maintains a protective lens when the shaded lens is opened, a hood with an auto-shading lens, or remove the hood and use a full-face shield.

Welding activities involve the use of fire or extreme heat to melt metal. Whenever fire is used, it must be controlled and contained. Welding activities are often performed on vessels that may once have contained flammable or explosive materials. Residues from those materials can catch fire or explode when a welder begins work on such a vessel. The following are fire and explosion prevention guidelines associated with welding:

- Never carry matches or gas-filled lighters in your pockets. Sparks can cause the matches to ignite or the lighter to explode, causing serious injury.
- Always comply with all site and/or employer requirements for a hot-work permit and a fire watch.
- Never use oxygen to blow dust or dirt off clothing. The oxygen can remain trapped in the fabric for a time. If a spark hits the oxygen in the fabric, the clothing can burn rapidly and violently.
- Make sure that any flammable material in the work area is moved or shielded by a fire-resistant covering.
- Approved fire extinguishers must be available before attempting any heating, welding, or cutting operations. Make sure the extinguisher is charged, the inspection tag is valid, and any individual that may be required to operate it knows how to do so.
- Never release a large amount of oxygen or use oxygen in place of compressed air. The presence of oxygen around flammable materials or sparks can cause rapid and uncontrolled combustion. Keep oxygen away from oil, grease and other petroleum products.
- Never release a large amount of fuel gas, especially acetylene. Methane and propane are heavier than air and tend to migrate to and concentrate in low areas. As a result, they can ignite at a considerable distance from the release point. Acetylene is lighter than air but is even more dangerous than methane; when mixed with air or oxygen, it will explode at much lower concentrations than any other common fuel gas.
- To prevent fires, maintain a neat and clean work area, and make sure that any metal scrap or slag is cold before disposal.

Before cutting or welding containers such as tanks or barrels, find out if they contained any explosive, hazardous, or flammable materials, including petroleum products, citrus products, or chemicals that decompose into toxic fumes when heated. Proper procedures for cutting or welding hazardous containers are described in the *American Welding Society (AWS) F4.1, Safe Practices for the Preparation of Containers and Piping for Welding and Cutting,* and *ANSI Z49.1.* As a standard practice, always clean and then fill any tanks or barrels with water, or purge them with a flow of inert gas such as nitrogen to displace any oxygen.

Containers must be cleaned by steam cleaning, flushing with water, or washing with detergent until all traces of the material have been removed.

After cleaning the container, fill it with water or a purging gas, such as carbon dioxide, argon, or nitrogen to displace the explosive fumes. Air, which contains oxygen, is displaced from inside the container by the water or inert gas. Without oxygen, combustion cannot take place.

A water-filled vessel is the best alternative. When using water, position the container to minimize the air space. When using an inert gas, provide a vent hole so the inert gas can push the air and other vapors out to the atmosphere. Keep in mind, though, that even these precautions do not guarantee the absence of flammable materials inside. For that reason, these types of activities should not be done without proper supervision and the use of proper testing methods.

GOING GREEN

Welding Water

To better protect the environment, send any water used in a welding process, or water used to clean tanks, to a waste treatment facility instead of simply allowing it to run into a storm drain or onto the ground.

Vapors and fumes tend to rise in the air from their sources. Welders often have to work above welding areas where fumes are being created. Welding fumes can be harmful. Good work area ventilation helps to remove the vapors and protect the welder. Always perform cutting or welding operations in a well-ventilated area. Cutting or welding operations involving zinc or cadmium materials or coatings result in toxic fumes. For long-term cutting or welding of such materials, always wear an approved full-face, supplied-air respirator (SAR) that uses breathing air supplied from outside the work area. For occasional, very short-term exposure, a HEPA-rated or metal fume filter may be used on a standard respirator.

1.2.2 Preparing the Welding Area

To practice welding, a welding table, bench, or stand is needed (*Figure 7*). The welding surface must be steel, and provisions must be made for placing welding coupons out of position.

To set up the area for welding, follow these steps:

Step 1 Make sure that the area is properly ventilated. Make use of doors, windows, and fans.

Step 2 Check the area for fire hazards. Remove any flammable materials before proceeding.

29112-15_F07.EPS

Figure 7 Welding station with vacuum system.

Step 3 Know the location of the nearest fire extinguisher. Do not proceed unless the extinguisher is charged and you know how to use it.

Step 4 Position a welding table near the welding machine, or vice versa.

Step 5 Set up welding curtains around the welding area.

1.2.3 Preparing the Weld Coupons

If possible, the weld coupons should be ⅜" (9.6 mm) thick carbon steel. If this size is not readily available, ¼" to ¾" (6.4 to 19.1 mm) thick steel can be used for practice welds. Open-root welds generally have a bevel angle of 30 degrees. Clean the steel plate prior to welding by using a wire brush or grinder to remove heavy mill scale or corrosion.

> **WARNING!**
>
> Be extremely careful when using wire brush attachments on powered tools. Always ensure that the rotational speed rating of the wire brush matches or exceeds the rated speed of the powered tool. Loose wires from the brush and debris tend to fly off and embed into anything or anyone near the work. Make sure that all appropriate PPE is worn when wire brushing or grinding with all tools, but be especially cautious with powered tools.

Prepare weld coupons to practice the following welds:

- *Stringer bead welds* – The coupons can be any size or shape that is easily handled.
- *Open-root V-groove welds* – As shown in *Figure 8*, cut the metal into 3" × 7" (7.6 cm × 17.8 cm) rectangles with a bevel of 30 degrees on one of the long sides. Grind up to a ⅛" (3.2 mm) root face on the bevel, or leave without a root face. Your instructor may have a preference.

> **NOTE**
>
> The welding codes allow the root face and opening on open-root welds to be 0" to ⅛" (0 to 3.2 mm). Adjust the root face and root opening as needed when you start your welding practices.

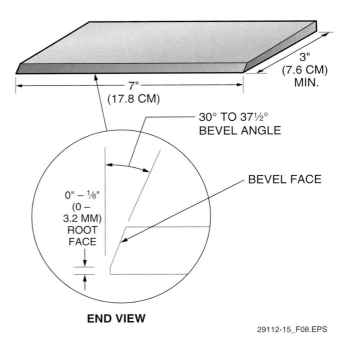

END VIEW

29112-15_F08.EPS

Figure 8 Metal cut for open-root V-groove weld coupons.

Follow these steps to prepare each V-groove weld coupon:

Step 1 Check the bevel face. There should be no dross and a 0" to ⅛" (0 to 3.2 mm) root face, and the bevel angle should be 30 degrees. An example of this is shown in *Figure 9*.

Step 2 Remove any burrs from the edges of the coupon.

Step 3 Align the beveled strips on a flat surface with a 0" to ⅛" (0 to 3.2 mm) root opening and tack weld them on each end. Use three to four ¼" (6.4 mm) long tack welds.

1.2.4 Electrodes

Obtain a small quantity of the electrodes to be used. For the welding exercises in this module, E6010 and E7018 electrodes will be used.

> **GOING GREEN**
>
> **Vapor Extraction Systems**
>
> To better protect the environment, perform any welding activities under a vapor extraction system that can filter the welding fumes before they reach the atmosphere.

For E6010 electrodes, use a whipping motion when depositing the stringer bead to control the weld puddle, then remove all slag between passes.

For E7018 electrodes, keep the following in mind:

- Remove only a small number of electrodes at a time from the oven because low-hydrogen electrodes absorb moisture quickly.
- Do not whip low-hydrogen electrodes.
- Maintain a short arc because a low-hydrogen weld relies on its slag to protect the molten metal. It does not have a heavy gaseous shield.
- Remove all slag between passes.
- Do not use low-hydrogen electrodes with chipped or missing flux.

For either E6010 or E7108 electrodes, observe the following:

- When running stringer beads, use a slight side-to-side motion and a small pause at the weld toe to tie-in and flatten the bead.
- When running weave beads, use a slight pause at the weld toes to tie-in to the base metal.
- Set the amperage in the lower portion of the suggested range for better puddle control when welding vertically.

Obtain only the electrodes to be used for a particular welding exercise at one time. Use some type of pouch or rod holder to store the electrodes in to prevent them from becoming damaged.

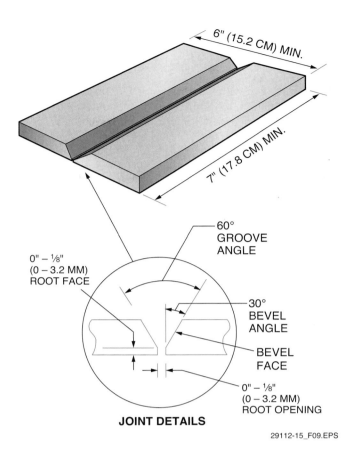

JOINT DETAILS

29112-15_F09.EPS

Figure 9 Open-root V-groove weld coupon.

Never store electrodes loose on a table. They may fall on the floor where they can become a tripping hazard or become damaged. A pail or other metal container must also be available to discard hot electrode stubs.

Recycle Metal

GOING GREEN

Steel for practice welding is expensive and difficult to obtain. Make every effort to conserve and not waste the material available. Reuse weld coupons until all surfaces have been welded upon by cutting the weld coupon apart and reusing the pieces. Use material that cannot be cut into weld coupons to practice running beads. When beveling plates for coupons, bevel both of the long edges to help conserve material.

Salvage any coupons or scrap metal that is no longer usable for welding practice and place it in a recycling bin. Almost all metals are recyclable and valuable. One technical school, for example, collected donated steel, used it for cutting and welding practice, and then sold all their recyclable metals to fund their school's annual picnic. The bottom line is that metal no longer usable for cutting or welding practice can be recycled and the proceeds are put to good use.

Electrode Housekeeping

A floor littered with electrode stubs is the site of an accident waiting to happen. Not only does throwing electrode stubs on the floor or ground create a hazard, it can also result in disciplinary action from the employer. In some cases, a worker may even lose his or her job for this infraction.

1.2.5 Preparing the Welding Machine

Welders can expect to find different types and makes of welding machines in welding shops and on jobs in the field. Many machines will have not have the operator manual with them. As a trainee, and even as a trained welder, you must be able to recognize the different types of welding machines and figure out how to safely set up and operate them. *Figure 10* shows typical welding machines.

Select a welding machine to use. Proceed with the following welding set-up steps:

Step 1 Verify that the welding machine can be used for DC (direct current) welding.

Step 2 Check the area for proper ventilation.

Step 3 Verify the location of the power disconnect for the welding machine.

Step 4 Set the polarity to direct current electrode positive (DCEP).

Step 5 Connect the clamp of the workpiece lead to the workpiece.

Running E7018 Electrodes

Moving an electrode back and forth makes a weave bead. Many different patterns can be used to make a weave bead, including circles, crescents, and zigzags. The pause at the edges will also flatten out the weld, giving it the proper profiles.

Step 6 Set the amperage for the electrode type and size to be used. Typical settings are:

Electrode	Size	Amperage
E6010	³⁄₃₂" (2.4 mm)	40A to 80A
E6010	⅛" (3.2 mm)	70A to 130A
E6010	⁵⁄₃₂" (4.0 mm)	90A to 165A
E7018	³⁄₃₂" (2.4 mm)	65A to 100A
E7018	⅛" (3.2 mm)	110A to 160A
E7018	⁵⁄₃₂" (4.0 mm)	130A to 200A

> **NOTE**
>
> Amperage recommendations can vary by manufacturer, position, current type, and electrode brand. Follow the WPS precisely when applicable; the WPS should always be the first source of weld information. Otherwise, consult the welding machine manufacturer's recommendations for the electrode being used.

Step 7 Check to be sure the electrode holder is not grounded to a conductive surface.

Step 8 Turn on the welding machine.

Lash

To ensure good welding results, remember LASH.

L *(length of arc)* – The distance between the electrode and the base metal (usually one times the electrode diameter).

A *(angle)* – Two angles are critical:

- Travel angle – The longitudinal angle of the electrode in relation to the axis of the weld joint
- Work angle – The traverse angle of the electrode in relation to the axis of the weld joint

S *(speed)* – Travel speed is measured in inches per minute (IPM). The width of the weld will determine if the travel speed is correct.

H *(heat)* – Controlled by the amperage setting and dependent upon the electrode diameter, base metal type, base metal thickness, and the welding position.

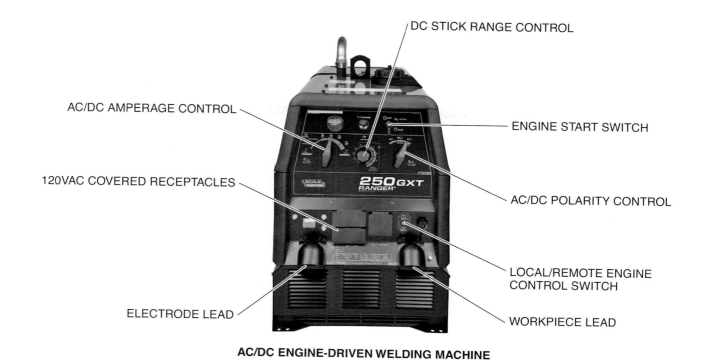

DC STICK RANGE CONTROL

AC/DC AMPERAGE CONTROL

ENGINE START SWITCH

120VAC COVERED RECEPTACLES

AC/DC POLARITY CONTROL

LOCAL/REMOTE ENGINE CONTROL SWITCH

ELECTRODE LEAD

WORKPIECE LEAD

AC/DC ENGINE-DRIVEN WELDING MACHINE

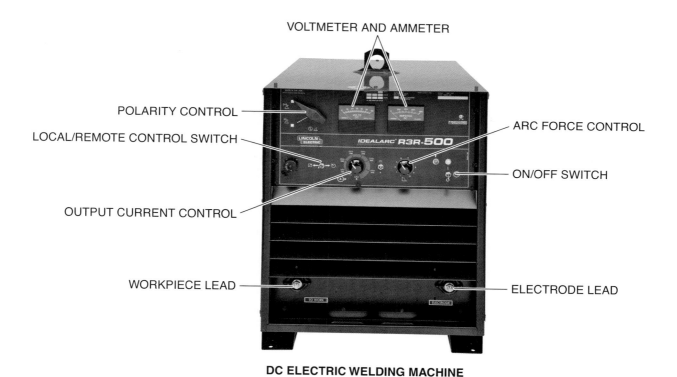

VOLTMETER AND AMMETER

POLARITY CONTROL

ARC FORCE CONTROL

LOCAL/REMOTE CONTROL SWITCH

ON/OFF SWITCH

OUTPUT CURRENT CONTROL

WORKPIECE LEAD

ELECTRODE LEAD

DC ELECTRIC WELDING MACHINE

29112-15_F10.EPS

Figure 10 AC/DC and DC-only welding machines.

Additional Resources

2014 Technical Training Guide. Current Edition. Cleveland, OH. USA: The Lincoln Electric Company. **www.lincolnelectric.com**

Welding Handbook. Current Edition. Miami, FL: The American Welding Society.

1.0.0 Section Review

1. The radius used to form the shape of a J- or U-groove is called the _____.
   - a. bevel radius
   - b. root radius
   - c. groove radius
   - d. face radius

2. Before welding on previously used containers such as tanks or barrels, it is best to clean and fill them with water or to purge them with _____.
   - a. compressed air
   - b. acetylene
   - c. oxygen
   - d. an inert gas

2.0.0 OPEN-ROOT V-GROOVE WELDS

Objective

Describe the technique required to produce various open V-groove welds.

a. Describe the technique required to produce open V-groove welds in the 1G and 2G positions.

b. Describe the technique required to produce open V-groove welds in the 3G and 4G positions.

Performance Tasks

1. Make open V-groove welds with E6010 and E7018 electrodes in the following positions:
   - Flat (1G) position
   - Horizontal (2G) position
   - Vertical (3G) position
   - Overhead (4G) position

Trade Term

Feather: The process of grinding a tack weld to a tapered edge.

The open-root V-groove weld is a common groove weld normally made on both plate and pipe. Practicing the open-root V-groove welds on plate will prepare the welder to make the more difficult pipe welds in the future.

The performance qualification requirements for this module for visual and destructive testing are based on the *ASME Boiler and Pressure Vessel Code, Section IX, Welder Certification Test.*

2.0.1 Root Pass

The most difficult part of making an open-root V-groove weld is the root pass. The root pass is made from the V-groove side of the joint and must have complete penetration, but not an excessive amount of root reinforcement. The penetration is controlled with a technique called "running a keyhole." A keyhole is a hole made when the root faces of two plates are melted away by the arc.

The molten metal flows to the back side of the keyhole, forming the weld. The keyhole should be about ⅛" to ³⁄₁₆" (3.2 to 4.8 mm) in diameter, and centered between the two coupons. If the keyhole is larger, there will be excess root reinforcement. If the keyhole closes up, there will be a lack of penetration. Root reinforcement should be flush to ⅛" (3.2 mm).

The keyhole is controlled by using fast-freeze E6010 electrodes and by whipping the electrode. Move the electrode forward about ³⁄₁₆" to ¼" (4.8 to 6.4 mm) and then back about ⅛" (3.2 mm) and pause. Use care not to increase the arc length when moving the electrode. If the keyhole starts to grow, decrease the pause and increase the forward length of the whip. If the keyhole starts to close up, pause slightly longer and decrease the length of the whip. Other factors that affect the keyhole size are the amperage setting, root face size, and root opening. All these factors can be adjusted within the parameters allowed by the WPS, welding codes, and/or your site quality procedures.

After the root pass has been run, it should be cleaned and inspected. Inspect the root face for excess buildup or undercut, as shown in *Figure 11*, which could trap slag when the filler pass is run. Remove excess buildup or undercut with a hand grinder by grinding the face of the root pass with the edge of a grinding disk. Use care not to grind through the root pass or widen the groove.

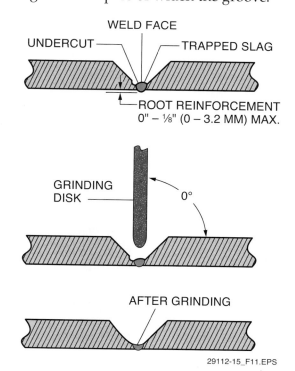

Figure 11 Grinding a root pass.

2.0.2 Groove Weld Positions

As illustrated in *Figure 12*, groove welds can be made in all positions. The weld position for plate is determined by the axis of the weld and the orientation of the workpiece. Groove weld positions for plate are flat, or 1G (groove); horizontal, or 2G; vertical, or 3G; and overhead, or 4G.

Grinding is required more often when low-hydrogen electrodes are used for the filler passes. This is because low-hydrogen electrodes have shallow penetrating characteristics and will not burn out slag inclusions as readily as E6010 electrodes. However, some structural codes do not allow the grinding of welds for any reason. In the 1G and 2G positions, the weld axis can be inclined up to 15 degrees. Any weld axis inclination for the other positions varies with the rotational position of the weld face as specified in AWS standards.

2.0.3 Acceptable and Unacceptable Groove Weld Profiles

Groove welds should be made with slight reinforcement and a gradual transition to the base metal at each toe. As shown in *Figure 13*, groove welds must not have excess reinforcement, underfill, excessive undercut, or overlap. If a groove weld has any of these defects, it is unacceptable.

2.1.0 Open V-Groove Welding – 1G and 2G Positions

The flat (1G) and horizontal (2G) positions are the easiest and best positions to begin practicing open-root V-groove welds.

2.1.1 Practicing Flat Open-Root V-Groove Welds (1G Position)

Practice flat open-root V-groove welds, as shown in *Figure 14*, using E6010 and E7018 electrodes. Use weave beads and keep the electrode angle 90 degrees to the plate surface (0-degree work angle) with a 10- to 15-degree drag angle. When using stringer beads, use a 10- to 15-degree drag angle, but adjust the electrode work angle to tie in and fill the bead. Pay particular attention at the termination of the weld to fill the crater. If the practice coupon gets too hot between passes, you may have to quench it in water.

> **WARNING!**
> Use pliers to handle the hot practice coupons. Wear gloves when quenching the practice coupon in water. Steam will rise off the coupon and can burn or scald unprotected hands.

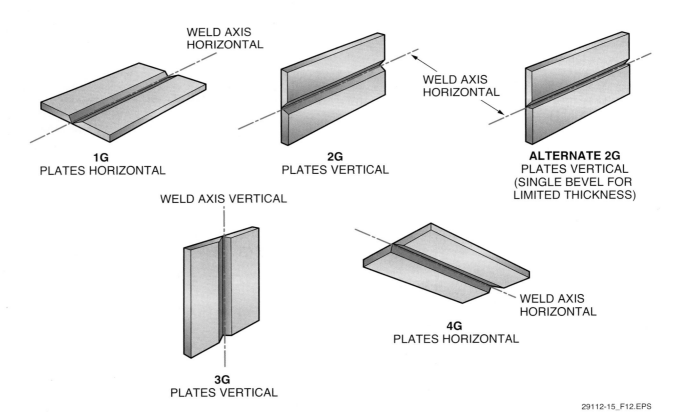

1G
PLATES HORIZONTAL

2G
PLATES VERTICAL

ALTERNATE 2G
PLATES VERTICAL
(SINGLE BEVEL FOR
LIMITED THICKNESS)

3G
PLATES VERTICAL

4G
PLATES HORIZONTAL

29112-15_F12.EPS

Figure 12 Open-root groove weld positions for plate.

NCCER – *Welding Level One* 29112-15

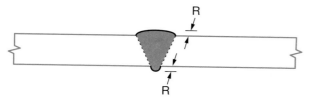

R = FACE AND ROOT REINFORCEMENT PER CODE
NOT TO EXCEED ⅛" (3.2 MM) MAX.

ACCEPTABLE WELD PROFILE

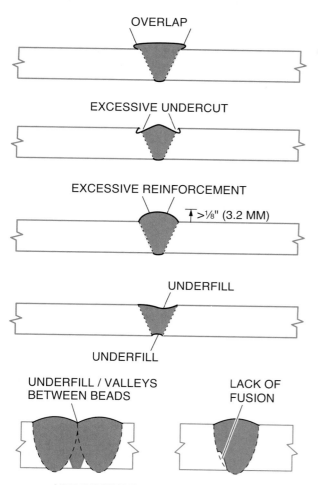

UNACCEPTABLE WELD PROFILES

29112-15_F13.EPS

Figure 13 Acceptable and unacceptable groove weld profiles.

Quenching is only done on practice coupons. Never cool test coupons or functional welds with water. Cooling with water can cause weld cracks and affect the mechanical properties of the base metal.

Follow these steps to practice open-root V-groove welds in the flat position:

Step 1 Tack weld the practice coupon together as explained earlier.

Step 2 Clean and feather the tack welds with the edge of a grinding wheel. Feathering the ends of the tack welds with a grinder will help to fuse the tack welds into the root pass.

Step 3 Place the weld coupon in a flat position.

Step 4 Run the root pass using E6010 electrodes.

Step 5 Remove the electrode stub from the electrode holder and discard as soon as you stop welding to prevent arcing to nearby surfaces, and place the stub in a proper container.

Step 6 Chip, clean, and brush the weld between all passes. Grind if required.

Step 7 Run the fill passes, cleaning the weld after each pass, to complete the weld using E7018 electrodes.

2.1.2 Horizontal Welds (2G Position)

Horizontal welds can be made on a vertically positioned flat surface or on a horizontal open-root V-groove. This section explains welding horizontal beads and making horizontal open-root V-groove welds.

Before welding an open-root V-groove in the horizontal position, practice running horizontal stringer beads by building a pad in the horizontal position (*Figure 15*). Use E6010 electrodes. The electrode angle should be a 10- to 15-degree drag angle and be perpendicular to the plate. To control the weld puddle, the electrode can be dropped slightly, but no more than 10 degrees.

Follow the WPS and Site Procedures

Refer to your site's WPS for specific requirements on groove welds. The information in this module is provided as a general guideline only. The site WPS or quality specifications must be followed for all welds. Check with your supervisor if you are unsure of the specifications for your application.

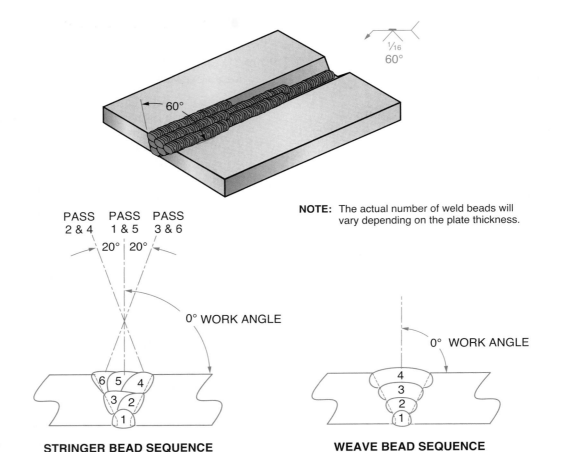

NOTE: The actual number of weld beads will vary depending on the plate thickness.

PASS 2 & 4   PASS 1 & 5   PASS 3 & 6
20° | 20°

0° WORK ANGLE

6 5 4
3 2
1

STRINGER BEAD SEQUENCE

0° WORK ANGLE

4
3
2
1

WEAVE BEAD SEQUENCE

29112-15_F14.EPS

Figure 14 Multipass 1G weld sequences and work angles.

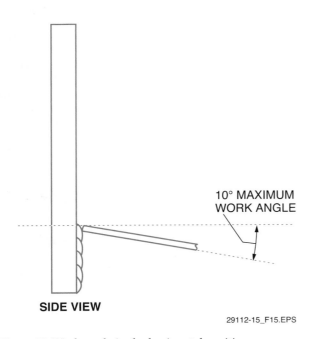

10° MAXIMUM WORK ANGLE

SIDE VIEW

29112-15_F15.EPS

Figure 15 Work angle in the horizontal position.

Follow these steps to practice welding beads in the horizontal position:

Step 1 Tack weld the flat plate welding coupon into the horizontal position.

Step 2 Clean and feather the tack welds with the edge of a grinding wheel. Feathering the ends of the tack welds with a grinder will help to fuse the tack welds into the root pass.

Step 3 Run the first pass along the bottom of the coupon using E6010 electrodes.

Step 4 Remove the electrode stub from the electrode holder and discard as soon as you stop welding to prevent arcing to nearby surfaces, and place the stub in a proper container.

Step 5 Chip, clean, and brush the weld between all passes.

Step 6 Continue running beads to complete the pad.

A horizontal open-root V-groove weld is shown in *Figure 16*. Practice horizontal open-root V-groove welds using E6010 and E7018 electrodes and stringer beads. For the electrode angles, use a 10- to 15-degree drag angle and adjust the work angle as required. Pay particular attention at the termination of the weld to fill the crater. Note that the weave bead is not recommended for the 2G position.

Follow these steps to practice welding open-root V-groove welds in the horizontal position:

Step 1 Tack weld the practice coupon together as explained earlier. Use the standard or alternate horizontal weld coupon as directed by your instructor.

Step 2 Clean and feather tack welds with the edge of a grinding wheel.

Step 3 Tack weld the flat plate weld coupon into the horizontal position.

Step 4 Run the root pass using E6010 electrodes.

Step 5 Remove the electrode stub from the electrode holder and discard as soon as you stop welding to prevent arcing to nearby surfaces, and place the stub in a proper container.

Step 6 Chip, clean, and brush the weld between all passes.

Step 7 Run the fill and cover passes, cleaning the weld after each pass, to complete the weld. Use E7018 electrodes and stringer beads.

2.2.0 Open V-Groove Welding – 3G and 4G Positions

Now that the open-root V-groove techniques in the 1G and 2G positions have been practiced, the more challenging 3G and 4G positions can be attempted.

Can You Spare Some Change?

A nickel and a dime make excellent gauges for fitting open-root V-groove weld coupons. The thickness of the nickel can be used to check the root face, while the dime is used to set the root opening.

2.2.1 *Vertical Welds (3G Position)*

Vertical welds can be made on a vertically positioned flat surface or on a vertical open-root V-groove, as shown in *Figure 17*. This section explains welding vertical beads and making vertical open-root V-groove welds.

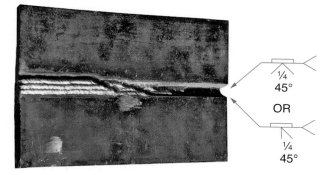

NOTE: The actual number of weld beads will vary depending on the plate thickness and electrode diameter.

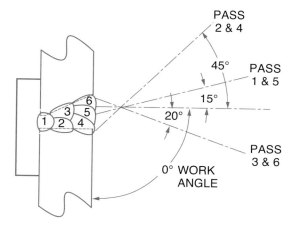

ALTERNATE JOINT REPRESENTATION

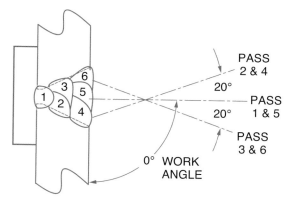

STANDARD JOINT REPRESENTATION

29112-15_F16.EPS

Figure 16 Multipass 2G weld sequences and work angles.

Before welding a joint in the vertical position, practice running vertical weave beads by building a pad in the vertical position (see *Figure 17*). Use E6010 electrodes. Set the amperage in the lower part of the allowable range for better puddle control. The electrode travel should be upward with a 0- to 10-degree push angle and perpendicular to the plate (0-degree work angle) for the first pass. A 0- to 10-degree work angle toward the previous bead should be used for subsequent beads. Use a side-to-side motion with a slight pause at each weld toe to tie-in and to control the weld puddle. The width of the weave beads should be no more than five times the electrode wire (metal core) diameter. Follow these steps to practice welding beads in the vertical position:

Step 1 Tack weld the flat plate welding coupon into the vertical position.

Step 2 Using a whipping motion, run the first pass along the vertical edge of the coupon.

Step 3 Remove the electrode stub from the electrode holder and discard as soon as you stop welding to prevent arcing to nearby surfaces, and place the stub in a proper container.

Step 4 Chip, clean, and brush the weld between all passes.

Step 5 Continue running beads and cleaning each pass to complete the pad.

An example of a vertical open-root V-groove weld is shown in *Figure 18*. Practice vertical open-root V-groove welds using E6010 for the root weld pass and E7018 electrodes and weave beads for the remaining passes. Set the amperage in the lower part of the range for better puddle control. Use an upward travel direction with a 0- to 10-degree push angle, but adjust the electrode work angle to tie-in the weld on one side or the other as needed. Pause slightly at each weld toe to penetrate and fill the crater to prevent undercut. Pay particular attention at the termination of the weld to fill the crater.

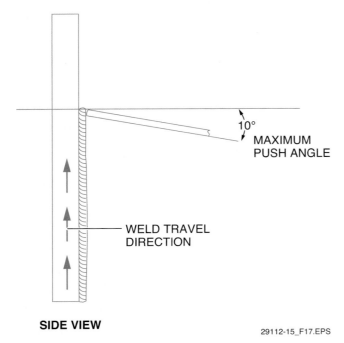

SIDE VIEW

29112-15_F17.EPS

Figure 17 Work angle and direction of travel in the vertical position.

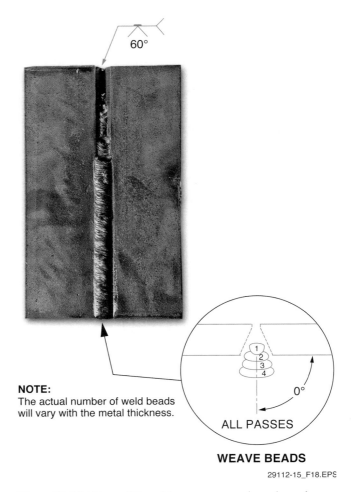

NOTE:
The actual number of weld beads will vary with the metal thickness.

WEAVE BEADS

29112-15_F18.EPS

Figure 18 Multipass 3G weld sequences and work angles.

Follow these steps to practice open-root V-groove welds in the vertical position:

Step 1 Tack weld the practice coupon together as explained earlier.

Step 2 lean and feather the tack welds with the edge of a grinder wheel.

Step 3 Tack weld the weld coupon into the vertical position.

Step 4 Run the root pass using E6010 electrodes.

Step 5 Remove the electrode stub from the electrode holder and discard as soon as you stop welding to prevent arcing to nearby surfaces, and place the stub in a proper container.

Step 6 Chip, clean, and brush the weld between all passes.

Step 7 Run the fill and cover passes with E7018 electrodes to complete the weld. Clean the weld between each pass.

2.2.2 Overhead Welds (4G Position)

Overhead welds, as shown in *Figure 19*, can be made on an overhead flat surface or on an overhead open-root V-groove.

Before welding a joint in the overhead position, practice running overhead stringer beads by building a pad in the overhead position. Use E6010 electrodes. The electrode work angle should be 90 degrees to the plate with a 10- to 15-degree drag angle for the first pass. Then for the following beads, the work angle should be 10 degrees toward the previous beads. Follow these steps to practice welding beads in the overhead position:

Step 1 Tack weld the flat plate welding coupon into the overhead position.

Step 2 Run the first pass along the edge of the coupon.

Step 3 Remove the electrode stub from the electrode holder and discard as soon as you stop welding to prevent arcing to nearby surfaces, and place the stub in a proper container.

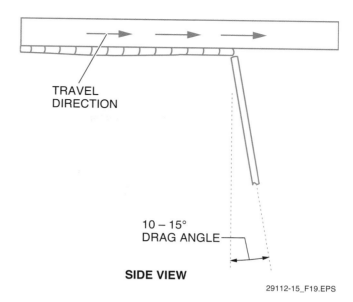

TRAVEL DIRECTION

10 – 15° DRAG ANGLE

SIDE VIEW

29112-15_F19.EPS

Figure 19 Building a pad in the overhead position.

Step 4 Chip, clean, and brush the weld between all passes.

Step 5 Continue running beads, cleaning the weld between passes, to complete the pad.

An overhead open-root V-groove weld is shown in *Figure 20*. Practice running overhead open-root V-groove welds using E6010 and E7018 electrodes and stringer beads. Use a 10- to 15-degree drag angle, but adjust the electrode work angle to tie-in the weld as needed. Pay particular attention at the termination of the weld to fill the crater. Follow these steps to practice welding open-root V-groove welds in the overhead position:

Step 1 Tack weld the practice coupon together as explained earlier.

Step 2 Clean and feather the tack welds with the edge of a grinder wheel.

Step 3 Tack weld the weld coupon into the overhead position.

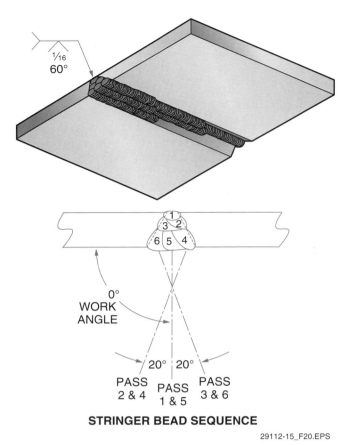

STRINGER BEAD SEQUENCE

29112-15_F20.EPS

Figure 20 Multipass 4G weld sequences and work angles.

Step 4 Run the root pass using E6010 electrodes.

Step 5 Remove the electrode stub from the electrode holder and discard as soon as you stop welding to prevent arcing to nearby surfaces, and place the stub in a proper container.

Step 6 Chip, clean, and brush the weld between all passes.

Step 7 Run the fill and cover passes with E7018 electrodes to complete the weld. Clean the weld after each pass.

Hold It Open

Before beginning the root pass, tack the opposite end of the coupons (the end opposite the starting point) together with a bead about ½" (≈13 mm) long. This will prevent the root opening from closing up at the opposite end as you begin welding.

Using Your Bifocals and Trifocals

Many welders have used what are known as "cheater lenses" in welding hoods for years to help sharpen their vision when the hood is down. A cheater lens is simply a magnification lens. Although many welders wear prescription glasses with bifocal or trifocal lenses, it was impractical to use the glasses due to the design of the hood. Adjustment of the hood's position in front of the face was nonexistent, and the lens size restricted vision.

Newer welding hoods, like the Viking model shown here, allow welders to make use of their bifocal and trifocal prescription glasses. The hood distance from the front of the face can be adjusted in the headgear. This allows the hood to be tilted down farther, even to the chest. The welder can then tilt the head back and forth slightly, using the different focal lengths of the lenses, without losing sight of the task. The significantly larger lens also ensures that the task at hand remains in the welder's field of view.

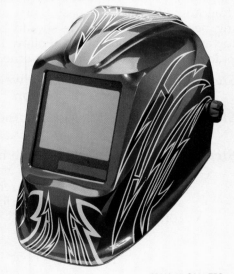

29112-15_SA01.EPS

Additional Resources

2014 Technical Training Guide. Current Edition. Cleveland, OH. USA: The Lincoln Electric Company. **www.lincolnelectric.com**

Welding Handbook. Current Edition. Miami, FL: The American Welding Society.

2.0.0 Section Review

1. Which of the following can be quenched without concern?

    a. Weld test coupons
    b. Functional welds
    c. Practice coupons
    d. Welds done in the 3G position

2. On an open-root V-groove weld in the 4G position, the electrode work angle should be _____.

    a. 30 degrees
    b. 45 degrees
    c. 90 degrees
    d. varied as needed to tie in

SUMMARY

The ability to make open-root V-groove welds on plate is an essential welding skill. A welder must possess this skill to perform welding jobs and to progress to more difficult pipe welding procedures. Making open-root V-groove welds requires the welder to properly and safely set up and prepare the welding equipment and the coupons. V-groove welds can be made in the flat, horizontal, vertical, and overhead positions. Practice these welds until you can consistently produce acceptable welds in all positions.

1. Which of the following statements is true?
   a. Groove weld styles include single-, double-, and quadruple-groove styles.
   b. Groove welds can be combined with fillet welds to maintain strength.
   c. The root opening dimension of a groove weld is determined by the specified work angle.
   d. Weld symbols on drawings always show both metric and imperial dimensions.

2. Open-root weld coupons generally have a bevel angle of _____.
   a. 10 degrees
   b. 15 degrees
   c. 22½ degrees
   d. 30 degrees

3. Welding codes typically allow the root face on open-root groove welds to be within a range of _____.
   a. ¹⁄₃₂" to ⁵⁄₃₂" (0.8 to 4.0 mm)
   b. ¹⁄₃₂" to ³⁄₁₆" (0.8 to 4.8 mm)
   c. ¹⁄₁₆" to ¼" (1.6 to 6.4 mm)
   d. 0" to ⅛" (0 to 3.2 mm)

4. When preparing to tack weld an open-root V-groove weld coupon, align the beveled strips on a flat surface with a _____.
   a. 0" to ⅛" (0 to 3.2 mm) root opening
   b. ³⁄₃₂" to ⁵⁄₃₂" (2.4 to 4.0 mm) root opening
   c. ¹⁄₁₆" to ¼" (1.6 to 6.4 mm) root opening
   d. ¹⁄₃₂" to ¹⁄₁₆" (0.8 to 1.6 mm) root opening

5. When making an open-root V-groove weld, the hole made when the two root faces of two plate edges are melted away by the arc is called a(n) _____.
   a. bead
   b. overlap
   c. keyhole
   d. root pass

6. When the plates are vertical and the axis of the weld is also vertical, the code for this position of groove weld is _____.
   a. 1G
   b. 2G
   c. 3G
   d. 4G

7. When making weave beads on flat open-root V-groove welds, keep the electrode work angle at 90 degrees to the plate surface with a _____.
   a. 5- to 10-degree drag angle
   b. 10- to 15-degree drag angle
   c. 15- to 20-degree drag angle
   d. 20- to 22-degree drag angle

8. To control the weld puddle when welding an open-root V-groove in the horizontal position, the electrode work angle can be _____.
   a. dropped up to 10 degrees
   b. raised up to 15 degrees
   c. raised up to 20 degrees
   d. dropped up to 30 degrees

9. When practicing horizontal open-root V-groove welds, pay particular attention at the termination of the weld to fill the _____.
   a. pad
   b. root
   c. crater
   d. puddle

10. For better puddle control when welding open-root V-groove welds in the 3G position, _____.
    a. set the amperage to the lower part of the allowable range
    b. change to a downward direction of travel
    c. increase the work angle to 25 degrees
    d. change from a push to a drag angle

Trade Terms Quiz

Fill in the blank with the correct term that you learned from your study of this module.

1. The portion of the edge of a part to be joined by a groove weld that has not been beveled or grooved; previously referred to as the land is a _____.

2. Grinding a tack weld to a tapered edge is called a _____.

Trade Terms

Feather
Root face

Appendix

PERFORMANCE ACCREDITATION TASKS

The American Welding Society (AWS) School Excelling through National Skills Standards Education (SENSE) program is a comprehensive set of minimum Standards and Guidelines for Welding Education programs. The following Performance Accreditation Tasks (PATs) are aligned with and designed around the SENSE program. The PATs correspond to and support the learning objectives in *AWS EG2.0, Guide for the Training and Qualification of Welding Personnel: Entry-Level Welder*.

Note that in order to satisfy all learning objectives in *AWS EG2.0*, the instructor must also use the PATs contained in the second level of the NCCER Welding curriculum.

PATs 1 through 4 correspond to *AWS EG2.0, Module 4 – Shielded Metal Arc Welding*, Key Indicators: 3, 4, and 6

PATs provide specific acceptable criteria for performance and help to ensure a true competency-based welding program for students.

The following tasks are designed to evaluate your ability to run open-root V-groove welds with SMAW equipment in all positions using E6010 and E7018 electrodes. Perform each task when you are instructed to do so by your instructor. As you complete each task, take it to your instructor for evaluation. Do not proceed to the next task until instructed to do so.

OPEN V-GROOVE WITH E6010 AND E7018 ELECTRODES IN THE FLAT POSITION

Using ³⁄₃₂" or ⅛" (2.4 or 3.2 mm) E6010 electrodes for the root pass and ³⁄₃₂" or ⅛" (2.4 or 3.2 mm) E7018 electrodes for the fill and cover passes, make an open-root V-groove weld on carbon steel plate in the flat position as indicated.

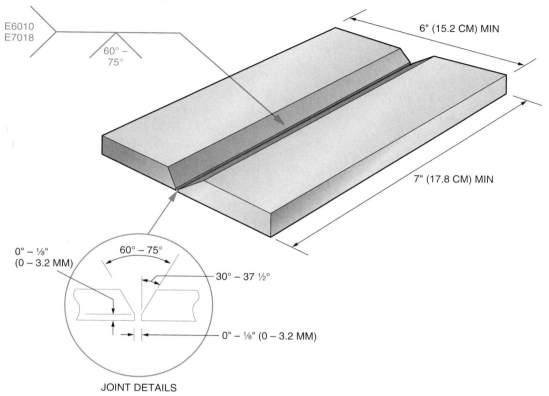

JOINT DETAILS

29112-15_A01.EPS

Criteria for Acceptance:

- No arc strikes outside the weld area
- Uniform rippled appearance on the bead face
- Craters and restarts filled to the full cross section of the weld
- Uniform weld size
- Acceptable weld profile in accordance with the *ASME Boiler and Pressure Vessel Code*
- Smooth transition with complete fusion at the toes of the weld
- Complete uniform root penetration at least flush with the base metal to a maximum buildup of ⅛" (3.2 mm)
- No porosity
- No excessive undercut
- No inclusions
- No cracks
- Acceptable guided bend test results (instructor option)

OPEN V-GROOVE WITH E6010 AND E7018 ELECTRODES IN THE HORIZONTAL POSITION

Using ³⁄₃₂" or ⅛" (2.4 or 3.2 mm) E6010 electrodes for the root pass and ³⁄₃₂" or ⅛" (2.4 or 3.2 mm) E7018 electrodes for the fill and cover passes, make an open-root V-groove weld on carbon steel plate in the horizontal position as shown.

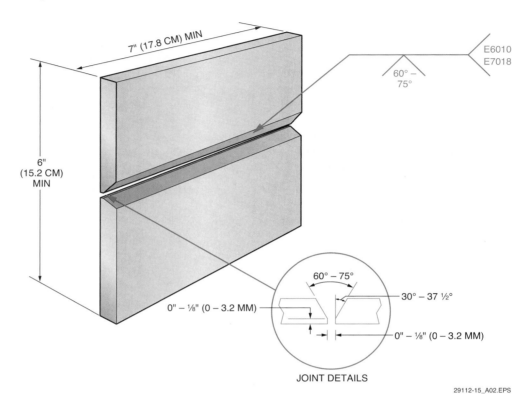

JOINT DETAILS

29112-15_A02.EPS

Criteria for Acceptance:

- No arc strikes outside the weld area   _____
- Uniform rippled appearance on the bead face   _____
- Craters and restarts filled to the full cross section of the weld   _____
- Uniform weld size   _____
- Acceptable weld profile in accordance with the *ASME Boiler and Pressure Vessel Code*   _____
- Complete uniform root penetration at least flush with the base metal to a maximum buildup of ⅛" (3.2 mm)   _____
- Smooth transition with complete fusion at the toes of the weld   _____
- No porosity   _____
- No excessive undercut   _____
- No inclusions   _____
- No cracks   _____
- Acceptable guided bend test results (instructor option)   _____

OPEN V-GROOVE WITH E6010 AND E7018 ELECTRODES IN THE VERTICAL POSITION

Using ³⁄₃₂" to ⅛" (2.4 or 3.2 mm) E6010 electrodes for the root pass and ³⁄₃₂" or ⅛" (2.4 or 3.2 mm) E7018 electrodes for the fill and cover passes, make an open-root V-groove weld on carbon steel plate in the vertical position as shown.

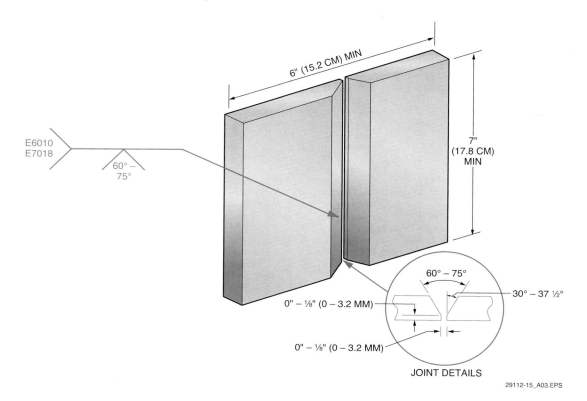

JOINT DETAILS

29112-15_A03.EPS

Criteria for Acceptance:

- No arc strikes outside the weld area
- Uniform rippled appearance on the bead face
- Craters and restarts filled to the full cross section of the weld
- Uniform weld size
- Acceptable weld profile in accordance with the *ASME Boiler and Pressure Vessel Code*
- Complete uniform root penetration at least flush with the base metal to a maximum buildup of ⅛" (3.2 mm)
- Smooth transition with complete fusion at the toes of the weld
- No porosity
- No excessive undercut
- No inclusions
- No cracks
- Acceptable guided bend test results (instructor option)

OPEN V-GROOVE WITH E6010 AND E7018 ELECTRODES IN THE OVERHEAD POSITION

Using ³⁄₃₂" to ⅛" (2.4 or 3.2 mm) E6010 electrodes for the root pass and ³⁄₃₂" or ⅛" (2.4 or 3.2 mm) E7018 electrodes for the fill and cover passes, make an open-root V-groove weld on carbon steel plate in the overhead position as indicated.

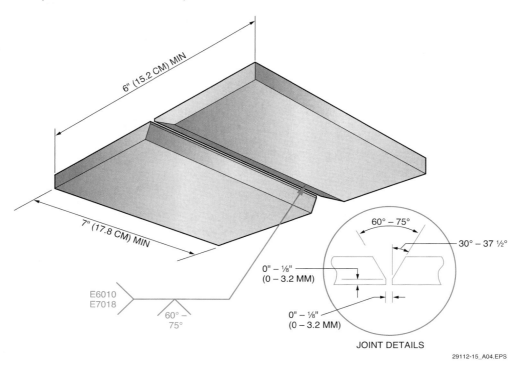

JOINT DETAILS

29112-15_A04.EPS

Criteria for Acceptance:

- No arc strikes outside the weld area   _____
- Uniform rippled appearance on the bead face   _____
- Craters and restarts filled to the full cross section of the weld   _____
- Uniform weld size   _____
- Acceptable weld profile in accordance with the *ASME Boiler and Pressure Vessel Code*   _____
- Complete uniform root penetration at least flush with the base metal to a maximum buildup of ⅛" (3.2 mm)   _____
- Smooth transition with complete fusion at the toes of the weld   _____
- No porosity   _____
- No excessive undercut   _____
- No inclusions   _____
- No cracks   _____
- Acceptable guided bend test results (instructor option)   _____

Trade Terms Introduced in This Module

Feather: The process of grinding a tack weld to a tapered edge.

Root face: The portion of the edge of a part to be joined by a groove weld that has not been beveled or grooved; previously referred to as the land.

Additional Resources

This module presents thorough resources for task training. The following resource material is suggested for further study.

2014 Technical Training Guide. Current Edition. Cleveland, OH. USA: The Lincoln Electric Company. **www.lincolnelectric.com**

Welding Handbook. Current Edition. Miami, FL: The American Welding Society.

Figure Credits

The Lincoln Electric Company, Cleveland, OH, USA, Module Opener, Figures 6, 7, 10, SA01

Topaz Publications, Inc., Figures 16 (photo), 18 (photo)

Section Review Answer Key

Answer	Section Reference	Objective
Section One		
1. a	1.1.3	1a
2. d	1.2.1	1b
Section Two		
1. c	2.1.1	2a
2. d	2.2.2	2b

NCCER CURRICULA — USER UPDATE

NCCER makes every effort to keep its textbooks up-to-date and free of technical errors. We appreciate your help in this process. If you find an error, a typographical mistake, or an inaccuracy in NCCER's curricula, please fill out this form (or a photocopy), or complete the online form at **www.nccer.org/olf**. Be sure to include the exact module ID number, page number, a detailed description, and your recommended correction. Your input will be brought to the attention of the Authoring Team. Thank you for your assistance.

Instructors – If you have an idea for improving this textbook, or have found that additional materials were necessary to teach this module effectively, please let us know so that we may present your suggestions to the Authoring Team.

NCCER Product Development and Revision
13614 Progress Blvd., Alachua, FL 32615

Email: curriculum@nccer.org
Online: www.nccer.org/olf

❏ Trainee Guide ❏ Lesson Plans ❏ Exam ❏ PowerPoints Other _____

Craft / Level: _____ Copyright Date: _____

Module ID Number / Title: _____

Section Number(s): _____

Description: _____

_____

_____

_____

_____

Recommended Correction: _____

_____

_____

_____

_____

Your Name: _____

Address: _____

_____

Email: _____ Phone: _____

Glossary

Addendum: Supplementary information typically used to describe corrections or revisions to documents.

Alloy: A metal that has had other elements added to it that substantially change its mechanical properties.

Ambient temperature: The room temperature of the atmosphere completely surrounding an object on all sides.

Amperage: The unit of measure used for the intensity of an electric current; often abbreviated to amp or amps.

Annealed: To free from internal stress by heating and gradually cooling.

Arc blow: The deflection of the arc from its intended path due to magnetic forces that develop from welding current flow.

Arc burn: Burn to the skin produced by brief exposure to intense radiant heat and ultraviolet light.

Arc strike: A discontinuity consisting of localized melting of the base metal or finished weld caused by the initiation of an arc. It remains visible as a result of striking the arc outside the area to be welded and can lead to rejection of an otherwise good weld.

Austenitic: Containing austenite (a solid solution of carbon or of carbon and other elements in a ferrous alloy), added through heating.

Backfire: A loud snap or pop as a torch flame is extinguished.

Backing: A weldable or nonweldable material used behind a root opening to allow defect-free welding at the open root of a joint.

Base metal: Metal to be welded, cut, or brazed.

Bonded: The permanent joining of metallic parts to form an electrically conductive path that will assure electrical continuity and the capacity to safely conduct any current likely to be imposed on it.

Brazing: A method of joining metal using heat and a filler metal with a melting point above 842°F (450°C). Unlike welding, the base metal is not melted during the brazing process.

Carbon steel: An alloy of iron combining iron and usually less than 1 percent carbon to provide hardness.

Carbon-graphite electrode: An electrode composed of a mixture of soft amorphous carbon and hard graphite carbon that may be coated with copper.

Carburizing flame: A flame burning with an excess amount of fuel; also called a reducing flame.

Castings: Something cast; any article that has been cast in a mold.

Code: A document that establishes the minimum requirements for a product or process. Codes can be, and often are, adopted as laws.

Coefficient of thermal expansion (linear): The change in length per unit length of material for a 1°C change in temperature.

Concentric cable system: A-CAC configuration in which a unique combination fitting is used to connect the torch cable to welding power in order to enable compressed air passage through the power conductor.

Condensation: The process in which atmospheric water vapor condenses (returns to its liquid state) on a cool surface.

Conductor: A material that will support the flow of an electrical current. Copper wire is the most common conductor.

Consumable insert: Preplaced filler metal that is completely fused into the root of the joint during welding, becoming part of the weld.

Defect: A discontinuity or imperfection that renders a part of the product or the entire product unable to meet minimum acceptable standards or specifications.

Discontinuity: A change or break in the shape or structure of a part that may or may not be considered a defect, depending on the code.

Distortion: The expansion and contraction of welded parts caused by the heating and subsequent cooling of the weld joint.

Drag angle: Describes the travel angle when the electrode is pointing in a direction opposite to the welding bead's progression.

Drag lines: The lines on the edge of the material that result from the travel of the cutting oxygen stream into, through, and out of the metal.

Dross: A waste byproduct of molten metal; the material (oxidized and molten metal) that is expelled from the kerf when cutting using a thermal process. It is sometimes called slag.

Ductile: Able to undergo change of form without breaking.

Ductility: Refers to the mechanical property of a material that allows it to be bent or shaped without breaking.

Duty cycle: The percentage of a ten-minute period that a welding machine can continuously produce its rated amperage without overheating; the percentage of time a plasma arc cutting machine can cut without overheating within a ten-minute period.

Electric arc: The flow of an electrical current across an air gap or gaseous space.

Electrically grounded: Connected to Earth or to some conducting body that serves in place of the earth.

Electrode: The point from which a welding arc is produced.

Embrittled: Metal that has been made brittle and that will tend to crack with little bending.

Essential variable: Items in a welding procedure specification (WPS) that cannot be changed without requalifying the WPS.

Feather: The process of grinding a tack weld to a tapered edge.

Ferritic: Steel containing less than 0.10 percent carbon and is magnetic. This steel can't be hardened via heat treatment.

Ferrous: Containing iron.

Ferrous metals: Metals containing iron.

Flash burn: Burns to the eyes sometimes called welder's flash; caused by exposure to intense radiant heat and ultraviolet light.

Flashback: The flame burning back into the tip, torch, hose, or regulator, causing a high-pitched whistling or hissing sound.

Flux: A material used to dissolve or prevent the formation of oxides and other undesirable substances on a weld joint; material used to prevent, dissolve, or facilitate the removal of oxides and other undesirable substances on a weld or base metal.

Fume plume: The fumes, gases, and particles from the consumables, base metal, and base metal coating during the welding process.

Galvanized steel: Carbon steel dipped in zinc to inhibit corrosion.

Gouging: The process of cutting a groove into a surface.

Governor: A device that limits engine speed.

Hardenable materials: Metals that have the ability to be made harder by heating and then cooling.

Heat-affected zone: The part of the base metal that has been altered by heating, but not melted by the heat.

Hermetically sealed: Having an airtight seal.

High-low: The discrepancy in the internal alignment of two sections of pipe.

Homogeneity: The quality or state of having a uniform structure or composition throughout.

Immediate danger to life and health (IDLH): As defined by OSHA, this term refers to an atmosphere that poses an immediate threat to life, would cause irreversible adverse health effects, or would impair an individual's ability to escape from a dangerous atmosphere.

Inclusion: Foreign matter introduced into and remaining in a weld.

Kerf: The gap produced by a cutting process.

Laminations: Cracks in the base metal formed when layers separate.

Level: A line on a horizontal axis parallel with the earth's surface (horizon). Also refers to the tool used to check if an object is level.

Load: The amount of force applied to a material or a structure.

Low-hydrogen electrode: An electrode specially manufactured to contain little or no moisture.

Malleable: Capable of being extended or shaped by hammering or by pressure from rollers.

Martensitic: Steel that shares some characteristics with ferritic, but has a higher levels of carbon, up to a full 1 percent. It can be tempered and hardened and is used where strength is more important than a resistance to oxidation.

Melt-through: Complete joint penetration.

Motor-generator: A combination device in which the motor turns the shaft of a generator, which in turn produces an AC voltage.

Neutral flame: A flame burning with correct proportions of fuel gas and oxygen.

Non-essential variable: Items in a welding procedure specification (WPS) that can be changed without requalifying the WPS.

Nonferrous metal: A metal, such as aluminum, copper, or brass, lacking sufficient quantities of iron to have any effect on its properties.

Notch toughness: The ability of a material to absorb energy in the presence of a flaw such as a notch; the ability of a material to resist breaking at points where stress is concentrated.

Oscillation: A repetitive and consistent side-to-side motion.

Oxide: The scale that forms on metal surfaces when they are exposed to oxygen or air containing oxygen.

Oxidize: To combine with oxygen, such as in burning (rapid oxidation) or rusting (slow oxidation).

Oxidizing flame: A flame burning with an excess amount of oxygen.

Peening: The mechanical working of metal using repeated hammer blows, typically using a ball-shaped hammer head such as that found on a ball peen hammer. Peening may be done for both metallurgical and appearance reasons.

Phase: In a three-phase power supply system, the sine waves of the voltage on each of the three separate conductors is displaced from each of the others by 120 degrees, although all conductors are carrying alternating current at the same frequency and are synchronized. Each of the three sine waves is referred to as a phase of the power source.

Pierce: To penetrate through metal plate with an oxyfuel cutting torch.

Piping porosity: A form of porosity having a length greater than its width and that is approximately perpendicular to the weld face.

Plasma: A fourth state of matter (not solid, liquid, or gas) created by heating a gas to such a high temperature that it boils electrons off the gas molecules (ionization).

Plumb: A line on a vertical axis perpendicular to the earth's surface (horizon).

Polarity: Refers to the direction of electrical current flow in a DC welding circuit; the condition of a system in which it has opposing physical properties at different points, such as an electric charge.

Porosity: Gas pockets, or voids in the weld metal.

Potential: The relative electrical voltage difference between two points of reference; the electromotive force or difference in potential between two points, expressed in volts.

Procedure qualification record (PQR): The document containing the results of the nondestructive and destructive testing required to qualify a WPS.

Purge gas: An inert gas such as nitrogen used to drive oxygen away from a weld site.

Push angle: Describes the travel angle when the electrode is pointing in the same direction as the welding bead's progression.

Quench: To cool suddenly by plunging into a liquid; to rapidly cool a hot component such as a freshly welded coupon. Note that only coupons should be quenched; real welds with a function should never be quenched.

Radiographic: Describes images made by passing X-rays or gamma rays through an object and recording the variations in density on photographic film.

Rectifier: An electronic device that converts AC voltage to DC voltage.

Residual stress: Stress remaining in a weldment as a result of heat.

Restart: The action of restarting a weld bead that is already in progress that requires the crater left from the termination of the previous arc.

Root face: A small flattened area on the end of a bevel for a groove weld; the portion of the edge of a part to be joined by a groove weld that has not been beveled or grooved; previously referred to as the land.

Root opening: The space between the base metal pieces at the bottom or root of the joint.

Shielding gas: A gas such as argon, helium, or carbon dioxide used to protect the welding electrode wire from contamination in GMAW and FCAW welding.

Soapstone: Soft, white stone used to mark metal.

Solenoid valve: A valve used to control the flow of gases or liquids that is opened or closed by the action of an energized electromagnet.

Specific heat per unit volume: The quantity of heat required to raise one unit mass of a substance by one unit degree.

Specification: A document that defines in detail the work to be performed or the materials to be used in a product or process.

Stainless steel: An iron-based alloy usually containing at least 11 percent chromium.

Standard: A document that defines how a code is to be implemented.

Step-down transformer: An electrical device that uses two wire coils of different sizes, referred to as the primary and secondary windings, to convert a higher voltage to a lower voltage through induction. Primary windings are connected to the primary power source; the secondary windings provide the reduced voltage to the process.

Stringer bead: A type of weld bead made without any significant weaving motion. With SMAW, stringer beads are not more than three times the diameter of the electrode.

Supplemental essential variable: The variables that must be considered when notch toughness requirements are invoked.

Surfacing: The application by welding, brazing, or thermal spraying of a layer of material to a surface to obtain desired properties or dimensions.

Tack weld: A weld made to hold parts of a weldment in proper alignment until the final weld is made.

Tempering: To impart strength or toughness to (steel or cast iron) by heating and cooling.

Tensile strength: The measure of the ability of a material to withstand a longitudinal stress without breaking.

Traceability: The ability to verify that a procedure has been followed by reviewing the documentation step-by-step.

Transducer: A device that converts one form of energy into another.

Ultraviolet (UV) radiation: Invisible rays capable of causing burns. UV rays from the sun are the causes of sunburn.

Underbead cracking: Cracking in the base metal near the weld, but under the surface.

Undercut: A defect that occurs when the cross-sectional thickness of the base metal is reduced at the edges of the weld. Excessive weld current can cause this as it causes the base metal at the edges to melt and drain into the weld puddle. A downward sloped edge toward the weld bead is the visual evidence.

Ventricular fibrillation: Irregular heart rhythm characterized by a rapid fluttering heartbeat.

Vertical welding: Welding with an upward or downward progression.

Washing: A term used to describe the process of cutting out bolts, rivets, previously welded pieces, or other projections from the metal surface.

Weathering steel: Steel alloy that, under specific conditions, is designed to form a very dense oxide layer on its outer surfaces, which retards further oxidation.

Weave bead: A type of weld bead made by transverse oscillation of the electrode in a particular pattern.

Weld axis: A straight line drawn through the center of a weld along its length.

Weld coupon: Metal pieces to be welded together for test or practice.

Welding procedure qualification: The demonstration that welds made following a specific process can meet prescribed standards.

Welding procedure specification (WPS): The document containing all the detailed methods and practices required to produce a sound weld.

Weldment: An assembly that is fastened together by welded joints.

Work angle: An angle less than 90 degrees between a line perpendicular to the major workpiece surface and a plane determined by the electrode axis and the weld axis. 0-degree work angle is common. In a T-joint or corner joint, the line is perpendicular to the non-butting member. The definition of work angle for a pipe weld is covered in a later module, as it differs from that of plate welding.

Wrought: Produced or shaped by beating with a hammer, as with iron.

Index